Semiconductor Devices
Basic Principles

Semiconductor Devices
Basic Principles

Jasprit Singh
University of Michigan

John Wiley & Sons, Inc.
New York / Chichester / Weinheim / Brisbane
Singapore / Toronto

EDITOR Bill Zobrist
MARKETING MANAGER Katherine Hepburn
ASSOCIATE PRODUCTION DIRECTOR Lucille Buonocore
SENIOR PRODUCTION EDITOR Monique Calello
SENIOR DESIGNER Karin Kincheloe

This book was set in LaTeX 10/12 Times Roman by Teresa Singh and printed and bound by Malloy Lithographing, Inc. The cover was printed by Phoenix Color Corporation.

This book is printed on acid-free paper.

The paper in this book was manufactured by a mill whose forest management programs include sustained yield harvesting of its timberlands. Sustained yield harvesting principles ensure that the numbers of trees cut each year does not exceed the amount of new growth.

To order books or for customer service call 1-800-CALL-WILEY (225-5945).

Library of Congress Cataloging-in-Publication Data

Singh, Jasprit.
    Semiconductor devices: basic principles / Jasprit Singh.
        p.cm.
    Includes bibliographical references and index.
    ISBN 0-471-36245-X (cloth : alk. Paper)
     1. Semiconductors. I. Title

    TK7871.85 .S5558 2000
    621.3815'2—dc21   00-038225

Printed in the United States of America

10 9 8 7 6 5 4 3 2 1

Preface

Solid state electronics technology has been moving by leaps and bounds over the last several decades. This has created new opportunities and challenges for the electrical engineer. Opportunities are apparent in the new industries that have been spawned by electronic and optoelectronic devices. These new industries rely not only on silicon technology, but also on gallium-arsenide, indium-gallium-arsenide, and a host of other semiconductors. Fine-line lithography, new semiconductor systems, and new device concepts are driving state-of-the-art devices. Herein also lies the challenge to electrical engineering students and electrical engineering textbooks. New technological advances also mean that a greater knowledge base is needed to understand and exploit the new devices.

How does an undergraduate textbook respond to the changes occurring in semiconductor technology? The study of semiconductor devices involves three ingredients: (i) physics of semiconductors; (ii) workings of semiconductor electronic devices; and (iii) operation and design of circuits. Most instructors have different views on how a semester of classes should be spread among these topics. This book addresses the first two topics, while pointing out some of the important issues in the third category.

Physics of Semiconductors: In order to understand the underlying physics upon which concepts such as effective mass, mobility, absorption coefficient, bandstructure, etc., are based, one needs one or more graduate-level courses. However, the path that leads us to these concepts can be explained to undergraduate students without all the mathematical rigor. The first three chapters of this textbook discuss the underlying physics of semiconductors using a minimum of mathematics and a minimum of quantum mechanics. I have tried to make the sections complete. By placing about 60 solved examples in these physics-related chapters, I hope that students can follow most topics fairly easily.

Semiconductor Electronic Devices: Semiconductor electronic devices form the core of most undergraduate courses. In this text, all of the important devices are covered in detail. About 100 device-related solved examples are included to illustrate the concepts discussed. In addition, the motivations for going from our "ideal device" towards the real-world devices of today and tomorrow are presented. While this book does not address electronic circuits, some important circuit requirements are discussed.

In each chapter I have also included a section called "A Bit of History." This section explores the history of the concepts developed in that chapter and provides a snapshot of the personalities involved and the challenges of the time. I feel it is important for the student of electrical engineering to feel connected through history to the devices that are an integral part of our modern life.

The figures, cover design, and the formatting of this book were done by Teresa Singh, my wife, who also provided the support without which this book would not have been possible. I am also indebted to my colleagues and the administration at the University of Michigan for providing the atmosphere and the physical resources to make book writing possible in these times of fast-paced research.

An Instructor's Manual is available to professors wishing to use this text. Please write on your department stationery for a copy of this manual to John Wiley and Sons.

I am grateful to my editor Bill Zobrist and his assistant Jenny Welter for their help. I am also grateful to Monique Calello for her help in producing the book. The comments by the following reviewers were extremely useful:
Winston Chan, University of Iowa,
Wai Tung Ng, University of Toronto,
Jacob B. Khurgin, Johns Hopkins University,
Dee-Son Pan, University of California, Los Angeles,
Chang Liu, University of Illinois, Urbana-Champaign,
Jorge Santiago-Aviles, University of Pennsylvania,
Gregory Lush, University of Texas at El Paso.
I am also grateful to Professors Stella Pang, Dimitris Pavlidis, George Haddad, and Pallab Bhattacharya for sharing photographs that were used in the book.

Jasprit Singh
singh@engin.umich.edu
www.eecs.umich.edu/~ singh

Contents

3 CARRIER DYNAMICS IN SEMICONDUCTORS 89

APPENDICES

Introduction

I.1 SOLID STATE ELECTRONICS IN MODERN LIFE

As children and as adults we are fascinated by the stories from mythologies and fairy tales. The laws of physics seem to be taking a rest as we read about doors that open by a magic word or demigods that can destroy life merely by looking at an object. If we now take a stroll through an exhibit of modern electronic products, we are amazed to see that many of those fantasies were, after all, not inconsistent with the laws of physics. Modern solid state electronics is rapidly turning fantasy into reality. Laser scanners, heat and chemical sensors, video cameras, fax machines, etc., are turning our lives into a magic land. The device scientist, the systems designer, and the software engineer have created a veritable "wish fountain" where one can ask for almost any kind of gadget and it is delivered.

Today's information-age technologies depend upon two general classes of electronic technologies—digital and analog. Digital technologies are based on Boolean algebra where all sorts of mathematical operations can be performed using a two-state system. The two-state system is represented by 0 and 1 and provides the basis for a binary number-based technology. The great advantage of digital technology is its high immunity to noise and the ease with which a vast array of information manipulation algorithms can be implemented.

While digital technology offers tremendous advantages in information manipulation (addition, subtraction, multiplication, etc.), the real world we live in is essentially analog in nature. The sounds we hear and produce, the sights we see, the forces we feel, etc., are continuously varying. For many applications, these continuous signals are discretized, i.e., converted into a digital representation (this process is called *analog-to-digital* or A/D conversion). After the information is manipulated, the final information may be reconverted from digital to analog (D/A conversion). Analog devices are needed to amplify small signals, produce signals (oscillators), shape signals, etc.

Digital Devices

A basic digital device needed for Boolean algebra is the inverter, shown in Fig. I.1a. In this device, an input (usually small) controls the output (large). When the input is low (representing a 0 state) the input is high (state 1) and when the input is high (state 1) the output is low (state 0). In an ideal inverter the output switches sharply about a value V_i, as shown in Fig. I.1a.

The basic inverter can be combined in circuits to produce various Boolean operations, such as AND, OR, NOR, etc., thus forming the basic building blocks for computation. The inverter is essentially a switch producing a state of high conductance and low conductance. Before the advent of semiconductor technology, mechanical relays and vacuum tubes were used for the implementation of digital logic. However, as we will see in this book, semiconductor devices have excellent qualifications. Not only do they provide the characteristics needed for the basic inverter, they are also inexpensive, reliable, and compact. In Fig. I.2 we show a schematic of how one starts from the basic inverter and builds the most complex digital circuitry.

Analog Devices
Analog devices form a critical component of information-processing technology. As noted earlier, the "natural" information of our world is not digital, but analog. Devices are needed to amplify signals that are very weak, generate signals, and modify other signals. A key need in such devices is gain, i.e., a small input signal should be able to control a large output signal, as shown schematically in Fig. I.1b. Used in an appropriate circuit, the high gain of an amplifier can be exploited not only to amplify and differentiate signals, but also to generate signals (i.e., as an oscillator).

In addition to devices that can act as an inverter (switch) and/or have high gain for amplification, another useful building block for information processing is a rectifying device with characteristics shown in Fig. I.1c. In this case the device is conducting if the bias is above a certain value, V_c, and is non-conducting otherwise. Such a device with strongly nonlinear characteristics can be used in a host of applications, such as for "mixing" different electrical signals and "shaping" signals, as well as in digital signal processing.

The input-output responses reviewed here and needed for information processing can be produced in a variety of systems. They can be generated from vacuum tubes, from electromechanical relays, even from certain kinds of plastics and polymers. However, while there are special niche areas where vacuum tubes are useful (e.g., in high-power microwave applications) and where electromechanical relays are used, semiconductor-based devices form the backbone of information-age technologies.

In this text we will see why semiconductors are the materials of choice for microelectronic technologies. We will see why metals and insulators—the other classes of materials—are relegated to passive though important roles in which semiconductors form the "active" materials for devices.

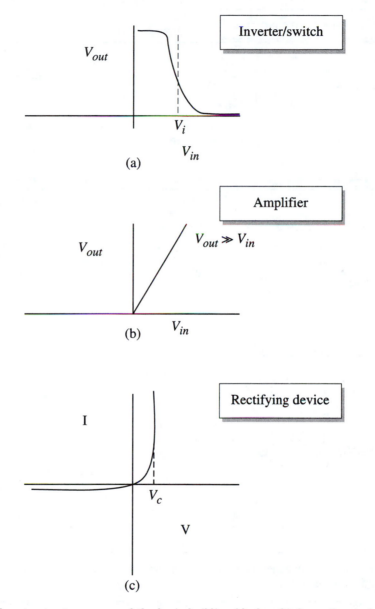

Figure I.1: Input-output response of the basic building blocks of information technology. (a) An inverter response, (b) an amplifier response, and (c) response of a rectifier.

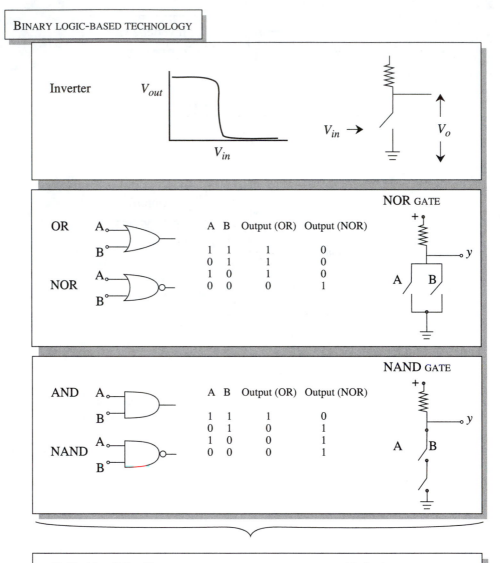

Figure I.2: A figure showing how, starting from the inverter, various building blocks of digital technology can be produced.

In this text, we will provide an understanding of the following issues:

• *Semiconductors have special properties that allow one to alter their conductivities from very low values to very high values*: We will examine why only semiconductors (not metals or insulators) have this unusual property.

• *Current transport in semiconductors can occur by two different kinds of particles— electrons and holes*: The density of electrons and holes can be altered in a semiconductor by doping and by applying a bias.

• *Semiconductor devices can be designed that have input-output relations to produce rectifying properties; inverters and amplifiers.* We will discuss the underlying basis of each of these devices. In particular, we will discuss *p-n* and Schottky diodes (rectifying devices), bipolar junction transistors, and field effect transistors (for digital and analog applications).

• *Semiconductor devices can be combined with other elements (resistors, capacitors and even other devices) to produce circuits on which modern information-processing chips are based.* We will briefly review the basic circuits for these building blocks.

If we examine the physics, technology, and computer science upon which modern information technologies are based we get an overview, shown in Table I.1. At the most basic level we have the building blocks of our universe—quantum electrons, protons, neutrons, mesons, etc. Starting from these one moves to higher levels, i.e., atoms, molecules, crystals, etc., as shown. Topics covered in this text are shown on a shaded background in Table I.1.

I.2 INFORMATION AGE: THE ARSENAL

The information age is influencing the way we work and play. At work the computer and communication revolution has enhanced productivity, while the Internet has brought world-wide information to our offices and homes. The rapid use of the Internet for electronic commerce is altering entire businesses. Entertainment has also been transformed by the internet and high-speed communication networks. Cable television, on-demand movies and shows, Internet and television shopping—the list is endless. The hardware that is driving this revolution is based on semiconductor devices—the subject of this book.

We can loosely divide the products of the microelectronic market into the categories shown in Fig. I.3. Let us take a brief overview of these "weapons" in the hardware arsenal.

Digital Signal Processing Systems (DSPs): These are an important set of tools in modern information systems. They are used for applications such as medical imaging, speech recognition, 3D graphics, output data visualization for complex problems, etc. They can be found in hardware equipment such as modems, hearing aids, cell

INFORMATION TECHNOLOGIES

Software: Operating systems, data management systems, graphic display packages...

Systems: Computers, communications...

Integrated circuit chips: Microprocessor, RAMs...

Circuits: Inverter, AND, OR... amplifier, oscillator...

Electronic devices: Input-output relations

Electrons in crystals: Band theory, transport theory

Solids: Crystals, non-crystals

Atoms, molecules

Electrons, protons, neutrons, mesons

Quarks

Table I.1: From fundamentals to applications—subjects that make up the basis of information technologies. Topics covered in this text are shaded.

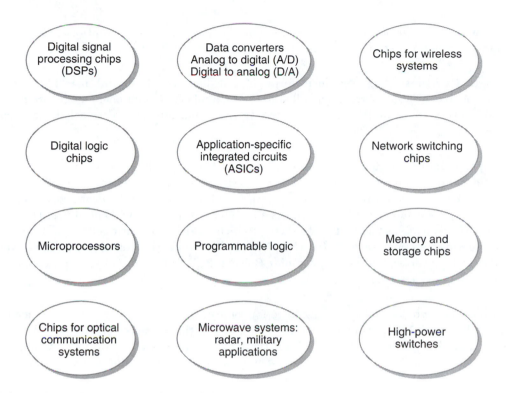

Figure I.3: "Tools of the trade" in the hardware used in modern information systems.

phones, pagers, digital cameras, control circuitry for heating/cooling, etc. The chips are specifically designed to implement certain mathematical algorithms for information manipulation.

Data Converter Chips: Most of the information with which we interface is analog in nature. The sounds we hear and produce, the wind pressure and temperature we feel, etc., are all examples of analog information. There is a continuous variation in the signal. However, from the point of view of information reception, manipulation, storage, transmission, etc., it is more efficient to use digital technology. Analog-to-digital (A/D) and digital-to-analog (D/A) systems are designed to convert rapidly from one domain to another while maintaining a certain resolution level. Such chips find applications in audio, graphics, communications, video, industrial control, disc drive control, etc.

Wireless Communications: Wireless systems are proving to be one of the biggest drivers of integrated circuit (IC) products today. It is argued that once the cost of using wireless services decreases, practically all systems will be connected by wireless systems. Printers, faxes, laptop computers, phones, etc., will no longer be hardwired to each other to communicate. Wireless systems are being driven by: (a) smaller, faster, and less power-hungry semiconductor chips; (b) DSPs with superior performance; (c) linear radio frequency (rf) components which reduce power consumption drastically, and (d) superior batteries that are lighter, smaller, and have a longer lifetime.

Logic Products: Logic products form the backbone of the digital age. They add, subtract, crunch numbers, and do all kinds of manipulations of numbers. Logic products are in every conceivable electronic system from fax machines to workstations and cell phones.

Application-Specific Integrated Circuits (ASICs): ASICs represent a growing trend in semiconductor technology. ICs are designed for specific customer applications, thus optimizing performance. Areas of wireless and wired telecommunications, networking, computers and consumer electronics are all benefiting from ASIC products. Designers have tremendous flexibility with these products, which have basic cells capable of implementing functions such as: (i) AND/NAND/OR/NOR; (ii) exclusive-OR/exclusive-NOR; (iii) inverter/buffers; (iv) complex Boolean logic functions; (v) multiplexers; (vi) clock signal generators, etc.

Wired Networks: A key driving force for semiconductor technology is the ability to provide broadband communication to homes and within business enterprises. Analog and digital chips are needed to control the traffic in local area networks (LANs) and wide area networks (WANs). Local area networks depend on the ethernet and a number of companies provide specific chips to deal with the networking challenges.

There are key needs for remote access to the Internet, corporate intranet, and WAN. Routers, asynchronous transfer mode (ATM) switches, remote access servers, asymmetrical digital subscriber line (ADSL) modems, very-high-bit-rate digital subscriber line (VDSL), etc., are central to modern information technology.

Microprocessors: These chips form the "brain" of the computer and are responsible for the ubiquity of the personal computer. Ever faster microprocessors are making desktop and even laptop computers do complex tasks, like image manipulations, in a fraction of the time taken by mainframe computers just a few years ago. Microprocessors are now being used not only in computers, but for machine control, automobiles, aircraft ... the list is endless.

Programmable Logic Devices: Chips that can be programmed to perform functions specific to a user task. Depending on specific needs, the user could use the same chip for multiple applications.

Memory and Storage Devices: The adage that there is never too much memory is perhaps truer than ever as chips with memory capacity of 64 MB and higher make their way into technology. Memory is as essential to technology as it is to us humans! Technology companies produce memories with exotic-sounding names: dynamic random-access memory, or DRAMs; first in first out (FIFO); video RAM (VRAM); synchronous DRAM (SDRAM); ultraviolet-erasable programmable read-only RAM (EPROM), etc.

Optical Communication Chips: A fiber optic communication revolution has been spawned by a new class of chips—chips that generate and manipulate optical signals. These chips convert signals into optical pulses, guide and switch these pulses, and finally, detect and switch the pulses.

Microwave Systems: Microwave systems are used in applications such as radar, remote sensing of terrestrial features, communication systems, and special military applications. While in digital applications most systems use semiconductors, in high-power microwave applications (radar, radio broadcasts, etc.), vacuum tubes still are devices of choice.

High-Power Switches: High-power switches are needed in factories, in power supplies and in power stations. As in high-power microwave systems, this is another area where semiconductor chips often are unable to compete with tube technology.

Semiconductor-based technologies are now firmly entrenched in the global economy. In Fig. I.4 we show the annual sales of semiconductor products as a function of time. We can see the rapid growth of this industry. What is even more important is the fact that semiconductor devices have spawned entirely new kinds of commercial and social interactions. The Internet is a prime example of this.

I.3 ROLE OF THIS BOOK

The role of this textbook is to prepare students at the undergraduate level so that they can appreciate (i) why semiconductors have such unique properties and are the materials of choice for devices; (ii) what material properties are required for developing superior devices; (iii) how various classes of semiconductor devices function; and (iv) what the current trends are in materials and devices.

Figure I.5 shows the topics covered in this book. The first three chapters of this book deal with the physics behind semiconductor devices. The level of physics needed to fully appreciate physical phenomena governing the devices is more complex than can be covered in an undergraduate textbook. Graduate-level courses on this level of physics would include: (a) bandstructure studies based on the tight-binding method or pseudo-potential method; (b) phonons in semiconductors; and (c) transport and optical processes based on matrix formulation of quantum mechanics and perturbation theory. Such depth is neither possible nor appropriate for an undergraduate book. Nevertheless, to understand semiconductor devices, the student must appreciate the concepts of effective mass, valence and conduction bands, bandgap, mobility, diffusion, etc.

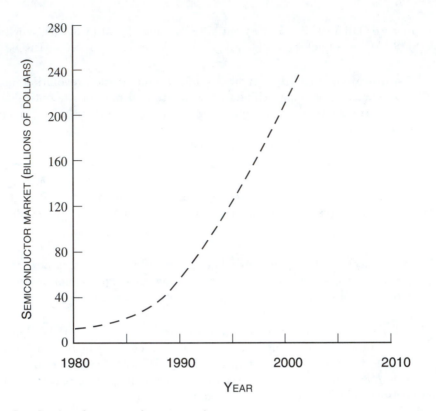

Figure I.4: Semiconductor market versus time.

The first three chapters present the physical basis of the concepts needed. The student is shown, through simple arguments, the path that leads to concepts such as effective mass, mobility, etc. The student will still have to make a leap of faith in accepting the concepts of mobility, absorption coefficient, etc. However, this leap of faith involves more of the mathematical framework than of the underlying physics. As the instructor and the reader will discover, the physics presented in the first three chapters is at the level of an undergraduate student with some background in mathematics.

Chapter 4 discusses basic material synthesis and device fabrication issues. This chapter can be assigned as reading material to the students. Chapters 5–10 cover electronic devices and form the bulk of this text. Chapter 11 discusses optoelectronic devices. Topics from this chapter could be covered in discussion sections once p-n diode theory is understood. Some of the important features of this text are listed below.

Solved Examples: The textbook has about 150 worked examples. This feature should be useful not only in a classroom setting, but also for practicing engineers and applied scientists who may want to learn more about modern semiconductor devices.

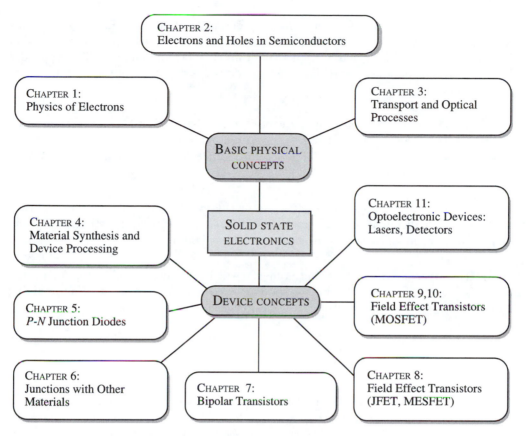

Figure I.5: A flow chart of topics covered in this text.

Units: The book uses SI units throughout. Many worked examples provide the units at each step. The student may notice the use of centimeters at some places and meters at others. Also, the energy unit is joules in some places and electron volts at others. These are to conform to standard practices and should not cause the student any difficulty.

End-of-Chapter Problems: There are several kinds of end-of-chapter problems. There are the usual problems that test the understanding of topics covered in the chapter. These are arranged according to the sections in which the relevant material is covered. In addition there are problems under the heading *Cumulative Chapter Problems*. The conceptual chapters of the text have several problems under this heading. They test the students' ability to integrate their knowledge. Finally, in device chapters there are problems under the heading *Design Problems*. These problems are important for the student to address. Usually the solutions to these problems are not unique and students are challenged to make intelligent guesses.

Relationship to SPICE: It is very important for students to relate their understanding of device physics to circuit design programs such as SPICE. To develop this relationship the book contains several sections (e.g., Section 5.7 for *p-n* diodes, Section 7.9 for bipolar transistors, and Section 10.4 for MOSFETs) and an appendix (Appendix D) where SPICE parameters are discussed and derived using basic physical understanding of devices.

A Bit of History: Several chapters contain brief historical studies of developments covered in the chapter. A good source for such studies is M. Eckert and H. Shubert, *Crystals, Electrons, Transistors: From Scholar's Study to Industrial Research*, (American Institute of Physics, New York, 1990).

It is important to note that some of the events presented in the historical sections may be disputed, since microelectronics and optoelectronics is still a developing area. Several companies are still litigating on patent issues.

Web Resources: Instructors and students will find many useful resources on the author's website: www.umich.edu/~singh. Particularly useful will be *pdf* files containing figures and tables.

I.4   GUIDELINES FOR INSTRUCTORS

Guidelines for the use of this text in introductory courses on semiconductor electronic devices are presented in Table I.2. A rough lecture schedule is also suggested for a semester with thirty-five one-hour lectures. These are, of course, only suggested guidelines and may benefit an instructor using the book for the first time. It may be useful for an instructor to use visual aids to avoid spending too much class time drawing complicated device structures or writing out long equations. The use of viewgraphs can be helpful and the emphasis can be kept on the real information to be conveyed. It is also very useful if the students can be provided a detailed lecture outline with a strong suggestion to read ahead.

<table>
<tr><td rowspan="7" style="font-size:2em">1</td>
<td>• Discussion of a simple classical model for Ohm's law and the failure of classical physics in explaining semiconductors</td><td>(1 lecture)</td></tr>
<tr><td>• Basic lattice structure of semiconductors</td><td>(1 lecture)</td></tr>
<tr><td>• Discussion of quantum-mechanical approach to physical phenomena</td><td>(1 lecture)</td></tr>
<tr><td>• The free electron problem; density of states concepts</td><td>(1 lecture)</td></tr>
<tr><td>• Electron statistics</td><td>(1 lecture)</td></tr>
<tr><td>• Electrons in a periodic structure; from atomic levels to bands; concept of k-vector</td><td>(1 lecture)</td></tr>
<tr><td>• Metals, semiconductors, insulators</td><td>(1 lecture)</td></tr>
</table>

1

- Discussion of a simple classical model for Ohm's law and the failure of classical physics in explaining semiconductors — (1 lecture)
- Basic lattice structure of semiconductors — (1 lecture)
- Discussion of quantum-mechanical approach to physical phenomena — (1 lecture)
- The free electron problem; density of states concepts — (1 lecture)
- Electron statistics — (1 lecture)
- Electrons in a periodic structure; from atomic levels to bands; concept of k-vector — (1 lecture)
- Metals, semiconductors, insulators — (1 lecture)

2

- Semiconductors; effective mass concept — (1 lecture)
- Concept of holes — (1 lecture)
- Discussion of important semiconductor bandstructures — (1 lecture)
- Intrinsic carrier concentration — (1 lecture)
- Concept of doping; carriers in doped semiconductors — (1-1/2 lectures)

3

- Concepts of scattering and mobility — (1 lecture)
- Drift transport (velocity-field relations, breakdown) — (1 lecture)
- Brief discussion of optical processes; quasi-Fermi level concepts — (1 lecture)
- Nonradiative recombination — (1/2 lecture)
- Continuity equation; drift diffusion — (1 lecture)

4

- *This chapter on material synthesis and device processing can be assigned as reading material at various stages of the course.*

5

- *P-N* junction equilibrium — (1 lecture)
- *P-N* junction under bias (issues related to light-emitting diodes; detectors can be briefly mentioned) — (2 lectures)
- Time response of a *p-n* diode — (1-2 lectures)

Table I.2: Suggested topics for an introductory course oriented towards basic electronic devices.

6	• Discussion of Schottky diode	(2 lectures)
	• Discussion of ohmic contacts	(1/2 lecture)

7	• Conceptual workings of a bipolar device	(1 lecture)
	• I-V characteristics for a bipolar device; BJT performance parameters	(2 lectures)
	• Issues in a real bipolar device	(reading assignment)
	• High-frequency/high-speed issues	(1 lecture)
	• Band tailoring and HBT	(1 lecture)
	• Advanced devices	(reading assignment)

8	• JFET, MESFET current-voltage characteristics	(3 lectures)
	• Issues in real devices	(reading assignment)
	• MODFET	(optional)
	• High-frequency/high-speed issues	(1 lecture)
	• Advanced device issues	(reading assignment)

9, 10	• MOS capacitor	(2 lectures)
	• MOSFET current-voltage characteristics	(1 lecture)
	• CMOS inverter	(1 lecture)
	• High-frequency/high-speed issues	(1 lecture)
	• Issues in real devices	(reading assignment)
	• CCD	(reading assignment)

11	*Section on LEDs, laser diodes, and detectors can be assigned to discussion sections after the p-n diode is covered.*

Table I.2: Suggested topics for an introductory course oriented towards basic electronic devices (continued).

Chapter

1

ELECTRONS IN SOLIDS

Chapter at a Glance

1.1 INTRODUCTION

Information-processing devices that have been responsible for the modern information age are based on one common theme. If an input is applied to them, a well-defined output results and the nature of the input-output relation is such that it can be used to carry out basic information-processing operations such as binary logic, amplification, etc. The input may be due to an electrical signal or an electromagnetic wave or a pressure wave, etc. The output may be a current pulse or a voltage pulse or light pulse, etc. This input-output relation is then exploited to design digital switches, amplifiers, memory devices, oscillators, lasers, etc. In order to understand and then exploit the input-output relation, we need to understand how electrons behave inside the material forming the electronic device.

All electronic and optoelectronic devices based on semiconductors exploit special properties electrons have inside semiconductors. The neutrons and protons that are present do not participate directly in any physical process, although they are essential and provide the chemistry that causes the material to be a semiconductor. The neutrons and protons are immobile, while the electrons, being very light particles in comparison, are "free" to move around and respond to external perturbations.

To understand the properties of semiconductors and their devices, we must understand the properties of electrons inside semiconductors. In particular, we should be able to understand two aspects of electronic properties: i) what are the energy, momentum, position, etc., of the electron inside a semiconductor; and ii) how do electrons respond to an external perturbation such as an electric field, magnetic field, electromagnetic field, etc.? In free space we know that electrons obey the classical "free" electron equations:

$$E = \frac{p^2}{2m_0}$$

$$\frac{d\mathbf{p}}{d\mathbf{t}} = \mathbf{F}_{\text{ext}} = -e\left(\mathbf{F} + \mathbf{v} \times \mathbf{H}\right)$$

where m_0, E, $\mathbf{p}$, $\mathbf{F}_{\text{ext}}$, and $-e$ are the electron mass, energy, momentum, force, and charge. The external forces include the effects of an electric field $\mathbf{F}$ and a magnetic field $\mathbf{H}$. Using such classical equations we are able to describe a number of important properties of electrons. For example, we can use these equations to design a cathode ray tube (like a TV or a computer screen) by predicting how electrons will strike the screen pixels. However, a number of important things cannot be understood with these classical equations. We cannot understand, for example, why metals like aluminum or copper have such a high conductivity while materials like glass or silicon or diamond have such poor conductivity. In fact, we cannot understand any significant aspect of semiconductor technology. To develop this understanding we need to use quantum mechanics. Once we know some basic properties of electrons in semiconductors through quantum mechanics we can then develop some simple *classical-looking* equations to describe semiconductors and their devices.

The first step in our understanding of semiconductors is to appreciate the basic concepts of quantum mechanics. In this chapter we will discuss some important properties of electrons using quantum mechanics. Before doing so, let us briefly discuss some classical attempts to explain how materials behave.

1.2 ELECTRONIC MATERIALS: FAILURE OF CLASSICAL PHYSICS

In an electrical engineer's world all materials are divided into three categories: (i) Metals—materials which have a very high conductivity or low resistivity to current flow; (ii) Insulators—materials which have very poor conductivity; and (iii) Semiconductors—materials whose conductivity falls in between metals and insulators. Each kind of material mentioned above finds important uses. For example, in an extension cord used to bring electricity from an outlet to, say, a hair dryer, the inner core is made of copper—a metal—while the outer cover is made from an insulator. The metal carries the current due to its low resistivity while the insulator prevents us from getting a shock.

An important law that is widely used in electrical engineering is Ohm's law, which tells us how current I flows in a material when a potential V is applied. The law has the form

$$I = \frac{V}{R} \tag{1.1}$$

This law was introduced by Ohm on an empirical basis, i.e., on the basis of observation with no *fundamental explanation*. Let us express Ohm's law in some other equivalent forms. If A is the area of the sample and L its length, the resistance of the sample is

$$R = \rho \frac{L}{A} = \frac{1}{\sigma} \frac{L}{A} \tag{1.2}$$

where ρ and σ are known as the resistivity and conductivity of the material respectively. We may write the current as

$$I = JA \tag{1.3}$$

where J is the current density. We may also write the electric field applied to the sample as

$$F = \frac{V}{L} \tag{1.4}$$

Ohm's law now becomes

$$J = \sigma F \tag{1.5}$$

Let us proceed further along this path. We have by the definition of current density

$$J = nev \tag{1.6}$$

where n is the density of the carriers carrying current, e is their charge, and v is the average velocity with which the carriers are moving in the direction of the field. We may also define a relation between velocity of the carriers and the applied field through a definition

$$v = \mu F \tag{1.7}$$

where μ is called the mobility of the carrriers. With this definition we get

$$\sigma = ne\mu \tag{1.8}$$

All of the equations we have given above are just different ways of expressing Ohm's law. Even though we have introduced carrier density and charge and mobility, however, *we have not derived Ohm's law*. The first attempt to derive Ohm's law was due to Drude at the turn of the 20th century. At this time there was no knowledge of quantum mechanics and Drude built a model based on the chemistry and classical physics known at that time. He exploited his understanding of atoms, electrons, and nuclei.

The picture of the solid used by Drude is shown schematically in Fig. 1.1. Solids are a collection of atoms. Each atom has a nucleus around which electrons are present. The total number of electrons in each atom is Z_a, which is the atomic number of the atom. Out of these electrons, Z_c are *valence* electrons in the outermost shell of the atom. These electrons are weakly bound to the atoms and are capable of carrying current when a field is applied. The remaining electrons are *core* electrons, which are strongly bound to the nucleus and don't participate in current flow.

We are now interested in the behavior of a very large number of negatively charged electrons moving through positively charged fixed ions. To appreciate the enormity of the problem let us examine the number of electrons involved. An element contains 6.022×10^{23} atoms per mole (Avogadro's number). If ρ is the density of the material, the number of moles per unit volume is ρ/A, where A is the atomic mass. We now assume that the number of electrons that are free to conduct current per atom is Z_c, the number of electrons in the outermost shell (i.e., the valence of the element) of the atom, as shown in Fig. 1.1. The electron density for the conduction electrons is now

$$n = 6.022 \times 10^{23} \frac{Z_c \rho}{A} \tag{1.9}$$

For most materials, this number is $\sim 10^{23}$ cm^{-3}! There is an enormously large density of free conduction electrons in the material. One can define an average radius r_s of a spherical volume per electron by

$$\frac{4\pi r_s^3}{3} = \frac{1}{n}$$

or

$$r_s = \left(\frac{3}{4\pi n}\right)^{1/3} \tag{1.10}$$

This radius is 1-2 Å for most materials!

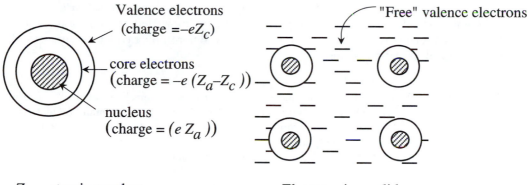

Figure 1.1: A conceptual picture of an atom showing the nucleus with charge eZ_a, the core electrons, and the valence electrons. In a solid the valence electrons are "free" and are capable of charge conduction. The electron concentration in solids is quite high.

Drude wanted to understand why the conductivity of different materials is different. For this purpose he used the model of atoms described above. He made the following assumptions for his model:

• A solid is made up of a collection of atoms.
• Each atom has electrons around the nucleus as shown in Fig. 1.1.
• The electrons can be divided into two categories—electrons that are *core* electrons and electrons that are in the outer shell of the atom and are known as *valence* electrons.
• The core electrons are strongly bound to the atom and don't participate in any current conduction.
• The outer shell valence electrons are weakly bound to the nucleus and in the solid become free to carry current.
• As the electrons move in the presence of an applied electric field they scatter from the nuclei and reach a terminal velocity (much like raindrops falling through air).

An outcome of the Drude model is that the conductivity of any material is proportional to the valence electrons per volume that are in the material. While this model worked for some metals, it failed to explain why conductivity of materials like diamond (C) or silicon (Si) is so small. It also failed to explain a number of other important experimental observations that were made shortly after Drude introduced his model. Some of these observations are:

• In some materials current appears to be carried by positively charged particles!
• The conductivity of some materials is changed by orders of magnitude by introduction of *tiny* amounts (say one part in a million or less) of impurities.
• In some materials Ohm's law is not obeyed at all.

These and many other observations simply could not be reconciled with Drude's or any other model based on classical physics. We now know that to understand these observations we have to invoke quantum mechanics. Before doing so let us understand some structural aspects of solids.

EXAMPLE 1.1 In our discussion we have argued that classical mechanics cannot explain why some materials have high conductivity (metals) and others have very small conductivity (insulators). The conductivity (σ), is proportional to the electron density n. Calculate the electron density in the metal Al and the semiconductors C and Si. Also calculate the density of electrons that can carry current according to the Drude model.

To calculate the electron density we use the following expression:

$$\text{electron density} = 6.023 \times 10^{23} \frac{\text{Number of electrons per atom} \times \rho}{A}$$

where ρ is the material's mass density, the first number in the equation above is Avogadro's number, and A is the atomic mass of the element. We have

$$A_{Al} = 27 \quad A_C = 12 \quad A_{Si} = 28$$

The density of Al is 2.7 gcm^{-3}, that of C is 3.515, and that of Si is 2.33. The valence of Al is 3 and that of C and Si is 4.

In this problem we have to calculate the electron density using two different values for the electron number per atom. In the first case we use the *total* number per atom to get

$$
\begin{aligned}
n_{Al} &= 6.023 \times 10^{23} \frac{13 \times 2.7}{27} = 7.83 \times 10^{23} \text{ cm}^{-3} \\
n_{Si} &= 6.023 \times 10^{23} \frac{14 \times 2.33}{28} = 7.02 \times 10^{23} \text{ cm}^{-3} \\
n_C &= 6.023 \times 10^{23} \frac{6 \times 3.515}{12} = 1.058 \times 10^{24} \text{ cm}^{-3}
\end{aligned}
$$

We see that these densities are quite close to each other.

Using Drude's approach we get

$$
\begin{aligned}
n_{Al} &= 6.023 \times 10^{23} \frac{3 \times 2.7}{27} = 1.8 \times 10^{23} \text{ cm}^{-3} \\
n_{Si} &= 6.023 \times 10^{23} \frac{4 \times 2.33}{28} = 2.0 \times 10^{23} \text{ cm}^{-3} \\
n_C &= 6.023 \times 10^{23} \frac{4 \times 3.515}{12} = 7.06 \times 10^{23} \text{ cm}^{-3}
\end{aligned}
$$

We see that with this model Si has a larger electron density than Al. This cannot be reconciled within classical physics with the fact that Al has a much greater conductivity than Si.

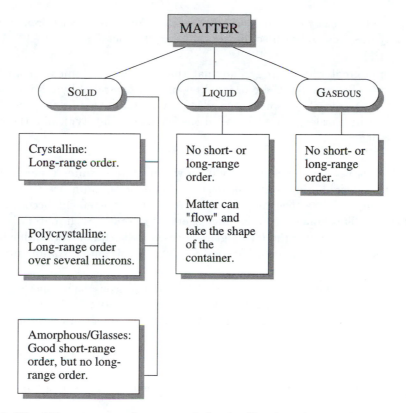

Figure 1.2: The different states of matter and the classification of materials according to the ordered nature of the arrangement of atoms.

1.3 STATES OF MATTER: ORDER AND CRYSTALS

The physical matter we observe in our universe can be characterized as solid, liquid, and gaseous, as shown in Fig. 1.2. One way to characterize these states of matter is by the density and the hardness of the material and the ability of the matter to flow. Thus a solid is usually dense with atomic spacing of a few angstroms (10^{-8} cm). A solid is also hard and has its own shape, in contrast to a liquid or a gas, which takes the shape of the container it is in.

For our purposes, a more useful way of characterizing materials is by the presence of order in the collection of atoms. While order can be defined precisely through mathematical expressions for the correlation functions for the arrangements of atoms, we will simply use a physical and intuitive definition of order. Let us imagine that we can actually look at the atoms inside a material. We may find that in some cases *the atoms are arranged in complete precision so that by knowing the position and species of a few atoms, we can predict the position and chemical nature of all the atoms in the sample.*

When this occurs, we say that the structure has long-range order. Such long-range order is found only in solids and such solids are called crystalline.

In some materials we may find that the precise arrangement of atoms exists over thousands of atoms, but these ordered regions are separated by a boundary across which there is no order. Such solids are called polycrystalline materials. The size of the region across which order persists is called the grain size and, typically, the grain size is a few microns.

There is yet another class of solids in which, if we examine a particular atom, we find that the neighboring atoms (nearest neighbors or second nearest neighbors) are precisely arranged, but as one moves further out, the arrangement becomes less and less predictable. This kind of order is called short-range order. Such materials are called amorphous materials. Often the term "glass" is used for such non-crystalline materials. Window glass or the glass used to have a drink is made of non-crystalline material.

Matter that falls in the category of liquid or gaseous phase has no short- (or long-) range order. The atoms have an average spacing between them determined by the density of the material, but there is no precise arrangement of the atoms.

In this text we will be dealing with semiconductor devices made from crystalline materials that have long-range order. While there are semiconductor devices that are made from amorphous materials, their performance is relatively very poor.

Crystals are made up of identical building blocks, the blocks being an atom or a group of atoms. The underlying periodicity of crystals is the key that controls the properties of the electrons inside the material. The intrinsic property of a crystal is that the environment around a given atom or atoms is exactly similar to the environment around another atom or atoms. The building block that is repeated infinitely to produce the crystal may be quite simple, such as a single atom for many metals (such as copper, gold, aluminum, etc.), or can be quite complex (as in some proteins). In man-made superlattices the number can be arbitrary. For almost all *"natural" semiconductors*, the building block consists of two atoms. To understand and define the crystal structure, two important concepts are introduced.

The *lattice* represents a set of points in space that form a periodic structure. Each point sees exactly the same environment. The lattice is by itself a mathematical abstraction. A building block of atoms called the *basis* is then attached to each lattice point, yielding the crystal structure.

An important property of a lattice is the ability to define three vectors $\mathbf{a}_1$, $\mathbf{a}_2$, $\mathbf{a}_3$ such that any lattice point $\mathbf{R}'$ can be obtained from any other lattice point $\mathbf{R}$ by a translation:

$$\mathbf{R}' = \mathbf{R} + m_1\mathbf{a}_1 + m_2\mathbf{a}_2 + m_3\mathbf{a}_3 \qquad (1.11)$$

where m_1, m_2, m_3 are integers. Such a lattice is called a Bravais lattice. The entire lattice can be generated by choosing all possible combinations of the integers m_1, m_2, m_3. The crystalline structure is now produced by attaching the basis to each of these lattice points:

$$\boxed{\text{lattice} + \text{basis} = \text{crystal structure}} \qquad (1.12)$$

The translation vectors $\mathbf{a}_1$, $\mathbf{a}_2$, and $\mathbf{a}_3$ are called primitive if the volume of the cell formed by them is the smallest possible. There is no unique way to choose the primitive vectors. One choice is to pick

$\mathbf{a}_1$ to be the shortest period of the lattice
$\mathbf{a}_2$ to be the shortest period not parallel to $\mathbf{a}_1$
$\mathbf{a}_3$ to be the shortest period not coplanar with $\mathbf{a}_1$ and $\mathbf{a}_2$

Figure 1.3 shows how one starts with a lattice and builds a crystal by placing a basis on each lattice site. The basic building block of the crystal is called the unit cell.

The various kinds of lattice structures possible in nature are described by the symmetry that describes their properties. We will focus on the cubic lattice, which is the underlying structure taken by all currently important semiconductors.

1.3.1 Cubic Lattices

There are three kinds of cubic lattices: simple cubic, body-centered cubic, and face-centered cubic.

Simple cubic: The simple cubic lattice is generated by the primitive vectors

$$a\hat{\mathbf{x}}, a\hat{\mathbf{y}}, a\hat{\mathbf{z}} \qquad (1.13)$$

where the $\hat{\mathbf{x}}, \hat{\mathbf{y}}, \hat{\mathbf{z}}$ are unit vectors. Figure 1.4 shows the basic cubic structure and the primitive vectors. None of the semiconductors found in nature have a simple cubic structure.

Body-centered cubic: The body-centered cubic (bcc) lattice can be generated from the simple cubic structure by placing an atom at the center of the cube as shown in Fig. 1.5.

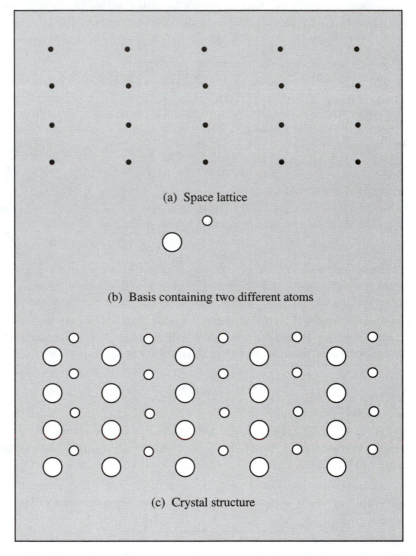

(a) Space lattice

(b) Basis containing two different atoms

(c) Crystal structure

Figure 1.3: The figure shows how a crystal structure is formed. One starts with a lattice. Then an arbitrarily complex basis consisting of atoms oriented in a particular manner to each other is placed at each lattice point in exactly the same order to form the crystal structure.

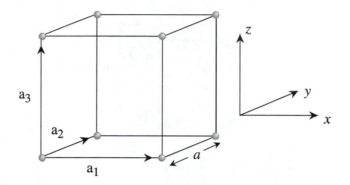

Figure 1.4: The simple cubic structure. The lattice points are simply the points at the edge of the cube of side a.

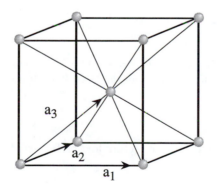

Figure 1.5: The body-centered cubic (bcc) lattice. Atoms are placed at the edges of a cube of side a and at the center of the cube. The crystal is formed by replicating the cube to fill space.

Face-centered cubic: The most important lattice for semiconductors is the *face-centered cubic* (fcc) Bravais lattice. To construct the face-centered cubic Bravais lattice, add to the simple cubic lattice an additional point in the center of each square face (Fig. 1.6).

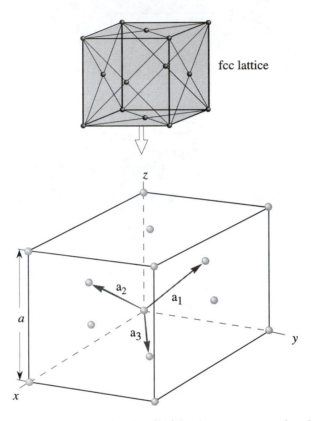

Figure 1.6: To create the face-centered cubic (fcc) lattice, atoms are placed at the edges of a cube of side a, as well as at the centers of all the faces. The figure also shows primitive vectors for the fcc lattice.

A symmetric set of primitive vectors for the face-centered cubic lattice is shown in Fig. 1.6 and is

$$\mathbf{a}_1 = \frac{a}{2}(\hat{\mathbf{y}} + \hat{\mathbf{z}}), \mathbf{a}_2 = \frac{a}{2}(\hat{\mathbf{z}} + \hat{\mathbf{x}}), \mathbf{a}_3 = \frac{a}{2}(\hat{\mathbf{x}} + \hat{\mathbf{y}}) \qquad (1.14)$$

where a is the cube edge and is called the lattice constant of the semiconductor. The face-centered cubic and body-centered cubic lattices are of great importance, since an enormous variety of solids crystallize in these forms with an atom (or ion) at each lattice site.

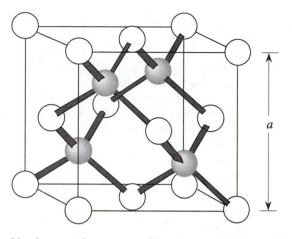

Figure 1.7: The zinc blende crystal structure. The structure consists of two interpenetrating fcc lattices, as discussed in the text.

1.3.2 Diamond and Zinc Blende Structures

Essentially all semiconductors of interest for electronics and optoelectronics have an underlying fcc lattice. However, they have two atoms per basis. The coordinates of the two basis atoms are

$$(000) \text{ and } \left(\frac{a}{4}, \frac{a}{4}, \frac{a}{4}\right) \tag{1.15}$$

Since each atom lies on its own fcc lattice, such a two-atom basis structure may be thought of as two interpenetrating fcc lattices, one displaced from the other by a translation along a body diagonal direction $(\frac{a}{4}\frac{a}{4}\frac{a}{4})$. The separation between the atoms is $\sqrt{3}a/4$.

Figure 1.7 gives details of this important structure. If the two atoms of the basis are identical, the structure is called diamond. Semiconductors such as Si, Ge, C, etc., fall in this category. If the two atoms are different, the structure is called the zinc blende structure. Semiconductors such as GaAs, AlAs, CdS, etc., fall in this category. Semiconductors with diamond structure are often called elemental semiconductors (although not all elemental semiconductors have a diamond structure), while the zinc blende semiconductors are called compound semiconductors. The compound semiconductors are also denoted by the position of the atoms in the periodic chart; e.g., GaAs, AlAs, InP are called III-V (three-five) semiconductors, while CdS, HgTe, CdTe, etc., are called II-VI (two-six) semiconductors. *The cube edge a is called the lattice constant of the crystal.* In a zinc blende crystal like GaAs there are four Ga atoms and four As atoms in each cube of volume a^3.

EXAMPLE 1.2 The lattice constant of silicon is 5.43 Å. Calculate the number of silicon

atoms in a cubic centimeter. Also calculate the number density of Ga atoms in GaAs, which
has a lattice constant of 5.65 Å.

Silicon has a diamond structure, which is made up of the fcc lattice with two atoms
on each lattice point. The fcc unit cube has a volume a^3. The cube has eight lattice sites at the
cube edges. However, each of these points is shared with eight other cubes. In addition, there
are six lattice points on the cube face centers. Each of these points is shared by two adjacent
cubes. Thus the number of points per cube of volume a^3 are

$$N(a^3) = \frac{8}{8} + \frac{6}{2} = 4$$

In silicon there are two silicon atoms per lattice point. The number density is, therefore,

$$N_{Si} = \frac{4 \times 2}{a^3} = \frac{4 \times 2}{(5.43 \times 10^{-8})^3} = 4.997 \times 10^{22} \text{ atoms/cm}^3$$

In GaAs, there is one Ga atom and one As atom per lattice point. The Ga atom density is,
therefore,

$$N_{Ga} = \frac{4}{a^3} = \frac{4}{(5.65 \times 10^{-8})^3} = 2.22 \times 10^{22} \text{ atoms/cm}^3$$

There are an equal number of As atoms.

EXAMPLE 1.3   In semiconductor technology, an Si device on a VLSI chip is one of the
smallest devices while a GaAs laser is one of the larger devices. Consider a Si device with
dimensions $(5 \times 2 \times 1)\mu m^3$ and a GaAs semiconductor laser with dimensions $(200 \times 10 \times 5)\mu m^3$.
Calculate the number of atoms in each device.

From Example 1.2, the number of Si atoms in the Si transistor are

$$N_{Si} = (5 \times 10^{22} \text{ atoms/cm}^3)(10 \times 10^{-12} \text{ cm}^3) = 5 \times 10^{11} \text{ atoms}$$

The number of Ga atoms in the GaAs laser are

$$N_{Ga} = (2.22 \times 10^{22})(10^4 \times 10^{-12}) = 2.22 \times 10^{14} \text{ atoms}$$

An equal number of As atoms are also present in the laser.

1.3.3   Notation for Planes and Points in a Lattice: Miller Indices

A simple scheme is used to describe lattice planes, directions, and points. For a plane,
we use the following procedure:

1) Define the x, y, z axes.
2) Take the intercepts of the plane along the axes in units of lattice constants.
3) Take the reciprocal of the intercepts and reduce them to the smallest
   integers, h, k, and l. The resulting numbers are called *Miller indices*.

The notation (hkl) denotes a family of parallel planes.

The notation {hkl} denotes a family of equivalent planes. For example, the planes {100},{010}, {001} are all equivalent in the cubic structure.

To denote directions, we use the smallest set of integers having the same ratio as the direction cosines of the direction. In a cubic system the Miller indices of a plane are the same as the direction perpendicular to the plane. The notation [ ] is used for a set of parallel directions. < > is used for a set of equivalent directions. Figure 1.8. shows some examples of the use of the Miller indices to define planes in semiconductors, based on the fcc lattice.

EXAMPLE 1.4 Calculate the surface density of Ga atoms on a Ga-terminated (001) GaAs surface.

In the (001) surfaces, the top atoms are either Ga or As, leading to the terminology Ga-terminated (or Ga-stabilized) and As-terminated (or As-stabilized), respectively. A square of area a^2 has four atoms on the edges of the square and one atom at the center of the square. The atoms on the square edges are shared by a total of four squares. The total number of atoms per square is

$$N(a^2) = \frac{4}{4} + 1 = 2$$

The surface density is then

$$N_{Ga} = \frac{2}{a^2} = \frac{2}{(5.65 \times 10^{-8})^2} = 6.26 \times 10^{14} \, \text{cm}^{-2}$$

EXAMPLE 1.5 Calculate the height of a GaAs monolayer in the (001) direction.

In the case of GaAs, a monolayer is defined as the combination of a Ga and As atomic layer. The monolayer distance in the (001) direction is simply

$$A_{m\ell} = \frac{a}{2} = \frac{5.65}{2} = 2.825 \, \text{Å}$$

1.4 COMING OF THE QUANTUM AGE

The classical description of the 1800s was quite adequate for describing complex phenomena, such as planetary motion, machine tools of various kinds, reflection, refraction and interference of light, etc. In fact, almost everything we could observe with our senses could be (at least superficially) examined by classical physics. At the beginning of the 20th century, some observations were made that started shaking the foundations of classical thinking and eventually led to quantum mechanics. Among these experiments

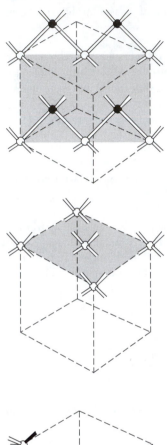

ATOMS ON THE (110) PLANE

Each atom has 4 bonds:
• 2 bonds in the (110) plane
• 1 bond connects each atom to adjacent (110) planes

⟹ Cleaving adjacent planes requires breaking 1 bond per atom

ATOMS ON THE (001) PLANE

2 bonds connect each atom to adjacent (001) plane

Atoms are either Ga or As in a GaAs crystal

⟹ Cleaving adjacent planes requires breaking 2 bonds per atom

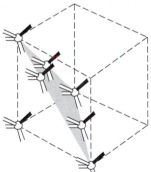

ATOMS ON THE (111) PLANE

Could be either Ga or As

1 bond connecting an adjacent plane on one side

3 bonds connecting an adjacent plane on the other side

Figure 1.8: Some important planes in the fcc lattice-based semiconductor structures along with their Miller indices. This figure also shows how many bonds connect adjacent planes. This number determines how easy or difficult it is to cleave the crystal along these planes by cutting the bonds joining the adjacent planes.

were: i) the photoelectric effect, which was explained by Einstein using the hypothesis that light consisted of particles called photons that had energy equal to $h\nu$ where h is Planck's constant ($h = 6.626 \times 10^{-34}$ J·s) and ν is the frequency of light; ii) radioactivity was discovered; iii) superconductivity was discovered; and iv) atomic spectra were measured, which suggested electrons behaved like waves.

From the point of view of solid state electronics, the observation of electronic levels in atoms was the most significant experiment. Understanding of the atomic levels through quantum mechanics eventually allowed us to understand band theory—the basis of modern electronic devices.

1.4.1 Particles Behaving as Waves: Atomic Spectra

A key area where classical physics broke down was the area of atomic physics. The most serious challenge for any of the pre-quantum mechanics models arose from the detailed measurements of the frequencies of the emitted and absorbed radiation by atoms. When atoms are excited (say, by electromagnetic radiation), they can absorb radiation. Once they absorb radiation, they can emit radiation as well. It was found experimentally that emitted and absorbed spectra from a species of atoms consisted of several series of sharp lines i.e., discrete frequencies, as shown in Fig. 1.9. It was possible to fit simple relations to the positions of these lines. For example, Johannes Balmer found that the emission wavelengths of hydrogen in the visible regime could be fit to the relation (the Balmer formula)

$$\lambda_n = 364.5 \frac{n^2}{n^2 - 4} \; nm; \quad n = 3, 4, \ldots \tag{1.16}$$

In fact, other groups of lines in H-spectra were fitted to other expressions. For example, we have the following:

Paschen Series:

$$\lambda_n = 820.1 \frac{n^2}{n^2 - 3^2} \; nm; \quad n = 4, 5, \ldots \tag{1.17}$$

Lyman Series:

$$\lambda_n = 91.35 \frac{n^2}{n^2 - 1} \; nm; \quad n = 2, 3, \ldots \tag{1.18}$$

Other atoms were found to have spectra that satisfied similar relations. In Fig. 1.9 we show series of lines observed in atomic spectra of hydrogen. The observation of atomic spectra showed that the electrons inside atoms occupy well-defined discrete energy levels that have a special relationship to each other.

According to classical physics, a particle can have a continuous range of energies, since energy is just the sum of kinetic and potential energy. The notion that a particle's energies could be "quantized," i.e., limited to certain discrete values, is inconceivable in

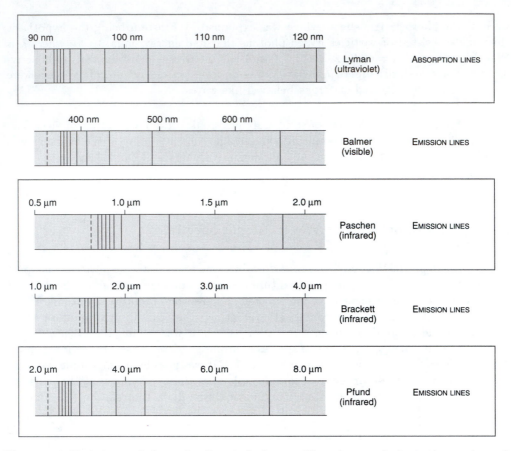

Figure 1.9: Emission and absorption lines in hydrogen. There is a regularity in the spacings of the spectral lines and the lines get closer as they reach the upper limit of each series (denoted by the dashed lines).

classical physics. It is clear that the classical picture of a particle (electron) had to be extended.

1.4.2   De Broglie's Hypothesis

It is known that when we have wave phenomena such as waves from a string instrument, the wavelengths or frequencies that are allowed are fixed. What if a particle also had a wave-like nature? In that case, perhaps the energy it could have would also be noncontinuous, as observed in the atomic spectra. A hypothesis was put forward by de Broglie according to which a particle was given a wave character by assigning a wavelength to

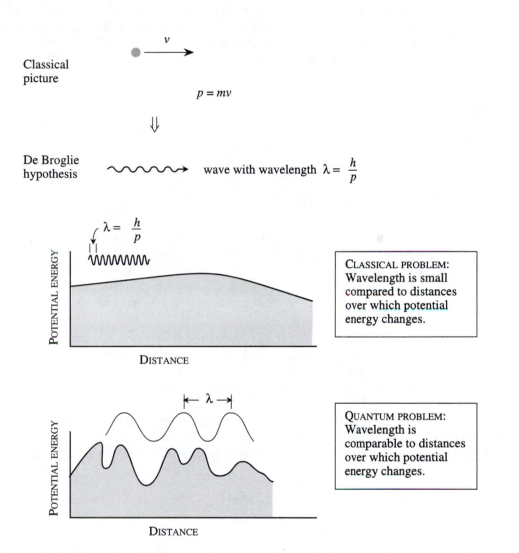

Figure 1.10: When does classical mechanics break down? The comparison between the de Broglie wavelength and scale of spatial variation in potential energy determines if quantum mechanics is needed.

it. The wavelength is given by

$$\lambda = \frac{h}{p} \qquad (1.19)$$

where p is the particle momentum. For most situations encountered by us in our daily life the wavelength of particles is extremely small. As shown in Fig. 1.10, *when the wavelength of a particle is small compared to the distance over which potential energy changes, classical concepts are quite valid.*

There are cases where the potential energy changes over distances comparable or smaller than the de Broglie wavelength of a particle. *In such cases the wave character of the particle is manifested and quantum mechanics is needed.* These ideas are illustrated schematically in Fig. 1.10.

1.5 SCHRÖDINGER EQUATION

We know that in order to describe the behavior of waves we need to solve an appropriate wave equation. The solutions of the wave equation (a differential equation) give us the *allowed* forms of the wave. The question that we now ask is the following: If particles have a wave-like behavior, what is the nature of the equation they satisfy? This question was first answered by Schrödinger and led to the equation bearing his name.

It is important to note that the Schrödinger equation describing the non-relativistic behavior of particles cannot be derived from any fundamental principles—just as Newton's equation cannot be derived. It is an equation that gives us solutions that can explain experimentally observed physical phenomena.

A clue to the form of the Schrödinger equation comes from the relation between a particle momentum and its wavelength as given by de Broglie. According to this relation, for a particle in free space the kinetic energy K is (replacing the momentum by wavelength, λ, or wavevector, k)

$$K = \frac{p^2}{2m} = \frac{\hbar^2 k^2}{2m}; \quad k = \frac{2\pi}{\lambda} \qquad (1.20)$$

If we examine the identity

$$K + U = E$$

where U is the potential energy and E is the total energy, we can write a *wave equation* for a wave with amplitude ψ:

$$(K + U)\psi = E\psi$$

In case the particle is free (i.e., the potential energy is zero) we have

$$K\psi = E\psi$$

Using a hint from Eqn. 1.19 for the form of the kinetic energy, we write the kinetic energy as an operator (say, for the one-dimensional case)

$$K = -\frac{\hbar^2}{2m}\frac{d^2}{dx^2}$$

We now see that the wave equation takes the form

$$-\frac{\hbar^2}{2m}\frac{d^2}{dx^2}\psi = E\psi$$

and the general solution to the equation is

$$\psi = Ae^{ikx} + Be^{-ikx}$$

The kinetic energy is

$$K = E = \frac{\hbar^2 k^2}{2m}$$

which is consistent with the relation we get from the de Broglie relation.

The Schrödinger equation is written by expressing the kinetic energy as a second-order operator, so that the identity

$$(K + U)\psi = E\psi$$

becomes

$$\boxed{\left[-\frac{\hbar^2}{2m}\nabla^2 + V(r)\right]\psi = E\psi} \tag{1.21}$$

To write the time dependence of the particle-wave we use the analogy for the phase of the particle-wave, which is

$$\psi(t) \sim e^{iwt}$$

We can write the time dependence of the wave as

$$\psi(t) \sim \exp\left(-\frac{iEt}{\hbar}\right) \tag{1.22}$$

so that the knowledge of ψ at any time allows one to predict the value of ψ at all times via the equation

$$\frac{\partial\psi}{\partial t} = \frac{-i}{\hbar}E\psi$$

or

$$\boxed{i\hbar\frac{\partial\psi}{\partial t} = E\psi} \tag{1.23}$$

Note that in this development, the quantity h or $\hbar$ has been introduced to define the proportionality between the energy and the particle-wave frequency. In classical physics, this quantity is assumed to be zero. Experiments carried out in the early 20th century showed that $\hbar$ was not zero but had a value of 1.055×10^{-34} Js.

It is important to emphasize that the derivation given above *does not constitute a proof for the Schrödinger equation*. Only experiments could determine the validity of such an extension.

In our "derivation" of the Schrödinger equation we see that the observable properties of the particle such as momentum and energy appear as *operators* operating on the particle wavefunction. From Eqns. 1.20 and 1.23 we see the operator form of these observables:

$$p_x \rightarrow -i\hbar \frac{\partial}{\partial x}$$

$$p_y \rightarrow -i\hbar \frac{\partial}{\partial y}$$

$$p_z \rightarrow -i\hbar \frac{\partial}{\partial z} \tag{1.24}$$

$$E \rightarrow i\hbar \frac{\partial}{\partial t} \tag{1.25}$$

This observation that physical observables are to be treated as operators is quite generic in the quantum description. The energy operator is called the *hamiltonian* in quantum mechanics.

EXAMPLE 1.6 Calculate the wavelength associated with a 1 eV (a) photon, (b) electron, and (c) neutron.

(a) The relation between the wavelength and the energy, E, of a photon is

$$\lambda_{ph} = \frac{hc}{E} = \frac{(6.6 \times 10^{-34} \text{ Js})(3 \times 10^8 \text{ m/s})}{(1.6 \times 10^{-19} \text{ J})} = 1.24 \times 10^{-6} \text{ m}$$

$$= 1.24 \ \mu\text{m}$$

(b) The relation between the wavelength and energy for an electron is $(k = 2\pi/\lambda)$

$$\lambda_e = \frac{h}{\sqrt{2m_0 E}} = \frac{6.6 \times 10^{-34} \text{ Js}}{[2(0.91 \times 10^{-30} \text{ kg})(1.6 \times 10^{-19} \text{ J})]^{1/2}}$$

$$= 12.3 \ \text{Å}$$

(c) For the neutron, using the same relation, we have

$$\lambda_n = \lambda_e \left(\frac{m_0}{m_n}\right)^{1/2} = \lambda_e \left(\frac{1}{1824}\right)^{1/2} = 0.28 \ \text{Å}$$

The wavelengths of different "particles" play an important role when these particles are used to "see" atomic phenomena in a variety of microscopic techniques.

1.5.1 The Wave Amplitude

When we solve a wave equation, say, describing electromagnetic waves, we get a solution that gives the amplitude of the fields and the frequency of vibration. Similarly, when we solve the Schrödinger equation for a given potential energy term, we get a set of solutions $\{E_n, \psi_n\}$ that give us the allowed energy E_n of the particle along with the wavefunction ψ_n.

Let us briefly recall the meaning of the wave amplitude $\phi(r, t)$ in the wave equations describing sound waves or light waves. The energy of the wave at a particular point in space and time is related to $|\phi(r, t)|^2$. It is "more likely" that the wave is found in regions where ϕ is large. Thus the quantity $|\phi(r, t)|^2$ represents some sort of probability function for the wave. A similar interpretation has to be developed for a wave ψ describing a particle of mass m. The interpretation has to be statistical in nature. It is thus natural to regard ψ as the measure of finding the particle at a particular point and space. It can be shown that the product of ψ and its complex conjugate ψ^* is the probability density

$$P(r, t) = \psi^*(r, t)\psi(r, t) = |\psi(r, t)|^2 \qquad (1.26)$$

Thus the probability of finding the particle in a region of volume $dx\ dy\ dz$ around the position r is simply

$$\boxed{P(r, t)\ dx\ dy\ dz = |\psi(r, t)|^2\ dx\ dy\ dz} \qquad (1.27)$$

Normalization of the Wavefunction
If we assume that the particle is confined to a certain volume V, we know that the probability of finding it somewhere in the region must be unity, so that one must have the following condition satisfied:

$$\int_V |\psi(r, t)|^2\ d^3r = 1 \qquad (1.28)$$

The volume over which the integral is carried out is sometimes arbitary and the coefficient of the wavefunction must be chosen to satisfy the normalization integral. The normalization factor does not change the fact that ψ is a solution of the Schrödinger equation due to the homogeneous (linear) form of the equation in ψ. Thus if ψ is a solution, and if A is a constant, then $A\psi$ is also a solution.

1.6   IMPORTANT QUANTUM PROBLEMS FOR SOLID STATE ELECTRONICS

In this section we will summarize the outcome of several important quantum problems that play a crucial role in our understanding of semiconductor devices. We will not discuss the mathematics behind the solutions. Just the results will be presented and their physical implications will be discussed.

1.6.1   Electrons in Free Space

We start with a discussion of the quantum mechanics of free electrons moving in a uniform (or zero) potential. It turns out that electrons inside semiconductors can be regarded as "free" electrons under proper conditions so that the concepts and mathematics developed here can be applied with simple modifications to semiconductors. Consider electrons in a uniform potential V_0.

The equation to be solved is

$$\left[\frac{-\hbar^2}{2m_o} \left(\frac{\partial^2}{\partial x^2} + \frac{\partial^2}{\partial y^2} + \frac{\partial^2}{\partial z^2} \right) + V_0 \right] \psi(r) = E\psi(r) \tag{1.29}$$

A general solution of this equation is

$$\psi(r) = \frac{1}{\sqrt{V}} e^{\pm ik \cdot r} \tag{1.30}$$

where the factor $\frac{1}{\sqrt{V}}$ arises because we wish to have one electron per volume V or

$$\int_V d^3r \mid \psi(r) \mid^2 = 1 \tag{1.31}$$

We assume that the volume V is a cube of side L.

The corresponding energy of the electron is obtained from Eqn. 1.29 as

$$E = \frac{\hbar^2 k^2}{2m_o} + V_0 \tag{1.32}$$

The momentum of the electron is (replacing p by the differential operator)

$$p\psi = -i\hbar \frac{\partial}{\partial \mathbf{r}} \psi \rightarrow \hbar \mathbf{k}\psi \text{ or } p = \hbar k \tag{1.33}$$

while the velocity is

$$\mathbf{v} = \frac{\hbar \mathbf{k}}{m_o} \tag{1.34}$$

The energy-momentum or *E-k* relation for the electrons is shown in Fig. 1.11a. Such a relation is called a *dispersion relation*. The allowed energies for the free electron form a continuous "band." We see from Eqn. 1.32 that the allowed "energy band" of the electron starts at the minimum energy V_0 and proceeds up in energy.

We will now define an important concept, density of states, i.e., density of allowed solutions. The concept of density of states is extremely powerful, and important physical properties such as optical absorption, transport, etc., are intimately dependent upon this concept. Density of states is the number of available electronic states *per unit volume per unit energy* around an energy E. If we denote the density of states by $N(E)$, the number of states in a unit volume in an energy interval dE around an energy E is $N(E)dE$. In Appendix B we have given a derivation for this function. We find that

$$\boxed{N(E) = \frac{\sqrt{2}m_o^{3/2}(E - V_0)^{1/2}}{\pi^2 \hbar^3}} \tag{1.35}$$

In Fig. 1.11b we plot the density of states function.

In free space the equation of motion for electrons is the same as in classical physics except the momentum is replaced by $\hbar k$. We have

$$\frac{\hbar dk}{dt} = \text{Force} \tag{1.36}$$

EXAMPLE 1.7 Calculate the density of states of electrons moving in zero potential at an energy of 0.1 eV.

The density of states in a 3D system is given by (E is the energy in joules)

$$\begin{aligned}
N(E) &= \frac{\sqrt{2}(m_o)^{3/2}E^{1/2}}{\pi^2 \hbar^3} \\
&= \frac{\sqrt{2}(0.91 \times 10^{-30} \text{ kg})(E^{1/2})}{\pi^2 (1.05 \times 10^{-34} \text{ Js})^3} \\
&= 1.07 \times 10^{56} E^{1/2} \text{ J}^{-1} \text{ m}^{-3}
\end{aligned}$$

Expressing E in eV and the density of states in the commonly used units of $\text{eV}^{-1}\text{cm}^{-3}$, we get ($E$ is now in units of eV)

$$\begin{aligned}
N(E) &= 1.07 \times 10^{56} \times (1.6 \times 10^{-19})^{3/2}(1.0 \times 10^{-6})E^{1/2} \\
&= 6.8 \times 10^{21} E^{1/2} \text{ eV}^{-1}\text{cm}^{-3} \\
\text{at } E &= 0.1 \text{ eV we get} \\
N(E) &= 2.15 \times 10^{21} \text{ eV}^{-1}\text{cm}^{-3}
\end{aligned}$$

EXAMPLE 1.8 Calculate the density of states of free electrons moving in a constant background potential of 2.0 eV. This background potential often arises in semiconductors, as we will see later.

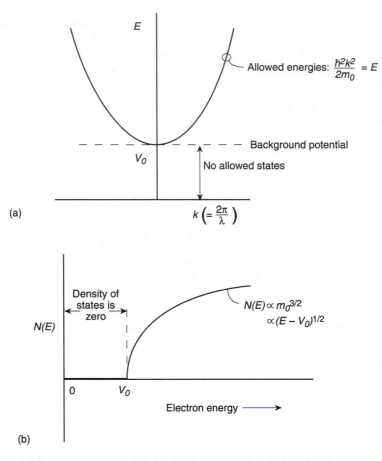

Figure 1.11: (a) Energy versus k ($\hbar k$ is the momentum) relation for electrons in a constant background potential V_0. (b) Energy dependence of density of states.

We have calculated the density of states assuming that the background potential is zero. We have for the constant background potential V_0,

$$E = V_0 + \frac{\hbar^2 k^2}{2m_0}$$

Thus we simply replace E by $E - V_0$ in our results. The density of states becomes

$$N(E) = \frac{\sqrt{2}m_0^{3/2}(E - V_0)^{1/2}}{\pi^2 \hbar^3}$$

Following the results derived in Example 1.7, we get (E is in eV)

$$N(E) = 6.8 \times 10^{21}(E - 2.0 \text{ eV})^{1/2} \quad \text{eV}^{-1}\text{cm}^{-3}$$

Note that, as shown in Fig. 1.12, when $E = 2.0$ eV, the density of states becomes zero. The density of states is zero for energies less than 2.0 eV. After that it is given by the expression above.

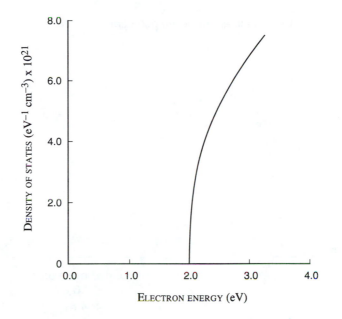

Figure 1.12: A plot of the density of states when the background potential is 2.0 eV (see Example 1.8).

1.6.2 Electrons in an Atom

In the previous subsection we have seen how electrons behave in free space in quantum mechanics. The result are quite similar to what we expect from classical physics except that we now talk about the electron as a wave with a certain wavevector k. When an electron is in an atom, it sees a coulombic interaction with the nucleus and other electrons. Only certain electronic energies are allowed and are schematically shown in Fig. 1.13. There are two regions of interest.

Free states: These states represent solutions where the electron is not confined near the atom and has so much kinetic energy that the attractive potential of the atom cannot hold it. All energies are allowed in this regime shown in Fig. 1.13.

Bound states: These are of greatest interest in atomic physics. Not all energies are allowed for the bound states, where the electron wavefunction is confined to ~1 Å around the nucleus. Only certain discrete energies are allowed, as shown in Fig. 1.13. The allowed energies are separated by regions that are *forbidden* for the electron. This concept of allowed energies and forbidden energy gaps is very central to semiconductor physics, as we shall see later.

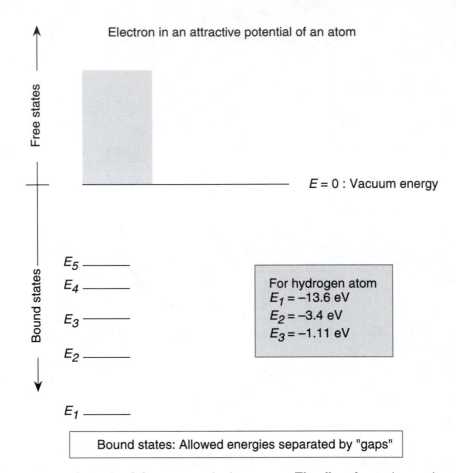

Figure 1.13: A schematic of electron energies in an atom. The allowed energies consist of free and bound states. The bound states consist of discrete allowed energies separated by "forbidden energy gaps."

In the case of the simplest atom—the hydrogen atom—the allowed bound states are given by (referring to the vacuum energy as zero)

$$E_n = \frac{-m_0 e^4}{2(4\pi\epsilon_0)^2 \hbar^2 n^2} = \frac{-13.6}{n^2} \ (\text{eV}) \ ; \ n = 1, 2, 3 \cdots \tag{1.37}$$

The lowest energy state is called the ground state (i.e., the state for which $n = 1$).

The energy at which the electron becomes just free is called the vacuum level energy E_{vac} and is a useful reference energy. Referred to the vacuum energy ($E_{vac} = 0$), the bound state energies are negative since the electrons are in an attractive potential.

1.6.3 Electrons in Crystalline Solids: Energy Bands

We have seen that quantum mechanics tells us that in an atom electrons are allowed to have certain energies that are separated by forbidden gaps. The electron is not allowed to occupy any energy in these gaps. What happens when atoms are brought together to form crystalline solids?

We have noted that in the bound states the electron wave is confined to ~ 1 Å around the nucleus. Let us consider a conceptual experiment where atoms of, say, silicon are brought together to form a crystal. We imagine that the lattice constant of the starting crystal is 1 cm and we start decreasing the spacing between atoms until we finally reach the actual silicon spacing.

As long as the neighboring atoms are more than a few angstroms away the electrons in each atom will see no influence of the neighbor and the electron states will look just as they would in an isolated atom. This is shown schematically on the left-hand side of Fig. 1.14. However, when spacing starts to reach 1-2 Å, electrons will sense the neighboring nuclei and be influenced by them. The result of these interactions is the following:

• The lower-energy core levels are relatively unaffected by the presence of neighboring atoms and remain quite similar to the levels in isolated atoms. This is shown schematically in Fig. 1.14.
• Electronic levels that have higher energies and whose wavefunctions are not as confined to the nucleus sense the presence of the neighbors and broaden into a "band" of allowed energies. Thus, instead of just one sharp energy level being allowed, a band of energies is allowed. Electrons can occupy any energy in these allowed bands.
• The allowed bands are separated by bandgaps, i.e., regions of energy that the electron cannot occupy.

We have seen that in free space electrons are described by the equations (V_0 is the uniform background potential)

$$\begin{aligned} \text{Energy}: \quad E(k) &= \frac{\hbar^2 k^2}{2m_0} + V_0 \\ \text{Momentum}: \quad \boldsymbol{p}(k) &= \hbar \boldsymbol{k} \\ \text{Equation of motion}: \quad \hbar \frac{d\boldsymbol{k}}{dt} &= \boldsymbol{F}_{\text{ext}} \end{aligned} \tag{1.38}$$

Inside an atom, electrons have discrete levels that are separated by forbidden gaps. If a small electric field is applied to isolated atoms, the electrons will not move, since the strong attractive potential of the nucleus will prevent them. The electron thus responds to the outside forces as if it had a very large mass. Of course, in reality the mass is always 9.1×10^{-31} kg.

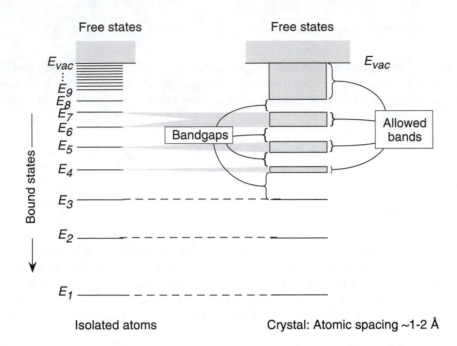

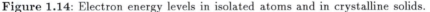

Figure 1.14: Electron energy levels in isolated atoms and in crystalline solids.

Within a crystal, we have seen that electrons have a series of allowed bands separated by bandgaps. Within each allowed band the electron is described by a k-vector, very much as in the free-space problem. As in free space we have an energy versus k-vector relation, only the relation is more complicated than the simple relation for free space.

Within each allowed band the electron behaves as if it were in free space, except it responds as if it had a different mass known as effective mass. This important property is not unlike what happens when we are in a swimming pool. If someone pushes us or pulls us from outside, our body responds as if the one "effective mass" had changed due to the buoyancy of water. Similarly, when the electron is in a periodic crystalline potential, it responds to the outside world as if it has a new effective mass. This mass is determined by the forces sensed by the electron inside the crystal.

The effective mass can be greater than or smaller than the free electron mass. It can even be negative, which represents the fact that when a force is applied, the electron energy decreases, instead of increasing. *The important point to note is that the real mass of the electron is always m_0. Due to the crystalline forces, the electron responds as if it had a new effective mass.*

In Fig. 1.15 we summarize the behavior of electrons inside crystals. As in free

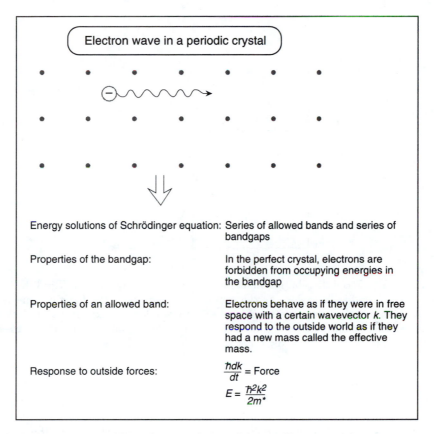

Figure 1.15: A summary of how electrons behave in crystalline materials.

space, the equation of motion in the perfect crystal is

$$\frac{\hbar d\boldsymbol{k}}{dt} = \text{Force} \tag{1.39}$$

and the relation between energy and effective momentum ($\hbar k$) is

$$E(k) = \frac{\hbar^2 k^2}{2m^*} + V_0 \tag{1.40}$$

where V_0 represents the starting energy of the allowed band. We will discuss this in more detail in the next chapter.

1.7 FILLING OF ELECTRONIC STATES: STATISTICS

In the previous section we discussed how the allowed electronic states are distributed in momentum and energy space. These electronic states result from our solution of the

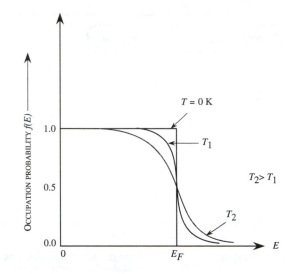

Figure 1.16: The Fermi distribution function at various temperatures. At $T = 0$ K, the Fermi function has a step-like behavior. As the temperature increases, the function smears out as shown. At high temperatures the function starts to develop an exponential tail that looks similar to the Boltzmann function. At the Fermi energy the occupation probability goes to 0.5. The Fermi energy does not have any other simple physical meaning attached to it.

Schrödinger equation. The presence of these states does not mean that electrons are actually present in these states.

When electrons are introduced in the system an important question arises: How do the electrons distribute themselves among the various allowed electronic states? Electrons have the property that at the most one electron can occupy an allowed state (this principle is called the Pauli exclusion principle). The distribution function $f(E)$ tells us the probability that an allowed level at energy E is occupied. This distribution function is given (at thermal equilibrium at temperature T) by the Fermi-Dirac distribution function, which is

$$f(E) = \frac{1}{1 + \exp \dfrac{E - E_F}{k_B T}} \tag{1.41}$$

The expression for electrons shows that $f(E)$ is always less than one, as shown in Fig. 1.16. Here E_F is called the Fermi level and it represents the energy where $f(E_F)$ becomes 1/2. The Fermi level is determined once the density of states and the electron density are known. If we have a case where $(E - E_F) \gg k_B T$, the unity in the denominator of the distribution function can be ignored. We get, in this case, the Maxwell-Boltzmann distribution function (widely used in semiconductor device physics)

$$f(E) = \exp - \left(\frac{E - E_F}{k_B T} \right) \tag{1.42}$$

The relation between Fermi energy and electron density is discussed next.

1.8 METALS, SEMICONDUCTORS, AND INSULATORS

We have discussed in Section 1.6 that the solution of the electron problem in a crystal gives an energy band picture that has regions of allowed bands separated by forbidden bandgaps. The question now arises: Which of these allowed states are occupied by electrons and which are unoccupied? Two important situations arise when we examine the electron occupation of allowed bands. In one case we have a situation where an allowed band is completely filled with electrons while the next allowed band is separated in energy by a gap E_g and is completely empty at 0 K. In a second case, the highest occupied band is only half full (or partially full). These situations are shown schematically in Fig. 1.17.

At this point a very important concept needs to be introduced. *When an allowed band is completely filled with electrons, the electrons in the band cannot conduct any current.* This concept is central to the special properties of semiconductors. A filled band is like a subway car completely filled with riders. There is no possibility of transport! One needs a few vacancies to cause transport. The electrons, being Fermions, cannot carry any net current in a filled band since an electron can only move into an empty state. One can imagine a net cancellation of the motion of electrons moving one way and those moving the other. Because of this effect, when we have a material in which a band is completely filled while the next allowed band is separated in energy and empty, the material has, in principle, infinite resistivity and is called an *insulator*. The material in which a band is only half full with electrons has a very low resistivity and is called a *metal*.

The distinction between a semiconductor and an insulator is not very sharp. Usually, if the bandgap between the filled and unfilled band is ≤ 3.0 eV, the material is called a *semiconductor*. At finite temperatures, some of the electrons in the filled band of the semiconductor are transferred to the upper unfilled band and this can lead to an increase in conductivity, as we shall see in the next chapter. *The band that is normally filled with electrons at 0 K in semiconductors is called the valence band while the upper unfilled band is called the conduction band.*

The energy difference between the vacuum level and the highest occupied electronic state in a metal is called the metal work function. The energy between the vacuum level and the bottom of the conduction band is called the electron affinity. These are shown in Fig. 1.17.

It is important to summarize a critical difference between metals and semiconductors. *The metals have a very high conductivity because of the very large number of electrons that can participate in current transport. It is difficult to alter the conductivity of metals in any simple manner as a result of this. On the other hand, semiconductors have zero conductivity at 0 K and quite low conductivity at finite temperatures, but it is possible to alter their conductivity by orders of magnitude. This is the key reason*

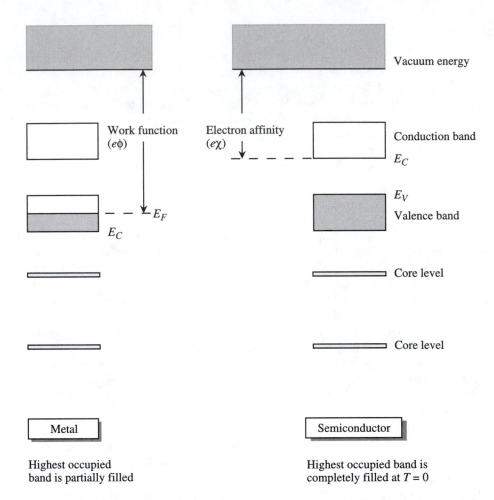

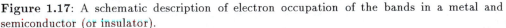

Figure 1.17: A schematic description of electron occupation of the bands in a metal and semiconductor (or insulator).

why semiconductors can be used for active devices while metals are relegated to being passive components like interconnects, contacts, etc.

1.8.1   Fermi Levels in Metals and Semiconductors

In the previous section we have discussed how electrons occupy available states in metals and semiconductors. Let us see how the Fermi distribution looks in metals and semiconductors (or insulators). A detailed examination of the distribution of electrons in semiconductors will be done in the next chapter. Here we merely want to contrast the differences in the position of the Fermi level.

In Fig. 1.18a we show once again the allowed bands and bandgaps in a typical metal. We can make the following observations:

- Several lower-lying bands are completely filled with electrons and are therefore incapable of carrying current. Any high-energy band(s) that is completely empty cannot carry current.
- Only one band that is partially filled (usually about half filled) can conduct current. The electrons can move because there are enough empty states to move into. A schematic of the allowed states and the Fermi function is shown in Fig. 1.18b. We can see that the Fermi level lies near the middle of the highest filled band.

Let us now move to the case of semiconductors. The case is shown schematically in Fig. 1.19. Once again, as in the case of the metal, there are some bands that are completely filled with electrons and cannot carry any current. As noted in the previous section, at zero temperature, the highest occupied band is completely filled and as a result, cannot carry any current. However, as the temperature is raised, the Fermi function smears, as shown in Fig. 1.19b, and a few empty states appear in the valence band while some states in the conduction band are filled with electrons. *As a result, both the valence band and the conduction band are able to carry current.*

In order to determine the Fermi level of an electronic system having a density n in a band that starts at, say, $E = E_0$, we note that

$$n = \int_{E_0}^{\infty} \frac{N(E)dE}{(\exp \frac{E - E_F}{k_B T}) + 1} \tag{1.43}$$

where $N(E)$ is the density of states and there are no allowed states for $E < E_0$. This integral is particularly simple to evaluate as 0 K, since at this temperature

$$\frac{1}{\exp(\frac{E - E_F}{k_B T}) + 1} = 1 \text{ if } E \le E_F$$

$$= 0 \text{ otherwise}$$

Thus

$$n = \int_{E_0}^{E_F} N(E)dE$$

Using the value of the density of states (see Eqn. 1.35), we find

$$n = \frac{\sqrt{2}m_o^{3/2}}{\pi^2 \hbar^3} \int_{E_0}^{E_F} (E - E_0)^{1/2} dE$$

$$= \frac{2\sqrt{2}m_o^{3/2}}{3\pi^2 \hbar^3} (E_F - E_0)^{3/2}$$

$$\tag{1.44}$$

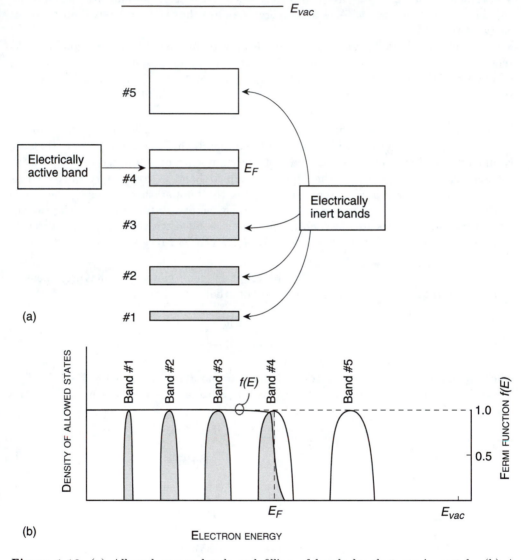

Figure 1.18: (a) Allowed energy bands and filling of bands by electrons in metals. (b) A schematic of allowed bands and density of states in a metal. In the example shown, Band #4 is the conduction band.

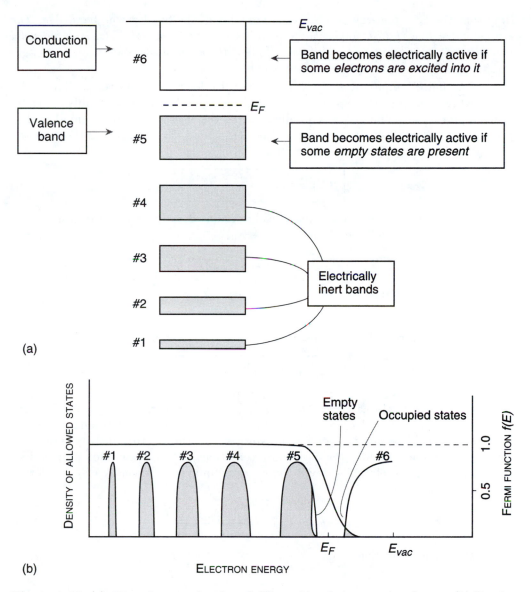

Figure 1.19: (a) Allowed energy bands and filling of bands in a semiconductor. (b) Density of allowed states in a semiconductor. Here, Band #5 is the valence band and Band #6 is the conduction band.

or

$$E_F - E_0 = \frac{\hbar^2}{2m_o}(3\pi^2 n)^{2/3} \tag{1.45}$$

Note that the Fermi energy is referred to the start of the band or the bandedge. *The expression in Eqn. 1.45 is applicable to metals such as copper, gold, etc. It is not valid in semiconductors because of additional effects that we will discuss later.*

At finite temperatures, it is not so easy to determine n in terms of E_F. For metals where n is large, the expression for E_F given above for 0 K is reasonably accurate. However, for semiconductors, n is usually small. If the electron density is small so that $f(E)$ is always small, we can represent the Fermi function by ignoring the 1 in the denominator. This gives the Boltzmann function

$$\boxed{f(E) = \exp\ -\left(\frac{E - E_F}{k_B T}\right)} \tag{1.46}$$

The electron density now can be analytically evaluated as (here also the start of the energy band or the bandedge is referred to as E_0)

$$\begin{aligned} n &= \int_{E_0}^{\infty} N(E)f(E)dE \\ &= N_c \exp\ \left(\frac{E_F - E_0}{k_B T}\right) \end{aligned} \tag{1.47}$$

where N_c is called the effective density of states and is given by

$$\boxed{N_c = 2\left(\frac{m_o}{2\pi\hbar^2}\right)^{3/2}(k_B T)^{3/2}} \tag{1.48}$$

The condition where the carrier density is small so that the carrier occupation function is small is called the non-degenerate condition. As noted above, the Boltzmann approximation can be used only for non-degenerate cases. If the electron density is high so that $f(E)$ approaches unity (the degenerate condition), an approximation proposed by Joyce and Dixon is useful. The Joyce-Dixon approximation is given by (E_F is referred to the bandedge energy)

$$E_F - E_0 = k_B T\left[\ln\frac{n}{N_c} + \frac{1}{\sqrt{8}}\frac{n}{N_c}\right] \tag{1.49}$$

where N_c is defined in Eqn. 1.48.

EXAMPLE 1.9 A particular metal has 10^{22} electrons per cubic centimeter. Calculate the Fermi level energy at 0 K, assuming that the allowed energy band starts at $E = 0$.

The Fermi energy at 0 K is given by

$$E_F = \frac{\hbar^2}{2m_o}(3\pi^2 n)^{2/3}$$

$$= \frac{(1.05 \times 10^{-34} \text{ Js})^2 \left[3\pi^2 (10^{28} \text{ m}^{-3})\right]^{2/3}}{2(0.91 \times 10^{-30} \text{ kg})} = 2.75 \times 10^{-19} \text{ J}$$

$$= 1.72 \text{ eV}$$

The highest energy electron thus has a large energy and is moving with a very large speed. In a classical system the electron energy would be $\sim \frac{3}{2} k_B T$, which would be zero at 0 K. The electron velocity will also be zero at 0 K in classical physics.

EXAMPLE 1.10 Calculate the Fermi level at at 77 K and 300 K for a case where the electron density is 10^{19} cm^{-3}. Use the Boltzmann statistics and the Joyce-Dixon approximation. Assume that the energy band starts at $E = 0$.

In the Boltzmann approximation the Fermi level is given by

$$E_F = k_B T \, \ell n \left(\frac{n}{N_c}\right)$$

and in the Joyce-Dixon approximation the Fermi level is

$$E_F = k_B T \left[\ell n \left(\frac{n}{N_c}\right) + \frac{1}{\sqrt{8}} \frac{n}{N_c}\right]$$

The effective density of electrons is

$$N_c(300K) = 2.56 \times 10^{19} \text{cm}^{-3}$$

$$N_c(77K) = N_c(300K) \left(\frac{77}{300}\right)^{3/2} = 3.34 \times 10^{18} \text{cm}^{-3}$$

At 77 K the Fermi levels are $(k_B T = 0.0067 \text{ eV})$

$$E_F(\text{Boltzmann}) = 0.0067 \left[\ell n \left(\frac{10^{19}}{3.34 \times 10^{18}}\right)\right] = 7.35 \text{ meV}$$

$$E_F(\text{Joyce-Dixon}) = 0.0067 \left[\ell n \left(\frac{10^{19}}{3.34 \times 10^{18}}\right) + \frac{10^{18}}{\sqrt{8}(3.34 \times 10^{18})}\right]$$

$$= 14.44 \text{ meV}$$

At 300 K we have

$$E_F(\text{Boltzmann}) = 0.026 \left[\ell n \left(\frac{10^{19}}{2.56 \times 10^{19}}\right)\right] = -24.4 \text{ meV}$$

$$E_F(\text{Joyce-Dixon}) = 0.026 \left[\ell n \left(\frac{10^{19}}{2.56 \times 10^{19}}\right) + \frac{10^{19}}{\sqrt{8}(2.56 \times 10^{19})}\right]$$

$$= -20.8 \text{ meV}$$

The relative error is much greater at low temperatures than at high temperatures. The reason is that at low temperatures the electron system has a higher occupancy (for a fixed carrier density) and the Boltzmann expression becomes less accurate.

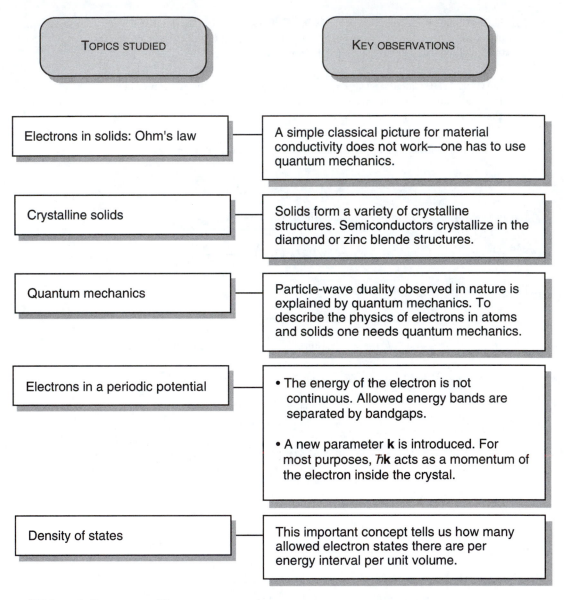

Table 1.1: Summary table.

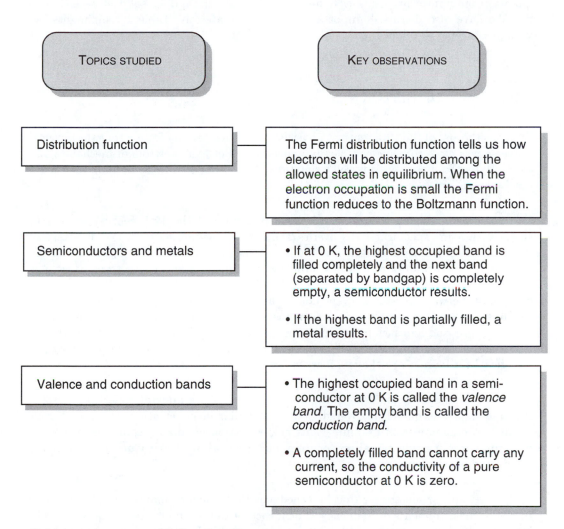

Topics studied	Key observations
Distribution function	The Fermi distribution function tells us how electrons will be distributed among the allowed states in equilibrium. When the electron occupation is small the Fermi function reduces to the Boltzmann function.
Semiconductors and metals	• If at 0 K, the highest occupied band is filled completely and the next band (separated by bandgap) is completely empty, a semiconductor results. • If the highest band is partially filled, a metal results.
Valence and conduction bands	• The highest occupied band in a semi-conductor at 0 K is called the *valence band*. The empty band is called the *conduction band*. • A completely filled band cannot carry any current, so the conductivity of a pure semiconductor at 0 K is zero.

Table 1.1: Summary table (continued).

1.9   SUMMARY

In this chapter we have seen that classical concepts are not applicable when it comes to understanding the behavior of electrons in solids. We have identified important quantum-mechanics problems that are critical in understanding solid state electronics. We have also discussed important crystalline materials. Table 1.1 highlights the various concepts developed.

1.10   A BIT OF HISTORY

Towards the end of the 19th century, physicists were increasingly feeling that the physical universe was close to being completely understood. Between Newtonian mechanics and electrodynamics based on Maxwell's equations, there appeared to be nothing that could not be explained.

The "know-it-all" attitude of classical physics of the 19th century started to shake with some spectacular discoveries and theories beginning with the discovery of X-rays by Wilhelm Conrad Röntgen in 1895. Röntgen was a professor of physics at the University of Wurtzburg and was experimenting with cathode ray (electrons as we now know them) tubes. He was amazed to find in November of 1895 that the screen near one such tube was fluorescing, even though the tube was completely covered with black cardboard. Clearly something mysterious (as a result, the term X-rays) was emanating from the tube and had such intense penetrating power that it could penetrate cardboard boxes and even human flesh. Unlike many other scientific discoveries, this effect was so "magical" for lay persons that its news spread throughout the world.

Working with uranium materials from his father's mineral collection in 1896, Antoine Henri Becquerel discovered that rays similar in penetration power to X-rays seemed to emanate from uranium. This new radiation emanated spontaneously from uranium without any external provocation. Further work by Marie Curie isolated many other "radioactive" materials.

An important trigger that launched new theoretical formalisms in physics came from the measurements of frequency dependence of radiation emitted from heated objects. The Imperial Institute of Physics and Technology in England was making measurements from glowing metals to develop a reliable light standard for testing new designs of lamps. The precise results showed discrepancies with the classical black-body radiation theory. Planck was in close contact with the scientists making these measurements and became interested in this problem. He could explain the spectral distribution in 1900 but only after assuming that energy is exchanged in "quanta." He introduced a new constant h (now called Planck's constant) which has no classical analog. Planck himself did not realize the significance of h and it was Einstein who in 1905 realized its significance. He introduced the notion of "light quanta" that had an energy of h times the light frequency.

Work by Arthur Compton in 1923 at Washington University (St. Louis) established that X-rays behave like "particles" with well defined energy quanta when they scatter from metals. The "Compton effect" showed that X-rays can act as particles while diffraction of X-rays from crystals showed that X-rays behave as waves. This wave-particle duality was an open challenge to physicists, who rapidly started developing theories to explain this key issue.

The 1920s were referred to as the "golden age" of physics by Sommerfeld. This was the time when quantum mechanics was born as a result of the work of many geniuses. The decade of 1923-1932 provided us with the theoretical foundations of quantum mechanics, which has had a tremendous impact on human life. By the mid-1930s questions of chemistry, metallurgy, crystallography, even biophysics were addressed by the new techniques of quantum mechanics. Electrical and optical properties of metals, semiconductors, and insulators could be explained in its basis. Of course, many detailed studies still continue, but the basic theories remain unchanged.

The developments in quantum theory come from two fronts: the matrix approach and the wave mechanics approach. The matrix approach was developed through the pioneering works of Max Born (1882-1970), Werner Heisenberg (1901-1976), Niels Bohr (1885-1962), and Wolfgang Pauli (1900-1958). The work in wave mechanics was due to Louis de Broglie and Erwin Schrödinger (1887-1961). The two approaches are different views of the physical world but provide identical results.

After de Broglie proposed a wave-particle duality for matter (not just waves), Schrödinger developed his wave mechanics. The issue of the amplitude ψ was, however, unclear. Max Born proposed that $|\psi|^2$ represented the probability of finding a particle at a particular point. Thus quantum mechanics incorporated a certain uncertainty, which many people at the time found disconcerting. Even Einstein had a difficult time accepting this "uncertain" view of the world. "God doesn't play dice," he said. Nevertheless, the theory was accepted because it explained all the experimental results.

Towards the beginning of the 20th century, ideas were developed to explain current and heat transport in metals. These ideas developed by Drude (see Chapter 4 for a discussion of the Drude model) were based on the concept that in metals a certain number of electrons (the valence electrons) are "free" to move. In insulators, it was believed that there were no "free" electrons. This explained the high conductivity of metals. However, this also suggested that the specific heat of metals should be much larger than that of insulators. This was not found to be so. This was a big dilemma for classical concepts.

The scientist who contributed perhaps the most to the understanding of electrons in solids was Arnold Sommerfeld. Sommerfeld's contributions were due not only to his own personal research, but also to his success at attracting and training superb students and disseminating his knowledge through travels and lectures. His students included Heisenberg, Pauli, Bethe, Debye, Ewald, London, Peierls, and his assistant was Laue. His travels included not only European cities, but far-flung countries such as India, Japan, China, and the United States.

Sommerfeld was an editor of a journal that published review articles on the state-of-the-art physics of the early 20th century. As a result, he was quite familiar with the problems being addressed in a wide area of physics. Sommerfeld's student, Pauli, had become a professor of physics at Zurich and had successfully applied Fermi-Dirac statistics to develop his exclusion principle (i.e., no two electrons can occupy the same state). Sommerfeld visited Pauli in 1927 and learned of his principle. He realized that the metal specific heat dilemma could be resolved if the Pauli principle was applied. Thus, even though metals have a large number of electrons, most of the electron states are occupied and cannot participate in heat conduction. He was thus the first to start applying the new quantum mechanics to solids.

Although Sommerfeld resolved an important dilemma in solid state physics, he could not explain why electrons are "free" in metals. In 1928, Felix Bloch successfully applied quantum mechanics to the Schrödinger equation in a periodic potential to show why electrons in a periodic crystal act *as if they are free*. This was a central discovery that launched the science of solid state physics.

The new concepts of quantum mechanics and solid state physics were not too welcome at many old-fashioned but powerful institutions. Sommerfeld was responsible for relentlessly pushing his students, such as Heisenberg, Pauli, Debye, etc., to important professorial positions at these institutions. As a result the new theories found entrance into a growing number of respected schools. Sommerfeld also traveled widely and wrote and persuaded others to write review articles to spread the gospel of quantum physics. Through his untiring effort, students around the world received exposure to the new physics.

Once Bloch's work was established, a rapid understanding of solid state crystals was developed. Scientists like Peierls, Bethe, Lothar, Wilson, Wigner, and Seitz made important contributions that provided the basic understanding of bandstructure, lattice vibrations, etc.

The advances in modern physics were centered around Germany until the early 1930s when the Nazi movement began to take a strong hold. People with Jewish heritage and political views not compatible with the movement were forced to emigrate from Germany. Within a few years, German science had lost its edge as a large number of scientists sought to escape the stifling climate. The United States was the biggest beneficiary of this unfortunate event. Britain also benefited considerably.

In addition to the theoretical developments of the '20s and '30s, important new developments took place that were driven by industrial needs for new materials. In Britain, the Bristol school was a pioneering force in introducing solid state physics as we know it today. The leadership was provided by Lennard-Jones and Nevill Mott, who were professors at Bristol. A large number of eminent visitors also provided impetus to the Bristol group, which was the first to recognize the importance of developing approximation schemes in addressing solid state problems.

In addition to the Bristol school, in Göttingen, Robert W. Pohl led an experimental group that laid the foundation of semiconductor physics. Semiconductor physics was considered "dirt physics" around the 1920s by most physicists. This was because of the tremendous changes that occurred in semiconductors when small impurities were introduced. Pohl and his co-workers were the pioneers who collected enormous amounts of data (often without fully understanding what was going on) that was invaluable in developing the physics of semiconductors.

As the new physics emerged, industries began to realize that use could be made of the knowledge for designing new instruments. The industries began to recruit scientists actively—an effort that paid off handsomely, especially in the area of semiconductors.

1.11 PROBLEMS

Section 1.2

Problem 1.1 Assume that the mobility of electrons in copper is 100 $cm^2/V \cdot s$. Calculate the conductivity and resistivity of the material using the Drude model. The atomic mass of copper is 63.55; the density is 8.96 gcm^{-3}; the valence is 1. What is the resistance of a copper wire with an area of 10^{-4} cm^2 and a length of 0.1 cm?

Problem 1.2 Calculate the density of valence electrons in silver. The atomic mass is 107.87; the density is 10.5 gcm^{-3} and the valence is 1. Compare the result to the valence electron density in silicon.

Problem 1.3 Assume that the Drude model works well to describe the conductivity of metals. Calculate the resistance of an aluminum interconnect used in IC technology with an area of 10^{-6} cm^2 and a length of 0.01 cm. The atomic mass of Al is 27; the density is 2.7 gcm^{-3} and the valence is 3.

Problem 1.4 The resistivity of an Al sample is found to be 2.0 $\mu\Omega$-cm. Using the Drude model, calculate the mobility of electrons in Al. Use the data needed from the previous problem.

Problem 1.5 Room-temperature resistivity of copper is found to be 1.5 $\mu\Omega$-cm. Calculate the mobility of electrons using the Drude model. The atomic mass of copper is 63.55; the density is 8.96 gcm^{-3}; the valence is 1. What is the average velocity of electrons in copper when a field of 100 V/cm is applied?

Section 1.3

Problem 1.6 (a) Find the angles between the tetrahedral bonds of a diamond lattice.
(b) What are the direction cosines of the nearest-neighbor bonds with the atom at the origin along the x, y, z axes?

Problem 1.7 Consider a semiconductor with the zinc blende structure (such as GaAs).
(a) Show that the (100) plane is made up of either cation (i.e., Ga) or anion (i.e., As) type atoms.
(b) Draw the positions of the atoms on a (110) plane.
(c) Show that there are two types of (111) surfaces: one where the surface atoms are bonded to three other atoms in the crystal, and another where the surface atoms are bonded to only one. These two types of surfaces are called the A and B surfaces, respectively.

Problem 1.8 Suppose that identical solid spheres are placed in space so that their centers lie on the atomic points of a crystal and the spheres on the neighboring sites touch each other. Assuming that the spheres have unit density, show that the density of such crystals is the following for the various crystal structures:

$$
\begin{aligned}
fcc &: \quad \sqrt{2}\pi/6 = 0.74 \\
bcc &: \quad \sqrt{3}\pi/8 = 0.68 \\
sc &: \quad \pi/6 = 0.52 \\
diamond &: \quad \sqrt{3}\pi/16 = 0.34
\end{aligned}
$$

Problem 1.9 Calculate the number density of In atoms in the semiconductor InAs.

Problem 1.10 Calculate the nearest-neighbor distance between atoms in Si, GaAs, and InAs crystals.

Problem 1.11 Calculate the surface number density of atoms on the (100), (110), and (111) surfaces of silicon.

Problem 1.12 The linear expansion coefficient α of Si is 2.6×10^{-6} K^{-1}. Calculate the number density of Si at 300 K and 77 K. Note that you can use the relation between the lattice constants at two temperatures T_1 and T_2,

$$
a(T_2) = a(T_1)\left[1 + \alpha(T_2 - T_1)\right]
$$

Problem 1.13 The linear expansion coefficient of GaAs is 6.86×10^{-6} K^{-1}. Calculate the lattice mismatch between Si and GaAs at 300 K and 600 K. Use the silicon expansion coefficient from the previous example.

Section 1.5

Problem 1.14 Consider an electron with a wavefunction given by

$$
\psi(z) = \sqrt{\frac{2}{W}} \cos\frac{\pi z}{W} ; \quad -\frac{W}{2} < z < \frac{W}{2}
$$

The wavefunction is zero everywhere else. Calculate the probability of finding the electron in the following regions: (a) between 0 and $W/4$; (b) between $W/4$ and $W/2$; (c) between $-W/2$ and $W/2$.

Problem 1.15 Consider an electron with a wavefunction given by

$$\psi(z) = \sqrt{\frac{2}{W}} \sin \frac{2\pi z}{W} \; ; \quad -\frac{W}{2} < z < \frac{W}{2}$$

The wavefunction is zero everywhere else. Calculate the probability of finding the electron in the following regions: (a) between 0 and $W/4$; (b) between $W/4$ and $W/2$; (c) between $-W/2$ and $W/2$.

Section 1.6

Problem 1.16 Plot the density of states for electrons that have an E-k relation

$$E = \frac{\hbar^2 k^2}{2m_0} - 2.0 \ \ \text{eV}$$

Plot the results from 0 to 3.0 eV.

Problem 1.17 Consider the previous problem. Calculate the number of allowed electron states per cm^3 between 0 and 2.5 eV.

Problem 1.18 Calculate the density of states for a particle that has an E-k relation given by

$$E = -\frac{\hbar^2 k^2}{2m_0}$$

Note the negative sign in the relation.

Problem 1.19 Consult the periodic table to see which group the following elements belong to: (a) Si; (b) Ga; (c) As; (d) Al; (e) Cu; (f) C.

Problem 1.20 Calculate the crystal momentum $\hbar k$ of an electron with energy 0.1 eV and an E-k relation

$$E = \frac{\hbar^2 k^2}{2m^*}; \quad m^* = 0.06 m_0$$

Problem 1.21 Calculate the density of states at 0.1 eV of electrons having a dispersion relation

$$E = \frac{\hbar^2 k^2}{2m^*}; \quad m^* = 0.1 m_0$$

Problem 1.22 In a particular periodic potential electrons are found to have the E-k relation

$$E = \frac{\hbar^2 k^2}{2m^*} - 5.0 \ \ \text{eV}; \quad m^* = 0.1 m_0$$

Plot the density of states for the electron between -6.0 eV and -4.0 eV.

Section 1.7

Problem 1.23 Show that if electrons obey a Maxwell-Boltzmann distribution function the average energy of the electrons is $\frac{3}{2} k_B T$.

Problem 1.24 Show that the Fermi level E_F has the property that the probability that an electron state ΔE above E_F is occupied is the same as the probability that a state ΔE below E_F is empty.

Section 1.8

Problem 1.25 Calculate the effective density of states N_c at 300 K and 77 K for free electrons ($m = m_o$). If the mass of the electron changes to $0.1 m_o$, calculate the effective density of states at the same temperatures.

Problem 1.26 In GaAs, electrons behave as if their mass is $m^* = 0.067 m_o$. If 10^{18} electrons per cm^3 are placed in the conduction band of GaAs, what is the position of the Fermi level at 0 K? Using the Joyce-Dixon approximation, calculate the position of the Fermi level at 300 K.

Problem 1.27 In copper and aluminum, each atom provides an electron for conduction. The density of copper is 8.92 g/cc and its atomic mass is 63.546. The corresponding values in aluminum are 2.702 g/cm^3 and 26.98. What are the positions of the Fermi level in copper and aluminum?

Problem 1.28 Silicon is a semiconductor with a filled valence band. The electrons coming into the valence band are all due to the valence electrons from the individual Si atoms making up the crystal. Calculate the density of the valence electrons. Density of Si is 2.33 gcm^{-3}; the atomic mass is 28 and valence is 4.

Problem 1.29 Al is a metal in which the conduction electrons are all due to the electrons in the outer shell (valence electrons) of the atom. Calculate the density of the electrons in the outermost half-filled band in Al. The atomic mass of Al is 27; the density is 2.7 gcm^{-3} and the valence is 3.

Problem 1.30 GaAs is a semiconductor with a lattice constant $a = 5.65$ Å. There are $a^3/4$ unit cells in the crystal. If each Ga atom provides 3 electrons for the valence band and each As atom gives 5 electrons, calculate the density of electrons in the valence band of GaAs.

Problem 1.31 Discuss, on physical grounds, why we have to worry only about electrons belonging on the outermost shell of atoms (the so-called valence electrons) when we discuss electrons in semiconductors. Why don't the deep-core-level electrons of the atom participate in carrying currents?

Problem 1.32 Consider copper, which is a metal in which each atom contributes 1 electron for conduction to the highest occupied band. Calculate the position of the Fermi level if the atomic mass of copper is 63.55 and the density is 8.96 gcm^{-3}. If the mobility of electrons is 100 cm^2/V·s and 30% of the total electrons participate in carrying current, calculate the conductivity of copper.

1.12   READING LIST

- **Quantum Mechanics**

    - R. L. Longini, *Introductory Quantum Mechanics for the Solid State* (Wiley Interscience, New York, 1970).

- R. P. Feynman, R. B. Leighton, and M. Sands, *The Feynman Lectures on Physics* (Addison-Wesley, Reading, Massachusetts, 1964).

- J. Singh, *Modern Physics for Engineers* (Wiley Interscience, New York, 1999).

- **Statistical Physics**

  - F. Reif, *Fundamentals of Statistical and Thermal Physics* (McGraw-Hill, New York, 1965).

  - M. W. Zemansky, *Heat and Thermodynamics* (McGraw-Hill, New York, 1968).

  - W. B. Joyce and R. W. Dixon, *Appl. Phys. Lett.* **31**, *354* (1977).

- **Crystal Structures**

  - J. M. Buerger, *Introduction to Crystal Geometry* (McGraw-Hill, New York, 1971).

  - M. Lax, *Symmetry Principles in Solid State and Molecular Physics* (Wiley, New York, 1974). Has a good description of the Brillouin zones of several structures in Appendix E.

  - J. F. Nye, *Physical Properties of Crystals* (Oxford University Press, Oxford, 1985).

  - F. C. Phillips, *An Introduction of Crystallography* (Wiley, New York, 1971).

Chapter

2

ELECTRONS IN
SEMICONDUCTORS

2.1 INTRODUCTION

We discussed in the previous chapter that a semiconductor is characterized by the fact that at zero Kelvin, the valence band is completely occupied (i.e., filled with electrons), while the conduction band is completely empty. We also saw that a completely full band does not conduct charge. How then does current flow in a semiconductor? In Section 1.2 we wrote an equation for conductivity of a material

$$\sigma = ne\mu$$

where, according to the classical Drude model, n is the electron density due to the valence electrons. We have to modify this equation to account for the fact that a filled band does not conduct current. In this chapter we will discuss what the proper extension of the above equation should be for semiconductors. Before doing this we will examine some details of semiconductor bandstructure.

2.2 BANDSTRUCTURE OF SEMICONDUCTORS

When the Schrödinger equation is solved for electrons in a semiconductor crystal (usually through a sophisticated computer algorithm), one obtains an energy versus effective momentum relation or an E-k relation. This relation is called the bandstructure of the semiconductor. Every semiconductor has its own unique bandstructure. For the purpose of understanding semiconductor devices we are interested in knowing the bandstructure in the conduction band and in the valence band. In particular, we are interested in what the E-k relation is near the top of the valence band and near the bottom of the conduction band.

2.2.1 Direct and Indirect Semiconductors: Effective Mass

The top of the valence band of most semiconductors occurs at $k=0$, i.e., at effective momentum equal to zero. A typical bandstructure of a semiconductor near the top of the valence band is shown in Fig. 2.1. We notice the presence of two different E-k curves near the valence bandedge. These curves or bands are labeled I and II in the figure and are called (for reasons that will be clear later) the heavy hole (HH) and the light hole (LH).

    The bottom of the conduction band in some semiconductors occurs at $k=0$. Such semiconductors are called direct bandgap materials. Semiconductors such as GaAs, InP, InGaAs, etc., are direct bandgap semiconductors. In other semiconductors, the bottom of the conduction band does not occur at the $k=0$ point but at certain other points. Such semiconductors are called indirect semiconductors. Examples are Si, Ge, AlAs, etc.

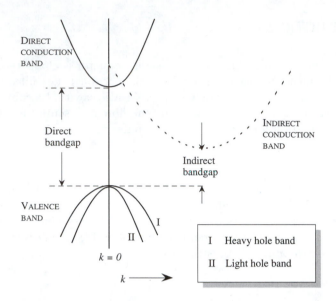

Figure 2.1: Schematic of the valence band, direct bandgap, and indirect bandgap conduction bands. The conduction band of the direct gap semiconductor is shown as the solid line while the conduction band of the indirect semiconductor is shown as the dashed line. Curves I and II in the valence band are called heavy hole and light hole states, respectively. The reason why valence band states are called holes will be discussed in Section 2.3.

An important outcome of the alignment of the bandedges in the valence band and the conduction band is that direct gap materials have a strong interaction with light. Indirect gap materials have a relatively weak interaction with light. This is a result of the law of momentum conservation and will be discussed in the next chapter.

When the conduction bandedge is at $k=0$, it is possible to represent the bandstructure by a simple relation of the form

$$E(k) = E_c + \frac{\hbar^2 k^2}{2m^*} \tag{2.1}$$

where E_c is the conduction bandedge, and the bandstructure is a simple parabola. The equation for the E-k relation looks like that of an electron in free space except that the *free electron mass m_0 is replaced by a new quantity m^**. The electron behaves as if its mass had changed to a different effective value. The mass m^* is called the effective mass and the electron responds to the outside world forces as if it had this mass. The conduction-band electron effective mass has a strong dependence on the value of the bandgap. The smaller the bandgap the smaller the mass, as shown in Fig. 2.2.

Silicon, the most important semiconductor, is an indirect semiconductor. The bottom of the conduction band does not occur at $k = 0$ but at six equivalent minima along the x-, y- and z-axis. The k-values at the minima are: $\frac{2\pi}{a}(0.85,0,0)$, $\frac{2\pi}{a}(0,0.85,0)$, $\frac{2\pi}{a}(0,0,0.85)$, and their inverses $\frac{2\pi}{a}(-0.85,0,0)$, $\frac{2\pi}{a}(0,-0.85,0)$, $\frac{2\pi}{a}(0,0,-0.85)$. The quantity a is the lattice constant ($= 5.43$ Å for Si).

At each of the six k-points given above, the conduction-band energy reaches a minimum value and as k moves away from these points, the energy rises. It is as if there were six *valleys* in the conduction band. The energy momentum relation in these valleys has the form given below.

- Valleys along the x-axis and $-x$-axis: $k_{0x} = \frac{2\pi}{a}(0.85,0,0)$ and $k_{0x} = \frac{2\pi}{a}(-0.85,0,0)$:

$$E(k) = E_c + \frac{\hbar^2}{2}\left[\frac{(k_x - k_{0x})^2}{m_l^*} + \frac{k_y^2 + k_z^2}{m_t^*}\right] \qquad (2.2)$$

- Valleys along the y-axis and $-y$-axis: $k_{0y} = \frac{2\pi}{a}(0,0.85,0)$ and $k_{0y} = \frac{2\pi}{a}(0,-0.85,0)$:

$$E(k) = E_c + \frac{\hbar^2}{2}\left[\frac{(k_y - k_{0y})^2}{m_l^*} + \frac{k_x^2 + k_z^2}{m_t^*}\right] \qquad (2.3)$$

- Valleys along the z-axis and $-z$-axis: $k_{0z} = \frac{2\pi}{a}(0,0,0.85)$ and $k_{0z} = \frac{2\pi}{a}(0,0,-0.85)$:

$$E(k) = E_c + \frac{\hbar^2}{2}\left[\frac{(k_z - k_{0z})^2}{m_l^*} + \frac{k_x^2 + k_y^2}{m_t^*}\right] \qquad (2.4)$$

The effective mass m_l^* is called the longitudinal mass and has a value of $0.98\ m_0$. The mass m_t^* is called the transverse mass and has a value of $0.19\ m_0$.

Near the top of the valence band, as noted earlier (see Fig. 2.1), there are two curves. The heavier mass band is called the heavy hole band (I in Fig. 2.1). The second lighter band is called light hole band (II in Fig. 2.1). The masses of the valence-band electrons are usually much heavier than those in the conduction-band and are also negative. *The reason we call the valence band states "holes" will be discussed in the next section.*

The energy-momentum relation in the valence band can be written as (here the quantities m_{hh}^* and m_{lh}^* are defined as positive values):

Heavy hole band:

$$E = E_v - \frac{\hbar^2 k^2}{2m_{hh}^*} \qquad (2.5)$$

Light hole band:

$$E = E_v - \frac{\hbar^2 k^2}{2m_{lh}^*} \qquad (2.6)$$

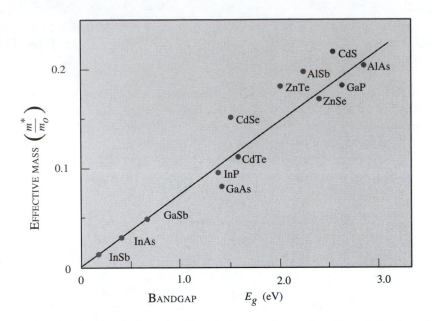

Figure 2.2: Conduction-band electron effective mass m^* as a function of the lowest direct gap E_g for various compound semiconductors. It is interesting to note that the effective mass decreases as the bandgap decreases.

In Section 1.6.1, we discussed the concept of the density of states. We use the same concept here except we use the appropriate effective mass instead of the free electron mass. If the E-k relation is of the form

$$E = E_c + \frac{\hbar^2 k^2}{2m^*}$$

which is appropriate for the conduction band, the density of states is (see Eqn. 1.35 and Appendix B)

$$N(E) = \frac{\sqrt{2}m^{*3/2}(E - E_c)^{1/2}}{\pi^2 \hbar^3} : \ E \geq E_c \tag{2.7}$$

For direct gap semiconductors, m^* is the effective mass discussed above. For a material like Si with different effective masses and six different conduction-band valleys, the conduction-band density of states mass to be used in the expression above is given by

$$m^*_{dos} = \left[6^{2/3} \left(m^*_l m^{*2}_t \right)^{1/3} \right] \tag{2.8}$$

If the E-k relation is of the form

$$E = E_v - \frac{\hbar^2 k^2}{2m^*}$$

which is relevant for the valence band, the density of states is

$$N(E) = \frac{\sqrt{2}m^{*3/2}(E_v - E)^{1/2}}{\pi^2 \hbar^3}; \ E \leq E_v \tag{2.9}$$

PROPERTY	SI	GAAS
Electron effective mass (m_0)	$m_l^* = 0.98$ $m_t^* = 0.19$ $m_{dos}^* = 1.08$ $m_\sigma^* = 0.26$	$m^* = 0.067$
Hole effective mass (m_0)	$m_{hh}^* = 0.49$ $m_{lh}^* = 0.16$ $m_{dos}^* = 0.55$	$m_{hh}^* = 0.45$ $m_{lh}^* = 0.08$ $m_{dos}^* = 0.47$
Bandgap (eV)	$1.17 - \dfrac{4.37 \times 10^{-4}\, T^2}{T + 636}$	$1.519 - \dfrac{5.4 \times 10^{-4}\, T^2}{T + 204}$
Electron affinity (eV)	4.01	4.07

For Si: m_{dos}^*: To be used in calculating density of states, position of Fermi level
m_σ^*: To be used in calculating response to electric field, e.g., in mobility

Table 2.1: Important properties of Si and GaAs.

Note that the valence-band density of states is zero for $E > E_v$. In the valence band (summing the density from the HH and LH bands);

$$
\begin{aligned}
N(E) &= \frac{\sqrt{2}m_{hh}^{*3/2}(E_v - E)^{1/2}}{\pi^2 \hbar^3} + \frac{\sqrt{2}m_{lh}^{*3/2}(E_v - E)^{1/2}}{\pi^2 \hbar^3} \\
&= \frac{\sqrt{2}m_{dos}^{*3/2}(E_v - E)^{1/2}}{\pi^2 \hbar^3}
\end{aligned} \tag{2.10}
$$

where the density of states mass for the valence band is given by

$$
m_{dos}^{*3/2} = (m_{hh}^{*3/2} + m_{lh}^{*3/2}) \tag{2.11}
$$

In Table 2.1 we list some important properties of Si and GaAs, the two most important semiconductors.

EXAMPLE 2.1 Calculate the k-value for an electron in the conduction band of GaAs having an energy of 0.1 eV (measured from the bandedge). Compare this to the case where the electron is in free space. The effective mass of electrons in GaAs is 0.067 m_0.

The k-value is given by

$$
k = \frac{\sqrt{2mE}}{\hbar}
$$

For GaAs the appropriate mass in the conduction band is 0.067 m_0. This gives

$$
k = \frac{[2(0.067 \times 9.1 \times 10^{-31}\ \text{kg})(0.1 \times 1.6 \times 10^{-19}\ \text{J})]^{1/2}}{1.05 \times 10^{-34}\ \text{Js}} = 4.2 \times 10^8\ \text{m}^{-1}
$$

In free space we get
$$k = 1.625 \times 10^9 \text{ m}^{-1}$$
The two values are quite different since the k-value in the crystal represents an *effective* momentum.

EXAMPLE 2.2 In Si we have the following relevant masses: $m_l^* = 0.98m_0$, $m_t^* = 0.19m_0$, $m_{hh}^* = 0.49m_0$ and $m_{lh}^* = 0.16m_0$. Calculate the density of states masses.

Using the definitions given in this section, we get for the conduction band
$$m_{dos}^* = 1.08 \ m_0$$
and for the valence band
$$m_{dos}^* = 0.55 \ m_0$$

2.3 HOLES IN SEMICONDUCTORS

As noted in the previous section, semiconductors are defined as materials in which the valence band is full of electrons and the conduction band is empty at 0 K. At finite temperatures some of the electrons leave the valence band and occupy the conduction band. The valence band is then left with some unoccupied states. Let us consider the situation shown in Fig. 2.3 where an electron with momentum k_e is missing from the valence band.

When all the valence band states are occupied, the sum over all wavevector states is zero, i.e.,
$$\sum \mathbf{k}_i = 0 = \sum_{\mathbf{k}_i \neq \mathbf{k}_e} \mathbf{k}_i + \mathbf{k}_e \tag{2.12}$$

This result is just an indication that there are as many positive k states occupied as negative. Now in the situation where the electron at wavevector $\mathbf{k}_e$ is missing, the total wavevector is
$$\sum_{\mathbf{k}_i \neq \mathbf{k}_e} \mathbf{k}_i = -\mathbf{k}_e$$

The missing state is called a hole and the wavevector of the system $-\mathbf{k}_e$ is attributed to it. It is important to note that the electron is missing from the state $\mathbf{k}_e$ and the momentum associated with the hole is at $-\mathbf{k}_e$. The position of the hole is depicted as that of the missing electron. But in reality the hole wavevector $\mathbf{k}_h$ is $-\mathbf{k}_e$, as shown in Fig. 2.3:
$$\mathbf{k}_h = -\mathbf{k}_e \tag{2.13}$$

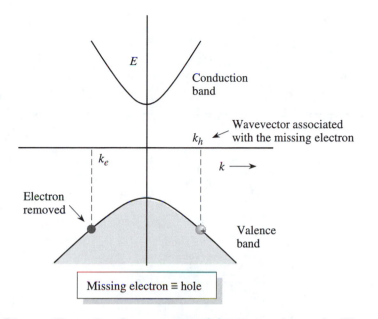

Figure 2.3: Diagram illustrating the wavevector of the missing electron k_e. The wavevector of the system with the missing electron is $-k_e$, which is associated with the hole.

Note that the hole is a representation for the valence band with a missing electron. As discussed earlier, if the electron is not missing, the valence-band electrons cannot carry any current. However, if an electron is missing, the current flow is allowed, as shown in Fig. 2.4. In Fig. 2.4a is shown a hole at $k = 0$, i.e., at point F. Now, if an electric field is applied, all the electrons move in the direction opposite to the electric field as shown. This results in the unoccupied state moving in the field direction as shown. *The hole thus responds as if it had a positive charge.* It therefore responds to external electric and magnetic fields $\mathbf{F}$ and $\mathbf{B}$, respectively, according to the equation of motion,

$$\hbar \frac{dk_h}{dt} = e\left[\mathbf{F} + v_h \times \mathbf{B}\right] \tag{2.14}$$

where $\hbar k_h$ and v_h are the momentum and velocity of the hole.

Thus, the equation of motion of holes is that of a particle with a *positive* charge e. It is important to recognize that we do not actually have a positively charged particle. The holes behave "as if" they had a positive charge. This occurs because all of the occupied states in the valence band respond to the external forces with a proper negative charge response, but the missing electron state responds with a positive charge response. The mass of the hole has a positive value, although the electron mass in its valence band is negative.

Because of the equation of motion discussed above, in an electric field the elec-

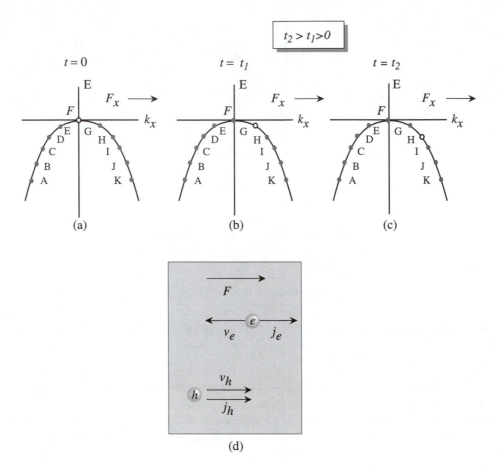

Figure 2.4: The movement of an empty electron state, i.e., a hole under an electric field. The electrons move in the direction opposite to the electric field so that the hole moves in the direction of the electric field, thus behaving as if it were positively charged, as shown in (a), (b), and (c). (d) The velocities and currents due to electrons and holes. The current flow is in the same direction, even though the electron and holes have opposite velocities. The electron effective mass in the valence band is negative, but the hole behaves as if it had a positive mass.

trons in the conduction band move in a direction opposite to the direction of holes in the valence band. However, because they effectively carry a positive charge, the direction of the current produced is the same, as shown in Fig. 2.4d.

When we discuss the conduction-band properties of semiconductors we talk about electrons, but when we discuss the valence-band properties, we talk about holes. This is because in the valence band, only the missing electrons or holes lead to charge transport and current flow.

In this chapter we have seen the words, "electrons behave as if ..." in a lot of places. The concepts of effective mass, effective momentum, holes with positive charge, etc., do not represent the real mass, momentum, charge, etc., of electrons. However, they take into account the effects of the background potential due to the atoms in the crystal. Thus by invoking these "effective" quantities one can forget about the background atoms and use very simple "free electron-like" equations to represent electrons inside semiconductors.

EXAMPLE 2.3 Calculate the energy of an electron and of a hole in the heavy hole band of a semiconductor at a k-value of 0.1 Å. The heavy hole mass is 0.5 m_0.

The electron energy in the valence band is

$$E_e = E_v - \frac{\hbar^2 k^2}{2m_{hh}^*}$$

After using the parameters given we get

$$E_e = E_v - 1.21 \times 10^{-20} \text{ J} = E_v - 0.0755 \text{ eV}$$

The hole energy is the inverse of the electron energy

$$E_h = E_v + 0.0755 \text{ eV}$$

Note that the deeper the electron is in the valence band, the greater the energy it takes to create a hole at that level. As a result, when referred to the valence bandedge, the electron energies in the valence band are negative while the hole energies are positive.

2.4 BANDSTRUCTURES OF SOME SEMICONDUCTORS

We will now examine special features of some semiconductors. Of particular interest are the bandedge properties, since they dominate the transport and optical properties. We will examine some important semiconductors in this section.

Silicon
Silicon is the unchallenged material of choice for electronic products. While the intrinsic bandstructure properties of Si are not outstanding, it seems to be almost chosen by nature for electronics applications. It is relatively easy to fabricate in pure form, has excellent processing properties, and has a high-quality native oxide that can act as an excellent isolation layer. The bandstructure of Si is shown in Fig. 2.5a. The bandgap is 1.1 eV, with the bottom of the conduction band occurring at $k = (0.85\frac{2\pi}{a}, 0, 0)$ and five equivalent points $(0, 0.85\frac{2\pi}{a}, 0); (0, 0, 0.85\frac{2\pi}{a}); (-0.85\frac{2\pi}{a}, 0, 0); (0, -0.85\frac{2\pi}{a}, 0); (0, 0, -0.85\frac{2\pi}{a})$, where a is the lattice constant (5.43 Å). The bandstructure near the conduction band is given by Eqns. 2.2, 2.3, and 2.4. The surfaces of constant energy in k-space are ellipsoid since the effective mass is different along the longitudinal and transverse directions, as

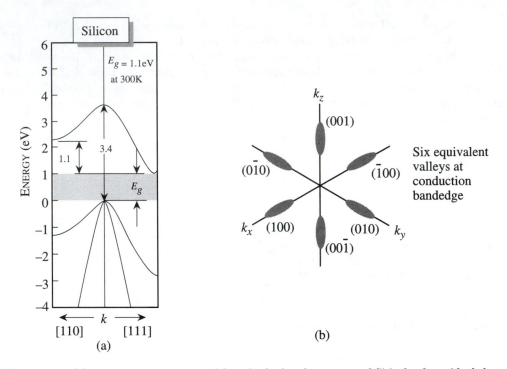

Figure 2.5: (a) Bandstructure of Si. Although the bandstructure of Si is far from ideal, having an indirect bandgap, high hole masses, and small spin-orbit splitting, processing related advantages make Si the premier semiconductor for consumer electronics. (b) Constant energy ellipsoids for Si conduction band. There are six equivalent valleys in Si at the bandedge.

discussed in Section 2.2. There are six valleys and six constant-energy surfaces, as shown in Fig. 2.5b.

Being an indirect material, Si has very poor optical properties and cannot be used to make lasers, as will be discussed in Chapter 3. The hole properties of Si are also quite poor, since the hole masses are quite large.

As we can see from this discussion, the electronic properties of Si are far from ideal. Several important considerations save the day for Si. Most important are the material properties, i.e., ease of processing and availability of SiO_2 as an insulator. Another important reason is that most electronic devices operate at very high electric fields ($E \gtrsim 20$ kV/cm). At such high fields, the response of the electrons to fields in almost all semiconductors is nearly the same.

GaAs
After Si, GaAs has become one of the most important semiconductors. Unlike Si, this semiconductor has its success based upon its superior electronic bandstructure. However,

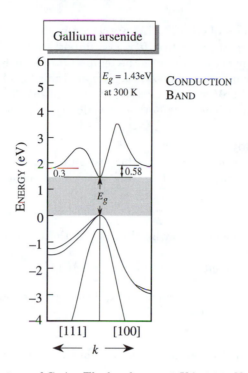

Figure 2.6: Bandstructure of GaAs. The bandgap at 0 K is 1.51 eV and at 300 K it is 1.43 eV. The bottom of the conduction band is at $k = (0,0,0)$, i.e., the Γ-point. The upper conduction band valleys are at the L-point.

it suffers from several drawbacks related to its material properties. GaAs does not have an oxide that can form a high quality insulator-semiconductor interface. It is much more difficult to process. In spite of these drawbacks, GaAs is making steady inroads into the technology.

The bandstructure of GaAs is shown in Fig. 2.6. The bandgap is direct, which is the chief attraction of GaAs. The direct bandgap ensures excellent optical properties of GaAs as well as superior electron transport in the conduction band. The bandedge E vs. $\mathbf{k}$ relation is quite isotropic, leading to spherical equal-energy surfaces. The bandstructure can be represented by the relation

$$E = E_c + \frac{\hbar^2 k^2}{2m^*}$$

with $m^* = 0.067\ m_0$.

The values of the hole masses are $m_{hh}^* = 0.45\ m_0$; $m_{lh}^* = 0.1\ m_0$. It may be noted that there are some uncertainties in the hole masses in most materials since it is very difficult to measure hole masses accurately.

The bandstructures of Ge and AlAs are shown in Fig. 2.7 along with brief comments about their important properties. It is also important to note that the bandgap of semiconductors, in general, decreases as the temperature increases. The temperature dependence of the bandgap for Si and GaAs is given in Table 2.1 above. The bandgap of GaAs, for example, is 1.51 eV at 0 K and 1.43 eV at room temperature. These changes have very important consequences for both electronic and optoelectronic devices.

Bandstructure of Alloys

It is possible to *mix* semiconductors to produce an alloy. If A and B are two semiconductors with lattice constants a_A and a_B, the lattice constant of an alloy $A_x B_{1-x}$ is given by Vegard's law

$$a_{all} = x a_A + (1-x) a_B \qquad (2.15)$$

The bandstructure of of the alloy may be obtained by the weighted average

$$E_{all}(k) = x E_A(k) + (1-x) E_B(k) \qquad (2.16)$$

Alloys play an important role in semiconductor technology since they allow one to adjust the bandgap of materials to a desired value.

EXAMPLE 2.4 Calculate the effective momentum of an electron in the conduction band of GaAs when the electron energy measured from the bandedge is 0.5 eV. Also calculate the momentum of a free electron in space with the same energy.

The energy versus momentum relation for electrons in GaAs is (measured from the bandedge)

$$E - E_c = \frac{\hbar^2 k^2}{2m^*}$$

where $m^* = 0.067 m_0$. This gives for the effective momentum

$$
\begin{aligned}
\hbar k = \sqrt{2m^*(E - E_c)} &= \left[2(0.067 \times 0.91 \times 10^{-30} \text{ kg})(0.5 \times 1.6 \times 10^{-19} \text{ J}) \right]^{1/2} \\
&= 9.83 \times 10^{-26} \text{ kg m s}^{-1}
\end{aligned}
$$

The corresponding wavevector is

$$k = 9.36 \times 10^8 \text{ m}^{-1} = 0.09 \text{ Å}^{-1}$$

Note that this is the electron energy measured from the conduction bandedge. The real energy of the electron measured from the vacuum level is, of course, negative, which means that the electrons are inside the semiconductor.

If a free electron has the energy 0.5 eV measured from the vacuum level, its momentum would be

$$p = \sqrt{2m_0 E} = 3.8 \times 10^{-25} \text{ kg m s}^{-1}$$

The two momenta are quite different since the effective crystal momentum is related to the electron energy inside the crystal measured from the conduction bandedge.

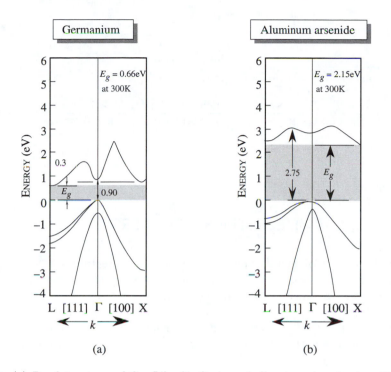

Figure 2.7: (a) Bandstructure of Ge. Like Si, Ge is an indirect semiconductor. The bottom of the conduction band occurs at the L-point. There are eight L-points, but only four of them are independent since the opposite points are connected to each other. The hole properties of Ge are the best of any semiconductor with extremely low hole masses. Ge was initially the semiconductor of choice, but, because of processing considerations, lost out to Si. (b) Bandstructure of AlAs. AlAs is an important III-V semiconductor because of its excellent lattice constant, matching GaAs. The material has indirect bandgap and is usually used as an AlGaAs alloy for barrier materials in GaAs/AlGaAs heterostructures.

EXAMPLE 2.5 A conduction-band electron in Si occupies the (100) valley and has a k-vector of $\frac{2\pi}{a}$ (0.95, 0.1, 0.0). Calculate the energy of the electron measured from the conduction bandedge.

In Si the bottom of the (100) valley occurs at a k-value of $\frac{2\pi}{a}$ (0.85, 0, 0). The change in the k-vector measured from the bottom of the conduction band is

$$\Delta k = \frac{2\pi}{a}(0.1, 0.1, 0.0)$$

Thus, we have ($a = 5.43$ Å)

$$k_\ell = k_x - k_{ox} = \frac{2\pi}{a} \times 0.1 = 1.157 \times 10^9 \text{ m}^{-1}$$

$$k_t = \frac{2\pi}{a} \times 0.1 = 1.157 \times 10^9 \text{ m}^{-1}$$

The energy of the electron is

$$E = \frac{\hbar^2}{2m_\ell^*} k_\ell^2 + \frac{\hbar^2}{2m_t^*} k_t^2$$

Using $m_\ell^* = 0.98m_0$; $m_t^* = 0.19m_0$, we get

$$E = \frac{(1.05 \times 10^{-34})^2(1.157 \times 10^9)^2}{2(0.98 \times 0.91 \times 10^{-30})} + \frac{(1.05 \times 10^{-34})^2(1.157 \times 10^9)^2}{2(0.19 \times 0.91 \times 10^{-30})} \quad = \quad 5.08 \times 10^{-20} \text{ J}$$

$$= \quad 0.318 \text{ eV}$$

2.5   MOBILE CARRIERS: INTRINSIC CARRIERS

From our discussion of metals and semiconductors in Section 1.8, we see that in a metal, current flows because of the electrons present in the highest (partially) filled band. This is shown schematically in Fig. 2.8a. The density of such electrons is very high ($\sim 10^{28}$ cm^{-3}). In a semiconductor, on the other hand, no current flows if the valence band is filled with electrons and the conduction band is empty of electrons. However, if somehow empty states or holes are created in the valence band by removing electrons, current can flow through the holes. Similarly, if electrons are placed in the conduction band, these electrons can carry current. This is shown schematically in Fig. 2.8b. If the density of electrons in the conduction band is n and that of holes in the valence band is p, the total mobile carrier density is $n + p$.

How do holes appear in the valence band and how do electrons occupy the conduction band? In a pure semiconductor, at finite temperatures, the thermal energy of the crystal knocks some electrons from the valence band into the conduction band. When this happens, an electron-hole pair is produced, as shown in Fig. 2.9a. In the pure semiconductor the electron density is equal to the hole density and is denoted by n_i and p_i, the subscript i standing for intrinsic. It is also possible to create holes in the valence band and electrons in the conduction band by using impurity atoms with special properties. This process is called doping and we will discuss it later in this chapter.

The intrinsic carrier concentration refers to the electrons (holes) present in the conduction band (valence band) of a pure semiconductor. The intrinsic carrier concentration depends upon the bandgap and temperature as well as the details of the bandedge masses. The conduction- and valence-band density of states are shown in Fig. 2.9 along with the position of a Fermi level. In Fig. 2.9b we show the case where the temperature is very low, so that there are very few electrons and holes. In Fig. 2.9c we show the case where temperature is high, so that there is a higher density of electrons and holes.

The concentration of electrons in the conduction band is (it would be useful for

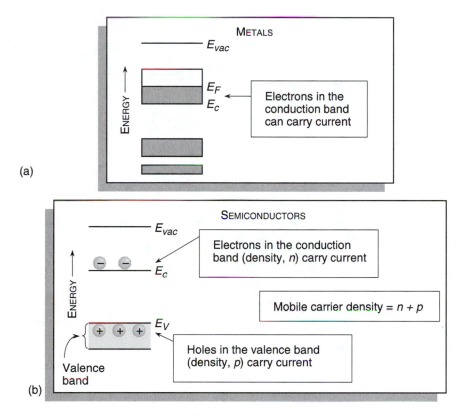

Figure 2.8: (a) A schematic showing allowed energy bands in electrons in a metal. The electrons occupying the highest partially occupied band are capable of carrying current. (b) A schematic showing the valence band and conduction band in a typical semiconductor. In semiconductors only electrons in the conduction band and holes in the valence band can carry current.

the reader to review the discussion in Section 1.6.1)

$$n = \int_{E_c}^{\infty} N_e(E) f(E) dE \qquad (2.17)$$

where $N_e(E)$ is the electron density of states near the conduction bandedge and $f(E)$ is the Fermi function. Using the appropriate expressions for N_e and f we get (the conduction band density of states starts at $E = E_c$, as shown in Fig. 2.9),

$$n = \frac{1}{2\pi^2} \left(\frac{2m_e^*}{\hbar^2}\right)^{3/2} \int_{E_c}^{\infty} \frac{(E - E_c)^{1/2} dE}{\exp\left(\frac{E - E_F}{k_B T}\right) + 1} \qquad (2.18)$$

If the Fermi level is far from the bandedge, then the unity in the denominator can be neglected. This approximation, called the Boltzmann approximation, is valid

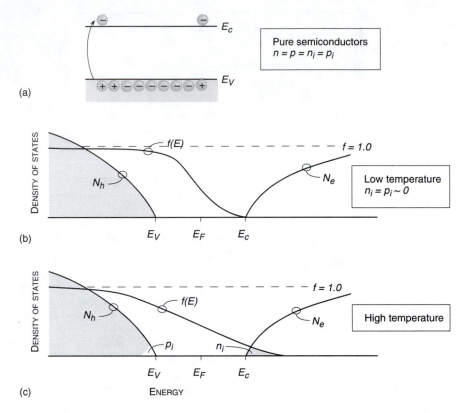

Figure 2.9: (a) A schematic showing that electron and hole densities are equal in a pure semiconductor. (b) Density of states and Fermi occupation function at low temperature. (c) Density of states and Fermi function at high temperatures when n_i and p_i became large.

when n is small ($\lesssim 10^{17}$ cm^{-3} for most semiconductors), and is usually valid for intrinsic concentrations. Then we get

$$
\begin{aligned}
n &= \frac{1}{2\pi^2}\left(\frac{2m_e^*}{\hbar^2}\right)^{3/2}\exp\left(\frac{E_F}{k_BT}\right)\int_{E_c}^{\infty}(E-E_c)^{1/2}\exp\left(-E/k_BT\right)dE \\
&= 2\left(\frac{m_e^* k_B T}{2\pi\hbar^2}\right)^{3/2}\exp\left[(E_F-E_c)/k_BT\right] \\
&= N_c\exp\left[(E_F-E_c)/k_BT\right]
\end{aligned}
\tag{2.19}
$$

where

$$
N_c = 2\left(\frac{m_e^* k_B T}{2\pi\hbar^2}\right)^{3/2}
\tag{2.20}
$$

N_c is known as the effective density of states at the conduction bandedge. Note that the units of the density of states N_e are eV^{-1} cm^{-3} while those of the effective density of states N_c are cm^{-3}. Note also that the effective mass to be used is the density of states effective mass.

The carrier concentration is known *when E_F is calculated*. To find the intrinsic carrier concentration requires finding the hole concentration p as well.

The hole distribution function f_h is given by (remember a hole is the absence of an electron)

$$f_h = 1 - f_e = 1 - \frac{1}{\exp\left(\frac{E-E_F}{k_BT}\right)+1} = \frac{1}{\exp\left[\frac{(E_F-E)}{k_BT}\right]+1}$$

$$\cong \exp -\left[\frac{(E_F-E)}{k_BT}\right] \tag{2.21}$$

The approximation is again based on our assumption that $E_F - E \gg k_BT$, which is a good approximation for pure semiconductors. Carrying out mathematics similar to that for electrons, we find that

$$p = 2\left(\frac{m_h^* k_BT}{2\pi\hbar^2}\right)^{3/2} \exp\left[(E_v - E_F)/k_BT\right]$$

$$= N_v \exp\left[(E_v - E_F)/k_BT\right] \tag{2.22}$$

where N_v is the effective density of states for the valence bandedge.

In Section 2.2 we have discussed the proper density of states mass that should be used to calculate N_c and N_v. In Example 2.9 we explicitly calculate these quantities for Si and GaAs.

In intrinsic semiconductors, the electron concentration is equal to the hole concentration since each electron in the conduction band leaves a hole in the valence band. If we multiply the electron and hole concentrations, we get

$$np = 4\left(\frac{k_BT}{2\pi\hbar^2}\right)^3 (m_e^* m_h^*)^{3/2} \exp\left(-E_g/k_BT\right) \tag{2.23}$$

We note that the product np is independent of the position of the Fermi level and is dependent only on the temperature and intrinsic properties of the semiconductor. This observation is called the law of mass action. If n increases, p must decrease, and vice versa. For the intrinsic case $n = n_i = p = p_i$, we have from the square root of the equation above,

$$n_i = p_i = 2\left(\frac{k_BT}{2\pi\hbar^2}\right)^{3/2} (m_e^* m_h^*)^{3/4} \exp\left(-E_g/2k_BT\right) \tag{2.24}$$

MATERIAL	CONDUCTION BAND EFFECTIVE DENSITY (N_C)	VALENCE BAND EFFECTIVE DENSITY (N_v)	INTRINSIC CARRIER CONCENTRATION ($n_i = p_i$)
Si (300 K)	2.78×10^{19} cm^{-3}	9.84×10^{18} cm^{-3}	1.5×10^{10} cm^{-3}
Ge (300 K)	1.04×10^{19} cm^{-3}	6.0×10^{18} cm^{-3}	2.33×10^{13} cm^{-3}
GaAs (300 K)	4.45×10^{17} cm^{-3}	7.72×10^{18} cm^{-3}	1.84×10^{6} cm^{-3}

Table 2.2: Effective densities and intrinsic carrier concentrations of Si, Ge, and GaAs. The numbers for intrinsic carrier densities are the accepted values even though they are smaller than the values obtained by using the equations derived in the text.

If we set $n = p$, we also obtain the Fermi level position using Eqns. 2.19 and 2.22. We denote the intrinsic Fermi level by E_{Fi}

$$\exp \ (2E_{Fi}/k_BT) = (m_h^*/m_e^*)^{3/2} \ \exp \ ((E_c + E_v)/k_BT)$$

or

$$\boxed{E_{Fi} = \frac{E_c + E_v}{2} + \frac{3}{4}k_BT \ln \left(m_h^*/m_e^*\right)} \tag{2.25}$$

Thus the Fermi level of an intrinsic material lies close to the midgap. Note that in calculating m_h^* and m_e^* the number of valleys and the sum of heavy and light hole states have to be included. See Example 2.10 for details.

In Table 2.2 we show the effective densities and intrinsic carrier concentrations in Si, Ge, and GaAs. The values given are those accepted from experiments. These values are lower than the ones we get by using the equations derived in this section. The reason for this difference is due to inaccuracies in carrier masses and the approximate nature of the analytical expressions.

We note that the carrier concentration increases exponentially as the bandgap decreases. Results for the intrinsic carrier concentrations for some semiconductors are shown in Fig. 2.10. The strong temperature dependence and bandgap dependence of intrinsic carrier concentration can be seen from this figure. In electronic devices where current has to be modulated by some means, the concentration of intrinsic carriers is fixed by the temperature and therefore is detrimental to device performance. Once the intrinsic carrier concentration increases to $\sim 10^{15}$ cm^{-3}, the material becomes unsuitable for electronic devices due to the high leakage current arising from the intrinsic carriers. A growing interest in high-bandgap semiconductors such as diamond (C), SiC, etc., is partly due to the potential applications of these materials for high-temperature devices where, due to their larger gap, the intrinsic carrier concentration remains low up to very high temperatures.

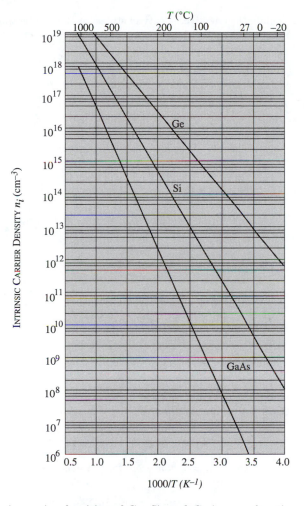

Figure 2.10: Intrinsic carrier densities of Ge, Si, and GaAs as a function of reciprocal temperature.

It is quite clear from the discussion above that pure semiconductors have a very low concentration of carriers that are mobile and can respond to an electric field to carry current. One must compare the room-temperature concentrations of $\sim 10^{11}$ cm^{-3} to the carrier concentrations of $\sim 10^{21}$ cm^{-3} for metals. Indeed, pure semiconductors would have little use by themselves. It was observed experimentally in the 1920s that the conductivity of semiconductors could change by several orders of magnitude if the semiconductor had some impurities. For some time these "impure" materials were looked upon with disdain by solid state physicists until the theory behind "doping" was understood. Now, of course, doping is exploited as the most versatile technique to modify the properties of semiconductors. This subject will be discussed in the next section.

EXAMPLE 2.9 Calculate the effective density of states for the conduction and valence bands of GaAs and Si at 300 K. Let us start with the GaAs conduction-band case. The effective density of states is

$$N_c = 2 \left(\frac{m_e^* k_B T}{2 \pi \hbar^2} \right)^{3/2}$$

Note that at 300 K, $k_B T = 26$ meV $= 4 \times 10^{-21}$ J.

$$
\begin{aligned}
N_c &= 2 \left(\frac{0.067 \times 0.91 \times 10^{-30} \ (\text{kg}) \times 4.16 \times 10^{-21} \ (\text{J})}{2 \times 3.1416 \times (1.05 \times 10^{-34} \ (\text{Js}))^2} \right)^{3/2} \text{m}^{-3} \\
&= 4.45 \times 10^{23} \ \text{m}^{-3} = 4.45 \times 10^{17} \ \text{cm}^{-3}
\end{aligned}
$$

In silicon, the density of states mass is to be used in the effective density of states. This is given by

$$m_{dos}^* = 6^{2/3} (0.98 \times 0.19 \times 0.19)^{1/3} \ m_0 = 1.08 \ m_0$$

The effective density of states becomes

$$
\begin{aligned}
N_c &= 2 \left(\frac{m_{dos}^* k_B T}{2 \pi \hbar^2} \right)^{3/2} \\
&= 2 \left(\frac{1.06 \times 0.91 \times 10^{-30} \ (\text{kg}) \times 4.16 \times 10^{-21} \ (\text{J})}{2 \times 3.1416 \times (1.05 \times 10^{-34} \ (\text{Js}))^2} \right)^{3/2} \text{m}^{-3} \\
&= 2.78 \times 10^{25} \ \text{m}^{-3} = 2.78 \times 10^{19} \ \text{cm}^{-3}
\end{aligned}
$$

One can see the large difference in the effective density between Si and GaAs.

In the case of the valence band, one has the heavy hole and light hole bands, both of which contribute to the effective density. The effective density is

$$N_v = 2 \left(m_{hh}^{3/2} + m_{\ell h}^{3/2} \right) \left(\frac{k_B T}{2 \pi \hbar^2} \right)^{3/2}$$

For GaAs we use $m_{hh} = 0.45 m_0, m_{\ell h} = 0.08 m_0$ and for Si we use $m_{hh} = 0.5 m_0, m_{\ell h} = 0.15 m_0$, to get

$$
\begin{aligned}
N_v(\text{GaAs}) &= 7.72 \times 10^{18} \text{cm}^{-3} \\
N_v(\text{Si}) &= 9.84 \times 10^{18} \text{cm}^{-3}
\end{aligned}
$$

EXAMPLE 2.10 Calculate the position of the intrinsic Fermi level in Si at 300 K.

The density of states effective mass of the combined six valleys of silicon is

$$m_{dos}^* = (6)^{2/3} \left(m_\ell^* \ m_t^2 \right)^{1/3} = 1.08 \ m_0$$

The density of states mass for the valence band is 0.55 m_0. The intrinsic Fermi level is given by (referring to the valence bandedge energy as zero)

$$
\begin{aligned}
E_{Fi} &= \frac{E_g}{2} + \frac{3}{4} k_B T \ell n \left(\frac{m_h^*}{m_e^*} \right) = \frac{E_g}{2} + \frac{3}{4}(0.026) \ell n \left(\frac{0.55}{1.08} \right) \\
&= \frac{E_g}{2} - (0.0132 \ \text{eV})
\end{aligned}
$$

The Fermi level is then 13.2 meV below the center of the mid-bandgap.

EXAMPLE 2.11 Calculate the intrinsic carrier concentration in InAs at 300 K and 600 K.

The bandgap of InAs is 0.35 eV and the electron mass is $0.027m_0$. The hole density of states mass is $0.4m_0$. The intrinsic concentration at 300 K is

$$
\begin{aligned}
n_i \;=\; p_i \;&=\; 2\left(\frac{k_B T}{2\pi\hbar^2}\right)^{3/2} (m_e^{*}\, m_h^{*})^{3/4} \exp\left(\frac{-E_g}{2k_B T}\right) \\
&=\; 2\left(\frac{(0.026)(1.6\times10^{-19})}{2\times3.1416\times(1.05\times10^{-34})^2}\right)^{3/2} \\
&\quad \left(0.027\times0.4\times(0.91\times10^{-30})^2\right)^{3/4}\; \exp\left(-\frac{0.35}{0.052}\right) \\
&=\; 1.025\times10^{21}\ \mathrm{m}^{-3} = 1.025\times10^{15}\,\mathrm{cm}^{-3}
\end{aligned}
$$

The concentration at 600 K becomes

$$
n_i(600\ \mathrm{K}) = 2.89\times10^{15}\,\mathrm{cm}^{-3}
$$

2.6   DOPING: DONORS AND ACCEPTORS

To alter the electron or hole density in a semiconductor, special impurities, called dopants, are introduced into a crystal. There are two kinds of dopants—donors, which donate an electron to the conduction band, and acceptors, which accept an electron from the valence band (thus creating a hole). To understand the donor (or acceptor) problem, we consider a donor atom on a crystal lattice site. The donor atom could be a pentavalent atom in silicon or a Si atom on a Ga site in GaAs. Thus it has one or more electrons in its outermost shell compared to the host atom it replaces. Focusing on the pentavalent atom in Si, four of the valence electrons of the donor atom behave as they would in a Si atom; the remaining fifth electron now sees a positively charged ion to which it is attracted, as shown in Fig. 2.11. The ion has a charge of unity and the attraction is simply a coulombic attraction suppressed by the dielectric constant of the material. The problem is now that of the hydrogen atom case discussed in Chapter 1 except that the electron mass is the effective mass at the bandedge. The attractive potential is (ϵ is the dielectric constant of the semiconductor, i.e., the product of ϵ_0 and the relative dielectric constant)

$$
U(r) = \frac{-e^2}{4\pi\epsilon r} \tag{2.26}
$$

This problem is now essentially the same as that of an electron in the hydrogen atom problem (see Section 1.6.2). The only difference is that the electron mass is m^{*} and the coulombic potential is reduced by ϵ_0/ϵ.

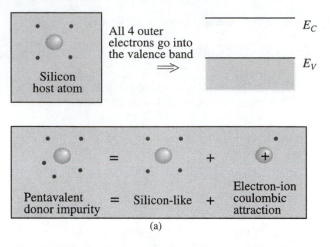

Figure 2.11: A schematic showing the approach to understanding donors in semiconductors. The donor problem is treated as the host atom problem together with a coulombic interaction term. The silicon atom has four "free" electrons per atom. All four electrons are contributed to the valence band at 0 K. The dopant has five electrons out of which four are contributed to the valence band, while the fifth one can be used for increasing electrons in the conduction band.

The lowest-energy solution for this problem is (see Eqn. 1.37)

$$
\begin{aligned}
E_d &= E_c - \frac{e^4 m_e^*}{2(4\pi\epsilon)^2 \hbar^2} \\
&= E_c - 13.6 \left(\frac{m^*}{m_o}\right) \left(\frac{\epsilon_o}{\epsilon}\right)^2 \text{ eV}
\end{aligned}
\tag{2.27}
$$

Note that in the hydrogen atom problem the electron level is measured from the vacuum energy level, which is taken as $E = 0$. *In the donor problem, the energy level is measured from the bandedge.* The ground state is shown schematically in Fig. 2.12 and the values of the donor energy are shown for some semiconductors in Table 2.3.

At this point we introduce another effective mass called the conductivity effective mass m_σ^*, which tells us how electrons respond to external potentials. This mass is used for donor energies as well as for charge transport in an electric field. For direct bandgap materials like GaAs, this is simply the effective mass. For materials like Si the conductivity mass is

$$
m_\sigma^* = 3 \left(\frac{2}{m_t^*} + \frac{1}{m_\ell^*}\right)^{-1}
\tag{2.28}
$$

In indirect bandgap materials the density of states mass is different from the conductivity mass. The density of states mass represents the properties of the electrons

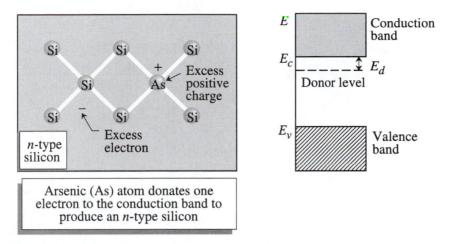

Figure 2.12: Charges associated with an arsenic impurity atom in silicon. Arsenic has five valence electrons, but silicon has only four valence electrons. Thus four electrons on the arsenic form tetrahedral covalent bonds similar to silicon, and the fifth electron is available for conduction. The arsenic atom is called a donor because when ionized it donates an electron to the conduction band.

at a constant-energy surface in the bandstructure. The conductivity mass, on the other hand, gives the response of electrons to an external potential.

According to the simple picture of the donor impurity discussed above, the donor energy levels depend only upon the host crystal (through ϵ and m^*) and *not* on the nature of the dopant. According to Eqn. 2.27, the donor energies for Ge, Si, and GaAs should be 0.006 V, 0.025, and 0.007 eV, respectively. However, as can be seen from Table 2.1, there is a small deviation from these numbers, depending upon the nature of the dopant. This difference occurs because of the simplicity of our model.

Another important class of intentional impurities is the acceptors. Just as donors are defect levels that are neutral when an electron occupies the defect level and positively charged when unoccupied, the acceptors are neutral when empty and negatively charged when occupied by an electron. The acceptor levels are produced when impurities that have a similar core potential as the atoms in the host lattice, but have one less electron in the outermost shell, are introduced in the crystal. Thus group III elements can form acceptors in Si or Ge, while Si could be an acceptor if it replaces As in GaAs. An example is shown for boron (B) in Si in Fig. 2.13.

The acceptor impurity potential could now be considered to be equivalent to a host atom potential together with a *negatively* charged coulombic potential. The "hole" (i.e., the absence of an electron) can then bind to the acceptor potential. The effective mass equation can again be used since only the top of the valence band contributes to the acceptor level.

Semiconductor	Impurity (Donor)	Donor Energy (meV)	Impurity (Acceptor)	Acceptor Energy (meV)
GaAs	Si	5.8	C	26
	Ge	6.0	Be	28
	S	6.0	Mg	28
	Sn	6.0	Si	35
Si	Li	33	B	45
	Sb	39	Al	67
	P	45	Ga	72
	As	54	In	160
Ge	Li	9.3	B	10
	Sb	9.6	Al	10
	P	12.0	Ga	11
	As	13.0	In	11

Table 2.3: Shallow level energies in some semiconductors. The values are all referred to as the energy *below* the conduction bandedge (for donors) and *above* the valence bandedge (for acceptors).

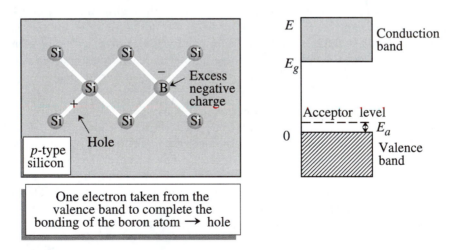

Figure 2.13: Boron has only three valence electrons; it can complete its tetrahedral bonds only by taking an electron from an Si-Si bond, leaving behind a hole in the silicon valence band. The positive hole is then available for conduction. The boron atom is called an acceptor because when ionized it accepts an electron from the valence band.

From the discussion above, it is clear that while in group IV semiconductors, the donor- or acceptor-like nature of an impurity is unambiguous, in compound semiconductors, the dopants can be "amphoteric." For example, Si can act as a donor in GaAs if it replaces a Ga atom while it can act as an acceptor if it replaces an As atom.

EXAMPLE 2.12 Calculate the donor and acceptor level energies in GaAs and Si. The shallow level energies are in general given by

$$E_d = E_c - 13.6(\text{eV}) \times \frac{m^*/m_0}{(\epsilon/\epsilon_0)^2}$$

The conduction-band effective mass in GaAs is $0.067m_0$ and $\epsilon = 13.2\epsilon_0$, and we get for the donor level

$$E_d(\text{GaAs}) = E_c - 5.2 \text{ meV}$$

The problem in silicon is a bit more complicated since the effective mass is not so simple. For donors we need to use the conductivity mass, which is given by

$$m_\sigma^* = \frac{3m_0}{\left(\frac{1}{m_\ell^*} + \frac{2}{m_t^*}\right)}$$

where m_ℓ^* and m_t^* are the longitudinal and transverse masses for the silicon conduction band. Using $m_\ell^* = 0.98$ and $m_t^* = 0.2m_0$, we get

$$m_\sigma^* = 0.26m_0$$

Using $\epsilon = 11.9\epsilon_0$, we get

$$E_d(\text{Si}) = 25 \text{ meV}$$

The acceptor problem is much more complicated due to the degeneracy of the heavy hole and light hole band. The simple hydrogen atom problem does not give very accurate results. However, a reasonable approximation is obtained by using the heavy hole mass ($\sim 0.45m_0$ for GaAs, $\sim 0.5m_0$ for Si), and we get

$$E_a(\text{Si}) \quad = \quad 48 \text{ meV}$$
$$E_a(\text{GaAs}) \quad \cong \quad 36 \text{ meV}$$

This is the energy above the valence bandedge. However, it must be noted that the use of the heavy hole mass is not strictly valid.

2.7 CARRIERS IN DOPED SEMICONDUCTORS

In the lowest energy state of the donor atom, the extra electron of the donor is trapped at the donor site and occupies the donor level E_d. Such an electron cannot carry any current and is not useful for changing the electronic properties of the semiconductor. At very low temperatures, the donor electrons are, indeed, tied to the donor sites and this effect is called carrier freezeout. At higher temperatures, however, the donor electron is *"ionized" and resides in the conduction band as a free electron. Such electrons can, of course, carry current and modify the electronic properties of the semiconductor.* The ionized donor atom is positively charged. Similarly, an ionized acceptor is negatively charged and contributes a hole to the valence band. Donors and acceptors are useful when they are ionized.

2.7.1 Extrinsic Carrier Density

We will now consider the case of materials that have a dominance of donors (n-type materials) or acceptors (p-type materials). We are interested in the free carrier densities n and p in these cases. Because of the doping, we no longer have the equality between electrons and holes, i.e.,

$$n - p \neq 0$$

However, the law of mass action still holds (the value of the product changes only at high doping levels, as will be discussed later)

$$np = \text{constant} = n_i^2 \tag{2.29}$$

Using Eqn. 2.19 for $n = n$ (Fermi energy is E_F) and $n = n_i$ (Fermi energy is E_{Fi}), we get

$$\boxed{\frac{n}{n_i} = e^{(E_F - E_{Fi})/k_B T}} \tag{2.30}$$

and similarly from Eqn. 2.22 (remember, $p_i = n_i$),

$$\boxed{\frac{p}{n_i} = e^{-(E_F - E_{Fi})/k_B T}} \tag{2.31}$$

When the semiconductor is doped n-type (p-type), the Fermi level moves towards the conduction (valence) bandedge. When the Fermi level approaches the bandedge, the Boltzmann approximation is not very good and the simple expressions relating the carrier concentration and the Fermi level are not very accurate. One approach for a more accurate relation is to use a tabulated function that gives the value of the integral for the carrier density. The carrier density is

$$n = \frac{1}{2\pi^2} \left(\frac{2m^*}{\hbar^2} \right)^{3/2} \int_{E_c}^{\infty} \frac{(E - E_c)^{1/2} dE}{\exp\left(\frac{E - E_F}{k_B T} \right) + 1} \tag{2.32}$$

In Section 1.8.1 we discussed the Joyce-Dixon approximation for the relation between the carrier concentration and the Fermi level. According to this relation, we have

$$E_F = E_c + k_B T \left[\ell n \ \frac{n}{N_c} + \frac{1}{\sqrt{8}} \ \frac{n}{N_c} \right] = E_v - k_B T \left[\ell n \ \frac{p}{N_v} + \frac{1}{\sqrt{8}} \ \frac{p}{N_v} \right] \quad (2.33)$$

This relation can be used to obtain the Fermi level if n is specified. Alternatively, it can be used to obtain n if E_F is known by solving for n iteratively. *If the term* $(n/\sqrt{8} \ N_c)$ *is ignored the result corresponds to the Boltzmann approximation. While we have given more accurate expressions for the carrier concentration, in all our calculations for electronic devices in later chapters we will assume that all the donors (acceptors) are ionized. Thus the electron concentration in n-type materials is the donor density* N_d. *We will also assume that the Boltzmann approximation is valid.*

EXAMPLE 2.13 A sample of GaAs has a free electron density of 10^{17} cm^{-3}. Calculate the position of the Fermi level using the Boltzmann approximation and the Joyce-Dixon approximation at 300 K.

In the Boltzmann approximation, the carrier concentration and the Fermi level are related by the following equation:

$$\begin{aligned} E_F &= E_c + k_B T \left[\ell n \frac{n}{N_c} \right] \\ &= E_c + 0.026 \left[\ell n \left(\frac{10^{17}}{4.45 \times 10^{17}} \right) \right] = E_c - 0.039 \text{ eV} \end{aligned}$$

The Fermi level is 39 meV below the conduction band. In the Joyce-Dixon approximation we have

$$\begin{aligned} E_F &= E_c + k_B T \left[\ell n \left(\frac{n}{N_c} \right) + \frac{1}{\sqrt{8}} \frac{n}{N_c} \right] \\ &= E_c + 0.026 \left[\ell n \left(\frac{10^{17}}{4.45 \times 10^{17}} \right) + \frac{10^{17}}{\sqrt{8}(4.45 \times 10^{17})} \right] \\ &= E_c - 0.039 + 0.002 = E_c - 0.037 \text{ eV} \end{aligned}$$

The error produced by using the Boltzmann approximation (compared to the more accurate Joyce-Dixon approximation) is 2 meV.

EXAMPLE 2.14 Assume that the Fermi level in silicon coincides with the conduction band-edge at 300 K. Calculate the electron carrier concentration using the Boltzmann approximation and the Joyce-Dixon approximation.

In the Boltzmann approximation, the carrier density is simply

$$n = N_c = 2.78 \times 10^{19} \text{cm}^{-3}$$

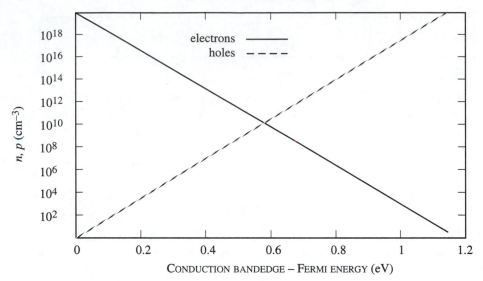

Figure 2.14: A plot of the electron and hole density in Si at 300 K as a function of the position of the Fermi level.

According to the Joyce-Dixon approximation, the carrier density is obtained from the solution of the equation

$$E_F = 0 = k_B T \left[\ell n \frac{n}{N_c} + \frac{n}{\sqrt{8}N_c} \right]$$

Solving by trial and error, we get

$$\frac{n}{N_c} = 0.76 \text{ or } n = 2.11 \times 10^{19} \text{cm}^{-3}$$

We see that the less accurate Boltzmann approximation gives a higher charge density.

EXAMPLE 2.15 Plot the electron density in the conduction band and the hole density in the valence band for silicon as the Fermi level moves from the valence bandedge to the conduction bandedge at 300 K. Use the Maxwell-Boltzmann approximation.

    The results are plotted in Fig. 2.14. Note that the Maxwell-Boltzmann approximation is not accurate as the Fermi level approaches either of the bandedges. *Will the real carrier density be less or greater than the approximate values plotted?*

2.7.2 Population of Impurity Levels: Carrier Freezeout

When we dope a material like Si with a pentavalent P atom, the following question arises. Will the extra fifth atom attach itself to the donor atom and occupy an energy

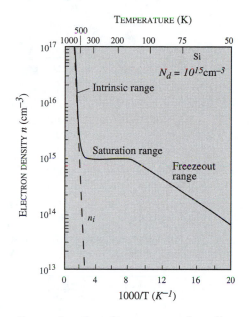

Figure 2.15: Electron density as a function of temperature for a Si sample with donor impurity concentration of 10^{15} cm^{-3}. It is preferable to operate devices in the saturation region where the free carrier density is approximately equal to the dopant density.

level E_d below the conduction band? Or will it "abandon" the donor atom and become a "free" electron in the conduction band? At low temperature the electrons are all confined to the donor atom. The free electron density goes to zero and this phenomenon is called carrier freezeout. As temperature increases, the fraction of "ionized" donors starts to increase until all of the donors are ionized and the free carrier density is equal to the donor density. This region is called the saturation region. Eventually, as the temperature is further raised, the carrier density starts to increase because of the intrinsic carrier density exceeding the donor density. These three regimes are shown in Fig. 2.15 for Si.

The full analysis of this problem is somewhat intricate and requires numerical techniques. We will simply provide an analytical approximation. Let us define the following quantities:

$$
\begin{aligned}
n &= \text{total free electrons in the conduction band} \\
n_d &= \text{electrons bound to the donors} \\
p &= \text{total free holes in the valence band} \\
p_a &= \text{holes bound to the acceptors}
\end{aligned}
$$

The fraction of electrons tied to the donor levels in a *n*-doped material with doping density N_d can be shown to be

$$\frac{n_d}{n + n_d} = \frac{1}{\frac{N_c}{2N_d} \exp\left[-\frac{(E_c - E_d)}{k_B T}\right] + 1} \tag{2.34}$$

Note that in this approximation the Fermi level does not appear in the ratio and the donor ionization energy $(E_c - E_d)$ and temperature determine the fraction of bound electrons.

At low temperatures, the ratio $n_d/(n + n_d)$ approaches unity, so that all the electrons are bound to the donors. This results in carrier freezeout. In general, the temperature dependence of the free carriers has the form shown in Fig. 2.15.

A similar treatment for a p-type material with acceptor doping N_a gives the ratio

$$\frac{p_a}{p + p_a} = \frac{1}{\frac{N_v}{4N_a} \exp\left[-\frac{(E_a - E_v)}{k_B T}\right] + 1} \tag{2.35}$$

We note that in our treatment of electronic devices operating at room temperature, unless otherwise stated, we will assume that $n = N_d$ for n-type materials and $p = N_a$ for p-type materials.

EXAMPLE 2.16   Consider a silicon sample doped with phosphorus at a doping density of 10^{16} cm^{-3}. Calculate the fraction of ionized donors at 300 K. How does this fraction change if the doping density is 10^{18} cm^{-3}?

We have for silicon for $N_d = 10^{16}$ cm^{-3} (donor binding energy is 45 meV)

$$\frac{n_d}{n + n_d} = \frac{1}{\frac{2.8 \times 10^{19}}{2(10^{16})} \exp\left(-\frac{0.045}{0.026}\right) + 1} = 0.004$$

Thus n_d is only 0.4% of the total electron concentration and almost all the donors are ionized.

For a doping level of 10^{18}, we get

$$\frac{n_d}{n + n_d} = 0.29$$

We see that in this case where the doping is heavy, only 71% of the dopants are ionized.

EXAMPLE 2.17.   A semiconductor is said to be n-type degenerate when the probability that the conduction bandedge electronic levels are occupied by electrons is close to unity. One can define a p-type degenerate semiconductor similarly. Assume a criterion that the Fermi level has to be $\sim 3k_B T$ into the band before the material can be called degenerate. What are the free electron densities in Si and GaAs before the semiconductors are n-type degenerate?

We will use the Joyce-Dixon approximation

$$E_F - E_c = k_B T \left[ln \frac{n}{N_c} + \frac{1}{\sqrt{8}} \frac{n}{N_c} \right]$$

Using $E_F - E_c = 3k_B T$, we get

$$\frac{n}{N_c} = 4.4$$

For Si this leads to a carrier density of

$$n = 4.4 \times 2.78 \times 10^{19} = 1.22 \times 10^{20} \, cm^{-3}$$

For GaAs the density for degeneracy is

$$n = 4.4 \times 4.45 \times 10^{17} = 1.96 \times 10^{18} \, cm^{-3}$$

It must be noted, however, that the condition for degeneracy used in this example is somewhat arbitrary.

2.7.3 Heavily Doped Semiconductors

In the theory discussed so far, we have made several important assumptions that are valid only when the doping levels are low: i) we have assumed that the bandstructure of the host crystal is not seriously perturbed and the bandedge states are still described by simple parabolic bands; ii) the dopants are assumed to be independent of each other and their potential is thus a simple coulombic potential. These assumptions become invalid as the doping levels become higher. When the average spacing of the impurity atoms reaches ~ 100 Å, the potential seen by the impurity electron is influenced by the neighboring impurities. In a sense this is like the problem of electrons in atoms. When the atoms are far apart, we get discrete atomic levels. However, when the atomic separation reaches a few angstroms, as in a crystal, we get electronic bands. At high doping levels we get impurity bands. Several other important effects occur at high doping levels. All these effects require us to abandon our simple picture that has worked well for low doping levels.

An important effect of heavy doping is the narrowing of the bandgap. This can have serious effects on performance of devices such as bipolar transistors. In silicon, if N_d is the donor density (cm^{-3}), the bandgap narrowing is given by a simple expression:

$$\Delta E_g \cong -22.5 \left(\frac{N_d}{10^{18}} \frac{300}{T(K)} \right)^{1/2} \, \text{meV} \tag{2.36}$$

This expression gives reasonable agreement with experiments at low doping levels. At high doping levels it overestimates the bandgap narrowing. However, due to its simplicity, we can use it to get an estimate of bandgap narrowing.

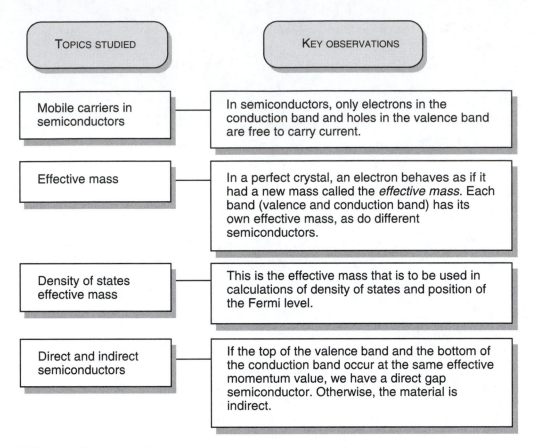

Table 2.4: Summary table.

2.8 SUMMARY

In this chapter we have discussed the very important concept of bandstructure and doping in semiconductors. Without doping, semiconductors are essentially useless for electronic devices. Thus a great deal of effort has focused on understanding the properties of dopants. The key findings of this chapter are summarized in Table 2.4.

2.9 A BIT OF HISTORY

As quantum mechanics developed, the earliest applications were geared toward isolated atoms in gases. The spectra of the atom provided test cases to which quantum-mechanical formalisms could be applied. In solid state, physicists focused more on metals, since the behavior of metals was not so strongly dependent upon the purity of the material. There was relatively little interest in semiconductors in which the smallest

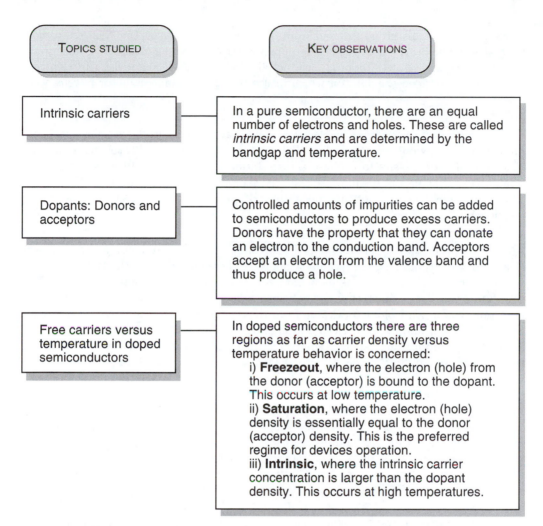

Table 2.4: Summary table (continued).

impurity levels seemed to have enormous effects on conductivity. The problems of this "dirt physics" (as referred to by Pauli) seemed rather uninteresting to the top physicists of the 1920s. During this time, however, Robert W. Pohl started his work on ionic crystals at Göttingen and for a while had no competition to his work. The Pohl school was an experimental school in the strictest sense of the word. The group primarily focused on the investigation of electrical conduction in non-metals. They documented the effects of impurities on the properties of the materials and thus laid the experimental foundations of semiconductor physics and, in particular, the doping process.

Pohl was surrounded in Göttingen by outstanding physicists like Max Born and James Franck, who were pioneers in the newly developing field of quantum mechanics.

However, he kept his group isolated from them. He also survived the Nazi era almost unscathed while other "new physics" groups were decimated. As a result, Pohl's group worked on crystal growth, doping, conductivity, and photoelectric effects in semiconductors and provided the data on which future theories of semiconductor physics were based.

The effect of "doping" on electrical conductivity was considered to be especially important, for obvious reasons. In fact, Pohl and Hilsch in 1938 used their understanding of semiconductor physics to propose a three-terminal current amplifier. They demonstrated the effect in a potassium bromide crystal and achieved an amplification of 100. The device, however, did not find any technical application because of the low level of technology of the time.

As properties of pure and doped semiconductors began to be understood, it was clear that there could be great technological use of these materials. In the 1930s the main technical use of non-metals was in the area of phosphors for scintillation counters and thin-film vapor-deposition techniques to reduce reflections from optical components. The real impetus for semiconductors came from the Second World War in connection with radar. There was an important need for diodes that could block current in one direction. In the U.S.A., important Defense Department research contracts were awarded to study the problem of building diode detectors. One such contract went in 1942 to Purdue University under the leadership of Karl Lark-Horovitz. Crystal detectors had been used since the early 1900s, but the devices were highly unreliable. Under the direction of Lark-Horovitz, the Purdue group decided to pursue germanium instead of then commonly used lead sulfide. A close collaboration was developed with the chemical industry and in 1942-43 high-quality germanium crystals (doped and undoped) started to become available. Essentially all the earlier electronic devices were produced in germanium. Ironically, germanium (named after Germany) provided the crystal detectors for the allied bombers on missions to blanket-bomb cities in Germany.

In addition to Purdue University, important research teams were formed at Bell Laboratories. The impetus of this research also came primarily from the Second World War. Bell Labs scientists were, of course, responsible for establishing the basis of today's solid state electronics and we will discuss these historical moments in later chapters.

2.10   PROBLEMS

Note: Please examine problems in the following Cumulative Problems section as well.

Section 2.2-2.3
Problem 2.1 The effective mass of a conduction-band electron in a semiconductor is

$0.1m_0$. Calculate the energy of this electron measured from the bottom of the conduction bandedge if the k-vector is 0.3 Å^{-1}. If the electron affinity of this semiconductor is 10 eV, calculate the energy of the electron measured from the vacuum level.

Problem 2.2 Plot the conduction-band and valence-band density of states in Si, Ge, and GaAs. Use the following data:

$$
\begin{aligned}
\text{Si} : m^*_{dos} \ (c-\text{band}) &= 1.08 \ m_0 \\
m^*_{hh} &= 0.49 \ m_0 \\
m^*_{\ell h} &= 0.16 \ m_0
\end{aligned}
$$

$$
\begin{aligned}
\text{Ge} : m^*_{dos} \ (c-\text{band}) &= 0.56 \ m_0 \\
m^*_{hh} &= 0.29 \ m_0 \\
m^*_{\ell h} &= 0.044 \ m_0
\end{aligned}
$$

$$
\begin{aligned}
\text{GaAs} : m^*_e &= m^*_{dos} = 0.067 \ m_0 \\
m^*_{hh} &= 0.5 \ m_0 \\
m^*_{\ell h} &= 0.08 \ m_0
\end{aligned}
$$

Problem 2.3 The wavevector of a conduction-band electron in GaAs is $k = (0.1, 0.1, 0.0)$ Å^{-1}. Calculate the energy of the electron measured from the conduction bandedge.

Problem 2.4 A conduction-band electron in silicon is in the (100) valley and has a k-vector of $\frac{2\pi}{a}(1.0, 0.1, 0.1)$. Calculate the energy of the electron measured from the conduction bandedge. Here a is the lattice constant of silicon.

Problem 2.5 Calculate the energies of electrons in the GaAs and InAs conduction-bands with k-vectors $(0.01, 0.01, 0.01)$ Å^{-1}. Refer the energies to the conduction band-edge values.

Problem 2.6 The bandstructure of GaAs conduction-band electrons is given by a simple parabolic relation ($m^* = 0.067 \ m_0$):

$$
E = \frac{\hbar^2 k^2}{2m^*}
$$

A better approximation is

$$
E(1 + \alpha E) = \frac{\hbar^2 k^2}{2m^*}
$$

where $\alpha = 0.67$ eV^{-1}. Calculate the difference in the energy of an electron using the two expressions at $k = 0.01$ Å^{-1} and $k = 0.1$ Å^{-1}.

Section 2.4

Problem 2.7 Calculate the effective density of states at the conduction and valence bands of Si and GaAs at 77 K, 300 K, and 500 K.

Problem 2.8 Calculate the intrinsic carrier concentration of Si, Ge and GaAs as a function of temperature from 4 K to 600 K. Assume that the bandgap is given by

$$E_g(T) = E_g(0) - \frac{\alpha T^2}{T + \beta}$$

where $E_g(0)$, α, and β are given by

Si   : $E_g(0) = 1.17$ eV; $\alpha = 4.37 \times 10^{-4}$ eVK^{-1}; $\beta = 636$ K

Ge   : $E_g(0) = 0.74$ eV; $\alpha = 4.77 \times 10^{-4}$ eVK^{-1}; $\beta = 235$ K

GaAs : $E_g(0) = 1.519$ eV; $\alpha = 5.4 \times 10^{-4}$ eVK^{-1}; $\beta = 204$ K

Problem 2.9 Estimate the intrinsic carrier concentration of diamond at 700 K (you can assume that the carrier masses are similar to those in Si). Compare the results with those for GaAs and Si. The result illustrates one reason why diamond is useful for high-temperature electronics.

Problem 2.10 An Si device is doped at 5×10^{16} cm^{-3}. Assume that the device can operate up to a temperature where the intrinsic carrier density is less than 10% of the total carrier density. What is the upper limit for the device operation?

Problem 2.11 Estimate the change in intrinsic carrier concentration with temperature, dn_i/dT, for InAs, Si, and GaAs near room temperature.

Problem 2.12 Calculate and plot the position of the intrinsic Fermi level in Si between 77 K and 500 K.

Section 2.5

Problem 2.13 A donor atom in a semiconductor has a donor energy of 0.045 eV below the conduction band. Assuming the simple hydrogenic model for donors, estimate the conduction bandedge mass.

Problem 2.14 Calculate the donor level energy for InAs.

Section 2.6

Problem 2.15 Calculate the density of electrons in a silicon conduction band if the Fermi level is 0.1 eV below the conduction band at 300 K. Compare the results by using the Boltzmann approximation and the Joyce-Dixon approximation.

Problem 2.16 In a GaAs sample at 300 K, the Fermi level coincides with the valence bandedge. Calculate the hole density using (a) the Boltzmann approximation and (b) the Joyce-Dixon approximation. Also calculate the electron density using the law of mass action.

Problem 2.17 The electron density in a silicon sample at 300 K is 10^{16} cm^{-3}. Calculate $E_c - E_F$ and the hole density using the Boltzmann approximation.

Problem 2.18 A GaAs sample is doped n-type at 5×10^{17} cm^{-3}. Assume that all the donors are ionized. What is the position of the Fermi level at 300 K?

Problem 2.19 Consider an n-type silicon with a donor energy 45 meV below the conduction band. The sample is doped at 10^{17} cm^{-3}. Calculate the temperature at which 90% of the donors are ionized.

Problem 2.20 Consider a GaAs sample doped at $N_d = 10^{17}$ cm^{-3}. The donor energy is 6 meV. Calculate the temperature at which 90% of the donors are ionized.

Problem 2.21 The shrinking of bandgap with doping is a serious problem in bipolar junction transistors, where one has to play an optimization game between band shrinkage and doping. Plot the band shrinkage in n-type Si as a function of doping up to a doping density of 5×10^{18} cm^{-3} at 300 K.

2.11 CUMULATIVE PROBLEMS: CHAPTERS 1-2

Problem 1 Calculate the minimum energy needed to remove an electron from the top of the valence band to free space (vacuum) in: (a) GaAs and (b) Si?

Problem 2 Calculate the position of the intrinsic Fermi level at 300 K with respect to the vacuum level in Si and GaAs.

Problem 3 A sample of Si is doped n-type at 10^{18} cm^{-3}. Assume that at room temperature, 90% of the donors are ionized. Calculate the position of the Fermi level referenced to: (a) the valence bandedge; (b) conduction bandedge; (c) intrinsic Fermi level position; and (d) vacuum level.

Problem 4 Calculate the n-type doping efficiency for phosphorus in silicon at room temperature for the following dopant densities: (a) $N_d = 10^{15}$ cm^{-3}; (b) $N_d = 10^{18}$ cm^{-3}; and (c) $N_d = 5 \times 10^{19}$ cm^{-3}. Assume that the donor level is 45 meV below the conduction band. Doping efficiency is the fraction of dopants that are ionized.

Problem 5 Consider pure silicon at room temperature. Calculate the ratio of mobile carriers to the *total electron density* in the material.

Problem 6 A Si sample is doped n-type at $N_d = 10^{18}$ cm^{-3}. Assume that all the donors are ionized. Calculate the ratio of mobile carriers to the density of *electrons in the valence band*. Assume that each Si atom gives four electrons to the valence band.

Problem 7 Calculate the total (i.e., not just the conduction band) density of electrons in GaAs, Si, and Ge. Also calculate the density of electrons in the valence band at 0 K. Assume that each atom gives four electrons to the valence band. Use a periodic table to find out the total number of electrons in each atom making up the crystal.

Problem 8 A 2 MeV α-particle hits a semiconductor sample. The entire energy of the α-particle is used up in producing electron-hole pairs by knocking electrons from the valence to the conduction band. Estimate the number of electron-hole pairs produced in Si and GaAs.

Problem 9 Calculate the average separation of dopants in Si when the doping density is: (a) 10^{16} cm^{-3}; (b) 10^{18} cm^{-3}; and (c) 10^{20} cm^{-3}.

Problem 10 Estimate the fraction of total electrons that are mobile in: (a) Ag; (b) Cu; (c) pure Si at 300 K; and (d) Si doped at 10^{18} cm^{-3} and with 90% dopant ionization.

2.12 READING LIST

- **General Bandstructure**

  - R. E. Hummel, *Electronic Properties of Materials—An Introduction for Engineers* (Springer, New York, 1985).
  - K. Seeger, *Semiconductor Physics: An Introduction* (Springer, Berlin, 1985).
  - M. S. Tyagi, *Introduction to Semiconductor Materials and Devices* (John Wiley and Sons, New York, 1991).
  - H. F. Wolf, *Semiconductors* (Wiley-Interscience, New York, 1971).

- **Intrinsic and Extrinsic Carriers**

  - J. S. Blakemore, *Semiconductor Statistics* (Pergamon Press, New York, 1962), reprinted by Dover, New York, 1988.
  - D. A. Neamen, *Semiconductor Physics and Devices: Basic Principles* (Irwin, Boston, 1997).
  - K. Seeger, *Semiconductor Physics: An Introduction* (Springer Berlin, 1985).
  - S. M. Sze, *Physics of Semiconductor Devices* (John Wiley and Sons, New York, 1981).
  - C. D. Thurmond, *J. Electrochem. Soc.* **122**, *1133* (1975).
  - M. S. Tyagi, *Introduction to Semiconductor Materials and Devices* (John Wiley and Sons, New York, 1991).

- **Heavily Doped Semiconductors**

  - H. C. Casey Jr. and F. Stern, *J. Appl. Phys.* **47**, *631* (1976).
  - S. T. Pantelides, A. Selloni, and R. Car, *Solid State Electron.* **28**, *17* (1985).
  - R. J. Van Overstraeten and R. P. Mertens, *Solid State Electron.* **30**, *1077* (1987).

Chapter

3

CARRIER DYNAMICS IN SEMICONDUCTORS

Chapter at a Glance

3.1 INTRODUCTION

We have seen that in semiconductors there are two kinds of carriers that can participate in transport—electrons in the conduction band and holes in the valence band. We now ask ourselves the following questions:

• If there is an electric field present in the semiconductor, how do the electrons and holes move?
• If there is a concentration gradient in the electron or hole density, how do the carriers respond?
• Do electrons in the conduction band *fall down* into the valence band and *recombine* with holes?
• Is it possible for electrons in the valence band to *jump up* into the conduction band? What causes such processes?
The answers to such questions provide us the understanding needed to design semiconductor devices.

Quantum mechanics tells us that in a perfect semiconductor there is *no scattering of electrons as they move through the complicated but periodic background potential of the ions.* The equation of motion of such "free" electrons is simply

$$\frac{\hbar d\mathbf{k}}{dt} = \mathbf{F}_{\text{ext}} \tag{3.1}$$

where $\hbar\mathbf{k}$ is the effective crystal momentum of the electrons. Thus, if there is no external force, the effective momentum of the electron $\hbar\mathbf{k}$ does not change. It is only the presence of imperfections that causes scattering of electrons in a semiconductor. In fact, all semiconductors have impurities and imperfections, either intentional or unintentional, and these in turn cause scattering of electrons. In Table 3.1 we show some of the important sources of scattering coming from a variety of imperfections in the semiconductor.

3.2 SCATTERING IN SEMICONDUCTORS

According to quantum mechanics, in a perfectly periodic potential, the electrons have a free electron-like behavior and no scattering occurs. In real semiconductors, due to the imperfections, some of which are listed in Table 3.1, electrons scatter as they move from one point to another. If, as shown in Fig. 3.1, one considers a beam of electrons moving initially with the same momentum, then due to the scattering processes, the momentum and energy will gradually lose coherence with the initial state values. The average time it takes to lose coherence or memory of the initial state properties is called the "relaxation time" or scattering time. Thus one can define a relaxation time for momentum (τ_{sc}) or velocity as shown in Fig. 3.1.

One of the most important scattering processes is due to ionized impurity scat-

Important Sources of Scattering in Semiconductors		
Ionized impurities	$\longrightarrow$	Due to dopants in the semiconductors
Phonons	$\longrightarrow$	Due to lattice vibrations at finite temperatures
Alloy	$\longrightarrow$	Random potential fluctuations in alloy semiconductors
Interface roughness	$\longrightarrow$	Important in heterostructure devices
Chemical impurities	$\longrightarrow$	Due to unintentional impurities

Table 3.1: Important sources of scattering in semiconductors. These sources cause scattering of electrons, which affects their transport properties.

tering. When a donor (acceptor) gives its extra electron (hole) to the semiconductor, one has an ionized impurity left behind. The electrons scatter from this ionized impurity.

At a finite temperature, the atoms in a crystal vibrate due to their thermal energies. This vibration causes local strain waves in the crystal that result in variations in the bandedge positions of the semiconductor. This is shown schematically in Fig. 3.2. The electrons scatter from these vibrations. As the temperature of the crystal is increased, the amplitude of vibration increases, causing an enhancement of the scattering rates.

In certain devices, such as the MOSFET, to be discussed later, electrons move close to an interface between a semiconductor and an insulator. Roughness at this interface also causes scattering.

In general, as the electron is moving in the semiconductor, it suffers scattering from all the various scattering processes. These processes are assumed to be unrelated. A total scattering rate may be defined as the sum

$$R_{Tot} = \sum_i R_i \tag{3.2}$$

where the sum is taken over all possible scattering processes. To a good approximation, this results in a total scattering time τ_{sc}, where

$$\frac{1}{\tau_{sc}} = \sum_i \frac{1}{\tau_{sc}^{(i)}} \tag{3.3}$$

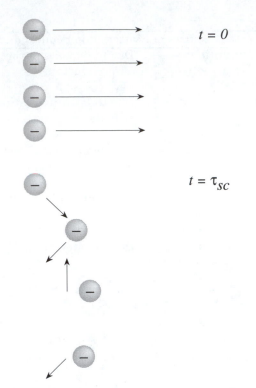

Figure 3.1: The effect of scattering on electron velocities. (a) A beam of electrons with constant velocity is considered at time $t = 0$; the arrow is the vector representing the velocity. (b) After a relaxation time (scattering time) τ_{sc}, the electrons have lost memory of their velocities.

and $\tau_{sc}^{(i)}$ is the scattering time of the electrons due to each individual scattering process. This rule for finding the total scattering time is called Mathieson's rule.

3.3 VELOCITY-ELECTRIC FIELD RELATIONS IN SEMICONDUCTORS

When an electron distribution is subjected to an electric field, the electrons tend to move in the field direction (opposite to the field $\mathbf{F}$) and gain velocity from the field. However, because of imperfections, they scatter in random directions. A steady state is established in which the electrons have some net drift velocity in the field direction.

In the absence of any applied field (i.e., under equilibrium), the electron distribution is given by the Fermi-Dirac distribution, as we have discussed before. Even though the electrons are moving, the motion is random and the average velocity in any

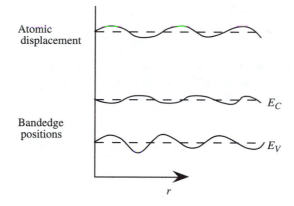

Figure 3.2: The effect of atomic displacement on bandedge energy levels in real space. The lattice vibrations cause spatial and time-dependent variations in the bandedges from which the electrons scatter.

direction is zero. When a field is applied and the electrons drift in the field direction, the distribution function changes. The new distribution is a function of the scattering rates and the field strength and is determined by solving an equation known as the Boltzmann transport equation. This is a difficult equation to solve, especially at high electric fields. The response of the electrons to the field can be represented by a velocity-field relation. We will briefly discuss the velocity-field relationships at low electric fields and moderately high electric fields.

3.3.1 Low Field Response: Mobility

At low electric fields, the macroscopic transport properties of the material (mobility, conductivity) can be related to the microscopic properties (scattering rate or relaxation time) by simple arguments. We will follow an approach developed by Drude at the turn of the century, except that at the time Drude did not know about the concept of effective mass and the source of scattering. In this approach we make the following assumptions derived from modifications to the Drude approach:

• The electrons in the semiconductor do not interact with each other. This approximation is called the independent electron approximation.
• Electrons suffer collisions from various scattering sources and the time τ_{sc} describes the mean time between successive collisions.
• *In betweeen collisions*, the electrons move according to the free electron equation

$$\frac{\hbar dk}{dt} = \mathbf{F}_{\text{ext}} \qquad (3.4)$$

After a collision, the electrons lose all their excess energy (on the average) so that the electron gas is essentially at thermal equilibrium. This assumption is really valid only at very low electric fields.

According to these assumptions, immediately after a collision the electron velocity is the same as that given by the thermal equilibrium conditions. This average velocity is thus zero, after collisions. The electron gains a velocity in between collisions, i.e., only for the time τ_{sc}.

This average velocity gain is then that of an electron with mass m^* traveling in a field $\boldsymbol{F}$ for a time τ_{sc}:

$$\boldsymbol{v}_{avg} = -\frac{e\boldsymbol{F}\tau_{sc}}{m^*} = \boldsymbol{v}_d \tag{3.5}$$

where v_d is the drift velocity. The current density is now

$$\mathbf{J} = -ne\mathbf{v}_d = \frac{ne^2\tau_{sc}}{m^*}\mathbf{F} \tag{3.6}$$

Comparing this with the Ohm's law result for conductivity σ

$$\mathbf{J} = \sigma\mathbf{F} \tag{3.7}$$

we have

$$\boxed{\sigma = \frac{ne^2\tau_{sc}}{m^*}} \tag{3.8}$$

The resistivity of the semiconductor is simply the inverse of the conductivity. From the definition of mobility μ, for electrons

$$\mathbf{v}_d = -\mu\mathbf{F}$$

Note that the electrons move in a direction opposite to the field while the holes move in the same direction. From Eqn. 3.5, the definition of mobility, we have

$$\boxed{\mu = \frac{e\tau_{sc}}{m^*}} \tag{3.9}$$

If both electrons and holes are present, the conductivity of the material becomes

$$\sigma = ne\mu_n + pe\mu_p \tag{3.10}$$

where μ_n and μ_p are the electron and hole mobilities and n and p are their densities.

Notice that the mobility has an explicit $\frac{1}{m^*}$ dependence in it. Additionally τ_{sc} also decreases with m^*. Thus the mobility has a strong dependence on the carrier mass.

| Semiconductor | Mobility at 300 K (cm²/V•s) | |
	Electrons	Holes
C	800	1200
Ge	3900	1900
Si	1500	450
α-SiC	400	50
GaSb	5000	850
GaAs	8500	400
GaP	110	75
InAs	33000	460
InP	4600	150
CdTe	1050	100

Table 3.2: Bandgap, electron, and hole mobilities of some semiconductors. Notice that narrow bandgap materials have superior mobility, since the effective mass is smaller.

In Table 3.2 we show the mobilities of several important semiconductors at room temperature. The results are shown for pure materials. If the semiconductors are doped, the mobility decreases. Note that Ge has the best hole mobility among all semiconductors.

EXAMPLE 3.1 The mobility of electrons in pure GaAs at 300 K is 8500 cm²/V·s. Calculate the relaxation time. If the GaAs sample is doped at $N_d = 10^{17}$ cm^{-3}, the mobility decreases to 5000 cm²/V·s. Calculate the relaxation time due to ionized impurity scattering.

The relaxation time is related to the mobility by

$$\tau_{sc}^{(1)} = \frac{m^*\mu}{e} = \frac{(0.067 \times 0.91 \times 10^{-30} \text{ kg})(8500 \times 10^{-4} \text{ m}^2/\text{V} \cdot \text{s})}{1.6 \times 10^{-19} \text{ C}}$$

$$= 3.24 \times 10^{-13} \text{ s}$$

If the ionized impurities are present, the time is

$$\tau_{sc}^{(2)} = \frac{m^*\mu}{e} = 1.9 \times 10^{-13} \text{ s}$$

According to Mathieson's rule, the impurity-related time $\tau_{sc}^{(imp)}$ is given by

$$\frac{1}{\tau_{sc}^{(2)}} = \frac{1}{\tau_{sc}^{(1)}} + \frac{1}{\tau_{sc}^{(imp)}}$$

which gives
$$\tau_{sc}^{(imp)} = 4.6 \times 10^{-13} s$$

EXAMPLE 3.2 The mobility of electrons in pure silicon at 300 K is 1500 cm^2/Vs. Calculate the time between scattering events using the conductivity effective mass.

The conductivity mass is given by

$$
\begin{aligned}
m_\sigma^* &= 3 \left(\frac{2}{m_t^*} + \frac{1}{m_\ell^*} \right)^{-1} \\
&= 3 \left(\frac{2}{0.19 m_o} + \frac{1}{0.98 m_o} \right)^{-1} = 0.26 m_o
\end{aligned}
$$

The scattering time is then

$$
\begin{aligned}
\tau_{sc} &= \frac{\mu m_\sigma^*}{e} = \frac{(0.26 \times 0.91 \times 10^{-30})(1500 \times 10^{-4})}{1.6 \times 10^{-19}} \\
&= 2.2 \times 10^{-13} \text{ s}
\end{aligned}
$$

EXAMPLE 3.3 Consider two semiconductor samples, one Si and one GaAs. Both materials are doped n-type at $N_d = 10^{17}$ cm^{-3}. Assume 50 percent of the donors are ionized at 300 K. Calculate the conductivity of the samples. Compare this conductivity to the conductivity of undoped samples.

You may assume the following values:

$$
\begin{aligned}
\mu_n(\text{Si}) &= 1000 \quad \text{cm}^2/\text{V} \cdot \text{s} \\
\mu_p(\text{Si}) &= 350 \quad \text{cm}^2/\text{V} \cdot \text{s} \\
\mu_n(\text{GaAs}) &= 8000 \quad \text{cm}^2/\text{V} \cdot \text{s} \\
\mu_p(\text{GaAs}) &= 400 \quad \text{cm}^2/\text{V} \cdot \text{s}
\end{aligned}
$$

In the doped semiconductors, the electron density is (50 percent of 10^{17} cm^{-3})

$$n_{doped} = 5 \times 10^{16} \text{ cm}^{-3}$$

and hole density can be found from

$$p_{doped} = \frac{n_i^2}{n_{doped}}$$

For silicon we have:
$$p_{doped} = \frac{2.25 \times 10^{20}}{5 \times 10^{16}} = 4.5 \times 10^3 \text{ cm}^{-3}$$

which is negligible for the conductivity calculation.

The conductivity is

$$\sigma_{doped} = n e \mu_n + p e \mu_p = 8 \quad (\Omega \text{ cm})^{-1}$$

In the case of undoped silicon we get ($n = n_i = p = 1.5 \times 10^{10}$ cm^{-3}):

$$\sigma_{undoped} = n_i e \mu_n + p_i e \mu_p = 3.24 \times 10^{-6} \ (\Omega \ cm)^{-1}$$

For GaAs we get

$$\sigma_{doped} = 5 \times 10^{16} \times 1.6 \times 10^{-19} \times 8000 = 64 \ (\Omega \ cm)^{-1}$$

For undoped GaAs we get ($n_i = 1.84 \times 10^6$ cm^{-3})

$$\sigma_{undoped} = n_i e \mu_n + p_i e \mu_p = 2.47 \times 10^{-9} \ (\Omega \ cm)^{-1}$$

You can see the very large difference in the conductivities of the doped and undoped samples. Also there is a large difference between GaAs and Si.

EXAMPLE 3.4 Consider a semiconductor in equilibrium in which the position of the Fermi level can be placed anywhere *within the bandgap.*

What are the maximum and minimum conductivity for Si and GaAs at 300 K? You can use the data given in the problem above.

The maximum carrier density occurs when the Fermi level coincides with the conduction bandedge if $N_c > N_v$ or with the valence bandedge if $N_v > N_c$. If $N_c > N_v$, the Boltzmann approximation gives

$$n_{max} = N_c$$

while if $N_v > N_c$, we get

$$p_{max} = N_v$$

This gives us for the maximum density: i) for Si, 2.78×10^{19} cm^{-3} ii) for GaAs, 7.72×10^{18} cm^{-3}. Based on these numbers we can calculate the maximum conducitvity:

For Si:

$$\sigma_{max} = 2.78 \times 10^{19} \times 1.6 \times 10^{-19} \times 1000 = 4.45 \times 10^3 \ (\Omega \ cm)^{-1}$$

For GaAs:

$$\sigma_{max} = 7.72 \times 10^{18} \times 1.6 \times 10^{-19} \times 400 = 4.9 \times 10^2 \ (\Omega \ cm)^{-1}$$

To find the minimum conductivity we need to find the minima of the expression

$$\begin{aligned} \sigma &= ne\mu_n + pe\mu_p \\ &= \frac{n_i^2}{p} e\mu_n + pe\mu_p \end{aligned}$$

To find the minimum we take the derivative with respect to p and equate the result to zero. This gives

$$p = n_i \sqrt{\frac{\mu_n}{\mu_p}}$$

This then gives for the minimum conductivity

$$\sigma_{min} = n_i e [\mu_n \sqrt{\frac{\mu_p}{\mu_n}} + \mu_p \sqrt{\frac{\mu_n}{\mu_p}}]$$

For Si this gives upon plugging in numbers

$$\sigma_{min} = 2.8 \times 10^{-6} \ (\Omega \ \text{cm})^{-1}$$

and for GaAs

$$\sigma_{min} = 1.05 \times 10^{-9} \ (\Omega \ \text{cm})^{-1}$$

Note that these values are lower than the values we get in the the previous problem for the undoped cases. This example shows the tremendous variation in conductivity that can be obtained in a semiconductor.

3.3.2   High Field Transport

In most electronic devices a significant portion of the electronic transport occurs under strong electric fields. This is especially true of field effect transistors. At such high fields ($F \sim 1- 100$ kV/cm) the electrons get "hot" and aquire a high average energy. The extra energy comes due to the strong electric fields. The drift velocities are also quite high. The description of electrons at such high electric fields is quite complex and requires either numerical techniques or computer simulations. We will only summarize the results.

At high electric field as the carriers gain energy from the field, they suffer greater rates of scattering, i.e., τ_{sc} decreases. The mobility thus starts to decrease. It is usual to represent the response of the carriers to the electric field by velocity-field relations. In Fig. 3.3 we show such relations for several semiconductors. At very high fields the drift velocity becomes saturated, i.e., becomes independent of the electric field. The drift velocity for carriers in most materials saturates to a value of $\sim 10^7$ cm/s. The fact that the velocity saturates is very important in understanding current flow in semiconductor devices.

EXAMPLE 3.5 The mobility of electrons in a semiconductor decreases as the electric field is increased. This is because the scattering rate increases as electrons become hotter due to the applied field. Calculate the relaxation time of electrons in silicon at 1 kV/cm and 100 kV/cm at 300 K.

The velocity of the silicon electrons at 1 kV/cm and 100 kV/cm is approximately 1.4×10^6 cm s and 1.0×10^7 cm/s, respectively, from the *v-F* curves given in Fig. 3.3. The mobilities are then

$$\mu(1 \ kV/cm) \quad = \quad \frac{v}{F} = 1400 \ \text{cm}^2/\text{V} \cdot \text{s}$$

$$\mu(100 \ kV/cm) \quad = \quad 100 \ \text{cm}^2/\text{V} \cdot \text{s}$$

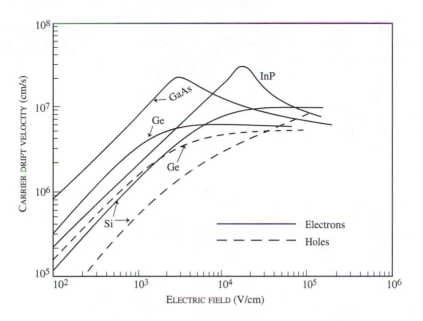

Figure 3.3: Velocity-field relations for several semiconductors at 300 K.

The corresponding relaxation times are

$$\tau_{sc}(1 \text{ kV/cm}) = \frac{(0.26 \times 0.91 \times 10^{-30} \text{ kg})(1400 \times 10^{-4} \text{ m}^2/\text{V})}{1.6 \times 10^{-19} \text{ C}} = 2.1 \times 10^{-13} \text{ s}$$

$$\tau_{sc}(100 \text{ kV/cm}) = \frac{(0.26 \times 0.91 \times 10^{-30})(100 \times 10^{-4})}{1.6 \times 10^{-19}} = 1.48 \times 10^{-14} \text{ s}$$

Thus the scattering rate has dramatically increased at the higher field.

EXAMPLE 3.6 The average electric field in a particular 2.0 μm GaAs device is 5 kV/cm. Calculate the transit time of an electron through the device a) if the low field mobility value of 8000 cm^2/V·s is used; b) if the saturation velocity value of 10^7 cm/s is used.

If the low field mobility is used, the average velocity of the electron is

$$v = \mu F = (8000 \text{ cm}^2/\text{Vs}) \times (5 \times 10^3 \text{ V/ cm}) = 4 \times 10^7 \text{ cm/s}$$

The transit time through the device becomes

$$\tau_{tr} = \frac{L}{v} = \frac{2.0 \times 10^{-4} \text{ cm}}{4 \times 10^7 \text{ cm/s}} = 5 \text{ ps}$$

The transit time, if the saturation velocity (which is the correct velocity value) is used, is

$$\tau_{tr} = \frac{L}{v} = \frac{2 \times 10^{-4}}{10^7} = 20 \text{ ps}$$

In our discussion on MESFET and MOSFET devices later in the text, we use a simple analytical model, and use the constant mobility model for electron velocity. As this example shows, this can cause an error in transit time.

3.4 VERY HIGH FIELD TRANSPORT: BREAKDOWN PHENOMENA

When the electric field becomes extremely high ($\gtrsim 100$ kV cm^{-1}), the semiconductor suffers a "breakdown" in which the current has a "runaway" behavior. The breakdown occurs due to carrier multiplication, which arises from the two sources discussed below. By carrier multiplication we mean that the number of electrons and holes that can participate in current flow increases. Of course, the total number of electrons is always conserved.

3.4.1 Impact Ionization or Avalanche Breakdown

In the transport considered in the previous sections, the electron (hole) remains in the same band during the transport. At very high electric fields, this does not hold true. In the impact ionization process shown schematically in Fig. 3.4, an electron which is "very hot" (i.e., has a very high energy due to the applied field) scatters with an electron in the valence band via coulombic interaction and knocks it into the conduction band. The initial electron must provide enough energy to bring the valence-band electron up into the conduction band. Thus the *initial electron should have energy slightly larger than the bandgap* (measured from the conduction-band minimum). In the final state we now have two electrons in the conduction band and one hole in the valence band. Thus the number of current carrying charges have multiplied, and the process is often called *avalanching*. Note that the same could happen to "hot holes" that could then trigger the avalanche.

For the purpose of device applications, the current in the device, once avalanching starts, is given by

$$\frac{dI(z)}{dz} = \alpha_{imp} I \tag{3.11}$$

where I is the current and α_{imp} represents the average rate of ionization per unit distance.

The coefficients α_{imp} for electrons and β_{imp} for holes depend upon the bandgap of the material in a very strong manner. This is because, as discussed above, the process can start only if the initial electron has a kinetic energy equal to a certain threshold (roughly equal to the bandgap). This is achieved for lower electric fields in narrow gap materials.

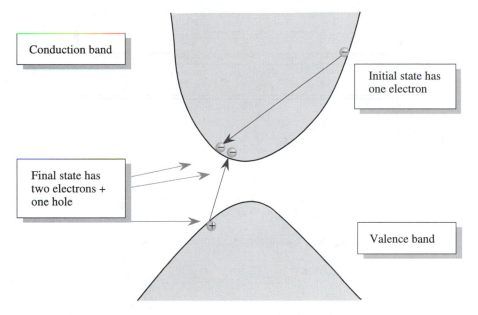

Figure 3.4: The impact ionization process where a high energy conduction-band electron scatters from a valence-band electron, producing two conduction-band electrons and a hole.

If the electric field is constant so that α_{imp} is constant, the number of times an initial electron will suffer impact ionization after traveling a distance x is

$$N(x) = \frac{I(x)}{I(0)} = \exp\left(\alpha_{imp}x\right) \tag{3.12}$$

A critical breakdown field F_{crit} is defined where α_{imp} or β_{imp} approaches 10^4 cm^{-1}. When α_{imp} (β_{imp}) approaches 10^4 cm^{-1}, there is about one impact ionization when a carrier travels a distance of one micron. Values of the critical field are given for several semiconductors in Table 3.3. The avalanche process places an important limitation on the power output of devices. Once the process starts, the current rapidly increases due to carrier multiplication and the control over the device is lost. The push for high-power devices is one of the reasons for research in large gap semiconductor devices. It must be noted that in certain devices such as avalanche photodetectors, the process is exploited for high gain detection. The process is also exploited in special microwave devices.

BREAKDOWN ELECTRIC FIELDS IN MATERIALS		
Material	Bandgap (eV)	Breakdown electric field (V/cm)
GaAs	1.43	4×10^5
Ge	0.664	10^5
InP	1.34	
Si	1.1	3×10^5
$In_{0.53}Ga_{0.47}As$	0.8	2×10^5
C	5.5	10^7
SiC	2.9	$2\text{-}3 \times 10^6$
SiO_2	9	$\sim 10^7$
Si_3N_4	5	$\sim 10^7$

Table 3.3: Breakdown electric fields in some materials.

3.4.2   Band-to-Band Tunneling: Zener Tunneling

In Appendix B we have discussed the quantum-mechanical tunneling theory. This theory
is invoked to understand another phenomenon responsible for high field breakdown.
Consider a semiconductor under a strong field as shown in Fig. 3.5a. At strong electric
fields, the electrons in the valence band can tunnel into an unoccupied state in the
conduction band. As the electron tunnels, it sees the potential profile shown in Fig. 3.5b.
The tunneling probability through the triangular barrier is given by (see Appendix B)

$$T = \exp\left(\frac{-4\sqrt{2m^*}E_g^{3/2}}{3e\hbar F}\right) \tag{3.13}$$

where F is the electric field in the semiconductor.

In narrow bandgap materials this band-to-band tunneling or Zener tunneling
can be very important. It is the basis of the Zener diode where the current is essentially
zero until the band-to-band tunneling starts and the current increases very sharply. A

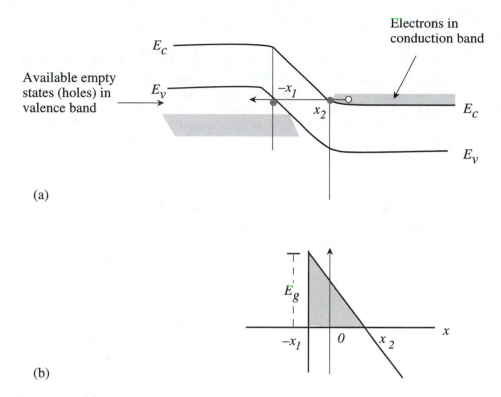

Figure 3.5: (a) A schematic showing the E-x and E-k diagram for a p-n junction. An electron in the conduction band can tunnel into an unoccupied state in the valence band or vice versa. (b) The potential profile seen by the electron during the tunneling process.

tunneling probability of $\sim 10^{-6}$ is necessary to start the breakdown process.

EXAMPLE 3.7 Calculate the band-to-band tunneling probability in GaAs and InAs at an applied electric field of 2×10^5 V/cm.

The exponent for the tunneling probability is (m^*(GaAs) $= 0.065$ m_0; m^*(InAs) ~ 0.02 m_0; E_g(GaAs) $= 1.5$ eV; E_g(InAs) $= 0.4$ eV) for GaAs

$$- \frac{4 \times (2 \times 0.065 \times 0.91 \times 10^{-30} \text{ kg})^{1/2}(1.5 \times 1.6 \times 10^{-19} \text{ J})^{3/2}}{3 \times (1.6 \times 10^{-19} \text{ C})(1.05 \times 10^{-34} \text{ Js})(2 \times 10^7 \text{ V/m})}$$

$$= -160$$

The tunneling probability is $\exp(-160) \cong 0$. For InAs the exponent turns out to be -12.5 and the tunneling probability is

$$T = \exp(-12.5) = 3.7 \times 10^{-6}$$

In InAs the band-to-band tunneling will start becoming very important if the field is $\sim 2 \times 10^5$ V/cm.

3.5   CARRIER TRANSPORT BY DIFFUSION

So far we have studied carrier transport due to drift in an electric field. It is intuitively easy to see why the force due to an electric field would cause electrons to move a certain way. However, there is another important transport mechanism that does not involve such direct forces. This is the diffusion process and is a wonderful phenomenon arising from thermodynamics. We depend upon this phenomenon when we smell a loaf of freshly baked bread. So do sharks, who can reputedly sense a few molecules of blood diffusing from a wound a mile away!

Whenever there is a gradient in the concentration of a species of mobile particles, the particles diffuse from the regions of high concentration to the regions of low concentration. This diffusion is due to the random motion of the particles.

In the case of electrons (or holes), as the particles move they suffer random collisions, as discussed in the previous section. The collision process can be described by the mean free path ℓ and the mean collision time τ_{sc}. The mean free path is the average distance the electron (hole) travels between successive collisions. These collisions are due to the various scattering processes that were discussed for the drift problem. In between the collisions the electrons move randomly with equal probability of moving in any direction (remember, there is no electric field). We are interested in finding out how the electrons move (diffuse) when there is a concentration gradient in space.

Consider a concentration profile $n(x, t)$ of electrons at time t as shown in Fig. 3.6. We are going to calculate the electron flux $\phi(x, t)$ across a plane $x = x_o$ at any instant of time. We consider a region of space a mean free path ℓ to each side of x_o, from which electrons can come across the $x = x_o$ boundary in time τ_{sc}. Electrons from regions further away will suffer collisions that will randomly change their direction. Since in the two regions labeled L and R in Fig. 3.6, the electrons move randomly, half of the electrons in region L will go across $x = x_o$ to the right and half in the region R will go across $x = x_o$ to the left in time τ_{sc}. The flux to the right is then

$$\phi_n(x, t) = \frac{(n_L - n_R)\ell}{2\tau_{sc}} \tag{3.14}$$

where n_L and n_R are the average carrier densities in the two regions. Since the two regions L and R are separated by the distance ℓ, we can write

$$n_L - n_R \cong -\frac{dn}{dx} \cdot \ell \tag{3.15}$$

Thus the net flux is

$$\boxed{\phi_n(x, t) = -\frac{\ell^2}{2\tau_{sc}} \frac{dn(x, t)}{dx} = -D_n \frac{dn(x, t)}{dx}} \tag{3.16}$$

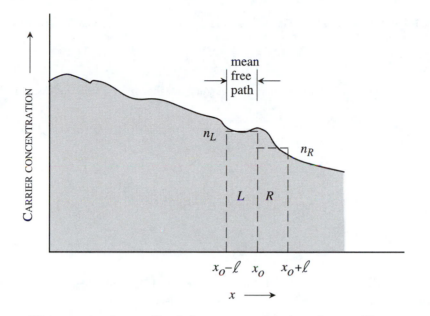

Figure 3.6: The concentration profile of electrons as a function of space. The terms n_L, n_R, L, and R are used to derive the diffusion law in the text. The distance ℓ is the mean free path for electrons, i.e., the distance they travel between collisions.

where D_n is called the diffusion coefficient of the electron system and clearly depends upon the scattering processes that control ℓ and the τ_{sc}. *Since the mean free path is essentially $v_{th}\tau_{sc}$, where v_{th} is the mean thermal speed, the diffusion coefficient depends upon the temperature as well.* In a similar manner, the hole diffusion coefficient gives the hole flux due to a hole density gradient

$$\phi_p(x,t) = -D_p \frac{dp(x,t)}{dx} \tag{3.17}$$

Because of this electron and hole flux, a current can flow in the structure that in the absence of an electric field is given by (current is just charge multiplied by particle flux)

$$
\begin{aligned}
\mathbf{J}_{tot}(diff) &= \mathbf{J}_n(diff) + \mathbf{J}_p(diff) \\
&= eD_n \frac{dn(x,t)}{dx} - eD_p \frac{dp(x,t)}{dx}
\end{aligned}
\tag{3.18}
$$

While both electrons and holes move in the direction of lower concentration of electrons and holes respectively, the currents they carry are opposite, since electrons are negatively charged while holes are positively charged.

EXAMPLE 3.8 In an n-type GaAs crystal at 300 K, the electron concentration varies along the x-direction as

$$n(x) = 10^{16} \, \exp \left(-\frac{x}{L} \right) \, \text{cm}^{-3} \qquad x > 0$$

where L is 1.0 μm. Calculate the diffusion current density at $x = 0$ if the electron diffusion coefficient is 220 cm^2/s.

The diffusion current density at $x = 0$ is

$$
\begin{aligned}
J_n(diff) &= eD_n \left. \frac{dn}{dx} \right|_{x=0} \\[2mm]
&= \left(1.6 \times 10^{-19} \, \text{C} \right) (220 \, \text{cm}^2/\text{s}) \left(\frac{10^{16} \, \text{cm}^{-3}}{10^{-4} \, \text{cm}} \right) \\[2mm]
&= 3.5 \text{kA/ cm}^2
\end{aligned}
$$

Note that in this problem the diffusion current of electrons changes with the position in space. Since the total current is constant in the absence of any source or sink of current, some other current must be present to compensate for the spatial change in electron diffusion current.

3.6 TRANSPORT BY DRIFT AND DIFFUSION: EINSTEIN'S RELATION

In many electronic devices the charge moves under the combined influence of electric fields and concentration gradients. The current is then given by

$$
\boxed{
\begin{aligned}
\mathbf{J}_n(x) &= e\mu_n n(x)\mathbf{F}(x) + eD_n \frac{\mathbf{d}n(x)}{\mathbf{d}x} \\[2mm]
\mathbf{J}_p(x) &= e\mu_p p(x)\mathbf{F}(x) - eD_p \frac{\mathbf{d}p(x)}{\mathbf{d}x}
\end{aligned}
}
\tag{3.19}
$$

In our discussion of the diffusion coefficient we indicated that it is controlled by essentially the same scattering mechanisms that control the mobility. We will now establish an important relationship between mobility and diffusion coefficients. To do so, let us examine the effect of electric fields on the energy bands of the semiconductor. We consider a case where a uniform electric field is applied as shown in Fig. 3.7a. The potential energy associated with the field is shown in Fig. 3.7b. There is a positive potential on the left-hand side in relation to the right-hand side. For a uniform electric field the potential energy is

$$U(x) = U(0) - eFx \tag{3.20}$$

The potential energy profile is shown in Fig. 3.7b. We now discuss how the electron energy band profile is displayed. The electron energy band includes the effect of negative charge of the electrons. The applied force is related to the potential energy by

$$\text{Force} = -\nabla U(x) \tag{3.21}$$

Thus, since the electron charge $-e$ is negative, the bands bend as shown in Fig. 3.7c according to the relation

$$E_c(x) = E_c(0) + eFx \tag{3.22}$$

Thus, if a positive potential is applied to the left of the material and a negative to the right, the energy bands will be lower on the left-hand side, as shown in Fig. 3.7c. The electrons drift downhill in the energy band picture and thus opposite to the field.

At equilibrium, the total electron and hole currents are individually zero and we have from Eqn. 3.19 for the electrons

$$\mathbf{F}(x) = -\frac{D_n}{\mu_n} \frac{1}{n(x)} \frac{dn(x)}{dx} \tag{3.23}$$

To obtain the derivative of carrier concentration, we write $n(x)$ in terms of the intrinsic Fermi level, E_{Fi}, which serves as a reference level, and the Fermi level in the semiconductor, $E_F(x)$. If we assume that the electron distribution is given by the Boltzmann distribution we have (see Eqn. 2.30)

$$n(x) = n_i \, \exp \left\{ -\left(\frac{E_{Fi} - E_F(x)}{k_B T} \right) \right\} \tag{3.24}$$

This gives

$$\frac{dn(x)}{dx} = \frac{n(x)}{k_B T} \left(-\frac{dE_{Fi}}{dx} + \frac{dE_F}{dx} \right) \tag{3.25}$$

At equilibrium, the Fermi level cannot vary spatially, otherwise the probability of finding electrons along a constant energy position will vary along the semiconductor. This would cause electrons at a given energy in a region where the probability is low to move to the same energy in a region where the probability is high. Since this is not allowed by definition of equilibrium conditions, i.e., no current is flowing, the Fermi level has to be uniform in space at equilibrium, or

$$\boxed{\frac{dE_F}{dx} = 0} \tag{3.26}$$

We then have from Eqns. 3.23 and 3.25 (using $F(x) = \frac{1}{e} \frac{dE_{Fi}}{dx}$)

$$\boxed{\frac{D_n}{\mu_n} = \frac{k_B T}{e}} \tag{3.27}$$

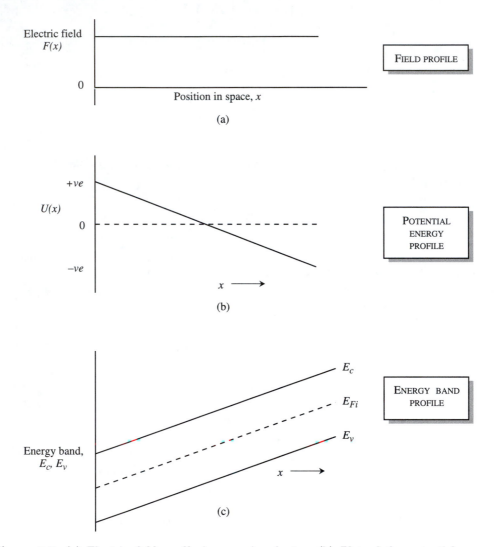

Figure 3.7: (a) Electric field profile in a semiconductor. (b) Plot of the potential energy associated with the electric field. (c) Electron energy band profile. The negative charge of the electron causes the energy band profile to have the opposite sign to the potential energy profile.

	D_n (cm^2/s)	D_p (cm^2/s)	μ_n (cm^2/V·s)	μ_p (cm^2/V·s)
Ge	100	50	3900	1900
Si	35	12.5	1350	480
GaAs	220	10	8500	400

Table 3.4: Low field mobility and diffusion coefficients for several semiconductors at room temperature. The Einstein relation is satisfied quite well.

which is the Einstein relation satisfied for electrons. A similar relation exists for the holes.

We list in Table 3.4 the mobilities and diffusion coefficients for a few semiconductors at room temperature. The Einstein relation is seen to be satisfied quite well.

EXAMPLE 3.9 Use the velocity-field relations for electrons in silicon to obtain the diffusion coefficient at an electric field of 1 kV/cm and 10 kV/cm at 300 K.

According to the v-F relations given in Fig. 3.3, the velocity of electrons in silicon is $\sim 1.4 \times 10^6$ cm/s and $\sim 7 \times 10^6$ cm/s at 1 kV/cm and 10 kV/cm. Using the Einstein relation, we have for the diffusion coefficient

$$D = \frac{\mu k_B T}{e} = \frac{v k_B T}{eF}$$

This gives

$$
\begin{aligned}
D(1kV/\ \mathrm{cm}^{-1}) &= \frac{(1.4 \times 10^4\, m/s)(0.026 \times 1.6 \times 10^{-19}\ \mathrm{J})}{(1.6 \times 10^{-19}\ \mathrm{C})(10^5\ \mathrm{V/m}^{-1})} \\
&= 3.64 \times 10^{-3}\ \mathrm{m}^2/s = 36.4\ \mathrm{cm}^2/s \\
D(10kV/\ \mathrm{cm}^{-1}) &= \frac{(7 \times 10^4\ \mathrm{m/s})(0.026 \times 1.6 \times 10^{-19}\ \mathrm{J})}{(1.6 \times 10^{-19}\ \mathrm{C})(10^6\ \mathrm{Vm}^{-1})} \\
&= 1.82 \times 10^{-3}\ \mathrm{m}^2/s = 18.2\ \mathrm{cm}^2/s
\end{aligned}
$$

The diffusion coefficient decreases with the field because of the higher scattering rate at higher fields.

3.7 CHARGE INJECTION AND QUASI-FERMI LEVELS

In Chapters 1 and 2 we discussed the distribution function of electrons among the various allowed levels. This function, given by the Fermi distribution function, is valid only under *equilibrium conditions*. This means that there is no net flow of external energy or of particles into the sample. In most real applications this condition is not satisfied. For example, light may be shining on the sample. Or a battery may inject electrons into the sample from one side and extract them from another. Under such conditions the material is not at equilibrium. If electrons and holes are injected into a semiconductor, either by external contacts or by optical excitation, the question arises: What kind of distribution function describes the electron and hole occupation? We know that in equilibrium the electron and hole occupation is represented by the Fermi function. A new function is needed to describe the system when the electrons and holes are not in equilibrium.

3.7.1   Quasi-Fermi Levels

In equilibrium the distribution of electrons and holes in the conduction and valence bands, respectively, is given by the Fermi function, which is defined once the Fermi level is known. Also the product of electrons and holes, np, is a constant. If excess electrons and holes are injected into the semiconductor, clearly the same function will not describe the occupation of states.

In most semiconductor devices, operating conditions are such that even though the system is not in equilibrium, we can assume that electrons are in equilibrium in the conduction band, while holes are in equilibrium in the valence band. In this case, the *quasi-equilibrium* electron and holes can be represented by an electron Fermi function f^e (with electron Fermi level) and a hole Fermi function f^h (with a *different* hole Fermi level). We now have

$$n = \int_{E_c}^{\infty} N_e(E) f^e(E) dE \tag{3.28}$$

$$p = \int_{-\infty}^{E_v} N_h(E) f^h(E) dE \tag{3.29}$$

where

$$f^e(E) = \frac{1}{\exp\left(\frac{E - E_{Fn}}{k_B T}\right) + 1} \tag{3.30}$$

and

$$f^h(E) = 1 - f^v(E) \quad = \quad 1 - \frac{1}{\exp\left(\frac{E - E_{Fp}}{k_B T}\right) + 1}$$

$$= \frac{1}{\exp\left(\frac{E_{Fp} - E}{k_B T}\right) + 1} \tag{3.31}$$

At equilibrium $E_{Fn} = E_{Fp}$. If excess electrons and holes are injected into the semiconductor, the electron Fermi level E_{Fn} moves towards the conduction band, while the hole Fermi level E_{Fp} moves towards the valence band. This is shown schematically in Fig. 3.8. The ability to define quasi-Fermi levels E_{Fn} and E_{Fp} provides a very powerful approach to solve non-equilibrium problems, which are, of course, of greatest interest in devices.

By defining separate Fermi levels for the electrons and holes, one can study the properties of excess carriers using the same relationship between Fermi level and carrier density as we developed for the equilibrium problem. Thus, in the Boltzmann approximation we have

$$
\begin{aligned}
n &= N_c \exp\left[\frac{(E_{Fn} - E_c)}{k_B T}\right] \\
p &= N_v \exp\left[\frac{(E_v - E_{Fp})}{k_B T}\right]
\end{aligned}
\tag{3.32}
$$

In the more accurate Joyce-Dixon approximation we have (see Eqn. 2.33)

$$
\begin{aligned}
E_{Fn} - E_c &= k_B T \left[\ell n \frac{n}{N_c} + \frac{n}{\sqrt{8}N_c}\right] \\
E_v - E_{Fp} &= k_B T \left[\ell n \frac{p}{N_v} + \frac{p}{\sqrt{8}N_v}\right]
\end{aligned}
\tag{3.33}
$$

Note the change of signs for the hole quasi-Fermi level expression.

EXAMPLE 3.10 Using Boltzmann statistics, calculate the position of the electron and hole quasi-Fermi levels when an *e-h* density of 10^{17} cm^{-3} is injected into pure (undoped) silicon at 300 K.

The electron and hole densities are related to the quasi-Fermi levels by the equations

$$
\begin{aligned}
n &= N_c \exp\left[(E_{Fn} - E_c)/k_B T\right] \\
p &= N_v \exp\left[(E_v - E_{Fp})/k_B T\right]
\end{aligned}
$$

Recall that when we were discussing carrier concentrations in *semiconductors in equilibrium*, we used similar equations, except E_{Fn} and E_{Fp} were the same. Here, since we are not at equilibrium, the two are different.

At room temperature for Si we have

$$
\begin{aligned}
N_c &= 2.8 \times 10^{19} \text{ cm}^{-3} \\
N_v &= 1.04 \times 10^{19} \text{ cm}^{-3}
\end{aligned}
$$

If $n = p = 10^{17}$ cm^{-3}, we obtain ($k_B T = 0.026$ eV)

$$
E_{Fn} = k_B T \, \ell n \left[\frac{n}{N_c}\right] + E_c
$$

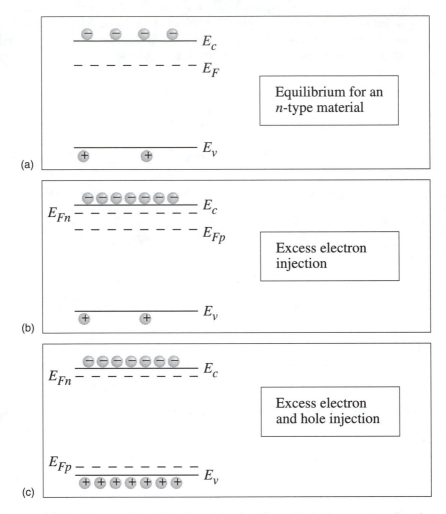

Figure 3.8: (a) Equilibrium Fermi level position in a hypothetical n-type semiconductor. (b) The positions of the quasi-Fermi levels for the case where excess electrons are injected in the conduction band. (c) The position of the quasi-Fermi levels when excess electrons and holes are injected.

$$
\begin{aligned}
&= E_c - 0.146 \text{ eV} \\
E_{Fp} &= E_v - k_B T \left[ln \, \frac{p}{N_v} \right] \\
&= E_v + 0.121 \text{ eV}
\end{aligned}
$$

Since in Si, the bandgap is $E_c - E_v = 1.1$eV, we have

$$
\begin{aligned}
E_{Fn} - E_{Fp} &= (E_c - E_v) - (0.146 + 0.121) \\
&= 1.1 - 0.267 = 0.833 \text{ eV}
\end{aligned}
$$

If we had injected only 10^{15} cm^{-3} electrons and holes, the difference in the quasi-Fermi levels would be

$$
\begin{aligned}
E_{Fn} - E_{Fp} &= (E_c - E_v) - (0.266 + 0.24) \\
&= 1.1 - 0.506 = 0.59 \text{ eV}
\end{aligned}
$$

Thus as the carrier injection is increased, the separation increases.

3.8   CARRIER GENERATION AND RECOMBINATION

Consider an experiment where a sample of pure silicon is taken from 0 K to 300 K. The electron density n will increase from 0 to the room-temperature intrinsic carrier concentration of 1.5×10^{10} cm^{-3}. How do the electrons get from the valence band into the conduction-band as the temperature is raised?

We know that at finite temperatures there is a certain thermal energy in the system. It is this energy that is responsible for exciting electrons from the conduction band to the valence band. Such a generation process is called *thermal generation*. We can also see that if electrons are continuously excited up from the valence band into the conduction band, there will be a build-up of free carriers. In order to reach an equilibrium concentration there has to be carrier *recombination* as well. Under equilibrium we have

$$
R_G = R_R \tag{3.34}
$$

where R_G is the generation rate and R_R is the carrier recombination rate.

In Fig. 3.9 we show a schematic description of carrier generation and recombination. Free carriers can be generated if an electron leaves the valence-band and goes to the conduction-band. They can also be generated if electrons leave a donor and go into the conduction-band. An electron from the valence-band going to an acceptor causes a hole to be generated. Reverse processes can also occur, as shown in Fig. 3.9.

Thermal energy is not the only source of carrier generation and recombination. There are several other processes that play an important role in exciting electrons from the valence to the conduction band and for electron hole recombination. One of the

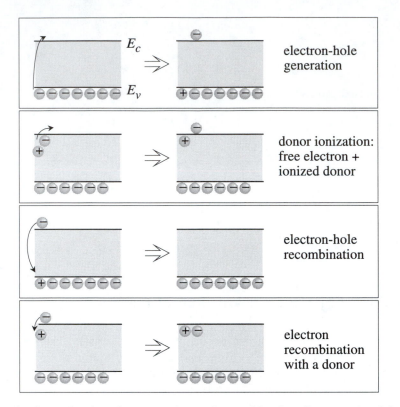

Figure 3.9: A schematic of carrier generation and recombination. Processes involving valence and conduction band directly are shown along with processes involving impurity levels.

most important mechanisms for carrier generation and recombination is absorption of light and emission of light.

3.8.1   Optical Processes in Semiconductors

When light shines on a semiconductor it can cause an electron in the valence band to go into the conduction band. This process generates electron-hole pairs. It is also possible for an electron and a hole to recombine and emit light. The interaction between light or photons and electrons in the semiconductor provides a rich variety of physical phenomena that are exploited for a fast growing field in solid state electronics—optoelectronics.

The most important optoelectronic interaction in semiconductors as far as devices are concerned is the band-to-band transition shown in Fig. 3.10. In the photon absorption process, a photon scatters an electron in the valence band, causing the electron to go into the conduction band. In the reverse process the electron in the conduction

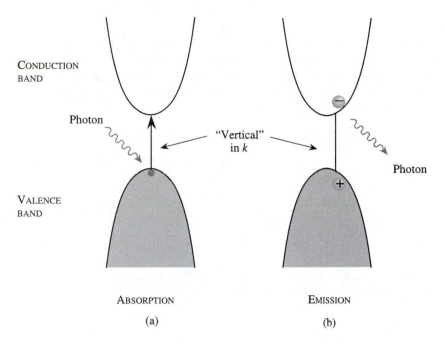

Figure 3.10: Band-to-band absorption in semiconductors. An electron in the valence band "absorbs" a photon and moves into the conduction band. In the reverse process, the electron recombines with a hole to emit a photon. Momentum conservation ensures that only vertical transitions are allowed.

band recombines with a hole in the valence band to generate a photon. These two processes are of obvious importance for light-detection and light-emission devices. The rates for the light-emission and absorption processes are determined by quantum mechanics. The processes involve the following issues:

• **Conservation of energy**: In the absorption and emission process we have for the initial and final energies of the electrons E_i and E_f

$$\text{absorption}: \quad E_f \;=\; E_i + \hbar\omega \tag{3.35}$$

$$\text{emission}: \quad E_f \;=\; E_i - \hbar\omega \tag{3.36}$$

where $\hbar\omega$ is the photon energy. Since the minimum energy difference between the conduction and valence band states is the bandgap E_g, *the photon energy must be larger than the bandgap*.

• **Conservation of momentum**: In addition to the energy conservation, one also needs to conserve the crystal momentum k for the electrons and the photon system. The photon k_{ph} value is given by

$$k_{ph} = \frac{2\pi}{\lambda} \tag{3.37}$$

Since 1 eV photons correspond to a wavelength of 1.24 μm, the k-values of relevance are $\sim 10^{-4}$ Å, which is essentially zero compared to the k-values for electrons. Thus k-conservation ensures that the initial and final electrons have the same k-value. Another way to say this is that *only "vertical" k transitions are allowed* in the bandstructure picture, as shown in Fig. 3.10.

Because of this constraint of k-conservation, in semiconductors where the valence band and conduction bandedges are at the same $k = 0$ value (the direct semiconductors), the optical transitions are quite strong. In indirect materials like Si, Ge, etc. the optical transitions are very weak and require the help of lattice vibrations to satisfy k-conservation. This makes a tremendous difference in the optical properties of these two kinds of materials. The transitions are very weak near the bandedges of indirect semiconductors.

Electromagnetic waves traveling through a medium like a semiconductor are described by Maxwell's equations, which show that the waves have a form given by the electric field vector dependence

$$\boldsymbol{F} = \boldsymbol{F}_o \, \exp \, \left\{ i\omega \left(\frac{n_r z}{c} - t \right) \right\} \, \exp \, \left(-\frac{\alpha z}{2} \right) \tag{3.38}$$

Here z is the propagation direction, ω the frequency, n_r the refractive index, and α the absorption coefficient of the medium. If α is zero, the wave propagates without attenuation with a velocity $\frac{c}{n_r}$. However, for nonzero α, the photon flux I ($\sim F^* F$) falls as

$$I(z) = I(0) \, \exp \, \{-\alpha z\} \tag{3.39}$$

In Fig. 3.11 we show the absorption coefficient for several direct and indirect bandgap semiconductors. Each photon that is absorbed creates an electron and a hole. If $\tilde{P}_{op}$ is the optical power density of light impinging on a semiconductor, the photon flux is

$$\Phi = \frac{\tilde{P}_{op}}{\hbar\omega}$$

and the electron-hole pair generation rate is

$$R_G = \alpha\Phi = \frac{\alpha \tilde{P}_{op}}{\hbar\omega} \tag{3.40}$$

As noted earlier, electrons in the conduction band can recombine with holes in the valence band to generate a photon. We will not discuss the rather complex details of electron-hole recombination. We will focus on a particular case that is of great importance in electronic devices such as p-n diodes or bipolar transistors. Consider an n-type (p-type) semiconductor in which $n \gg p$ ($p \gg n$). Assume that an excess hole

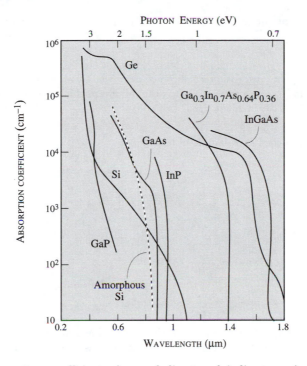

Figure 3.11: Absorption coefficient of several direct and indirect semiconductors. For the direct gap material, the absorption coefficient is very strong once the photon energy exceeds the bandgap. For indirect materials the absorption coefficient rises much more gradually. Once the photon energy is more than the direct gap, the absorption coefficient increases rapidly.

(electron) density δp (δn) is injected by some means. These excess minority carriers will recombine with the majority carriers with a rate given by

- Excess holes in n-type material:

$$R_R = \frac{\delta p}{\tau_p^r} \qquad (3.41)$$

- Excess electrons in p-type material:

$$R_R = \frac{\delta n}{\tau_n^r} \qquad (3.42)$$

Here τ_n^r and τ_p^r are called the electron and hole radiative lifetimes. They represent the time a typical carrier survives before recombining and emitting a photon. These times as defined above are also called minority carrier radiative lifetimes and for simplicity are denoted by τ_r.

In direct gap semiconductors the radiative lifetime is about 1 ns for heavily doped semiconductors. In indirect gap semiconductor the minority carrier lifetimes are two to three orders of magnitude larger. This is due to the k-conservation restriction discussed earlier.

EXAMPLE 3.11 The absorption coefficients near the bandedges of GaAs and Si are $\sim$ 10^4 cm^{-1} and 10^3 cm^{-1}, respectively. What is the minimum thickness of a sample in each case that can absorb 90% of the incident light?

The light absorbed in a sample of length L is

$$\frac{I_{abs}}{I_{inc}} = 1 - \exp(-\alpha L)$$

or

$$L = = \frac{1}{\alpha} \, \ell n \left(1 - \frac{I_{abs}}{I_{inc}}\right)$$

Using $\frac{I_{abs}}{I_{inc}}$ equal to 0.9, we get

$$L(\text{GaAs}) = -\frac{1}{10^4} \, \ell n \, (0.1) = 2.3 \times 10^{-4} \text{ cm}$$

$$= 2.3 \ \mu\text{m}$$

$$L(\text{Si}) = -\frac{1}{10^3} \, \ell n \, (0.1) = 23 \ \mu\text{m}$$

Thus, an Si detector requires a very thick active absorption layer in order to function. In an optical detector light is absorbed to create electrons and holes. These are collected and generate a signal.

EXAMPLE 3.12 Optical radiation with a power density of 1.0 kW/cm^2 impinges on GaAs. The photon energy is 1.5 eV and the absorption coefficient is 3×10^3 cm^{-1}. Calculate the carrier generation rate at the surface of the sample and at a depth of 10.0 μm.

At the surface the carrier generation rate is

$$R_G(0) = \frac{(3 \times 10^3 \ \text{cm}^{-1})(10^3 \ \text{W cm}^{-2})}{(1.5 \times 1.6 \times 10^{-19} \ \text{J})}$$

$$= 1.25 \times 10^{25} \text{ cm}^{-3}\text{s}^{-1}$$

At the depth of 10.0 μm we get

$$R_G(10.0 \ \mu\text{m}) = \frac{(3 \times 10^3 \ \text{cm}^{-1})(10^3 \ \text{W cm}^{-2})(e^{-3})}{(1.5 \times 1.6 \times 10^{-19} \ \text{J})}$$

$$= 6.22 \times 10^{23} \ \text{cm}^{-3} \ \text{s}^{-1}$$

3.8.2   Nonradiative Recombination

In perfect semiconductors, there are no allowed electronic states in the bandgap region. However, in real semiconductors there are always intentional or unintentional impurities

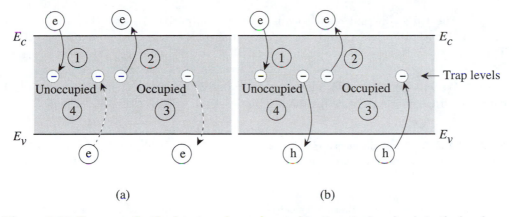

Figure 3.12: Processes that lead to trapping and recombination via deep levels in the bandgap region (dashed line). Processes 1 and 2 in (a) represent trapping and emission of electrons, while 3 and 4 represent the same for holes. The *e-h* recombination is shown in part (b).

that produce electronic levels that are in the bandgap. These impurity levels can arise from chemical impurities or from native defects such as a vacancy.

The bandgap levels are states in which the electron is "localized" in a finite space near the defect, unlike the usual *free* states, which represent the valence- and conduction-band states and in which electrons are free to move in space. As the free electrons move in the allowed bands, they can be trapped by the defects, as shown in Fig. 3.12. Some defects can allow the recombination of an electron and hole without emitting a photon, as we saw in the previous section. This nonradiative recombination competes with radiative recombination, discussed in the previous subsection. In a laser the nonradiative recombination is not desirable, but it is purposely increased in bipolar devices (discussed later) to increase device speed. We will briefly discuss the nonradiative processes involving a midgap level with density N_t. Such levels are called traps.

A trap state can be assigned a capture cross-section σ so that physically if a carrier comes within this area around a trap, it will be captured. If v_{th} is the velocity of the carrier, the rate at which the carrier encounters a trap is

$$r = N_t \sigma v_{th} = \frac{1}{\tau_{nr}} \tag{3.43}$$

where we have defined a lifetime or time it takes for a carrier to be captured by a trap as τ_{nr}. This non-radiative recombination process is known as Shockley–Read–Hall or SRH recombination. If we make the simplifying assumptions: i) the trap levels are at midgap level; and ii) $np \gg n_i^2$ under the injection conditions, the recombination rate can be shown to be

$$R_R = \frac{np}{\tau_{nr}(n + p)} \tag{3.44}$$

The time constant τ_{nr} depends upon the impurity density, the cross-section associated with the defect, and the electron thermal velocity, as seen in Eqn. 3.43. Typically the cross-sections are in the range 10^{-13} to 10^{-15} cm^2.

We will see in Chapter 5 how the nonradiative recombination leads to non-ideal current-voltage characteristics for the p-n diode.

EXAMPLE 3.13 A silicon sample has an impurity level of 10^{15} cm^{-3}. These impurities create a midgap level with a cross-section of 10^{-14} cm^2. Calculate the electron trapping time at 300 K and 77 K.

We can obtain the thermal velocities of the electrons by using the relation

$$\frac{1}{2}m^*v_{th}^2 = \frac{3}{2}k_BT$$

This gives

$$v_{th}(300\ K) = 2 \times 10^7 \text{ cm/s}$$
$$v_{th}(77\ K) = 1 \times 10^7 \text{ cm/s}$$

The electron trapping time is then

$$\tau_{nr}(300\ K) = \frac{1}{10^{15} \times 2 \times 10^7 \times 10^{-14}} = 5 \times 10^{-9} \text{ s}$$
$$\tau_{nr}(77\ K) = 10^{15} \times 1 \times 10^7 \times 10^{-14} = 10^{-8} \text{ s}$$

In silicon the e-h radiative recombination by emission of photon is of the range of 1 ms to 1 μs so that the nonradiative (trap-related) lifetime is much shorter for this example.

3.9 CONTINUITY EQUATION: DIFFUSION LENGTH

In our discussion of charge transport in Sections 3.3-3.6, we considered the drift and diffusion processes through the semiconductor, without worrying about the electron-hole recombination. The recombination process removes the electrons and holes and thus alters the charge transport picture. Similarly, the generation process adds additional electrons and holes. To describe the transport and recombination of injected electrons and holes we develop a continuity equation for the problem. The carrier recombination process is critical for any device that involves both electron and hole flow (e.g., p-n diode, bipolar junction transistor).

If we consider a volume of space in which charge transport and recombination is taking place, we have the simple equality (see Fig. 3.13a)

Rate of particle flow = Particle flow rate due to current −

Particle loss rate due to recombination + Particle gain due to generation.

Continuity equation $\Longrightarrow$ Conservation of electrons and hole densities

Total rate of particle flow = Particle flow due to current – Loss due to
recombination rate = Gain due to generation rate

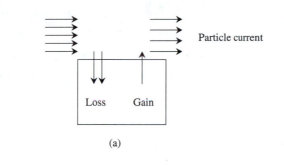

(a)

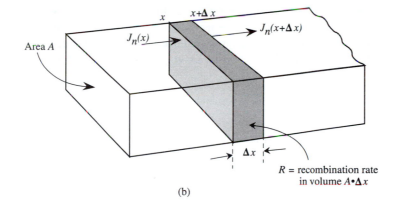

(b)

Figure 3.13: (a) A conceptual description of the continuity equation. (b) Geometry used to set up the current continuity equation.

This equality simply is a statement of conservations of particles. If δn is the excess carrier density in the region, the recombination rate R in the volume $A \cdot \Delta x$ shown in Fig. 3.13 may be written approximately as

$$R = \frac{\delta n}{\tau_n} \cdot A \cdot \Delta x \tag{3.45}$$

where τ_n is the electron recombination time per excess particle due to both the radiative and the nonradiative components. The particle flow rate into the same volume due to the current J_n is given by the difference of particle current coming into the region and the particle current leaving the region,

$$\left[\frac{J_n(x)}{(-e)} - \frac{J_n(x + \Delta x)}{(-e)} \right] A \cong \frac{1}{e} \frac{\partial J_n(x)}{\partial x} \Delta x \cdot A \tag{3.46}$$

If G is the generation rate per unit volume, the generation rate in the volume $A \cdot \Delta x$ is $GA\Delta x$. The total rate of electron build up in the volume $A \cdot \Delta x$ is then

$$A \cdot \Delta x \left[\frac{\partial n(x,t)}{\partial t} \equiv \frac{\partial \delta n}{\partial t} = \frac{1}{e} \frac{\partial J_n(x)}{\partial x} - \frac{\delta n}{\tau_n} + G \right] \qquad (3.47)$$

The same considerations apply to holes. We thus have, for the electrons and holes, the continuity equations (note the sign difference in the particle current density for electrons and holes)

$$\frac{\partial \delta n}{\partial t} = \frac{1}{e} \frac{\partial J_n(x)}{\partial x} - \frac{\delta n}{\tau_n} + G \qquad (3.48)$$

$$\frac{\partial \delta p}{\partial t} = -\frac{1}{e} \frac{\partial J_p(x)}{\partial x} - \frac{\delta p}{\tau_p} + G \qquad (3.49)$$

It is interesting to examine these equations for the special case where the current is being carried only by the diffusion process and there is no generation rate. This situation usually occurs when we consider transport in *p-n* junction diodes or in bipolar transistors and there is no light excitation to cause *e-h* generation. Writing the diffusion currents as (see Eqn. 3.18)

$$J_n(diff) = eD_n \frac{\partial \delta n}{\partial x} \qquad (3.50)$$

$$J_p(diff) = -eD_p \frac{\partial \delta p}{\partial x} \qquad (3.51)$$

we have

$$\frac{\partial \delta n}{\partial t} = D_n \frac{\partial^2 \delta n}{\partial x^2} - \frac{\delta n}{\tau_n} \qquad (3.52)$$

$$\frac{\partial \delta p}{\partial t} = D_p \frac{\partial^2 \delta p}{\partial x^2} - \frac{\delta p}{\tau_p} \qquad (3.53)$$

These equations will be used when we discuss the transient time responses of the *p-n* diodes and bipolar transistors. These equations are also used to study the steady-state charge profile in these devices. In steady state we have (the time derivative is zero)

$$\boxed{\begin{aligned} \frac{d^2 \delta n}{dx^2} &= \frac{\delta n}{D_n \tau_n} = \frac{\delta n}{L_n^2} \\ \frac{d^2 \delta p}{dx^2} &= \frac{\delta p}{D_p \tau_p} = \frac{\delta p}{L_p^2} \end{aligned}} \qquad (3.54)$$

where $L_n(L_p)$ defined as $D_n\tau_n(D_p\tau_p)$ are called the diffusion lengths for reasons that will become clear below.

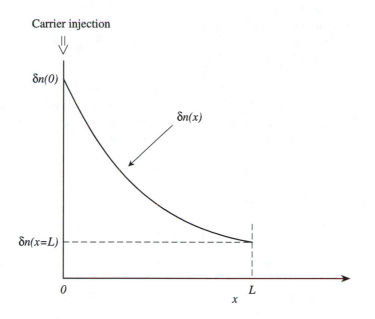

Figure 3.14: Electrons are injected at $x = 0$ into a sample. At $x = 0$, a fixed carrier concentration is maintained. The figure shows how the excess carriers decay into the semiconductor.

Consider first the case where, due to an external injection mechanism, an excess electron density $\delta n(0)$ is maintained at the semiconductor edge $x = 0$, as shown in Fig. 3.14. At some point L in the semiconductor the excess carrier density is fixed at $\delta(L)$. We are interested in finding out how the excess density varies with position. The general solution of the second-order differential Eqn. 3.54 is

$$\delta n(x) = A_1 e^{x/L_n} + A_2 e^{-x/L_n}$$

Using the boundary conditions at $x = 0$ and $x = L$, we find, after solving two simultaneous equations, that the coefficients A_1 and A_2 are

$$A_1 = \frac{\delta n(L) - \delta n(0)e^{-L/L_n}}{e^{L/L_n} - e^{-L/L_n}}$$

$$A_2 = \frac{\delta n(0)e^{L/L_n} - \delta n(L)}{e^{L/L_n} - e^{-L/L_n}} \qquad (3.55)$$

This gives for the excess carrier concentration

$$\delta n(x) = \frac{\delta n(0) \sinh\left(\frac{L-x}{L_n}\right) + \delta n(L) \sinh\left(\frac{x}{L_n}\right)}{\sinh\left(\frac{L}{L_n}\right)} \qquad (3.56)$$

Let us examine this result for two important cases:
(i) $L \gg L_n$ and $\delta n(L) = 0$: In this case the semiconductor sample is much longer than

L_n. This happens in the case of the *long p-n* diode, which will be discussed in Chapter 5. The carriers are injected at the origin and the excess density decays to zero deep in the semiconductor. For this case we have

$$\delta n(x) = \delta(0)e^{-x/L_n} \tag{3.57}$$

Thus the carrier density simply decays exponentially into the semiconductor.
(ii) $L \ll L_n$: This case is very important in discussing the operation of bipolar transistors and narrow *p-n* diodes. Using the small x expansion for $\sinh(x)$

$$\sinh(x) = x + \frac{x^3}{3!} + \frac{x^5}{5!} + \dots$$

and retaining only the first-order terms we get

$$\delta n(x) = \delta n(0) - \frac{x\,[\delta n(0) - \delta n(L)]}{L} \tag{3.58}$$

i.e., in this case the carrier density goes *linearly* from one boundary value to the other.

Note that once the carrier density is known the diffusion current can be simply obtained by taking its derivative.

Let us examine the case where excess carriers are injected into a thick semiconductor sample. As the excess carriers diffuse away into the semiconductor they recombine. The diffusion length L_n represents the distance over which the injected carrier density falls to $1/e$ of its original value. It also represents the average distance an electron will diffuse before it recombines with a hole. This can be seen as follows.

The probability that an electron survives up to point x without recombination is, from Eqn. 3.57,

$$\boxed{\frac{\delta n(x)}{\delta n(0)} = e^{-x/L_n}} \tag{3.59}$$

The probability that it recombines in a distance Δx is

$$\frac{\delta n(x) - \delta n(x + \Delta x)}{\delta n(x)} = -\frac{\Delta x}{\delta n(x)} \frac{d\delta n(x)}{dx} = \frac{1}{L_n}\Delta x \tag{3.60}$$

where we have expanded $\delta n(x + \Delta x)$ in terms of $\delta n(x)$ and the first derivative of δn. Thus the probability that the electron survives up to a point x and then recombines is

$$P(x)\Delta x = \frac{1}{L_n}e^{-x/L_n}\Delta x \tag{3.61}$$

Thus the average distance an electron can move and then recombine is

$$
\begin{aligned}
< x > \;&= \int_o^\infty x P(x)dx = \int_0^\infty \frac{xe^{-x/L_n}}{L_n}dx \\
&= \; L_n
\end{aligned}
\tag{3.62}
$$

This average distance $(= \sqrt{D_n \tau_n})$ depends upon the recombination time and the diffusion constant in the material. In the derivations of this section, we used a simple form of recombination rate

$$R = \frac{\delta n}{\tau_n} \tag{3.63}$$

where τ_n is given in terms of the radiative and nonradiative rates as

$$\frac{1}{\tau_n} = \frac{1}{\tau_r} + \frac{1}{\tau_{nr}} \tag{3.64}$$

The simple $\delta n / \tau_n$ form is valid, for example, for minority carrier recombination $(p \gg n)$. These equations are therefore used widely to discuss minority carrier injection.

EXAMPLE 3.14 In a p-type GaAs sample, electrons are injected from a contact. If the minority carrier mobility is 4000 cm^2/V·s at 300 K, calculate the diffusion length for the electrons. The recombination time is 0.6 ns.

The diffusion constant is given by the Einstein relation

$$
\begin{aligned}
D_n &= \frac{\mu \; k_B T}{e} \\
&= \frac{(0.4 \text{ m}^2/\text{V} \cdot s) \times (1.38 \times 10^{-23} (\text{J/K}) \times 300 \; K)}{(1.6 \times 10^{-19} \text{ C})} \\
&= 1.04 \times 10^{-2} \text{m}^2/\text{s}
\end{aligned}
$$

The diffusion length is

$$L_n = \sqrt{D_n \tau}$$

Using the recombination time value given, we get

$$
\begin{aligned}
L_n &= 2.5 \times 10^{-6} \text{ m} \\
&= 2.5 \; \mu\text{m}
\end{aligned}
$$

EXAMPLE 3.15 Consider a Si sample of length L. The diffusion coefficient for electrons is 25 cm^{2}s^{-1} and the electron lifetime is 0.01 μs. An excess electron concentration is maintained at $x = 0$ and $x = L$. The excess concentrations are:

$$\delta n(x = 0) = 2.0 \times 10^{18} \quad \text{cm}^{-3}; \quad \delta n(x = L) = -1.0 \times 10^{14} \quad \text{cm}^{-3}$$

Calculate and plot the excess electron distribution from $x = 0$ to $x = L$. Do the calculations for the following values of L:

$$
\begin{aligned}
L &= 10.0 \; \mu\text{m} \\
L &= 5.0 \; \mu\text{m} \\
L &= 1.0 \; \mu\text{m} \\
L &= 0.5 \; \mu\text{m}
\end{aligned}
$$

Note that the excess electron distribution starts out being nonlinear in space for the long structure, but *becomes linear between the two boundary values for the short structure.*

To solve this problem we need to use the general solution of the continuity equation, which tells us what the excess carrier distribution is in a semiconductor. The diffusion length is

$$L_n = (2.5 \times 10^{-7})^{1/2} = 5.0 \times 10^{-4} \text{ cm}$$

For a very long structure $(L \gg L_n)$ we see from Eqn. 3.55 that

$$A_2 = \delta n(0); \quad A_1 = 0$$

For $L = 10 \ \mu$m we have:

$$A_1 = -3.7329 \times 10^{16} \text{ cm}^{-3}; \quad A_2 = 2.0373 \times 10^{18} \text{ cm}^{-3}$$

For $L = 0.5 \ \mu$m we have:

$$A_1 = -9.0338 \times 10^{18} \text{ cm}^{-3}; \quad A_2 = 1.1034 \times 10^{19} \text{ cm}^{-3}$$

Other values for different values of L can be similarly calculated. Once we know A_1, A_2, the values of $\delta n(x)$ can be calculated.

In Fig. 3.15 we show the density carrier dependence for $L = 10.0 \ \mu$m and $L = 0.5 \ \mu$m. We see that for the shorter length, the carrier density falls off linearly.

EXAMPLE 3.16 An experiment is carried out at 300 K on n-type Si doped at $N_d = 10^{17}$ cm^{-3}. The conductivity is found to be 10.0 $(\Omega \text{ cm})^{-1}$.

When light with a certain intensity shines on the material the conductivity changes to 11.0 $(\Omega \text{ cm})^{-1}$. The light is turned off at time 0 and it is found that at time 1.0 μs the conductivity is 10.5 $(\Omega \text{ cm})^{-1}$. The light-induced excess conductivity is found to decay as

$$\delta\sigma = \delta\sigma(0) \exp\left(-\frac{t}{\tau}\right)$$

where τ is the carrier lifetime. Calculate the following:
• What fraction of the donors is ionized?
• What is the diffusion length of holes in this material?

Use the following data:

$$\mu_n = 1100 \ \text{cm}^2/\text{V} \cdot \text{s} : \quad \mu_p = 400 \ \text{cm}^2/\text{V} \cdot \text{s}$$

To answer the first part we find the carrier concentration in the n-type material

$$n = \frac{\sigma}{e\mu} = \frac{10(\Omega \text{ cm})^{-1}}{1.6 \times 10^{-19} \text{ C} \times 1100 \text{ cm}^2/\text{V} \cdot \text{s}} = 5.68 \times 10^{16} \ \text{cm}^{-3}$$

Thus 56.8% of the donors are ionized.

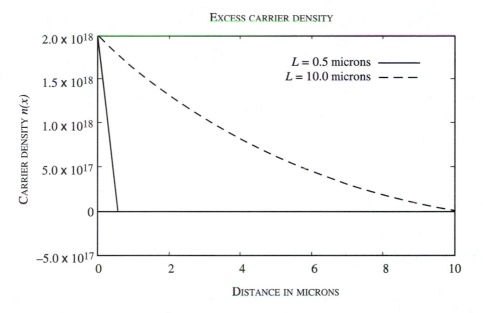

EXCESS CARRIER DENSITY

Figure 3.15: Carrier concentration as a function of position for Example 3.15.

To find the diffusion length we need to find the hole lifetime in the sample. We note that the presence of light changes the conductivity by 1.0 $(\Omega\ cm)^{-1}$ and once light is turned off the excess drops by 50% in a microsecond. We have

$$e^{-(1.0\ \mu s)/\tau} = 0.5$$

or

$$\tau = -\frac{1}{ln(0.5)} = 1.44\ \mu s$$

The diffusion coefficient is

$$D_p = \frac{\mu k_B T}{e} = 10.4\ \ cm^2/s$$

This gives for the diffusion length

$$L_p = \sqrt{D_p \tau} = \sqrt{10.4 \times 1.44 \times 10^{-6}} = 38.69\ \mu m$$

EXAMPLE 3.17 Consider a structure in which the hole density at $x = 0$ is $\delta p(0)$ and at $x = L$ it is $\delta p(L) = 0$. If $L \ll L_p$, estimate the transit time for holes to travel the distance L.

We can write the current due to the hole flow as (A is the area of the sample)

$$eAD_p \frac{dp}{dx} = \frac{Q_p}{\tau_t}$$

where Q_p is the total charge between $x = 0$ and $x = L$ and τ_t is the transit time. For $L \ll L_p$ we know that the carrier density changes linearly (see Example 3.15) and

$$\frac{dp}{dx} = \frac{\delta p(0)}{L}$$

and the charge is

$$Q_p = eAL\frac{(\delta p(0) - \delta p(L))}{2} = \frac{eAL\delta p(0)}{2}$$

This gives, for the transit time,

$$\tau_t = \frac{L^2}{2D_p}$$

This is a very useful result and is used to understand switching response of narrow *p-n* diodes and bipolar transistors.

3.10 SUMMARY

In this chapter we have discussed the basic physical phenomena upon which electronic and optoelectronic devices are based. All devices involve some physical response to external perturbations. These perturbations are usually electric fields or electromagnetic fields. The device performance depends upon how electrons respond to these external stimuli. Summary table 3.5 highlights the concepts discussed in this chapter.

3.11 PROBLEMS

Note: Please examine problems in the following Cumulative Problems section as well.

Sections 3.2-3.3

Problem 3.1 Why do mobilities of semiconductors decrease with an increase in temperature?

Problem 3.2 The mean free path of an electron (average speed times the scattering time) is the average distance an electron travels between two consecutive scattering events. Calculate the mean free path for silicon at room temperature. How many lattice constants (5.43 Å) is the mean free path?

Problem 3.3 Consider a sample of GaAs with electron effective mass of $0.067m_o$. If an electric field of 1 kV/cm is applied, what is the drift velocity produced if

a) $\tau_{sc} = 10^{-13}$ s

b) $\tau_{sc} = 10^{-12}$ s

c) $\tau_{sc} = 10^{-11}$ s?

How does the drift velocity compare to the average thermal speed of the electrons at room temperature?

Problem 3.4 Assume that at room temperature the electron mobility in Si is 1300 cm^2/

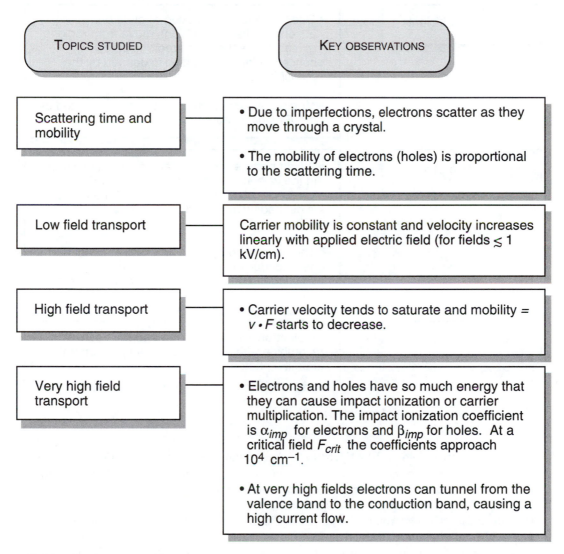

Table 3.5: Summary table.

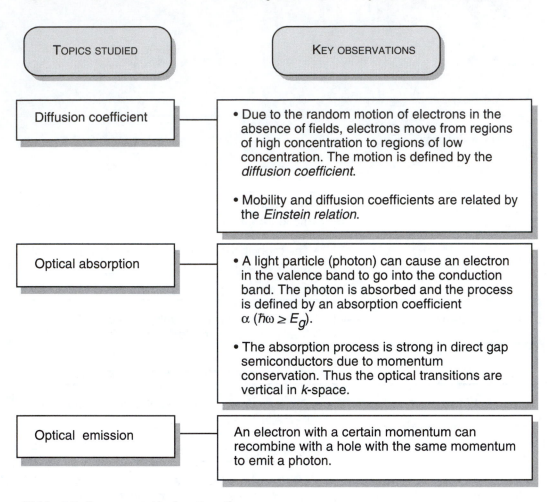

Table 3.5: Summary table (continued).

V·s. If an electric field of 100 V/cm is applied, what is the excess energy of the electrons? How does it compare with the thermal energy? If you assume that the mobility is unchanged, how does the same comparison work out at a field of 5 kV/cm? (Excess energy $= \frac{1}{2}m^*v_d^2$ where v_d is the drift velocity.)

Problem 3.5 The electron mobility of Si at 300 K is 1400 cm^2/V·s. Calculate the mean free path and the energy gained in a mean free path at an electric field of 1 kV/cm. Assume that the mean free path $= v_{th} \cdot \tau_{sc}$, where v_{th} is the thermal velocity of the electron ($v_{th} \sim 2.0 \times 10^7$ cm/s).

Problem 3.6 The mobility of electrons in the material InAs is $\sim 35{,}000$ cm^2/V·s at 300K compared to a mobility of 1400 cm^2/V·s for silicon. Calculate the scattering times in the two semiconductors. The electron masses are 0.02 m_0 and 0.26 m_0 for InAs and Si, respectively.

Problem 3.7 Use the velocity-field relations for Si and GaAs to calculate the transit

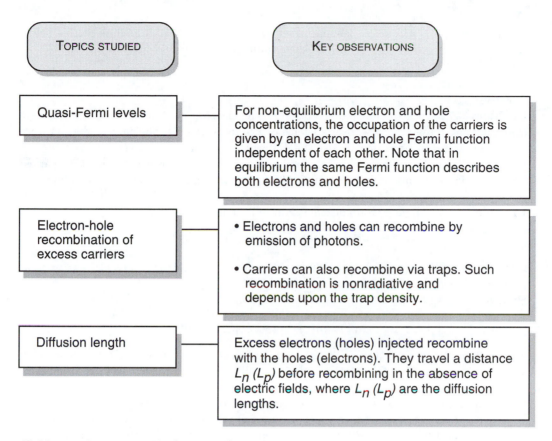

TOPICS STUDIED	KEY OBSERVATIONS
Quasi-Fermi levels	For non-equilibrium electron and hole concentrations, the occupation of the carriers is given by an electron and hole Fermi function independent of each other. Note that in equilibrium the same Fermi function describes both electrons and holes.
Electron-hole recombination of excess carriers	• Electrons and holes can recombine by emission of photons. • Carriers can also recombine via traps. Such recombination is nonradiative and depends upon the trap density.
Diffusion length	Excess electrons (holes) injected recombine with the holes (electrons). They travel a distance L_n (L_p) before recombining in the absence of electric fields, where L_n (L_p) are the diffusion lengths.

Table 3.5: Summary table (continued).

time of electrons in a 1.0 μm region for a field of 1 kVcm^{-1} and 50 kVcm^{-1}.

Problem 3.8 The velocity of electrons in silicon remains $\sim 1 \times 10^7$ cm s^{-1} between 50 kVcm^{-1} and 200 kVcm^{-1}. Estimate the scattering times at these two electric fields.

Section 3.4

Problem 3.9 The power output of a device depends upon the maximum voltage that the device can tolerate before impact-ionization-generated carriers become significant (say 10% excess carriers). Consider a device of length L over which a potential V drops uniformly. What is the maximum voltage that can be tolerated by an Si and a diamond device for $L = 2$ μm and $L = 0.5$ μm? Use the values of the critical fields given in this chapter.

Problem 3.10 An electron in a silicon device is injected in a region where the field is 500 kV cm^{-1}. The length of this region is 1.0 μm. Calculate the average number of impact ionization events that occur for the incident electron.

Problem 3.11 In GaAs impact ionization starts to be significant for electrons if the electric field is 350 kVcm^{-1}. Calculate the probability for Zener tunneling at this field. Estimate the electric field where the Zener tunneling probability approaches 10^{-6}.

Section 3.5

Problem 3.12 In a silicon sample at 300 K, the electron concentration drops linearly from 10^{18} cm^{-3} to 10^{16} cm^{-3} over a length of 2.0 μm. Calculate the current density due to the electron diffusion current. Use the diffusion constant values given in this chapter.

Problem 3.13 In a GaAs sample, it is known that the electron concentration varies linearly. The diffusion current density at 300 K is found to be 100 A/cm^2. Calculate the slope of the electron concentration.

Problem 3.14 The electron concentration in a Si sample is given by

$$n(x) = n(0)\exp(-x/L_n); \quad x > 0$$

with $n(0) = 10^{18}$ cm^{-3} and $L_n = 3.0$ μm. Calculate the diffusion current density as a function of position if $D_n = 35$ cm^2/s.

Problem 3.15 A silicon sample has the following electron density:

$$n(x) = n(0)\exp(-x/L_n) \quad x > 0$$

with $n(0) = 10^{18}$ cm^{-3} and $L_n = 2.0$ μm. There is also a uniform electric field of 2 kV/cm in the material. Calculate the drift and diffusion current densities as a function of position. The diffusion coefficient is 30 cm^2/s and the mobility is 1000 cm^2/V·s.

Problem 3.16 In a GaAs sample the electrons are moving under an electric field of 5 kV cm^{-1} and the carrier concentration is uniform at 10^{16} cm^{-3}. The electron velocity is the saturated velocity of 10^7 cm/s. Calculate the drift current density. If a diffusion current has to have the same magnitude, calculate the concentration gradient needed. Assume a diffusion coefficient of 100 cm^2/s.

Section 3.7

Problem 3.17 In a GaAs sample at 300 K, equal concentrations of electrons and holes are injected. If the carrier density is $n = p = 10^{17}$ cm^{-3}, calculate the electron and hole Fermi levels using the Boltzmann and Joyce-Dixon approximations.

Problem 3.18 In a p-type GaAs doped at $N_a = 10^{18}$ cm^{-3}, electrons are injected to produce a minority carrier concentration of 10^{15} cm^{-3}. What is the rate of photon emission assuming that all *e-h* recombination is due to photon emission ? What is the optical output power? The photon energy is $\hbar\omega = 1.41$ eV and the radiative lifetime is 1.0 ns.

Problem 3.19 Calculate the electron carrier density needed to push the electron Fermi level to the conduction bandedge in GaAs. Also calculate the hole density needed to push the hole Fermi level to the valence bandedge. Calculate the results for 300 K and 77 K.

Problem 3.20 The optical absorption coefficient in silicon is low until the photon energies reach ∼3.5 eV. Then it increases rapidly. Explain this feature on the basis of the momentum conservation rule.

Problem 3.21 Explain why the nonradiative *e-h* recombination time has a very strong temperature dependence.

Problem 3.22 Consider a pure silicon sample at 300 K. Light of intensity 10 W/cm^2 shines on the sample. The absorption coefficient is 10^3 cm^{-1} and the electron-hole recombination time is 10^{-7} s. Calculate the positions of the quasi-Fermi levels in the material.

Problem 3.23 A photodetector uses pure silicon as its active region. Calculate the *dark conductivity* of the detector (i.e., conductivity when no light is shining on the detector). Light with intensity 10^{-3} W/cm^2 shines on the device. Calculate the conductivity in presence of light.

$$
\begin{aligned}
\mu_n &= 1000 \text{ cm}^2/\text{V} \cdot \text{s} \\
\mu_p &= 400 \text{ cm}^2/\text{V} \cdot \text{s} \\
\alpha &= 10^3 \text{ cm}^{-1} \\
\tau_r &= 10^{-7} \text{ s}
\end{aligned}
$$

Problem 3.24 The radiative lifetime of a GaAs sample is 1.0 ns. The sample has a defect at the midgap with a capture cross-section of 10^{-15} cm^2. At what defect concentration does the nonradiative lifetime become equal to the radiative lifetime at (a) 77 K and (b) 300 K?

Section 3.9

Problem 3.25 Electrons are injected into a *p*-type silicon sample at 300 K. The electron-hole radiative lifetime is 1 μs. The sample also has midgap traps with a cross-section of 10^{-15} cm^2 and a density of 10^{16} cm^{-3}. Calculate the diffusion length for the electrons if the diffusion coefficient is 30 cm^2/s.

Problem 3.26 Assume that silicon has a midgap impurity level with a cross-section of 10^{-14} cm^2. The radiative lifetime is given to be 1 μs at 300 K. Calculate the maximum impurity concentration that will ensure that $\tau_r < \tau_{nr}$.

Problem 3.27 When holes are injected into an *n*-type ohmic contact, they decay within a few hundred angstroms. Thus one can assume that the minority charge density goes to zero at an ohmic contact. Discuss the underlying physical reasons for this boundary condition.

Problem 3.28 Electrons are injected into *p*-type GaAs at 300 K. The radiative lifetime for the electrons is 2 ns. The material has 10^{15} impurities with a cross-section of 10^{-14} cm^2. Calculate the distance the injected minority charge will travel before 50% of the electrons recombine with holes. The diffusion coefficient is 100 cm^2/s.

Problem 3.29 Electrons are injected into *p*-type silicon at $x = 0$. Calculate the fraction of electrons that recombine within a distance L where L is given by (a) 1.0 μm, (b) 2.0 μm, and (c) 10.0 μm. The diffusion coefficient is 30 cm^2/s and the minority carrier lifetime is 10^{-7}s.

3.12 CUMULATIVE PROBLEMS: CHAPTERS 1-3

Problem 1 A silicon sample doped n-type at 10^{18} cm^{-3} is found to have a resistance of 10 Ω. The sample has an area of 10^{-6} cm^2 and a length of 10 μm. Calculate the doping efficiency of the sample. The electron mobility is 800 cm^2/V·s.

Problem 2 A thin-film resistor is to be made from a GaAs film doped n-type. The resistor is to have a value of 1 kΩ. The resistor dimensions are to be: area = 10^{-6} cm^2 and length = 100 μm. The doping efficiency is known to be 90%. The mobility of electrons is 8000 cm^2/V·s. What doping is needed for this resistor?

Problem 3 Consider a sample of silicon at 300 K. A dc field of 1.0 kV/cm is applied and the drift velocity is measured to be 5.0×10^5 cm/s for the electrons and 1.0×10^5 cm/s for the holes. Calculate the resistance of a sample of silicon in which the Fermi level is 0.6 eV below the conduction band. The area of the sample is 10^{-4} cm^2 and the length is 10.0μm.

Problem 4 Consider an undoped Si sample with area 10^{-4} cm^2 and length 1.0 mm. Calculate the resistance of the sample at 300 K. It is found that when light with a certain intensity shines on the sample, the resistance decreases by a factor of 10^6. Calculate the electron and hole quasi-Fermi levels for this case.

$$\mu_n = 1000 \text{ cm}^2/\text{V} \cdot \text{s}; \quad \mu_p = 300 \text{ cm}^2/\text{V} \cdot \text{s}$$

Problem 5 Consider an n-type Si sample with a conductivity of 0.1 $(\Omega\text{cm})^{-1}$ at 300 K. Calculate the Fermi level in the material. Also calculate the Fermi level in a p-type material with the same conductivity.

$$\mu_n = 1000 \text{ cm}^2/\text{V} \cdot \text{s}; \quad \mu_p = 300 \text{ cm}^2/\text{V} \cdot \text{s}$$

Problem 6 A crystalline material is made from an element with atomic mass 22.98 and atomic number 11. The density of the crystal is 0.97 g/cm^3. The material has the following allowed bands; the allowed states per band are also given (N is the number density of atoms):

1st allowed band : -24.0 eV $\leq E \leq -23.9$ eV; $2N$ allowed states

2nd allowed band : -20.0 eV $\leq E \leq -19.8$ eV; $2N$ allowed states

3rd allowed band : -15.0 eV $\leq E \leq -14.5$ eV; $6N$ allowed states

4th allowed band : -12.0 eV $\leq E \leq -5.0$ eV; $2N$ allowed states

The mobility of electrons is 100 cm^2/V·s. Assume that the electron effective mass is m_0. Calculate (a) conductivity of the material and (b) position of the Fermi level at 0 K.

Problem 7 The mobility of electrons in Si, GaAs and InAs is found to be 1100, 8500 and 20000 cm^2/V·s, respectively, at room temperature. Estimate the maximum length of a semiconductor channel in which electrons will move *without* scattering in each of the three semiconductors under a field of 5 kV/cm. Note that for this to occur, the transit time has to be less than the scattering time.

3.13 READING LIST

- **Transport**

  - R. A. Smith, *Semiconductors* (Cambridge University Press, London, 1978).
  - R. B. Adler, A. C. Smith, and R. L. Longini, *Introduction to Semiconductor Physics* (Wiley, New York, 1969).
  - J. R. Haynes and W. Shockley, *Phys. Rev.* **81**, *835* (1951).
  - S. M. Sze, *Physics of Semiconductor Devices* (Wiley, New York, 1981).
  - D. L. Rode, *Low Field Electron Transport in Semiconductors and Semimetals* (edited by R. K. Willardson and A. C. Beer, Academic Press, New York, 1975), vol. 10.

- **General High Field Transport**

  - M. Lundstrom, *Fundamentals of Carrier Transport* (Modular Series on Solid State Devices, edited by G. W. Neudeck and R. F. Pierret, Addison-Wesley, Reading, Massachusetts, 1990), vol. X.

- **Optical Processes in Semiconductors**

  - J. I. Pankove, *Optical Processes in Semiconductors* (Prentice-Hall, Englewood Cliffs, NJ, 1971).
  - F. Stern, *Elementary Theory of the Optical Properties of Solids* (Academic Press, New York, 1963), vol. 15.
  - J. Singh, *Modern Physics for Engineers* (Wiley-Interscience, New York, 1999).

Chapter

4

PROCESSING OF DEVICES: A REVIEW

4.1 INTRODUCTION

In the first three chapters of this text we have discussed important properties of electrons (and holes) in semiconductors. How can we exploit these properties to design devices that can then be used for information processing? This will be the subject of the remainder of this text. In this chapter we will provide an overview of the technology involved in the fabrication of semiconductor devices. It is suggested that this chapter be used as a reading assignment for students.

In the previous chapters we have examined the following key questions: i) what is the mobile carrier density in a semiconductor and how does the Fermi level position control this density? ii) How do electrons and holes move in the presence of an applied electric field? iii) How does current flow in the presence of a carrier concentration gradient? From our discussions we know that the mobile carrier density can be greatly altered by somehow altering the position of the Fermi level. We also know that current flow can be greatly altered by changing the gradient of electron or hole concentrations. Thus, in principle, the conductivity of a semiconductor can be changed from a very low value (or OFF state) to a very high value (ON state). Such an ability is essential for making information-processing devices. It is important to note that semiconductor devices are not simply made from a piece of homogeneous semiconductor alone. They need several different components. A device may contain n-type and p-type regions. It may have an insulator and several metallic contacts. To be able to accommodate all these requirements a great deal of technology is needed. We will now survey this technology.

4.2 SEMICONDUCTOR CRYSTAL GROWTH

This textbook deals with the electronic devices that are designed to provide the key components of the information age. The devices we have addressed are based primarily on semiconductor crystals. In this section we will discuss how semiconductors are grown.

4.2.1 Bulk Crystal Growth

Bulk crystal growth techniques are used mainly to produce substrates on which devices are eventually fabricated. While for some semiconductors like Si and GaAs (and to some extent for InP), bulk crystal growth techniques are highly matured, for most other semiconductors it is difficult to obtain high-quality, large-area substrates. The aim of the bulk crystal growth techniques is to produce single-crystal boules with as large a diameter as possible and with as few defects as possible. In Si the boule diameters have reached 30 cm with boule lengths approaching 100 cm. Large-size substrates ensure low-cost device production.

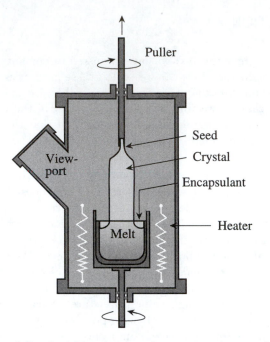

Figure 4.1: Schematic of Czocharlski-style crystal grower used to produce substrate ingots. The approach is widely used for Si, GaAs, and InP.

To grow the boules from which substrates are obtained, one starts out with a purified form of the elements that are to make up the crystal. One important technique that is used is the Czochralski (CZ) technique. In the Czochralski technique shown in Fig. 4.1, the melt of the charge, a high-quality polycrystalline material, is held in a vertical crucible. The top surface of the melt is just barely above the melting temperature. A seed crystal is then lowered into the melt and slowly withdrawn. As the heat from the melt flows up the seed, the melt surface cools and the crystal begins to grow. The seed is rotated about its axis to produce a roughly circular cross-section crystal. The rotation inhibits the natural tendency of the crystal to grow along certain orientations to produce a faceted crystal. The CZ technique is widely employed for Si, GaAs, and InP and produces long ingots (boules) with very good circular cross-section.

A second bulk crystal growth technique involves a charge of material loaded in a quartz container. The charge may be composed of either high-quality polycrystalline material or carefully measured quantities of elements that make up a compound crystal. The container, called a "boat, " is heated till the charge melts and wets the seed crystal. The seed is then used to crystallize the melt by slowly lowering the boat temperature, starting from the seed end. In the Bridgeman approach, the boat is kept stationary while the furnace temperature is temporally varied to form the crystal. This is shown in Fig. 4.2a. In the gradient-freeze approach the boat is pushed into a furnace (to melt the charge) and slowly pulled out, as shown in Fig. 4.2b.

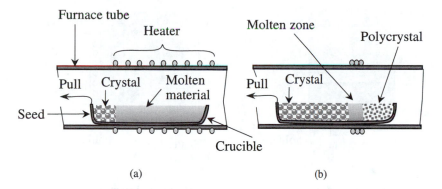

Figure 4.2: Crystal growing from the melt in a crucible: (a) solidification from one end of the melt (horizontal Bridgeman method); (b) melting and solidification in a moving zone.

The easiest approach for the boat technique is to use a horizontal boat. However, the boule that is produced has a D-shaped form. To produce circular cross-sections vertical configurations have now been developed for GaAs and InP.

In addition to producing high-purity bulk crystals, the techniques discussed above are also responsible for producing crystals with specified electrical properties. Later we will see that impurities or dopants must be added to semiconductors to alter their conductivities. For such applications carefully measured impurities are added in the melt.

The availability of high-quality substrates is essential to any device technology. Other than the three materials mentioned above (Si, GaAs, InP), the substrate fabrication of semiconductors is still in its infancy. Since epitaxial growth techniques used for devices require close lattice matching between the substrate and the overlayer, non-availability of substrates can seriously hinder the progress of a material technology. This is, for example, one of the reasons for slow progress in large bandgap semiconductor technology necessary for high-power high-temperature electronic devices and short wavelength semiconductor lasers.

4.2.2 Epitaxial Crystal Growth

The substrates that result once the bulk semiconductor boule is sliced and lapped are almost never used directly for devices. Invariably an additional epitaxial layer or epilayer is grown that may be a few microns in thickness. The word *epitaxy* comes from the Greek word "epi" (upon) and "taxis" (order), meaning the ordered continuation of the substrate crystal. Epitaxial growth techniques have a very slow growth rate (as low as a monolayer per second for some techniques), which allows one to control very

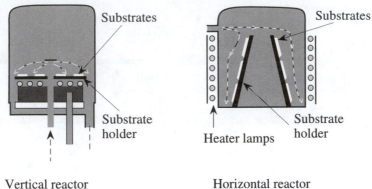

Vertical reactor   Horizontal reactor

Figure 4.3: Reactors for VPE growth. The substrate temperature must be maintained uniformly over the area. This is achieved by lamp heating.

accurately the dimensions in the growth direction. In fact, in techniques like molecular beam epitaxy (MBE) and metal organic chemical vapor deposition (MOCVD), one can achieve monolayer (~ 3 Å) control in the growth direction.

Vapor Phase Epitaxy (VPE)
A large class of epitaxial techniques relies on delivering the components that form the crystal from a gaseous environment. If one has molecular species in a gaseous form with partial pressure P, the rate at which molecules impinge upon a substrate is given by

$$F = \frac{P}{\sqrt{2\pi m k_B T}} \sim \frac{3.5 \times 10^{22} P(torr)}{\sqrt{m(g)T(K)}} \text{ mol/cm}^2 \text{ s}$$

where m is the molecular weight and T the cell temperature. For most crystals the surface density of atoms is $\sim 7 \times 10^{14}$ cm^{-2}. If the atoms or molecules impinging from the vapor can be deposited on the substrate in an ordered manner, epitaxial crystal growth can take place.

The VPE technique is used mainly for homoepitaxy, i.e., growth of a single component. As an example of the technique, consider the VPE of Si. The Si-containing reactant silane (SiH_4) or dichlorosilane (SiH_2Cl_2) or trichlorosilane ($SiHCl_3$) or silicon tetrachloride ($SiCl_4$) is diluted in hydrogen and introduced into a reactor in which heated substrates are placed as shown in Fig. 4.3. Chemical reactions at the substrate surface cause crystal growth to occur.

An important consideration in VPE is safety-related, since hydrogen, which is produced in the deposition, can explode in contact with any oxygen. Also, almost all the reactants are highly toxic.

VPE can be used for other semiconductors as well by choosing different appropriate reactant gases. The reactants used are quite similar to those employed in the MOCVD technique discussed below.

Molecular Beam Epitaxy (MBE)

Molecular beam epitaxy (MBE) is one of the most important epitaxial techniques as far as heterostructure physics and devices are concerned. Almost every semiconductor (other than a few very large bandgap semiconductors) has been grown by this technique. MBE is a high-vacuum technique ($\sim 10^{-11}$ torr vacuum when fully pumped down) in which crucibles containing a variety of elemental charges are placed in the growth chamber. The elements contained in the crucibles make up the components of the crystal to be grown as well as the dopants that may be used. When a crucible is heated, atoms or molecules of the charge are evaporated, and these travel in straight lines to impinge on a heated substrate.

The growth rate in MBE is $\sim$1.0 monolayer per second and this slow rate, coupled with shutters placed in front of the crucibles, allows one to switch the composition of the growing crystal with monolayer control. However, to do so, the growth conditions have to be adjusted so that growth occurs in the monolayer by monolayer mode rather than by three-dimensional island formation. This requires that the atoms impinging on the substrate have enough kinetics to reach an atomically flat profile. Thus the substrate temperature has to be maintained at a point where it is high enough to provide enough surface migration to the incorporating atoms, but not so high as to cause entropy-controlled defects.

Since no chemical reactions occur in MBE, the growth is the simplest of all epitaxial techniques and is quite controllable. However, since the growth involves high vacuum, leaks can be a major problem. The growth chamber walls are usually cooled by liquid N_2 to ensure high vacuum and to prevent atoms/molecules from coming off from the chamber walls. MBE is a relatively safe technique and has become the technique of choice for the testing of almost all new ideas on heterostructure physics.

Metal Organic Chemical Vapor Deposition (MOCVD)

Metal organic chemical vapor deposition (MOCVD) is another important growth technique widely used for heteroepitaxy. Like MBE, it is capable of producing monolayer abrupt interfaces between semiconductors. Unlike MBE, the gases that are used in MOCVD are not made of single elements, but are complex molecules that contain elements like Ga or As to form the crystal. Thus the growth depends upon the chemical reactions occurring at the heated substrate surface. For example, in the growth of GaAs one often uses triethyl gallium and arsine and the crystal growth depends upon the following reaction:

$$\mathrm{Ga(CH_3)_3 + AsH_3 \rightleftharpoons GaAs + 3CH_4}$$

There are several varieties of MOCVD reactors. In atmospheric MOCVD the growth chamber is essentially at atmospheric pressure. One needs large amounts of gases for growth in this case, although one does not have the problems associated with vacuum generation. In low-pressure MOCVD the growth chamber pressure is kept low.

4.3 LITHOGRAPHY

The success of solid state electronics and optoelectronics is due to device fabrication techniques that can produce extemely complicated devices with high yield. A variety of active and passive devices can be fabricated on the same wafer using lateral patterning techniques based on lithography. Since crystal growth processes do not produce any controlled lateral variations in material properties, lithographic techniques are needed to alter the lateral properties of the wafer. Advances in lithographic techniques are one of the most important driving forces for advances in device performance. The lithography process involves taking a certain design created in a computer or by an artist and transferring it onto the wafer. A number of steps are involved in this process. While new advances in lithographic techniques are introducing continual changes in the individual steps, the following discussion will provide an overview of the current state of the art.

4.3.1 Photoresist Coating

In order to transfer an image to a wafer, the surface of the wafer has to be made sensitive to light. To make the wafer (which is usually covered by a thin oxide film or some other dielectric passivation material) sensitive to an image, a photoresist is spread on the wafer by a process called spin coating. For the resist to be reliable it must satisfy three criteria: i) it must have good bonding to the substrate; ii) its thickness must be uniform; and iii) the thickness should be reliably controlled over different wafer runs.

Spin-coating has emerged as the most reliable technique for photoresist application. As shown in Fig. 4.4, a small puddle of the resist is applied to the center of the wafer, which is held to a spindle by a vacuum chuck. The spindle is now spun at a rate of 2000-8000 rpm for 10 to 60 seconds. During the first couple of seconds of spinning, most of the resist is thrown off and carefully drained away. The remaining resist forms a thin layer whose thickness is controlled by the spin speed. The thickness of the resists is usually in the range of 0.7 to 1.0 μm. After the resist is applied, it is soft-baked at 90 to 100 C to improve adhesion to the oxide.

Once the resist is ready, it is exposed to an optical image through a mask (to be discussed next) for a certain exposure time, as shown in Fig. 4.5. The resist is then developed by washing it in a solvent that dissolves away the regions of higher solubility. A resist that becomes more soluble when exposed to illumination is called positive. Its image is identical to the opaque image on the mask plate. A resist that loses solubility when illuminated is called negative.

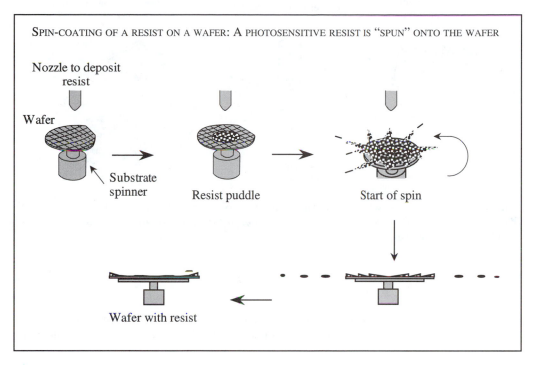

Figure 4.4: The process of spin-coating a wafer starts the lithographic process.

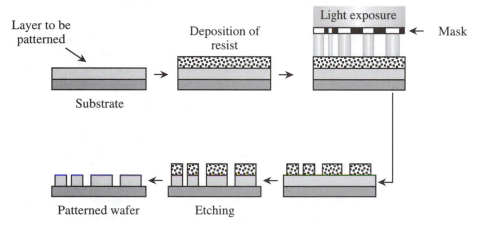

Figure 4.5: The exposure and etching process that allows one to transfer a pattern to the wafer.

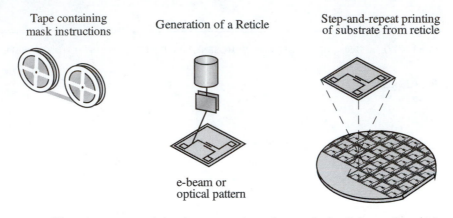

Tape containing   Generation of a Reticle   Step-and-repeat printing
mask instructions   of substrate from reticle

e-beam or
optical pattern

Figure 4.6: The processes used in the generation of a mask for lithography. (After W.S. Ruska, *Microelectronic Processing, An Introduction to the Manufacture of Integrated Circuits*, McGraw-Hill, New York (1988).)

4.3.2   Mask Generation and Image Transfer

A mask allows one to transfer a circuit pattern onto the sensitive resist deposited on the wafer. As shown in Fig. 4.6, computer-aided design (CAD) software programs allow one to directly generate a design tape which is then analyzed and "written" onto a mask plate. The pattern on the final mask plate can be generated optically or by electron beam writing. The electron beam (e-beam) produces much finer features and can be controlled with greater precision. The mask plate, which is later used to repeatedly transfer patterns to different wafers, must have good mechanical and thermal properties. Quartz or borosilicate glasses are often used for ultraviolet light lithography. The regions of the mask that are to be opaque are covered with metallic chrome or iron oxide.

Usually only the basic building block of the device or circuit is produced first on the mask. This single pattern is called a reticle, and a step-and-repeat camera (or stepper) is used to create the entire pattern from the reticle. The reticle can be used to generate a wafer-size mask plate or, in some cases, may be directly used to generate the entire pattern on the wafer by the stepper, as shown in Fig. 4.6.

Once the mask plate is made and the resist is deposited, the next step is to transfer the pattern on the mask to the wafer. For feature sizes greater than 0.25 μm, one uses optical equipment for the pattern transfer. The limit of $\sim$ 0.25 μm is determined by the wavelength of the light available. Most materials become opaque to light once the wavelength goes below $\sim$ 1000 Å. If electromagnetic radiation is to be used, one must go down to X-ray lithography with $\lambda \sim$ 40-80 Å. Another way to decrease feature size is by use of e-beam lithography.

The patterns to be printed in microelectronics can be classified as windows and lines. The windows may be used to allow doping diffusion or to make connections with a conductor through an insulator. The conducting interconnects and resistors are usually formed into line patterns. The imaging process involves transferring an ideal image to the mask and then from the mask to the wafer. Once the image is transferred, the resist is developed and etching is carried out. In all these steps, the ideal pattern in the designer's mind is gradually lost. This occurs even if the design considerations are well within the image-transferring technology.

4.4 DOPING OF SEMICONDUCTORS

We have seen that doping is a critical part of semiconductor technology. Every semiconductor device has some region that is doped n-type or p-type. Once a dopant is selected, two important hurdles are to be overcome: i) getting the dopant into the crystal and ii) *activating* the dopant, which may involve getting the dopant to the proper crystal site and removing any damage resulting from the doping process.

Three techniques are used to introduce dopant atoms into the crystal: i) epitaxial doping; ii) doping by diffusion; and iii) doping by ion implantation. We will briefly discuss these approaches. Once the dopant is in the crystal, thermal annealing is usually needed to remove the damage that may have been produced and to get the dopant into the correct lattice sites.

4.4.1 Epitaxial Doping

In epitaxial doping the dopant atoms are included in the flux of gases being used for crystal growth. As a result, this is the most controllable and precise doping technique. In this method the doping density and the placement of the dopants is very precise. It is also possible to switch from n-type dopants to p-type dopants and produce an almost atomically abrupt junction.

Epitaxial doping has some drawbacks. The doping process is quite expensive compared to other approaches. Also, in most microelectronic devices the use of *planar* technology requires that n and p regions be on the same plane. In epitaxial doping it is very difficult to produce any lateral variation in doping on the wafer. As a result, even when epitaxial doping is used, another doping process may be needed to produce the final device.

4.4.2   Doping by Diffusion

In Chapter 3 we discussed how carrier transport occurs by diffusion. A similar process can be exploited to dope semiconductors. The approach is carried out in two steps. During the first step, known as predeposition, the wafer is placed in a furnace in which inert gases carry the dopant, which is deposited on the wafer. In the second step, known as the drive-in, the wafer is heated for a period of time to allow the deposited dopants to diffuse into the wafer. During the second step no more dopants are added to the wafer.

In the drive-in process the dopants move from the high-concentration region (near the surface) to the low-concentration region (in the bulk). This produces a concentration variation as a function of time as shown in Fig. 4.7. The doping profile produced by diffusion is not as abrupt as that produced by epitaxial doping.

If the diffusion occurs on a sample that has opposite kind of dopants already in place, a junction is produced. The position of the junction is given by the point where the diffused doping density reaches the value present in the bulk.

4.4.3   Ion Implantation

The most important doping technique used in microelectronics industry is ion implantation. In this approach dopant ions are given a pre-chosen energy by accelerating them by an electric field. The high-energy beam (ranging in energies from a few keV to a few hundred keV) then impinges on the wafer as shown in Fig. 4.8.

When an energetic ion beam penetrates a semiconductor it suffers a series of collisions with the atoms and the electrons in the solid. These collisions are statistical in nature and on an average the ion loses its energy at a depth or range R_p from the surface. The ion distribution typically has a Gaussian distribution as shown in Fig. 4.9. The dopant profile can be controlled by the energy of the incoming particles. Higher-energy particles have a deeper range. Thus by controlling the initial energy one can control the position of the dopants in the semiconductor.

As shown in Fig. 4.9, the ion-implanted doping profile is defined by a full width at half maximum (FWHM), Δx_p, which defines how much spatial spread is present in the dopant distribution. Typically Δx_p is one-tenth of R_p. Due to the collisions with the host semiconductor, the ion implantation process causes a lot of damage in the implanted wafer. Vacancies, interstitial atoms or even locally non-crystalline regions are produced. This damage has to be removed before the dopants become activated. The defects are annealed by keeping the wafer at an elevated temperature for 20 to 30 minutes. In some applications regions between devices are implanted to make them electrically *dead* so that the devices are isolated.

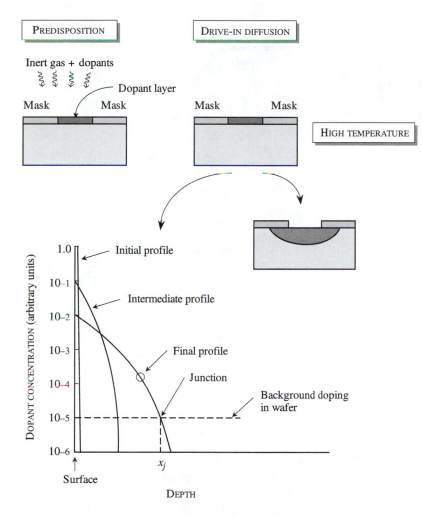

Figure 4.7: A schematic of doping by diffusion. The predeposition and drive-in diffusion are shown along with dopant concentration.

To produce lateral variation in doping, ion implantation is done through masks. This allows *n*-type doping in some regions and *p*-type doping in others.

4.5 ETCHING

Once an image is transferred from a mask to a wafer, one has to remove or etch material from selected regions to form the final device. Specially designed etchants allow one to remove material in a selective manner once the resist has been patterned. The choice and control of the etching process are crucial if the features in the resist film are to

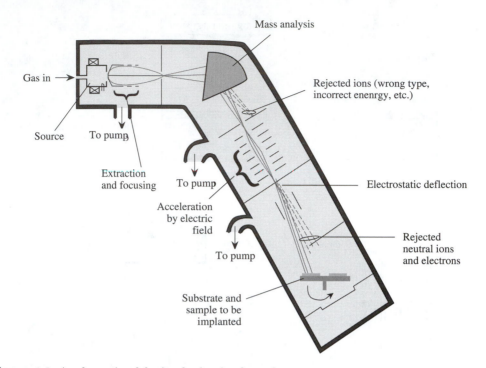

Figure 4.8: A schematic of doping by ion implantation.

become a part of the substrate.

An ideal etching process must be able to remove a layer of material from the region where there is no resist. The etchant should not attack the resist, nor should it penetrate under the resist, causing undercuts. It should also attack only one layer and should be self-limiting (e.g., should etch SiO_2 and not Si, etc.). A number of techniques have been developed to carry out etching. These are discussed below. Once etching is finished, one must remove the resist material either by other etchants or by the "lift-off" technique. We will now briefly review the etching approaches.

4.5.1   Wet Chemical Etching

The simplest and most commonly employed etching technique is wet chemical etching, in which the wafer is simply soaked in a liquid chemical that dissolves away the semiconductor. The etching process involves a chemical reaction in which the elements of the film to be etched react with the etching solution. The reaction products can then be rinsed away. The rate of the etching is proportional to the etch time and is usually isotropic. The isotropic nature of wet etching is not very suitable for devices where very

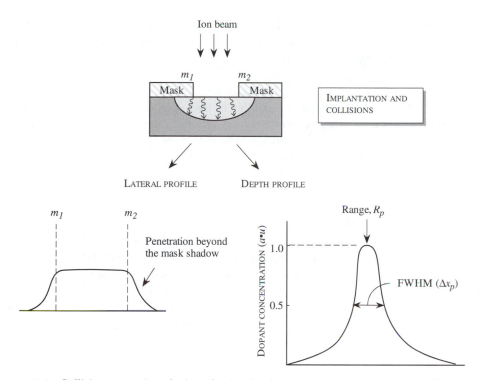

Figure 4.9: Collisions occuring during the ion implantation process are shown. The lateral and depth profile of the dopants is also shown.

sharp sidewalls are to be produced.

The etch rate depends upon the concentration of the chemicals used in the etchant and the temperature of the solution. The etchants are usually either acids (hydrofluoric acid–HF; nitric acid–HNO$_3$; acetic acid–H$_4$C$_2$O$_2$; sulfuric acid–H$_2$SO$_4$; phosphoric acid–H$_3$PO$_4$) or alkaline solutions of ammonium hydroxide (NH$_4$OH). The etchants are usually diluted in water according to well established recipes. In many devices, an important film to be etched is SiO$_2$. This film is etched with HF solutions and the ease and control of this process is one of the reasons for the success of the Si technology. Usually the HF is buffered with NH$_4$F to produce buffered oxide etch (BOE).

An important film that is often used for passivation or device protection is silicon nitride (Si$_3$N$_4$). This film can also be etched by BOE, but the rate is much slower than the SiO$_2$ rate. Thus, if a Si$_3$N$_4$ film is on top of an SiO$_2$ film and only the Si$_3$N$_4$ film is to be etched, one cannot use BOE. In such cases H$_3$PO$_4$ is used.

Since aluminum is commonly used for metal interconnects, one often has to etch it. A mixture of phosphoric acid, nitric acid, and acetic acid is usually used if the underlying crystal is Si. However, since this etchant attacks GaAs, one uses hydrochloric acid for etching Al over GaAs substrates.

Silicon and polycrystalline silicon can be etched by HF-HNO_3 mixtures, while GaAs is etched by using bromine in methanol or hydrogen peroxide mixed with sulfuric acid in water. A number of selective etches have also been developed for the heterostructure technology. While chemical etching is simple and inexpensive, it is not compatible with technology that demands deep anisotropic etching.

4.5.2   Plasma Etching

Plasma etching resolves one of the main problems with wet chemical etching, viz. feature size control. It also provides efficient etching for a wide variety of films including those that are difficult to etch by wet chemicals. The plasma is produced by passing an rf electrical discharge through a gas at a low pressure. The rf discharge creates ions and electrons. The ions can be used to interact with the elements in the substrate and cause etching.

The ions, being charged particles, can be accelerated in the electric field and be made to bombard the substrate with controlled energy. If the ion energy is large, the ions simply sputter off atoms from the surface in a rather unselective manner. However, this provides extremely anisotropic etching with almost no undercutting effects. At low energies the ions can cause chemical reactions at the surface and cause removal of atoms selectively.

In typical plasmas one introduces fluorine- or chlorine-containing gases. The fluorocarbons (e.g., CF_4) or silicon tetrafluoride (SiF_4) and silicon tetrachloride ($SiCl_4$) are often used for the plasma. Once the plasma is formed, fluorine (or chlorine) ions are produced that cause the etching process.

4.5.3   Reactive Ion Beam Etching (RIBE)

An important tool in microelectronic technology is ion-implantation, which is widely used to dope semiconductors. The implanter accelerates ions to a prechosen energy and shoots them into the semiconductor by controlling the energy. The depth at which the ions are embedded can be controlled. The ion-implanter can also be used for etching by using appropriate ions and it provides focusing of the ion beam. At low energies, the ions can be used to selectively etch very small feature sizes. The advantages are the same as for plasma etching.

4.5.4 Ion Beam Milling

Ion beam milling is another application of ions in "chiseling" off material from a substrate. Ion milling requires a focused beam of ions of energy high enough that the ions can physically knock out atoms from the film. The focus is on removing the atoms physically rather than through a chemical reaction. The process is thus highly directional and can produce extremely anisotropically etched structures. Ion milling is particularly advantageous if the etching involves small patterns, very steep walls, or materials that are relatively inert and cannot be etched by chemical reactions.

Ion milling is dominated by geometric effects as shown in Fig. 4.10. In Fig. 4.10a we show a substrate with a resist in the process of being etched by a perpendicular ion beam. The impact of the beam causes the substrate material to fly off randomly. Some of the material can get redeposited on the etched sides, forming "ears." Ions rebounding off the edges can cause "trenches" to be formed around the resist pattern. These effects are controlled by impinging the beam at a slight angle and rotating the substrate. The resolution of ion beam milling is controlled by the ion-beam spot size and can reach 0.1 μm or less.

4.6 PUTTING IT ALL TOGETHER

Semiconductor devices and chips are some of the most complex systems that humans build. In high-density memory chips, tens of millions of transistors are packed into a thumbnail-size Si chip. In microprocessors, thousands of logic elements—electronic buses, registers, and memory devices—are coupled by an intricate maze of interconnects. In microwave chips, transistors, inductors, capacitors, and resistors are built and connected together. In optoelectronic chips semiconductor lasers, transistor drivers, and modulators have to perform in unison. Semiconductors are also *micromachined* to produce tiny whiskers for oscillators, cantilevers for pressure sensors, thin pipes to separate molecules, and hundreds of other sensors and activators. The remarkable aspect of microelectronics is that not only are the chips very complex, but the features on them are extremely tiny. A view of microelectronic chips under a microscope reveals a panorama reminiscent of an aerial view of a large city. The smallest feature sizes on modern chips are as small as 0.1 μm (lateral dimension). In the depth profile of a chip one sees features as thin as 20 Å—just a few monolayers.

Figures 4.11–4.15 show some of the remarkable device structures that modern technology is capable of producing.

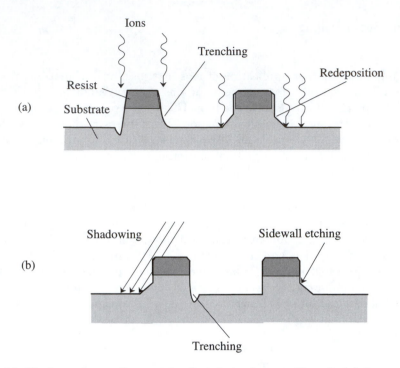

Figure 4.10: The importance of geometric effects in ion beam milling. In (a) the perpendicular incident beam can produce trenching effects as well as redeposition, causing sidewall "ears;" (b) if the beam comes at an oblique angle and the substrate is rotated, the trenching and "ear" formation can be balanced.

4.7   READING LIST

- **Microelectronic Processing**

    – J. W. Mayer and S. S. Lau, *Electronic Materials Science for Integrated Circuits in Si and GaAs* (McGraw-Hill, New York, 1990).

    – W. S. Ruska, *Microelectronic Processing: An Introduction to the Manufacture of Integrated Circuits* (McGraw-Hill, New York, 1988).

    – P. K. L. Yu and P. C. Chan, *Introduction to GaAs Technology* (ed. C. Wang, Wiley, New York, 1988).

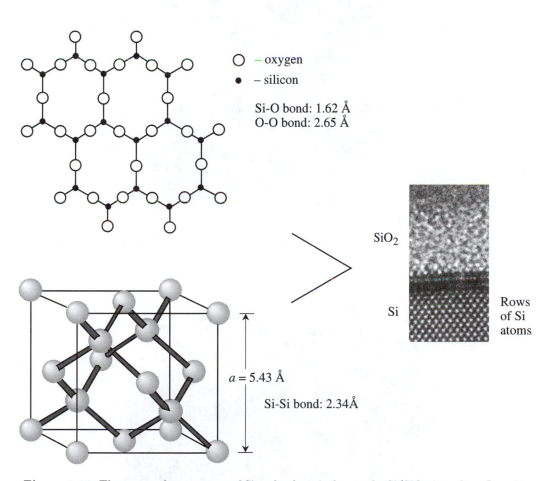

○ – oxygen
● – silicon

Si-O bond: 1.62 Å
O-O bond: 2.65 Å

SiO$_2$

Si

Rows
of Si
atoms

$a = 5.43$ Å

Si-Si bond: 2.34Å

Figure 4.11: The tremendous success of Si technology is due to the Si/SiO$_2$ interface. In spite of the very different crystal structure of Si and SiO$_2$, the interface is extremely sharp, as shown in the TEM picture in this figure.

Figure 4.12: This transmission electron microscope picture shows the precision with which semiconductor compositions can be altered by epitaxial growth techniques. Individual semiconductor layers as thin as 10 Å can be produced.

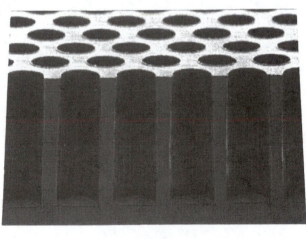

10 μm

Figure 4.13: Scanning electron micrographs of InP via holes that are 30 μm wide and 92 μm deep etched with vertical profile and smooth surface morphology. The picture shows the great precision achievable in modern etching technologies.

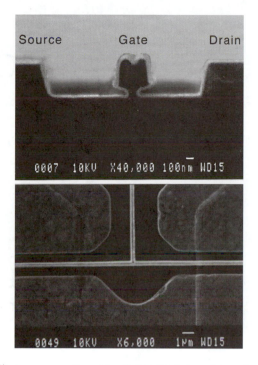

Figure 4.14: (top half) A cross-section of a field effect transistor showing the gate, source, and drain. The gate width is only 0.1 μm. (lower half) A planar view of the same device.

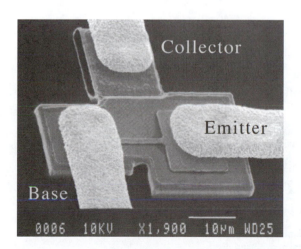

Figure 4.15: A high-resolution picture of a bipolar transistor with 1 μm emitter fingers.

Chapter

5

JUNCTIONS IN SEMICONDUCTORS: *P-N* DIODES

5.1 DEVICE DEMANDS

In the first three chapters of this book we examined the basic physical phenomena that control the properties of semiconductors. In the remainder of this book we will see how these special properties can be exploited to produce electronic devices. Before going into the details of how different devices operate, it is important to pause and examine what devices are supposed to do.

As noted in the Introduction, we live in the information age where the generation and manipulation of information is critical to the wellbeing of society. Semiconductor devices play a key role in this generation and manipulation process. These devices are responsible for the following key functions.

Information Reception/Detection
The devices should be able to detect information that may have originated from a faraway star, a supersonic jet, or a voice thousands of kilometers away. Semiconductor diodes, transistors, and photodetectors serve the role of detecting signals of various kinds.

Information Amplification
Very often the information received by a device is very weak and must be amplified to be useful. The device must have "gain" to be able to do this. This means that a weak signal input should result in a large output signal with minimal distortion. Bipolar and field effect transistors, to be discussed in Chapters 7, 8, and 9, are extremely useful devices because they are capable of large gain.

Information Manipulation
Semiconductor devices are often called "intelligent" because they are able to manipulate information. This manipulation may involve addition, multiplication, etc. or logic decisions such as AND, OR, etc. To be able to carry out such intelligent steps, devices must be able to have special input-output relations (e.g., input is high, output is low and vice versa for digital applications). The devices must also have gain and a nonlinear response so that small noise levels in the input do not cause errors in the output. Additionally, there should be *input-output isolation* so that the output does not affect the incoming information.

Information Storage
An important role of devices is to store information for future use. This is an area where semiconductors are most challenged by other technologies. Technologies based on plastic discs (the compact disc with optoelectronic reading), magnetic tapes, and magnetic discs still dominate the memory market. However, semiconductor memories based on transistors are much faster and the rapid decrease in the cost is making them increasingly important.

Information Generation
Another important demand on devices is to be able to generate information. Often this involves taking a stream of incoming information and generating an outgoing stream of electronic or optical information. Microwave devices and semiconductor lasers are examples of semiconductor devices that are being used to generate information.

Information Display
An important demand from devices is to display information in the form of a picture that a human being can understand. Displays are becoming increasingly important because of the vast amounts of information that are now being produced. A rough estimate suggests that over a trillion bits of information are created every day (mainly through the television and Internet industry)! Semiconductor devices, such as light-emitting diodes, are important devices in the display market.

We now start our journey into how semiconductor devices operate. As the reader covers each device, it may be useful to come back to this section and see how the device fits into the demands of the information age.

5.1.1   Need for Junctions

We have seen how the electrical properties of a semiconductor material can be altered by changing the doping in the material. While this is, of course, essential for semiconductor devices, it is not enough to make a device. An essential feature of an electronic device must be the ability to alter the "state" of the device. The state of the device may involve the conductivity of the material, for example. How can one rapidly alter the current flowing through a material?

Consider a flexible tube through which water is flowing at some rate. How can you modify the flow of water? An easy way would be to modify the area of the "flow channel" by applying pressure on the tube with your thumb. Can this be done for a semiconductor device? Yes, although one cannot, of course, physically deform a bar of semiconductor. Such tasks can be accomplished by semiconductor junctions.

One of the important requirements of good devices is nonlinear response. An important class of nonlinear responses is a rectifying response in which the flow is "easy" in one direction but "difficult" in another. A simple semiconductor (doped or undoped) does not have any such direction selectivity. In fact, a simple piece of a semiconductor cannot accomplish most of the information processing requirements of devices. So what good are semiconductors? We will see that semiconductors become useful when, instead of having a *uniform chemical composition*, they have *spatially nonuniform compositions*. This includes having *junctions at certain regions in space*. These junctions allow us to conceive of and build the devices that drive the information age.

In this and the next chapter we will discuss the properties of several kinds of junctions. A most important junction is the *p-n* junction, in which the nature of dopants is altered across a boundary to create a region that is *p*-type next to a region that is *n*-type. This junction has rectifying properties and can be used to produce strong nonlinear effects.

Another important class of junctions involves semiconductors and metals. These junctions are important in allowing charge flow in and out of the semiconductors, say from batteries. When used for such purposes, these junctions are called ohmic contacts. Additionally, certain kinds of metal-semiconductor junctions called Schottky barriers can be used to electrically control the flow area of a semiconductor film.

Another class of junctions involves an insulator or a large gap semiconductor. In the case of Si technology the insulator junction involves oxides or nitrides of Si. For GaAs technology one often uses AlGaAs, a material with bandgap larger than GaAs. These junctions provide isolation of the input signal from the output signal and also allow modulation of the channel conductivity.

The junctions we are about to discuss occasionally form important devices by themselves. More often, though, they are used along with other junctions to make up the device structure. In this chapter we will discuss the various properties of *p-n* diodes and some of their applications.

5.2 UNBIASED *P-N* JUNCTION

As noted in the previous section, the *p-n* junction is one of the most important junctions in solid-state electronics. The junction is used as a device in applications as rectifiers, waveform shapers, lasers, detectors, etc. Additionally, it forms the key ingredient of the bipolar transistor, which is one of the most important electronic devices. We are interested in the following aspects of this junction: i) What are the carrier distributions for electrons and holes in the material? ii) What are the physical processes responsible for current flow in the structure when a bias is applied?

In Chapter 4 we discussed how semiconductors are doped. We will assume in our analysis that the *p-n* junction is abrupt, even though this is really true only for epitaxially grown junctions. Let us first discuss the properties of the junction in the absence of any external bias. Obviously there is no current flowing in the diode (the word used for a *p-n* junction). However, does this mean that the electrons and holes are not feeling any forces to cause charge transport? We will see that, in fact, there is a lot of activity in the carriers, even though there is no net current.

In Fig. 5.1a we show schematically the *p-* and *n*-type semiconductors without

forming a junction between them. We have shown the positions of the Fermi level, the
conduction and valence bands, and the vacuum energy level. The electron affinity $e\chi$,
defined as the energy difference between the conduction band and vacuum level, is also
shown along with the work function ($e\phi_{sp}$ or $e\phi_{sn}$). The work function represents the
energy required to remove an electron from the semiconductor to the "free" vacuum
level and is the difference between the vacuum level and the Fermi level.

What happens when the p- and n-type materials are made to form a junction
and there is no externally applied field? We know that in the absence of any applied bias,
there is no current in the system. We know also from Chapter 3 (see Section 3.5) that
in the absence of any current flow, the Fermi level is uniform throughout the structure.

This gives the schematic view of the junction shown in Fig. 5.1b. Three regions
can be identified:

i) The p-type region at the top left where the material is neutral and the bands
are flat. The density of acceptors exactly balances the density of holes;

ii) The n-type region at the top right where again the material is neutral and
the density of immobile donors exactly balances the free electron density;

iii) The depletion region where the bands are bent and a field exists that has
swept out the mobile carriers, leaving behind negatively charged acceptors in the p-
region and positively charged donors in the n-region, as shown in Fig. 5.1b.

In the depletion region, which extends a distance W_p in the p-region and a distance W_n in the n-region, an electric field exists. Any electrons or holes in the depletion
region are swept away by this field. Thus a drift current exists that counterbalances the
diffusion current, which arises because of the difference in electron and hole densities
across the junction. To describe the junction properties we need to know the width of
the depletion region, the charge distribution of electrons and holes, and the electric field.
While in principle this problem is simple, analytical results can be obtained only if we
make some simplifying assumptions:

i) The physical junction is abrupt and each side is uniformly doped. This condition can be relaxed but then one needs to use numerical techniques to solve the problem.

ii) While the mobile charge density in the depletion region is not zero (otherwise
there will be no current flow possible), it is much smaller than the background fixed
charges. Thus to *solve Poisson equations we will assume that the mobile carrier density
is zero*. Of course, this approximation, called the depletion approximation, will *not be
used to calculate current flow*;

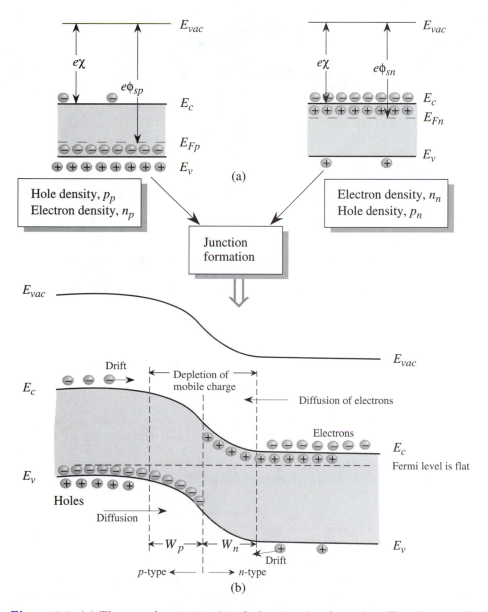

Figure 5.1: (a) The p- and n-type regions before junction formation. The electron affinity $e\chi$ and work functions $e\phi_{sp}$ and $e\phi_{sn}$ are shown along with the Fermi levels. (b) A schematic of the junction and the band profile showing the vacuum level and the semiconductor bands. The Fermi level is "flat" in the absence of current flow.

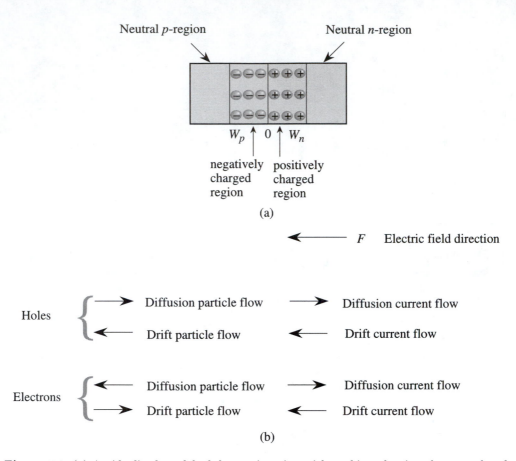

Figure 5.2: (a) An idealized model of the *p-n* junction without bias, showing the neutral and depletion areas. (b) A schematic showing various current and particle flow components in the *p-n* diode at equilibrium. For electrons, the current flow is in the direction opposite to that of the particle flow. Electrons that enter the depletion region from the *p*-side and holes that enter the depletion region from the *n*-side are swept away and are the source of the drift components.

iii) The transition between the bulk neutral *n*- or *p*-type region and the depletion region is abrupt, as shown in Fig. 5.2a. While in reality the transition is gradual, detailed calculations show that this is not such a bad approximation.

In order to understand the diode properties, let us first identify all the current components flowing in the device. There is the electron drift current and electron diffusion current as well as the hole drift and hole diffusion current, as shown in Fig. 5.2b. When there is no applied bias, these currents cancel each other *individually*. Let us consider these current components. The hole current density is

$$J_p(x) = e \left[\underbrace{\mu_p p(x) F(x)}_{\text{drift}} - \underbrace{D_p \frac{dp(x)}{dx}}_{\text{diffusion}} \right] = 0 \qquad (5.1)$$

Note that the mobility μ_p is actually field-dependent and, as discussed in Chapter 3, *the drift current density in the depletion region is really $ep(x)v_s(x)$ and independent of F,* where v_s is the saturated velocity. This is because, as we shall see later, *the field in the depletion region is very large even under equilibrium.* Similarly, the diffusion coefficient is not constant but is field-dependent. The ratio of μ_p and D_p is given by the Einstein relation (Section 3.6)

$$\frac{\mu_p}{D_p} = \frac{e}{k_B T} \qquad (5.2)$$

As shown in Fig. 5.3, as a result of bringing the p and n type semiconductors, a *built-in voltage,* V_{bi}, is produced between the n and the p side of the structure. Referring to Fig. 5.3, the built-in voltage is given by

$$eV_{bi} = E_g - (E_c - E_F)_n - (E_F - E_v)_p$$

where the subscripts n and p refer to the n-side and p-side of the device. We know that (see Eqn. 2.19)

$$(E_c - E_F)_n = -k_B T \ell n(\frac{n_n}{N_c})$$

where n_n is the electron density on the n-side of the device. Assuming that all of the donors are ionized,

$$n_n = N_d$$

Similarly (see Eqn. 2.22),

$$(E_F - E_v)_p = -k_B T \ell n(\frac{p_p}{N_v})$$

where p_p is the hole density on the p-side and is given by

$$p_p = N_a$$

We now have

$$eV_{bi} = E_g + k_B T \ell n(\frac{n_n p_p}{N_c N_v})$$

Using the relation (Eqn. 2.24)

$$n_i^2 = N_c N_v \exp\left(-\frac{E_g}{k_B T}\right)$$

we get

$$V_{bi} = \frac{k_B T}{e} \ell n(\frac{n_n p_p}{n_i^2}) \qquad (5.3)$$

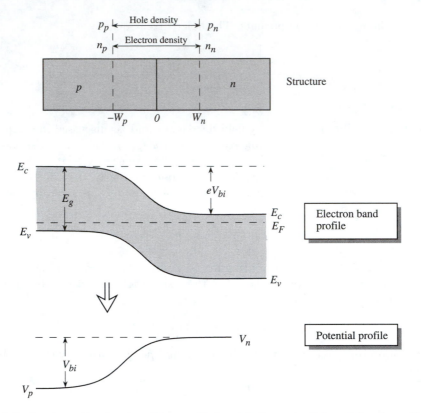

Figure 5.3: A schematic showing the p-n diode and the potential and band profiles. The voltage V_{bi} is the built-in potential at equilibrium. The expressions derived in the text can be extended to the cases where an external potential is added to V_{bi}.

If n_n and n_p are the electron densities in the n-type and p-type regions, the law of mass action (i.e., the product np is constant) tells us that

$$n_n p_n = n_p p_p = n_i^2 \tag{5.4}$$

Thus, the contact potential or built-in potential, $V_{bi} = V_n - V_p$, is

$$\boxed{V_{bi} = \frac{k_B T}{e} \; \ell n \; \frac{p_p}{p_n}} \tag{5.5}$$

or

$$\boxed{V_{bi} = \frac{k_B T}{e} \; \ell n \; \frac{n_n}{n_p}} \tag{5.6}$$

We can thus write the following equivalent expressions:

$$\frac{p_p}{p_n} = e^{eV_{bi}/k_B T} = \frac{n_n}{n_p} \tag{5.7}$$

In this relation V_{bi} is the built-in voltage in the absence of any external bias. Under the approximations discussed later, a similar relation holds when an external bias V is applied to alter V_{bi} to $V_{bi} + V$, and will be used when we calculate the effect of external potentials on the current flow.

Let us now calculate the width of the depletion region for the diode under no applied bias. The calculation in the presence of a bias V will follow the same approach and V_{bi} will simply be replaced by $V_{bi} - V$, the total potential across the junction. In the depletion region, the mobile carrier density of electrons and holes is very small compared to the fixed background charge. This leads to the depletion approximation discussed earlier. With this approximation, there is a region of negative charge (due to acceptors) extending from the junction to the point W_p on the p-side and a region of positive charge (due to donors) extending the junction to W_n. The total negative and positive charge have the same magnitude, so that we have the equality

$$A\, W_p N_a = A\, W_n N_d \qquad (5.8)$$

where A is the cross-section of the p-n structure and N_a and N_d are the uniform doping densities for the acceptors and donors.

Let us solve the Poisson equation for the field profile in the structure. The equations to consider (in the depletion approximation) are

$$\frac{d^2V(x)}{dx^2} = 0 \qquad -\infty < x < -W_p \qquad (5.9)$$

$$\frac{d^2V(x)}{dx^2} = \frac{eN_a}{\epsilon} \qquad -W_p < x < 0 \qquad (5.10)$$

$$\frac{d^2V(x)}{dx^2} = -\frac{eN_d}{\epsilon} \qquad 0 < x < W_n \qquad (5.11)$$

$$\frac{d^2V(x)}{dx^2} = 0 \qquad W_n < x < \infty \qquad (5.12)$$

The first of these equations has a simple solution ($F(x)$ is the electric field):

$$\frac{dV(x)}{dx} = -F(x) = C_1 \qquad -\infty < x < -W_p \qquad (5.13)$$

where C_1 is a constant that is zero since there is no electric field in the neutral region. The potential here is

$$V(x) = V_p \qquad -\infty < x < -W_p \qquad (5.14)$$

where V_p is defined as the potential in the neutral p-type region.

The solution in the negatively charged p-side depletion region is

$$F(x) = \frac{-dV}{dx} = \frac{(-)eN_a x}{\epsilon} + C_2 \qquad -W_p < x < 0 \qquad (5.15)$$

and since the field $F(x)$ is zero at $x = -W_p$, we have

$$C_2 = -\frac{eN_aW_p}{\epsilon} \tag{5.16}$$

Thus the electric field in the p-side of the depletion region is

$$F(x) = -\frac{eN_ax}{\epsilon} - \frac{eN_aW_p}{\epsilon} \qquad -W_p < x < 0 \tag{5.17}$$

The electric field reaches a peak value at $x = 0$. The potential is given by integrating Eqn. 5.17,

$$V(x) = \frac{eN_ax^2}{2\epsilon} + \frac{eN_aW_px}{\epsilon} + C_3 \tag{5.18}$$

The constant C_3 is obtained by equating $V(-W_p) = V_p$. This gives for the potential

$$V(x) = \frac{eN_ax^2}{2\epsilon} + \frac{eN_aW_px}{\epsilon} + \frac{eN_aW_p^2}{2\epsilon} + V_p \qquad -W_p < x < 0 \tag{5.19}$$

Carrying out a similar analysis for the n-side of the depletion region and n-side of the neutral region, we use the conditions

$$
\begin{aligned}
V(x) &= V_n & W_n < x < \infty \\
F(x) &= 0 &
\end{aligned}
\tag{5.20}
$$

where V_n is the potential at the neutral n-side. The electric field and potential on the n-side is

$$F(x) = \frac{eN_dx}{\epsilon} - \frac{eN_dW_n}{\epsilon} \qquad 0 < x < W_n \tag{5.21}$$

$$V(x) = -\frac{eN_dx^2}{2\epsilon} + \frac{eN_dW_nx}{\epsilon} - \frac{eN_dW_n^2}{2\epsilon} + V_n \qquad 0 < x < W_n \tag{5.22}$$

From this discussion, the potential difference between points $-W_p$ and 0 is

$$V(0) - V(-W_p) = \frac{eN_aW_p^2}{2\epsilon} \tag{5.23}$$

Similarly, the potential difference between the points W_n and 0 is

$$V(W_n) - V(0) = \frac{eN_dW_n^2}{2\epsilon} \tag{5.24}$$

Thus the built-in potential is

$$V(W_n) - V(-W_p) = V_{bi} = \frac{eN_dW_n^2}{2\epsilon} + \frac{eN_aW_p^2}{2\epsilon} \tag{5.25}$$

Recalling that the charge neutrality gives us

$$N_d W_n = N_a W_p \qquad (5.26)$$

we get the following set of relationships:

$$W_p(V_{bi}) = \left\{ \frac{2\epsilon V_{bi}}{e} \left[\frac{N_d}{N_a(N_a + N_d)} \right] \right\}^{1/2} \qquad (5.27)$$

$$W_n(V_{bi}) = \left\{ \frac{2\epsilon V_{bi}}{e} \left[\frac{N_a}{N_d(N_a + N_d)} \right] \right\}^{1/2} \qquad (5.28)$$

$$W(V_{bi}) = W_p(V_{bi}) + W_n(V_{bi}) = \left(W_n^2(V_{bi}) + W_p^2(V_{bi}) + 2W_n(V_{bi})W_p(V_{bi}) \right)^{1/2}$$

$$W(V_{bi}) = \left[\frac{2\epsilon V_{bi}}{e} \left(\frac{N_a + N_d}{N_a N_d} \right) \right]^{1/2} \qquad (5.29)$$

From these discussions, we can draw the following important conclusions about the diode:

• The electric field in the depletion region peaks at the junction and decreases linearly towards the depletion region edges.
• The potential drop in the depletion region has a quadratic form.

We remind ourselves that this procedure can be extended to find the electric fields, potential, and depletion widths for arbitrary values of V_p and V_n under certain approximations to be discussed next. Thus we can use these equations directly when the diode is under external bias V, by simply replacing V_{bi} by $V_{bi} + V$. The applied bias can increase or decrease the total potential, as will be discussed later.

The results of the calculations carried out above are shown schematically in Fig. 5.4. Shown are the charge density and the electric field profiles. Notice that the electric field is nonuniform in the depletion region, peaking at the junction with a peak value (the sign of the field simply reflects the fact that in our study the field points towards the negative x-axis)

$$F_m = -\frac{e N_d W_n}{\epsilon} = -\frac{e N_a W_p}{\epsilon} \qquad (5.30)$$

Notice that the depletion in the p- and n-sides can be quite different. If $N_a \gg N_d$, the depletion width W_p is much smaller than W_n. Thus a very strong field exists over a very narrow region in the heavily doped side of the junction. *In such abrupt junctions (p^+n or n^+p) the depletion region exists primarily on the lightly doped side.*

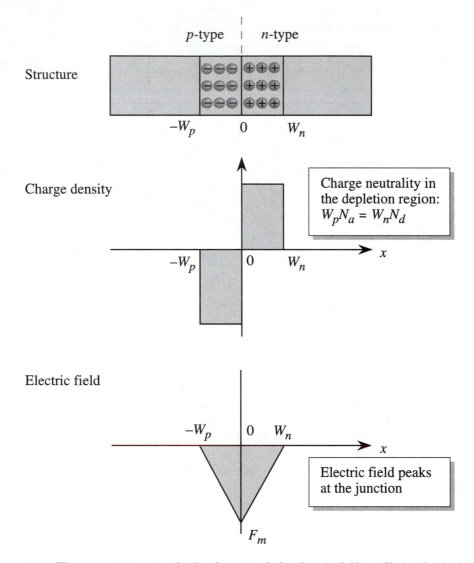

Figure 5.4: The *p-n* structure, with the charge and the electric field profile in the depletion region. Note that in the depletion approximation there is no charge or electric field outside the depletion region. The electric field peaks at the junction as shown.

In the depletion approximation we have assumed that the mobile carrier density is zero. However, this is not strictly true. Since we know the potential profile we can evaluate the mobile carrier density and the drift and diffusion currents. We will do this in the general case where an external bias is applied.

EXAMPLE 5.1 A silicon diode is fabricated by starting with an n-type ($N_d = 10^{16}$ cm^{-3}) substrate, into which indium is diffused to form as a p-type region doped at 10^{18} cm^{-3}. Assuming that an abrupt $p^+ - n$ junction is formed by the diffusion process, i) calculate the Fermi level positions in the p- and n-regions; ii) determine the contact potential in the diode; iii) calculate the depletion widths on the p- and n-side.

To find the positions of the Fermi levels, we can use either of the following equivalent relations. We assume complete ionization of the dopants. For the n-side we have

$$n_n = n_i \exp\left[\frac{-(E_{Fi} - E_{Fn})}{k_B T}\right]$$

or

$$n_n = N_c \exp\left[\frac{-(E_c - E_{Fn})}{k_B T}\right]$$

For the p-side we have

$$p_p = n_i \exp\left[\frac{(E_{Fi} - E_{Fp})}{k_B T}\right]$$

or

$$p_p = N_v \exp\left[\frac{(E_v - E_{Fp})}{k_B T}\right]$$

where $n_n = N_d; p_p = N_a$ and n_i, N_c, N_v are the intrinsic carrier concentration, conduction-band effective density of states, and valence-band effective density of states. Using the effective density of states relations, we have ($N_c = 2.8 \times 10^{19}$cm^{-3}; $N_v = 1 \times 10^{19}$cm^{-3} at 300 K)

$$
\begin{aligned}
E_{Fn} &= E_c + k_B T \, \ell n \, \frac{n_n}{N_c} \\
&= E_c - (0.026 \text{ eV}) \times 7.937 \\
&= E_c - 0.206 \text{ eV} \\
E_{Fp} &= E_v - k_B T \, \ell n \, \frac{p_p}{N_v} \\
&= E_v + (0.026 \text{ eV}) \times 2.3 \\
&= E_v + 0.06 \text{ eV}
\end{aligned}
$$

The built-in potential is given by

$$
\begin{aligned}
eV_{bi} &= E_g - 0.06 - 0.206 \text{ eV} \\
&= 1.1 - 0.06 - 0.206 \\
&= 0.834 \text{ eV}
\end{aligned}
$$

The depletion width on the p-side is given by

$$W_p(V_{bi}) = \left\{ \frac{2\epsilon V_{bi}}{e} \left[\frac{N_d}{N_a(N_a + N_d)} \right] \right\}^{1/2}$$

$$\cong \left\{ \frac{2 \times (11.9 \times 8.84 \times 10^{-12} \ \text{F/m}) \times 0.834 \ (\text{volts})}{(1.6 \times 10^{-19} \ \text{C})} \right.$$

$$\left. \times \frac{10^{22} \ \text{m}^{-3}}{10^{24} \ \text{m}^{-3} \times (1.01 \times 10^{24} \ \text{m}^{-3})} \right\}^{1/2}$$

$$= \quad 3.2 \times 10^{-9} \ \text{m} = 32 \ \text{Å}$$

The depletion width on the n-side is 100 times longer:

$$W_n(V_o) = 0.32 \ \mu\text{m}$$

This is an example of how abrupt the depletion region is on the heavily doped p-side.

5.3 *P-N* JUNCTION UNDER BIAS

Let us now consider the situation where an external potential is applied across the p and n regions. In the presence of the applied bias, the balance between the drift and diffusion currents will no longer exist and a net current will flow. In general, one needs a numerical treatment to understand the behavior of the p-n diode under bias. However, under some simplifying assumptions, one can use the formalism of the previous section to study the biased diode. These approximations are found to be valid as long as the current flow is not too large. We make the following approximations:

• The diode structure can be described by "quasi-neutral" n- and p-regions and a depletion region. We will see later that in forward bias, minority charge is injected into the quasi-neutral regions. However, we will assume that the injected density is quite small compared to the majority density.

• We assume that in the depletion region, the electron and hole distributions are essentially described by a Boltzmann distribution and that the concept of a quasi-Fermi level (see Section 3.7) is valid for electrons and holes. The quasi-Fermi levels for the electrons and holes extend from the quasi-neutral regions as shown in Fig. 5.5. This approximation is again valid if the current levels are small.

• Across the depletion region the mobile carrier density is low, and therefore the external potential drops mainly across this region. This is a generally applicable assumption and depends upon the fact that since current flow is continuous, wherever the mobile charge density is low, there has to be a high electric field.

In Fig. 5.5 we show the schematic profiles of the depletion region, potential profile, and the band profiles in equilibrium, forward bias, and reverse bias. When a forward bias V_f is applied, the p-side is at a positive potential with respect to the n-side. In the reverse bias case, the p-side is at a negative potential $-V_r$ with respect to the n-side. Remember that the way we plot the energy bands includes the negative electron charge so the energy bands have the opposite sign to the potential profile.

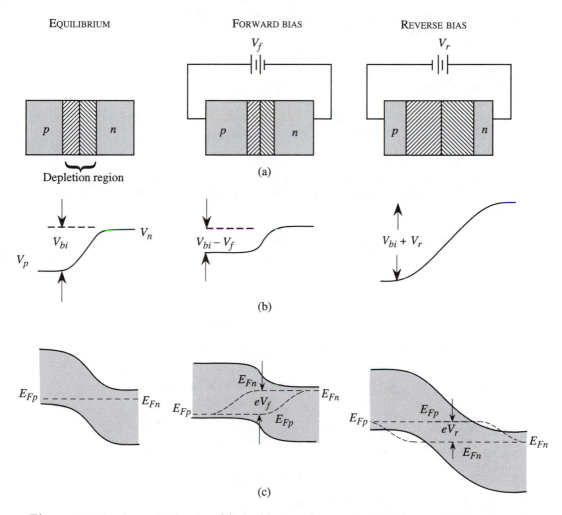

Figure 5.5: A schematic showing (a) the biasing of a *p-n* diode in the equilibrium, forward, and reverse bias cases; (b) the voltage profile, and (c) the energy band profiles. In forward bias, the potential across the junction decreases, while in reverse bias it increases. The quasi-Fermi levels are shown in the depletion region.

In the forward bias case, the potential difference between the n- and p-side is (applied bias $V = V_f$)

$$V_{Tot} = V_{bi} - V = V_{bi} - V_f \qquad (5.31)$$

while for the reverse bias case it is (the applied potential V is negative, $V = -V_r$, where V_r has a positive value)

$$V_{Tot} = V_{bi} - V = V_{bi} + V_r \qquad (5.32)$$

Under the approximations given above, the equations for electric field profile, potential profile, and depletion widths we calculated in the previous section are directly applicable except that V_{bi} is replaced by V_{Tot}. Thus the depletion width and the peak electric field at the junction decrease under forward bias, while they increase under reverse bias, as can be seen from Eqns. 5.29 and 5.30 if V_{bi} is replaced by V_{Tot}.

EXAMPLE 5.2 Consider a 20 μm diameter p-n diode fabricated in silicon. The donor density is 10^{16} cm^{-3} and the acceptor density is 10^{18} cm^{-3}. Calculate the following in this diode at 300 K:

i) The depletion widths and the electric field profile under reverse biases of 0, 2, 5, and 10 V, and under a forward bias of 0.5 V.

ii) What are the charges in the depletion region for these biases?

Let us start with calculating V_{bi}, the built-in potential for the diode:

$$V_{bi} = \frac{k_B T}{e} \ell n \frac{p_p}{p_n}$$

For our diode $p_p = 10^{18}$ cm^{-3}. The value of p_n is obtained by the law of mass action:

$$p_n = \frac{n_i^2}{n_n} = \frac{(1.5 \times 10^{10} \text{ cm}^{-3})^2}{10^{16} \text{ cm}^{-3}} = 2.25 \times 10^4 \text{ cm}^{-3}$$

Thus ($k_B T$ at 300 K = 0.026 eV)

$$V_{bi} = 0.026 \text{ V} \times 31.4 = 0.817 \text{ V}$$

The depletion widths are now (as in Example 5.1)

$$V_r = 0 \quad : \quad W_p(0.817 \text{ V}) = 32 \text{ Å}$$
$$W_n(0.817 \text{ V}) = 0.32 \ \mu\text{m}$$

$$V_r = 2 \text{ V} \quad : \quad W_p(2.817 \text{ V}) = 32 \text{ Å} \times \left(\frac{2.817}{0.817}\right)^{1/2} = 58.9 \text{ Å}$$
$$W_n(2.817 \text{ V}) = 0.589 \ \mu\text{m}$$

$$V_r = 5 \text{ V} \quad : \quad W_p(5.817 \text{ V}) = 84.5 \text{ Å}$$
$$W_n(5.817 \text{ V}) = 0.845 \ \mu\text{m}$$

$$V_r = 10 \text{ V} \quad : \quad W_p(10.817 \text{ V}) = 115.1 \text{ Å}$$

$$W_n(10.817 \text{ V}) = 1.151 \text{ } \mu m$$

$$V_f = 0.5 \text{ V} \quad : \quad W_p(0.317 \text{ V}) = 32 \text{ Å} \times \left(\frac{0.317}{0.817}\right)^{1/2} = 20 \text{ Å}$$

$$W_n(0.317 \text{ V}) = 0.20 \text{ } \mu m$$

The peak fields in the diode are given by

$$F_m = -\frac{e N_d W_n}{\epsilon}$$

$$F_m(V_r = 0 \text{ V}) = -\frac{\left(1.6 \times 10^{-19} \text{ C}\right)\left(10^{22} \text{ m}^{-3}\right)\left(0.32 \times 10^{-6} \text{ m}\right)}{\left(11.9 \times 8.84 \times 10^{-12} \text{ F/m}\right)}$$

$$= -4.95 \times 10^6 \text{ V/m} = -4.95 \times 10^4 \text{ V/cm}$$

$$F_m(V_r = 2 \text{ V}) = -4.95 \times 10^4 \times \frac{W_n(0.817 \text{ V})}{W_n(0.317)} = -9.11 \times 10^4 \text{ V/cm}$$

$$F_m(V_r = 5 \text{ V}) = -1.3 \times 10^5 \text{ V/cm}$$

$$F_m(V_r = 10 \text{ V}) = -1.78 \times 10^5 \text{ V/cm}$$

$$F_m(V_f = 0.5 \text{ V}) = -3.14 \times 10^4 \text{ V/cm}$$

We can see that at a reverse bias of $\sim$10 V, the peak field is beginning to approach the breakdown field for Si, which is around 3×10^5 V/cm.

The charge in the n- or p-side depletion region is

$$Q = e N_d W_n A = e N_a W_p A$$

where A is the area of the diode.

$$Q(V_r = 0) = \left(1.6 \times 10^{-19} \text{ C}\right) \times \left(10^{22} \text{ m}^{-3}\right) \times \left(0.32 \times 10^{-6} \text{ m}\right) \times \pi \left(10 \times 10^{-6} \text{ m}\right)^2$$

$$= 1.61 \times 10^{-13} \text{ C}$$

The charge can be obtained at other biases simply by using the various depletion widths.

EXAMPLE 5.3 Consider the diode discussed in Example 5.2. Calculate the average field in the depletion region at the four reverse-bias values considered. Calculate the velocity of the electrons at these average fields using the velocity-field results given in Chapter 3 or Appendix B. What can be said about the change in the drift components of the diode current with the change in bias?

We will define the average field in the depletion region as $\frac{F_m}{2}$. The average field and the electron velocities are

$$\begin{cases} \frac{F_m}{2}(V_r = 0 \text{ V}) = -2.5 \times 10^4 \text{ V/cm} \\ v \cong 1.0 \times 10^7 \text{ cm/s} \end{cases}$$

$$\begin{cases} \frac{F_m}{2}(V_r = 2 \text{ V}) = -4.56 \times 10^4 \text{ V/cm} \\ v \cong 1.0 \times 10^7 \text{ cm/s} \end{cases}$$

$$\begin{cases} \frac{F_m}{2}(V_r = 5 \text{ V}) = -6.5 \times 10^4 \text{ V/cm} \\ v \cong 1 \times 10^7 \text{ cm/s} \end{cases}$$

One can see that at all these reverse biases and also at the forward bias considered the carrier density velocity remains unchanged. Thus the drift current flowing in the depletion region is not affected by the bias conditions in a *p-n* diode.

5.3.1   Charge Injection and Current Flow

We will now examine how the applied bias changes the various current components in the *p-n* diode. The presence of the bias increases or decreases the electric field in the depletion region. However, under moderate external bias, the *electric field in the depletion region is always higher than the field for carrier velocity saturation* ($F \gtrsim 10\text{kV cm}^{-1}$). *Thus the change in electric field does not alter the drift part of the electron or hole current in the depletion region. Regardless of the bias, electrons or holes that come into the depletion region are swept out and contribute to the same current independent of the field.*

The situation is quite different for the diffusion current. Remember that the diffusion current depends upon the gradient of the carrier density. As the potential profile is greatly altered by the applied bias, the carrier profile changes accordingly, greatly affecting the diffusion current. Let us evaluate the mobile carrier densities across the depletion region. Recall that at no applied bias we have the relation

$$\frac{p_p}{p_n} = e^{eV_{bi}/k_B T} \tag{5.33}$$

In the presence of the applied bias, under the assumptions of quasi-equilibrium, the same mathematics used in the equilibrium case leads to the conditions

$$\frac{p(-W_p)}{p(W_n)} = e^{e(V_{bi}-V)/k_B T} \tag{5.34}$$

We assume that the injection of mobile carriers is small (low-level injection) so that the majority carrier densities are essentially unchanged because of injection, i.e., $p(-W_p) = p_p$. Taking the ratio of Eqns. 5.33 and 5.34 we get (V has a positive value for forward bias and a negative value for reverse bias)

$$\frac{p(W_n)}{p_n} = e^{eV/k_B T} \tag{5.35}$$

where we have assumed that $p(-W_p) = p_p$. This equation suggests that the hole minority carrier density at the edge of the *n*-side depletion region can be increased dramatically if one applies a forward bias. Conversely, in reverse bias, this injection is greatly reduced. This is simply a consequence of the hole distribution being described by a Boltzmann distribution. Only those holes from the *p*-side that can overcome the potential barrier $V_{bi} - V$ can be injected over into the *n*-side, as shown in Fig. 5.6. *In forward bias, this barrier decreases so that more charge can be injected across the barrier. In reverse bias, the barrier increases so that less charge can be injected.*

A similar consideration gives, for the electrons injected as a function of applied bias,

$$\frac{n(-W_p)}{n_p} = e^{eV/k_BT} \tag{5.36}$$

The *excess carriers injected* across the depletion regions are

$$\Delta p_n = p(W_n) - p_n = p_n(e^{eV/k_BT} - 1) \tag{5.37}$$

$$\Delta n_p = n(-W_p) - n_p = n_p(e^{eV/k_BT} - 1) \tag{5.38}$$

From our discussions in Section 3.9, we know that the excess minority carriers that are introduced will decay into the majority region due to recombination with the majority carriers. The decay is simply given by the appropriate diffusion lengths (L_p for holes, L_n for electrons). Thus the carrier densities of the minority carriers outside the depletion region are (see Section 3.9)

$$
\begin{aligned}
\delta p(x) &= \Delta p_n e^{(-(x-W_n)/L_p)} \\
&= p_n \left(e^{eV/k_BT} - 1\right) e^{-(x-W_n)/L_p} \qquad x > W_n
\end{aligned} \tag{5.39}
$$

$$
\begin{aligned}
\delta n(x) &= \Delta n_p e^{(x+W_p)/L_n} \\
&= n_p \left(e^{eV/k_BT} - 1\right) e^{(x+W_p)/L_n}
\end{aligned}
\left\{
\begin{array}{l}
x < -W_p \\
x \text{ is negative} \\
W_p \text{ is positive}
\end{array}
\right. \tag{5.40}
$$

We are now in a position to calculate the current in the diode as a function of the applied bias. We emphasize once again that we are assuming that the drift current is unaffected by the bias because of the assumption that the electric field in the depletion region is always larger than the field at which velocity saturates for the electrons and the holes. *Thus for the net current we need only consider the excess electrons and holes injected as given by Eqns. 5.39 and 5.40.*

The diffusion current due to holes in the *n*-type material is

$$I_p(x) = -eAD_p \frac{d(\delta p(x))}{dx} = eA \frac{D_p}{L_p}(\delta p(x)) \qquad x > W_n \tag{5.41}$$

where we have used Eqn. 5.39 to take the derivative of the excess hole density.

The hole current injected into the *n*-side is proportional to the excess hole density at a particular point. The total hole current injected into the *n*-side is given by the current at $x = W_n$ (after using the value of $\delta p(x = W_n)$ from Eqn. 5.39)

$$I_p(W_n) = e \frac{AD_p}{L_p} p_n \left(e^{eV/k_BT} - 1\right) \tag{5.42}$$

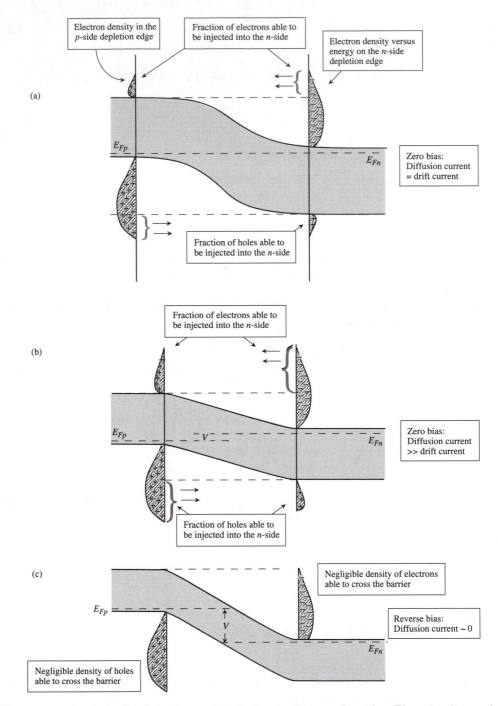

Figure 5.6: A schematic of the charge distribution in the n- and p-sides. The minority carrier injection (electrons from n-side to p-side or holes from p-side to n-side) is controlled by the applied bias as shown.

Similarly, the total electron current injected into the *p*-side region is given by

$$I_n(-W_p) = \frac{eAD_n}{L_n} \, n_p \left(e^{eV/k_BT} - 1\right) \tag{5.43}$$

We assume initially that in the ideal diode there is no recombination of the electron and hole injected currents in the depletion region. Thus the total current can be simply obtained by adding the hole current injected across W_n and electron current injected across $-W_p$. The diode current is then

$$
\begin{aligned}
I(V) &= I_p(W_n) + I_n(-W_p) \\[2mm]
&= eA\left[\frac{D_p}{L_p}p_n + \frac{D_n}{L_n}n_p\right]\left(e^{eV/k_BT} - 1\right) \\[2mm]
I(V) &= I_o\left(e^{eV/k_BT} - 1\right)
\end{aligned}
\tag{5.44}
$$

This equation, called the diode equation, gives us the current through a *p-n* junction under forward $(V > 0)$ and reverse bias $(V < 0)$. Under reverse bias, the current simply goes towards the value $-I_o$, where

$$\boxed{I_o = eA\left(\frac{D_p p_n}{L_p} + \frac{D_n n_p}{L_n}\right)} \tag{5.45}$$

Under forward bias the current increases exponentially with the applied forward bias. This strong asymmetry in the diode current is what makes the *p-n* diode attractive for many applications.

5.3.2 Minority and Majority Currents

In the calculations for the diode current we have simply calculated the excess minority carrier diffusion current. This current was evaluated at its peak value at the edges of the depletion region. However, we can also see from Eqn. 5.44 that the diffusion current decreases rapidly in the majority region because of recombination. As the holes recombine with electrons in the *n*-region, an equal number of electrons are injected into the region. These electrons provide a drift current in the *n*-side to exactly balance the hole current that is lost through recombination. The drift part of the minority carrier is negligible because of the relatively low carrier density and very small electric field in the "neutral" region. Let us consider the hole diffusion current in the *n*-type region. This current is, from Eqn. 5.41, using the value of $\delta p(x)$ from Eqn. 5.39,

$$I_p(x) = e\,A\,\frac{D_p}{L_p}\,p_n\,e^{-((x-W_n)/L_p)}\left(e^{eV/k_BT} - 1\right);\, x > W_n \tag{5.46}$$

The total current is, from Eqn. 5.44:

$$I = e \, A \left(\frac{D_p}{L_p} \, p_n + \frac{D_n}{L_n} \, n_p \right) \left(e^{eV/k_BT} - 1 \right) \tag{5.47}$$

Thus the electron current diode in the region is

$$I_n(x) \;\; = \;\; I - I_p(x) \qquad\qquad\qquad x > W_n$$

$$= \;\; eA \left[\frac{D_p}{L_p} \left(1 - e^{-(x-W_n)/L_p} \right) p_n \right.$$

$$\left. + \frac{D_n}{L_n} n_p \right] \left(e^{eV/k_BT} - 1 \right) \tag{5.48}$$

As the hole current decreases from W_n into the n-side, the electron current increases correspondingly to maintain a constant current. A similar situation exists on the p-side region. As the electron injection current decays, the hole current compensates. The general picture shown schematically in Fig. 5.7 emerges from these discussions.

We see from the discussions of this section that the current flow through the simple p-n diode has some very interesting properties. We have not the simple linear Ohm's law-type behavior, but a strongly nonlinear and rectifying behavior. The current as shown in Fig. 5.8 saturates to a value I_o given by Eqn. 5.45 when a reverse bias is applied. Since this value is quite small, the diode is essentially nonconducting. On the other hand, when a positive bias is applied, the diode current increases exponentially and the diode becomes strongly conducting. The forward bias voltage at which the diode current density becomes significant ($\sim 10^3$ Acm^{-2}) is called the cut-in voltage. This voltage is ~ 0.8V for Si diodes and ~ 1.2 V for GaAs diodes.

EXAMPLE 5.4 Consider an ideal diode model for a silicon p-n diode with $N_d = 10^{16}$ cm^{-3} and $N_a = 10^{18}$ cm^{-3}. The diode area is 10^{-3} cm^2.

The transport properties of the diode are given by the following values at 300 K:

$$n - \text{side} \begin{cases} \mu_p = 300 \text{ cm}^2 \text{ V}^{-1} \text{ s}^{-1}; & \mu_n = 1300 \text{ cm}^2 \text{ V}^{-1} \text{ s}^{-1} \\ D_p = 7.8 \text{ cm}^2 \text{ s}^{-1}; & D_n = 33 \text{ cm}^2 \text{ s}^{-1} \end{cases}$$

$$p - \text{side} \begin{cases} \mu_p = 100 \text{ cm}^2 \text{ V}^{-1} \text{ s}^{-1}; & \mu_n = 280 \text{ cm}^2 \text{ V}^{-1} \text{ s}^{-1} \\ D_p = 2.6 \text{ cm}^2 \text{ s}^{-1}; & D_n = 7.3 \text{ cm}^2 \text{ s}^{-1} \end{cases}$$

(Note that the mobility is a lot lower in the heavily doped p-side because of the increased ionized impurity scattering.) Assume that $\tau_n = \tau_p = 10^{-6}$ s. Calculate the diode current.

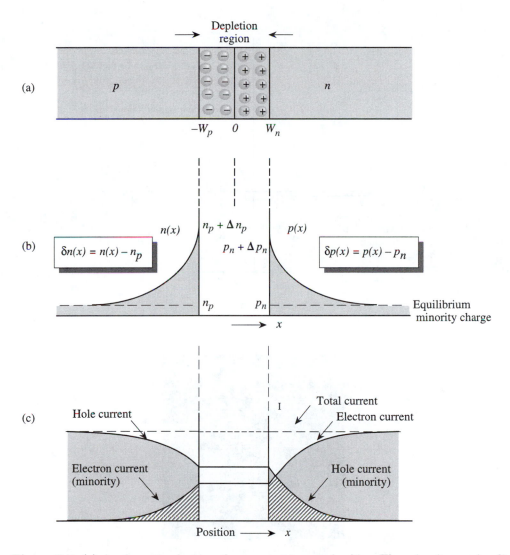

Figure 5.7: (a) A schematic showing the *p-n* structure under bias. The minority carrier distribution is shown in (b) while the form of the minority and majority current in the *n*-side is shown in (c). The minority carrier decays to essentially a zero value in a long diode. However, if the diode is narrow, the minority carrier continues to be significant up to the contact region, where it rapidly goes to zero. As the minority current (e.g., the hole current on the *n*-side) decays into the neutral region, the majority current (e.g., electron current on the *n*-side) increases so that the total current is uniform in the device.

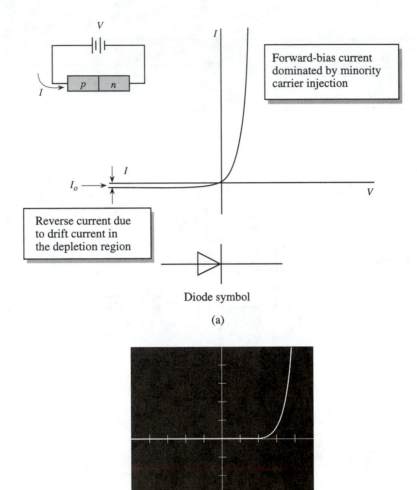

Diode symbol

(a)

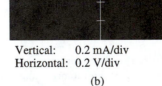

Vertical: 0.2 mA/div
Horizontal: 0.2 V/div

(b)

Figure 5.8: (a) The highly nonlinear and rectifying I-V current of the *p-n* diode. The strong nonlinear response makes the diode a very important device for a number of applications. (b) A typical experimental current-voltage relation measured for a Si diode.

We need to calculate the minority carrier diffusion length to obtain the diode current. The hole diffusion length on the n-side is

$$L_p = \sqrt{D_p \tau_p} = \left(7.8 \text{ cm}^2 \text{ s}^{-1} \times 1.0 \times 10^{-6} \text{ s}\right)^{1/2} = 2.79 \times 10^{-3} \text{ cm}$$

The electron diffusion length on the p-side is

$$L_n = \sqrt{D_n \tau_n} = \left(7.3 \times 10^{-6}\right)^{1/2} = 2.7 \times 10^{-3} \text{ cm}$$

The prefactor I_o of the current is now

$$I_o = eA \left[\frac{D_p}{L_p} p_n + \frac{D_n}{L_n} n_p\right]$$

where

$$p_n = \frac{n_i^2}{n_n}; n_p = \frac{n_i^2}{p_p}$$

Using $n_i = 1.5 \times 10^{10} \text{cm}^{-3}$ and assuming that the dopants are fully ionized, we get

$$I_o = \left(1.6 \times 10^{-19} \text{ C}\right)\left(10^{-7} \text{ m}^2\right)\left[\frac{7.8 \times 10^{-4}\,(\text{m}^2 \text{ s}^{-1})}{2.79 \times 10^{-5}\,(\text{m})} \times \left(2.25 \times 10^{10} \text{ m}^{-3}\right)\right.$$

$$\left. + \frac{7.3 \times 10^{-4}\,(\text{m}^2 \text{ s}^{-1})}{2.7 \times 10^{-5}\,(\text{m})} \times 2.25 \times 10^8 \left(\text{m}^{-3}\right)\right]$$

$$I_o = 1.0 \times 10^{-14} \text{ A}$$

Note that the second term corresponding to the electron injection into the heavily doped p-side does not contribute much to the saturation current. The current is thus dominated by the hole current (the first term) component. Thus by proper doping one can ensure that the current is carried by either electron injection or hole injection. This is exploited in the design of bipolar transistors, light-emitting diodes, as discussed in Examples 5.6 and 5.7.

The diode current is simply

$$I(V) = I_o \left[\exp\left(\frac{eV}{k_B T}\right) - 1\right]$$

EXAMPLE 5.5 We will define a p-n diode to be "turned on" when the current density reaches 10^3 A/cm^2 (this is an approximate criterion). Calculate the turn-on or *cut-in* voltage for a GaAs and a Si p-n diode with following parameters (same for both diodes):

$$N_d = N_a = 10^{17} \text{ cm}^{-3}$$
$$\tau_n = \tau_p = 10^{-8} \text{ s}$$

Use Table 3.2 for diffusion coefficients. Assume that the diodes are long and $T = 300$ K.

We use the ideal diode equation to calculate the current density. For Si the prefactor in the current equation is

$$J_0 = e\left(\frac{12.5 \text{ cm}^2/s \times 2.25 \times 10^3 \text{ cm}^{-3}}{3.5 \times 10^{-4} \text{ cm}} + \frac{35 \text{ cm}^2/s \times 2.25 \times 10^3 \text{ cm}^{-3}}{5.91 \times 10^{-4} \text{ cm}}\right)$$

$$= 3.418 \times 10^{-11} \text{ A/cm}^2$$

The forward bias needed to have a current density of 10^3 A/cm^2 at room temperature is then 0.8 V.

For GaAs the prefactor is similarly

$$J_0 = e\left(\frac{10 \text{ cm}^2/\text{s} \times 3.38 \times 10^{-5} \text{ cm}^{-3}}{3.16 \times 10^{-4} \text{ cm}} + \frac{220 \text{ cm}^2/\text{s} \times 3.38 \times 10^{-5} \text{ cm}^{-3}}{14.83 \times 10^{-4} \text{ cm}}\right)$$

$$= 9.73 \times 10^{-19} \text{ A/cm}^2$$

In this case the forward bias needed is 1.26 V.

EXAMPLE 5.6 An important use of a forward-biased p-n diode is as an emitter in a bipolar transistor (discussed in Chapter 7). In the emitter it is desirable that the current be injected via only one kind of charge. The diode efficiency is thus defined as (J_n is the current density carried by electron injection into the p-side)

$$\gamma_{inj} = \frac{J_n}{J_{Tot}}$$

Consider a GaAs p-n diode with the following parameters:

Electron diffusion coefficient,	D_n =	30 cm^2/s
Hole diffusion coefficient,	D_p =	15 cm^2/s
p-side doping,	N_a =	5×10^{16} cm^{-3}
n-side doping,	N_d =	5×10^{17} cm^{-3}
Electron minority carrier time,	τ_n =	10^{-8} s
Hole minority carrier time,	τ_p =	10^{-7} s

Calculate the diode injection efficiency (this is called the emitter efficiency in a bipolar transistor).

The minority charge density is

$$n_p = \frac{n_i^2}{N_a} = \frac{(1.84 \times 10^6 \text{ cm}^{-3})^2}{5 \times 10^{16} \text{ cm}^{-3}} = 6.77 \times 10^{-5} \text{ cm}^{-3}$$

$$p_n = \frac{n_i^2}{N_d} = \frac{(1.84 \times 10^6 \text{ cm}^{-3})^2}{5 \times 10^{17} \text{ cm}^{-3}} = 6.77 \times 10^{-6} \text{ cm}^{-3}$$

The diffusion lengths are

$$L_n = \sqrt{D_n \tau_n} = \left[(30 \text{ cm}^2 \text{ s}^{-1})(10^{-8} s)\right]^{1/2} = 5.47 \ \mu\text{m}$$

$$L_p = \sqrt{D_p \tau_p} = \left[(15)(10^{-7})\right]^{1/2} = 12.25 \ \mu\text{m}$$

The efficiency becomes

$$\gamma_{inj} = \frac{\frac{e D_n n_p}{L_n}}{\frac{e D_n n_p}{L_n} + \frac{e D_p p_n}{L_p}} = 0.98$$

As can be seen, the efficiency increases by making N_d larger than N_a.

EXAMPLE 5.7 A forward-biased pn^+ GaAs diode is used as an LED at 300 K. Assume that all the electrons injected into the top p region combine with holes to generate photons.

Calculate the photon generation rate and optical power coming from the LED if the device has the parameters of the diode in Example 5.6, and is forward-biased at 1.0 V. The diode area is 0.1 mm^2.

We will assume that all the electrons that are injected into the upper (closer to the surface) p region combine with the holes to produce photons that have an energy equal to the bandgap of GaAs (1.43 eV).

The electron current is (using results from the previous example)

$$I_n = \frac{eAD_n n_p}{L_n}\left[\exp\left(\frac{eV_f}{k_BT}\right) - 1\right]$$
$$= \frac{(1.6\times 10^{-19}\text{ C})(0.1\times 10^{-2}\text{ cm}^2)(30\text{ cm}^2/\text{s})(6.77\times 10^{-5}\text{ cm}^{-3})}{(5.47\times 10^{-4}\text{ cm})}\left[\exp\left(\frac{1}{0.026}\right) - 1\right]$$
$$= 0.03\text{ mA}$$

Each electron generates a photon so that the photons emitted per second are (remember that the current divided by charge gives the number of electrons passing per second)

$$\frac{I_n}{e} = \frac{(0.03\times 10^{-3}\text{ A})}{1.6\times 10^{-19}\text{ C}} = 1.9\times 10^{14}\text{ s}^{-1}$$

Each photon has an energy of 1.43 eV and the optical power is

$$\text{Power} = (1.9\times 10^{14}\text{ s}^{-1})(1.43\text{ eV})(1.6\times 10^{-19}\text{ J/eV})$$
$$= 4.3\times 10^{-5}\text{ W}$$

It may be noted that in real LEDs not all the photons emitted are able to emerge from the diode. A significant fraction are reflected back from the GaAs-air interface and are thus lost.

EXAMPLE 5.8 An important application of p-n (or p-i-n) diodes is their use as detectors of optical radiation. The optical signal creates electron-hole pairs that are collected as a current if the pairs are in the depletion region and regions within the diffusion lengths of the depletion region. Consider the following parameters of a silicon p-n diode at 300 K:

Diode area,	A	$= 10^4\ \mu\text{m}^2$
p-side doping,	N_a	$= 2\times 10^{16}\text{ cm}^{-3}$
n-side doping	N_d	$= 10^{16}\text{ cm}^{-3}$
Applied reverse bias,	V_r	$= 15$ V
Electron diffusion constant,	D_n	$= 20\text{ cm}^2/\text{s}$
Hole diffusion constant,	D_p	$= 12\text{ cm}^2/\text{s}$
Electron minority carrier lifetime,	τ_n	$= 10^{-8}$ s
Hole minority carrier lifetime,	τ_p	$= 10^{-8}$ s

Calculate the photocurrent of the device if the e-h pairs are generated from an optical signal at a rate $G_L = 10^{22}\text{ cm}^{-3}\text{s}^{-1}$ and the photocurrent is $eAG_L(W + L_n + L_p)$.

We need to calculate W, L_n, and L_p for the problem. The electron diffusion length is

$$L_n = \sqrt{D_n\tau_n} = \left[(20\text{ cm}^2/\text{s})(10^{-8}\text{ s})\right]^{1/2} = 4.5\ \mu\text{m}$$

The hole diffusion length is

$$L_p = \sqrt{D_p \tau_p} = \left[(12)(10^{-8})\right]^{1/2} = 3.4 \ \mu m$$

To calculate the depletion width we need to calculate the built-in voltage V_{bi} as

$$V_{bi} = \frac{k_B T}{e} \ ln \left(\frac{N_a N_d}{n_i^2} \right) = 0.026 \ ln \ \left(\frac{(2 \times 10^{16})(10^{16})}{(1.5 \times 10^{10})^2} \right) = 0.715 \ V$$

The depletion width is

$$
\begin{aligned}
W &= \left\{ \frac{2\epsilon}{e} \left(\frac{N_a + N_d}{N_a N_d} \right) (V_{bi} + V_R) \right\}^{1/2} \\
&= \left\{ \frac{2(11.9)(8.85 \times 10^{-14})}{(1.6 \times 10^{-19})} \left(\frac{(2 \times 10^{16} + 10^{16})}{(2 \times 10^{16})(10^{16})} \right) (15.715) \right\}^{1/2} = 1.75 \ \mu m
\end{aligned}
$$

The photocurrent becomes

$$
\begin{aligned}
I_L &= eAG_L(W + L_n + L_p) = (1.6 \times 10^{-19} \ C)(10^4 \times 10^{-8} \ cm^2)(10^{22} \ cm^{-3}s^{-1}) \\
&\quad (1.75 \times 10^{-4} \ cm + 4.5 \times 10^{-4} \ cm + 3.46 \times 10^{-4} \ cm) \\
&= 0.155 \ mA
\end{aligned}
$$

From the above example it can be seen that the photocurrent can be increased by increasing the depletion width.

5.3.3   The Case of a Narrow Diode

In the discussions so far we have assumed that the neutral sections of the diode are much wider than the diffusion lengths of the electrons and holes. In many devices, especially in the bipolar transistors to be discussed in the next chapter, one side of the diode is narrower than the minority carrier diffusion length. In this case we cannot assume that the injected excess minority carrier density will simply decay exponentially as $\exp\{-(x - W_n)/L_p\}$ (for holes). In fact, for the narrow diode one has to consider the *ohmic boundary conditions where at the contacts the minority carrier density goes to zero.*

Let us consider the geometry shown in Fig. 5.9 where the diode extends a distance $W_{\ell n}$ and $W_{\ell p}$ as shown in the n- and p-sides. We still assume that there is no recombination in the depletion region, i.e., $W_n \gg L_p$ and $W_p \gg L_n$. In Example 3.15 of Section 3.9, we discussed how the carrier concentration varies if the regions $|W_{\ell n} - W_n|$ and $|W_{\ell p} - W_p|$ are much smaller than the diffusion lengths L_p and L_n, respectively. We showed there that the injected minority current goes from its value at the depletion edge towards zero at the contact in a linear manner. The hole current injected across

W_n becomes (note that $\delta p_n(W_{\ell n}) = 0$)

$$
\begin{aligned}
I_p(W_n) &= -eAD_p \frac{d(\delta p(x))}{dx} = -eAD_p \left[\frac{\delta p_n(W_n) - \delta p_n(W_{\ell n})}{W_{\ell n} - W_n} \right] \\
&= \frac{-eAD_p}{W_{\ell n} - W_n} p_n \left[\exp\left(\frac{eV}{k_B T} \right) - 1 \right]
\end{aligned}
\tag{5.49}
$$

A similar expression results in this linear approximation for the electron current. The net effect is that the prefactor of the diode current changes (i.e., the term L_n or L_p in the denominator is replaced by a smaller term $(W_{\ell n} - W_n)$ or $(|W_{\ell p} - W_p|)$). The prefactor becomes

$$
\boxed{ I_o = eA \left(\frac{D_p p_n}{(W_{\ell n} - W_n)} + \frac{D_n n_p}{(|W_{\ell p} - W_p|)} \right) }
\tag{5.50}
$$

The narrow diode therefore has a higher saturation current than a long diode. The advantage of the narrow diode lies in its superior time-dependent response—a topic we will consider later.

EXAMPLE 5.9 Consider the p-n diode in Example 5.4. In that example we examined the prefactor of the diode current using the long diode conditions. Calculate the prefactor for the case of a short diode in which both the n- and p-side widths are 5.0 μm.

From Example 5.4 we saw that the diffusion lengths for the carriers are

$$
\begin{aligned}
L_p &= 27.9 \ \mu\text{m} \\
L_n &= 27.0 \ \mu\text{m}
\end{aligned}
$$

The depletion widths for a p-n diode were calculated in Example 5.2. The zero-bias depletion widths are

$$
\begin{aligned}
W_p &= 32 \ \text{Å} \\
W_n &= 0.32 \ \mu\text{m}
\end{aligned}
$$

For a value of $W_{\ell n} = W_{\ell p} = 5.0 \ \mu$m, the prefactor is found to increase by a factor of 5.6. Thus, as the diode gets narrower, the prefactor starts to increase.

5.4   REAL DIODES: CONSEQUENCES OF DEFECTS

In the calculations above we assumed that the semiconductor is perfect, i.e., there are no defects and associated bandgap states that may lead to trapping, recombination, or generation terms. In Section 3.8 we discussed the effects of bandgap states produced by defects. In our analysis of the ideal diode, we have assumed that the electrons and holes injected across the depletion region barrier are not able to recombine with each other.

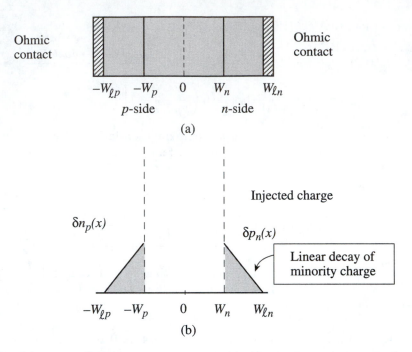

Figure 5.9: (a) A schematic of the narrow p-n diode with ohmic contacts at the boundaries. The effect of the ohmic contacts is to force the minority carrier density to go to zero at the boundary, by causing the minority particle to recombine "instantly" with the majority carrier. (b) The injected charge distribution. A reasonable approximation to the narrow diode problem is that the injected charge linearly falls to zero between W_n and $W_{\ell n}$ or $-W_p$ and $-W_{\ell p}$.

Only when they enter the neutral regions are they able to recombine with the majority carriers. This recombination in the neutral region is described via the diffusion lengths L_n and L_p that appear in the expression for I_o.

In a real diode, a number of sources may lead to bandgap states. The states may arise if the material quality is not very pure so that there are chemical impurities present. The doping process itself can cause defects such as vacancies, interstitials, etc. Let us assume that the density of such deep level states is N_t. We will assume that the deep level is at the center of the bandgap.

In Section 3.8 we addressed this problem and showed how the Shockley–Hall–Read recombination comes about. The electron-hole recombination rate per unit volume is given by (see Eqn. 3.50; we denote τ_{nr} of Eqn. 3.50 by τ)

$$R_t = \frac{np}{\tau(n + p)} \tag{5.51}$$

where n and p are the electron-hole concentrations and the recombination time τ is given by (see Eqn. 3.48)

$$\tau = \frac{1}{N_t v_{th} \sigma} \tag{5.52}$$

where v_{th} is the thermal velocity of the electron (assumed the same for the hole) and σ is the capture cross-section of the trap for the electron or hole. As electrons and holes enter the depletion region, one possible way they can cross the region without overcoming the potential barrier is to recombine with each other. This leads to an additional flow of charged particles. This current, called the generation-recombination current, must be added to the current calculated so far.

Note that at the edges of the depletion width we have from Eqns. 5.35 and 5.36 (using the relation $n_n p_n = n_i^2 = p_p n_p$)

$$n(-W_p)p_p(-W_p) = n_i^2 \, \exp \, \left(\frac{eV}{k_B T} \right) = p(W_n)n_n(W_n) \tag{5.53}$$

We assume the np product remains constant in the depletion region as well. This can be shown to be the case by detailed numerical calculations. We thus get (using $n \cong p$ for the maximum recombination rate) from Eqns. 5.51 and 5.53

$$R_t \cong \frac{n}{2\tau} \cong \frac{n_i}{2\tau} \, \exp \, \left(\frac{eV}{2k_B T} \right) \tag{5.54}$$

The recombination current is now simply (current is equal to charge times volume times rate)

$$
\begin{aligned}
I_R \; &= \; eAWR_t = \frac{eAWn_i}{2\tau} \, \exp \, \left(\frac{eV}{2k_B T} \right) \\[2mm]
&= \; I_{GR}^o \, \exp \, \left(\frac{eV}{2k_B T} \right)
\end{aligned}
\tag{5.55}
$$

where W is the depletion width. *At zero applied bias, a generation current of I_G balances out the recombination current.*

The generation-recombination current has an exponential dependence on the voltage as well, but the exponent is different. The generation-recombination current is

$$
\begin{aligned}
I_{GR} \; &= \; I_R - I_G = I_R - I_R(V=0) \\[2mm]
&= \; I_{GR}^o \, \left[\exp \, \left(\frac{eV}{2k_B T} \right) - 1 \right]
\end{aligned}
\tag{5.56}
$$

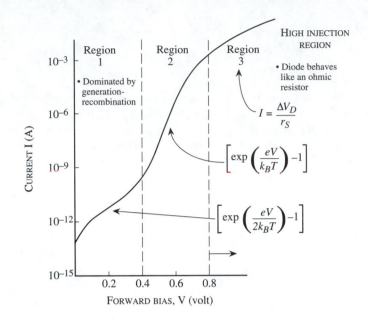

Figure 5.10: A log plot of the diode current versus forward bias. At low biases, the recombination effects are quite pronounced, while at higher biases the slope becomes closer to unity. At still higher biases the behavior becomes more ohmic, as discussed in Section 5.5.1.

The total device current now becomes

$$I \;=\; I_o \left[\exp\left(\frac{eV}{k_B T} \right) - 1 \right] + I_{GR}^o \left[\exp\left(\frac{eV}{2k_B T} \right) - 1 \right]$$

or

$$I \;\cong\; I_S \left[\exp\left(\frac{eV}{n k_B T} \right) - 1 \right] \tag{5.57}$$

where n is called the *diode ideality factor*. The prefactor I_{GR}^o can be much larger than I_o for real devices. Thus at low applied voltages the diode current is often dominated by the second term. However, as the applied bias increases, the diffusion current starts to dominate. We thus have two regions in the forward I-V characteristics of the diode, as shown in Fig. 5.10.

EXAMPLE 5.10   Consider the *p-n* diodes examined in Example 5.4. In that example, the diode prefactor was calculated assuming that there is no recombination in the depletion region. Calculate the effect of the generation-recombination current assuming a lifetime of 10^{-6} s.

The prefactor of the generation-recombination current is

$$I_{GR}^o = \frac{e A W n_i}{2\tau}$$

At zero applied bias, we know from the solution to Example 5.2 that $W = 0.32\ \mu$m. This gives

$$I_{GR}^o \;=\; \frac{\left(1.6 \times 10^{-19}\ \text{C}\right)\left(10^{-7}\ \text{m}^2\right)\left(0.32 \times 10^{-6}\ \text{m}\right)\left(1.5 \times 10^{16}\ \text{m}^{-3}\right)}{2\left(10^{-6}\ \text{s}\right)}$$

$$=\; 3.84 \times 10^{-11}\ \text{A}$$

and

$$I_{GR} \;=\; I_{GR}^o \left[\exp\left(\frac{eV}{2k_B T}\right) - 1\right]$$

We can see that the generation-recombination prefactor is much larger than the prefactor due to the diffusion term. Thus the reverse current will be dominated by the generation-recombination effects.

In forward bias, the diffusion current is initially much smaller than the generation-recombination term. However, at higher forward bias the diffusion current will start to dominate. For example, using the results obtained in Example 5.4, we see that at a forward bias of 0.2 V, the diffusion current is 2×10^{-11} A, while the generation-recombination current is 1.8×10^{-9} A. At a forward bias of 0.6 V, the diffusion current is 9.9×10^{-5} A, while the generation-recombination current is 3.9×10^{-6} A.

EXAMPLE 5.11 Consider a Si long *p-n* diode with the following parameters:

$$
\begin{aligned}
n\text{-side doping} &= 10^{17}\ \text{cm}^{-3}\\
p\text{-side doping} &= 10^{17}\ \text{cm}^{-3}\\
\text{Minority carrier lifetime } \tau_n &= \tau_p = 10^{-7}\ \text{s}\\
\text{Electron diffusion constant} &= 30\ \text{cm}^2/\text{s}\\
\text{Hole diffusion constant} &= 10\ \text{cm}^2/\text{s}\\
\text{Diode area} &= 10^{-4}\ \text{cm}^2\\
\text{Carrier lifetime in the depletion region} &= 10^{-8}\ \text{s}
\end{aligned}
$$

Calculate the diode current at a forward bias of 0.5 V and 0.6 V at 300 K. What is the ideality factor of the diode in this range?

For this diode structure we have the following:

$$
\begin{aligned}
n_p &= 2.25 \times 10^3\ \text{cm}^{-3}\\
p_n &= 2.25 \times 10^3\ \text{cm}^{-3}\\
L_n &= 17.32\ \mu\text{m}\\
L_p &= 10.0\ \mu\text{m}\\
V_{bi} &= 0.817\text{V}
\end{aligned}
$$

The prefactor in the ideal diode equation is

$$I_0 \;=\; e\left(\frac{D_p p_n}{L_p} + \frac{D_n n_p}{L_n}\right)$$

$$=\; 9.83 \times 10^{-16}$$

The prefactor to the recombination-generation current is

$$I_{GR}^0 = \frac{eAWn_i}{2\tau}$$

where τ is the lifetime in the depletion region.

The depletion width at a forward bias of 0.5 V is found to be

$$W(0.5 \text{ V}) = 9.2 \times 10^{-6} \text{ cm}$$

The depletion width at 0.6 V forward bias is

$$W(0.6 \text{ V}) = 7.6 \times 10^{-6} \text{ cm}$$

The prefactors to the recombination-generation current is

$$I_{GR}^0(0.5 \text{ V}) = 1.1 \times 10^{-10} \text{ A}$$

$$I_{GR}^0(0.6 \text{ V}) = 9.1 \times 10^{-11} \text{ A}$$

The current is now

$$
\begin{aligned}
I(0.5 \text{ V}) &= 9.83 \times 10^{-16} e^{\frac{0.5}{0.026}} + 1.1 \times 10^{-10} e^{\frac{0.5}{0.052}} \\
&= 2.2 \times 10^{-7} + 1.66 \times 10^{-6} \text{ A} \\
&= 1.88 \times 10^{-6} \text{ A}
\end{aligned}
$$

and

$$
\begin{aligned}
I(0.6 \text{ V}) &= 9.83 \times 10^{-16} e^{\frac{0.6}{0.026}} + 9.1 \times 10^{-11} e^{\frac{0.6}{0.052}} \\
&= 1.034 \times 10^{-5} + 1.1 \times 10^{-5} \text{ A} \\
&= 2.13 \times 10^{-5} \text{ A}
\end{aligned}
$$

We can write the diode current as

$$I = I_S \exp\left(\frac{eV}{nk_BT}\right)$$

Thus

$$\frac{I(V_2)}{I(V_2)} = \exp\left(\frac{e(V_2 - V_1)}{nk_BT}\right)$$

Using this relation, we find that

$$n = 1.6$$

5.5   HIGH-VOLTAGE EFFECTS IN DIODES

In the derivations discussed so far for the diode properties, we have made several important assumptions. Some have been made explicitly while some are implicit in the equations we have used for current as a function of electric field. Many of these assumptions break down at large applied voltages.

5.5.1 Forward Bias: High Injection Region

One of our assumptions was that the injection of minority carriers was at a fairly low level, so that essentially no voltage dropped across the bulk of the structure. All the voltage was assumed to be dropping over the depletion region. However, as the forward bias is increased, the injection level increases and eventually the injected minority carrier density becomes comparable to the majority carrier density. When this happens, an increasingly larger fraction of the external bias drops across the undepleted region. The diode current will then stop growing exponentially with the applied voltage, but will tend to saturate, as shown as Region 3 in Fig. 5.10. The minority carriers that are injected move not only under diffusion effects, but also under the electric field that is now present in the undepleted region. As the forward bias increases, the devices start to have more of an ohmic behavior, where the current-voltage relation is given by a simple linear expression. The current is now controlled by the resistance of the n- and p-type regions as well as the contact resistance. It must be noted, however, that at the high current densities involved, the device may heat and suffer burnout.

5.5.2 Reverse Bias: Impact Ionization

In Chapter 3 (Section 3.4) we discussed the transport of carriers under very high electric fields and the impact ionization phenomena. Under very high electric fields, the electron (for example) acquires so much energy that it can scatter from an electron in the valence band, knocking it into the conduction band. The final result is that instead of one initial electron to carry current, we have two electrons in the conduction band and one hole in the valence band. This results in current multiplication and the initial current reverse bias I_o becomes

$$I_o' = M(V)I_o \qquad (5.58)$$

where M is a factor that depends upon the impact ionization rate.

In Chapter 3 we discussed the impact ionization coefficients α_{imp} and β_{imp} for the electrons and holes. These coefficients depend upon the material bandgap and the applied electric field. One can define a critical electric field F_{crit} for a given semiconductor at which the impact ionization rate becomes large. At F_{crit}, α_{imp} or β_{imp} approaches $\sim 10^4$ cm^{-1}; values of F_{crit} for different materials are given in Appendix B. The value of 10^4 cm^{-1} is chosen so that an impact ionization occurs over a micron of distance, which represents a typical device dimension of modern devices.

In the previous sections we discussed the maximum electric field in a p-n junction as a function of applied bias. Once the applied bias becomes so large that $F_m \cong F_{crit}$, the impact ionization process starts to become dominant and the current shows a runaway behavior. If we consider a one-sided abrupt p^+n junction, the depletion width is essentially in the n-side and the values of F_m and W are, from Eqns. 5.30 and

5.29,

$$F_m = \frac{eN_dW}{\epsilon}; \; W \cong \left[\frac{2\epsilon V_r}{eN_d}\right]^{1/2} \tag{5.59}$$

where we have neglected the built-in field V_{bi} for the depletion width. The breakdown field V_{BD} is given by the reverse bias at which F_m becomes F_{crit}. This gives from Eqn. 5.29

$$\boxed{V_{BD} = \frac{\epsilon F_{crit}^2}{2eN_d}} \tag{5.60}$$

The breakdown fields for some semiconductors are given in Chapter 3.

5.5.3 Reverse Bias: Zener Breakdown

We will now discuss another source of breakdown at high reverse voltages. This breakdown mechanism is due to the quantum-mechanical process of tunneling. The tunneling process, discussed in Section 3.3.2, allows electrons in the valence band to tunnel into the conduction band and vice versa. As the electric field is increased (reverse-biased), the effective barrier that an electron in the valence band has to overcome to go to the conduction band starts to decrease. Once this tunneling probability becomes significant, there are so many free carriers that the diode effectively shorts out.

In Fig. 5.11 we show the band profile in a reverse-biased *p-n* junction. The diode is heavily doped so that the Fermi level on the *n*-side and the Fermi level on the *p*-side are in the conduction and valence bands, respectively. The heavy doping ensures that electrons in the conduction band can tunnel into "available" empty states in the valence band. We can see that electrons in the conduction band are aligned in energy with the states in the valence band. However, a typical electron sees a potential barrier between points x_2 and x_1, as shown in Fig. 5.11b. The tunneling probability is given under such conditions by the relation (see Eqn. 3.13)

$$T \approx \exp\left(-\frac{4\sqrt{2m^*}E_g^{3/2}}{3e\hbar F}\right) \tag{5.61}$$

where E_g is the bandgap of the semiconductor, m^* is the reduced mass of the electron-hole system, and F is the field in the depletion region.

Zener tunneling, as this phenomenon is called, depends upon the depletion greater as well as the applied bias. The greater the width, the lower the tunneling. The depletion width can be controlled by the doping density. If the junction is made from heavily doped materials, the Zener tunneling can start at a reverse bias of V_z, which could be as low as a few tenths of a volt. The voltage across the junction is then clamped at V_z, and the current is controlled by the external circuit as shown in Fig.

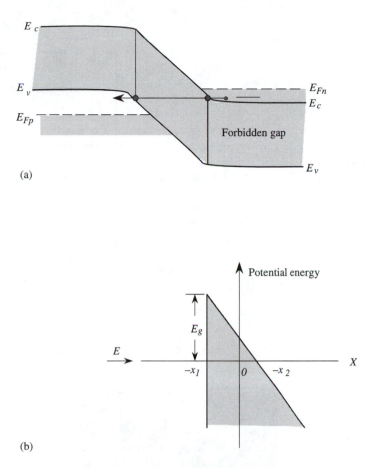

(a)

(b)

Figure 5.11: (a) A schematic showing the band diagram for a reverse-biased p-n junction. An electron in the valence band can tunnel into an unoccupied state in the conduction band or vice versa. (b) The potential profile seen by the electron during the tunneling process.

5.12. This clamping property provides a very useful application for the Zener diodes. If V_z is breakdown voltage (due to impact ionization or Zener breakdown), the current for reverse bias voltages greater than V_z is

$$I = \frac{|V - V_z|}{R_L} \tag{5.62}$$

In this section we have discussed two important sources for the reverse-biased diode to break down. This breakdown does not have to be catastrophic for the device if the external circuit is properly designed so that the current flow is not excessive. In a particular diode, which of the two processes will dominate the breakdown will be determined by the device design, which includes the diode width, doping levels, and, of course, the semiconductor material being used.

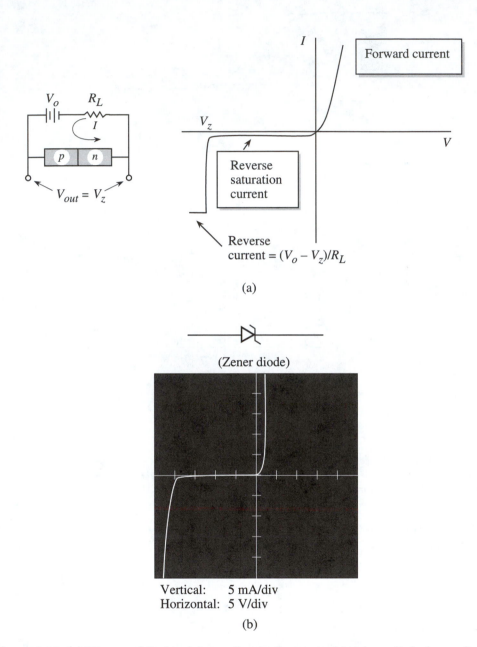

(a)

(Zener diode)

Vertical: 5 mA/div
Horizontal: 5 V/div

(b)

Figure 5.12: (a) The use of the breakdown effect in the reverse-biased *p-n* diode for a voltage-clamping circuit. The current saturates to a value determined by the external circuit resistor, while the output voltage is clamped at the diode breakdown voltage. The circuit is thus very useful as a voltage regulator. (b) A typical measurement for a Zener diode.

EXAMPLE 5.12 A silicon p^+n diode has a doping of $N_a = 10^{19}$ cm^{-3}, $N_d = 10^{16}$ cm^{-3}. Calculate the 300 K breakdown voltage of this diode. If a diode with the same ϵ/N_d value were to be made from diamond, calculate the breakdown voltage.

The critical fields of silicon and diamond are (at a doping of 10^{16} cm^{-3}) $\sim 4 \times 10^5$ V/cm and 10^7 V/cm. The breakdown field is

$$V_{BD}(Si) = \frac{\epsilon(F_{crit})^2}{2eN_d} = \frac{(11.9)(8.85 \times 10^{-14} \text{F/cm})(4 \times 10^5 \text{ V/cm})^2}{2(1.6 \times 10^{-19} \text{ C})(10^{16} \text{ cm}^{-3})} = 51.7 \text{ V}$$

The breakdown for diamond is

$$V_{BD}(C) = 51.7 \times \left(\frac{10^7}{4 \times 10^5}\right)^2 = 32.3 \text{ kV!}$$

One can see the tremendous potential of diamond for high-power applications where the device must operate under high applied potentials. At present, however, diamond-based diodes are not commercially available.

5.6 MODULATION AND SWITCHING: AC RESPONSE

So far in this chapter we have discussed the dc characteristics of the *p-n* diode. It is important to note that many applications of diodes will involve transient or ac properties of the diode. The transient properties of the diode are usually not very appealing, especially for high-speed applications. This is one of that reasons that diodes have been replaced by transistors and Schottky diodes (to be discussed later) in many applications.

The *p-n* diode is a minority carrier device, i.e., it involves injection of electrons into a *p*-type region and holes into an *n*-side region. In forward-bias conditions where the diode is in a conducting state, the current is due to the minority charge injection. In Fig. 5.13, we show the minority charge (hole) distribution in the *n*-side of a forward-biased *p-n* diode. If this diode is to be switched, this excess charge must be removed. The device time response, therefore, depends upon how fast one can alter the minority charge that has been injected. In Fig. 5.13b we show how the minority charge can be extracted. As noted in this figure, one can speed up the process either by introducing defects that speed up the recombination or by using very narrow diodes. Both these approaches have problems. A high defect density causes non-ideal diode behavior and a narrow diode has a large reverse-bias current.

Minority charge extraction is a problem for forward-biased diodes. For the reverse-biased case, where no minority charge injection occurs, the device speed can be quite high and is dominated by the device RC time constant. Let us examine the response of the *p-n* diode to large and small signals.

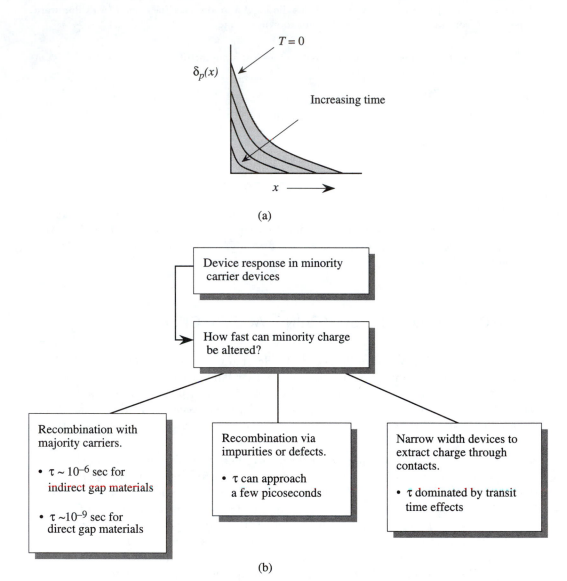

Figure 5.13: (a) The minority hole distribution in a forward-biased *p-n* diode. If the diode is to be switched to a reverse-biased state, the excess holes have to be extracted. This time is controlled by the minority carrier lifetime. (b) A schematic of what controls device response of minority-carrier-based devices. Three approaches used to speed up the device response are described.

5.6.1 Small-Signal Equivalent Circuit of a Diode

In order to analyze and design circuits using *p-n* diodes, it is essential to extract information on the diode small-signal capacitance and resistance. The diode capacitance arises from two distinct regions of charge: i) The junction capacitance arises from the depletion region where there is a dipole of fixed positive and negative charge; and ii) The diffusion capacitance is due to the region outside the depletion region where minority carrier injection has introduced charges. Under reverse-biased conditions, there are essentially no injected carriers and the junction capacitance dominates. The diffusion capacitance due to injected carriers dominates under forward-bias conditions. The capacitance is in general defined by the relation

$$C = \left| \frac{dQ}{dV} \right|$$

The junction depletion width of the diode is

$$W = \left[\frac{2\epsilon(V_{bi} - V)}{e} \left(\frac{N_a + N_d}{N_a N_d} \right) \right]^{1/2} \tag{5.63}$$

The depletion region charge is

$$|Q| = eA \, W_n N_d = eA \, W_p N_a \tag{5.64}$$

where we showed earlier (see Eqns. 5.27 through 5.29)

$$W_n = \frac{N_a}{N_a + N_d} W; \; W_p = \frac{N_d}{N_a + N_d} W \tag{5.65}$$

Thus

$$|Q| = \frac{eA \, N_a N_d}{N_d + N_a} W = A \left[2e\epsilon(V_{bi} - V) \frac{N_d N_a}{N_d + N_a} \right]^{1/2} \tag{5.66}$$

The junction capacitance is then

$$\begin{aligned} C_j &= \left| \frac{dQ}{dV} \right| = \frac{A}{2} \left[\frac{2e\epsilon}{(V_{bi} - V)} \frac{N_a N_d}{N_a + N_d} \right]^{1/2} \\ &= \frac{A\epsilon}{W} = \frac{C_{jo}}{(1 - \frac{V}{V_{bi}})^{1/2}} \end{aligned} \tag{5.67}$$

where C_{jo} is the capacitance at zero applied bias. It is important to note that the capacitance is dependent upon the applied voltage. This voltage-dependent diode capacitor is called a varactor and is very useful in numerous applications. For example, one can tune the frequency of a resonant cavity electronically by changing the capacitance.

In real diodes, the doping in the n-side and p-side usually does not switch abruptly. In such cases, the depletion capacitance of the diode is written as

$$C_j = \frac{C_{jo}}{\left(1 - \frac{V}{V_{bi}}\right)^m} \tag{5.68}$$

where C_{jo} is the capacitance when $V = 0$ and m is a parameter called the grading parameter. For abrupt junctions, $m = 1/2$ as can be seen in Eqn. 5.67. For linearly graded junctions (see Problem 5.15), $m = 1/3$.

For the forward-biased diode, the injected charge density is quite large and can dominate the capacitance. The injected hole charge is (see Eqn. 5.42 for the forward bias hole current; remember that charge is $I\tau_p$ and use $\tau_p D_p = L_p^2$; we also ignore 1 in the forward-bias state)

$$Q_p = I\tau_p = eA\ L_p p_n\ e^{eV/k_B T} \tag{5.69}$$

The corresponding capacitance is

$$C_{diff} = \frac{dQ_p}{dV} = \frac{e^2}{k_B T} A\ L_p p_n\ e^{eV/k_B T} = \frac{e}{k_B T} I\tau_p \tag{5.70}$$

For small-signal ac response, one can define the ac conductance of the diode as

$$\boxed{G_s = \frac{dI}{dV} = \frac{e}{k_B T} I(V)} \tag{5.71}$$

from the definition of the $I(V)$ function. At room temperature the conductance is (r_s is the diode resistance)

$$\boxed{G_s = \frac{1}{r_s} = \frac{I(\text{mA})}{25.86} \Omega^{-1}} \tag{5.72}$$

Consider now a p-n diode that is forward-biased at some voltage V_{dc}, as shown in Fig. 5.14a. If an ac signal is now applied to the diode, the current output changes as shown schematically. We are interested in the ac impedance presented by the diode to the small signal excitation.

The equivalent circuit of the diode is shown in Fig. 5.14b and consists of the diode resistance r_s ($= G_s^{-1}$), the junction capacitance, and the diffusion capacitance. In the forward-bias condition, the diffusion capacitance will dominate and we have the following relation between the current i_s and the applied voltage signal v_s:

$$i_s = G_s v_s + C_{diff} \frac{dv_s}{dt} \tag{5.73}$$

If we assume an input voltage with frequency ω ($v_s \sim v_s^o \exp(j\omega t)$), we get

$$i_s = G_s v_s + j\omega C_{diff} v_s \tag{5.74}$$

and the admittance of the diode becomes

$$y = \frac{i_s}{v_s} = G_s + j\omega C_{diff} \tag{5.75}$$

It is important to note here that the diffusion capacitance we calculated (in Eqn. 5.70) was based on the *total minority charge injected across the junction*. In a small signal response, *not all of the minority charge is modulated through the junction*. Some of the charge simply recombines in the neutral region. Thus the real diffusion capacitance of the small-signal description is

$$\boxed{C_{diff} = K \frac{e}{k_B T} I \tau_p} \tag{5.76}$$

where K is a factor which is 1/2 for long base diodes and 2/3 for narrow base devices.

Using a value of $K = 1/2$, we get the admittance

$$\boxed{y = \frac{eI}{k_B T} + \frac{j\omega e I \tau_p}{2 k_B T}} \tag{5.77}$$

In Fig. 5.14b we show the equivalent circuit of a packaged diode where we have the additional series resistance R_s associated with the diode n- and p-type neutral regions and a capacitance C_p associated with the diode packaging. As discussed, at forward bias the diffusion capacitance dominates, while at reverse bias the junction capacitance is dominant.

EXAMPLE 5.13 The $\frac{1}{C^2}$ versus applied voltage relation in a silicon $p^+ - n - n^+$ junction diode is measured to have a form shown in Fig. 5.15. Calculate the thickness of the n-region, the built-in voltage, and the N_a and N_d concentrations in the p^+ and n regions. The diode area is 10^{-3} cm^2. Also calculate the width of the n-region.

The junction capacitance is related to the applied bias by the equation

$$\frac{1}{C^2} = \frac{2(V_{bi} - V)}{A^2 e \epsilon N_{eff}}; N_{eff} = \frac{N_a N_d}{N_a + N_d}$$

The intercept of the $\frac{1}{C^2}$ vs. V plot on the voltage axis is 0.68 V, which is the built-in voltage V_{bi}.

The slope of the $\frac{1}{C^2}$ vs. V plot gives

$$\frac{d\left(\frac{1}{C^2}\right)}{dV} = 2.1 \times 10^{23} \, F^{-2} V^{-1} = \frac{2}{A^2 e \epsilon N_{eff}}$$

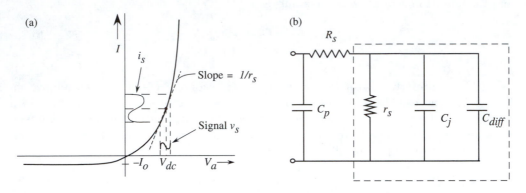

Figure 5.14: (a) A p-n diode is biased at a dc voltage V_{dc} and a small signal modulation is applied to it. (b) The equivalent circuit of a forward-biased diode. At reverse bias the diffusion capacitance vanishes. The diode resistance is given by the differential slope of the I-V characteristics $(r_s = G_s^{-1})$. The resistance R_s is the series resistance of the neutral n- and p- regions, and the capacitance C_p is the capacitance associated with the diode packaging.

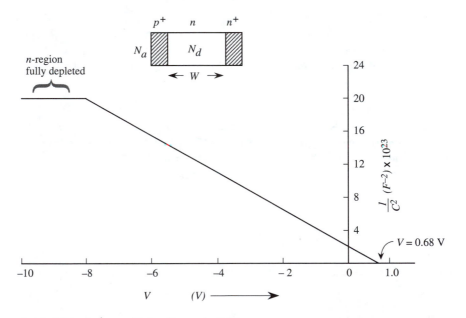

Figure 5.15: Plot of $\frac{1}{C^2}$ vs. V for Example 5.13.

This gives for N_{eff}

$$N_{eff} = \frac{2}{2.1 \times 10^{23} \times 10^{-14}(\text{m}^2) \times 1.6 \times 10^{-19} \text{ (C)} \times (8.84 \times 10^{-12} \times 11.9(\text{ F/m}))}$$
$$= 5.64 \times 10^{19} \text{ m}^{-3} = 5.64 \times 10^{13} \text{ cm}^{-3}$$

The built-in voltage is given by

$$V_{bi} = \frac{k_B T}{e} \ln \left(\frac{N_a N_d}{n_i^2} \right)$$

so that at $kT = 0.026$ eV ($T = 300$ K)

$$N_a N_d = \exp \left(\frac{0.65}{0.026} \right) \times 2.25 \times 10^{20} \text{ cm}^{-6}$$
$$= 5.1 \times 10^{31} \text{ cm}^{-6}$$

Using the definition of N_{eff}, we have from our previous results

$$\frac{N_a N_d}{N_a + N_d} = 5.64 \times 10^{13} \text{ cm}^{-3}$$

This gives us

$$N_d = 5 \times 10^{13} \text{ cm}^{-3}$$
$$N_a = 9.04 \times 10^{17} \text{ cm}^{-3}$$

From the $\frac{1}{C^2}$ vs. V plot we see that the lightly doped n-region becomes depleted at an applied reverse-bias value of 8 V. At this point the capacitance is $7 \times 10^{-13} F$ and we can obtain the width of the n-region as

$$W = \frac{\epsilon A}{C} = \frac{(8.84 \times 10^{-12} \text{ F/m}) \times (11.9 \times 10^{-7} \text{ m}^2)}{7 \times 10^{-13} \text{ F}}$$
$$= 1.48 \times 10^{-5} m = 1.48 \times 10^{-3} \text{ cm}$$

EXAMPLE 5.14 Consider a long base $p^+ n$ diode that is biased to carry a forward current of 1 mA. The junction capacitance is 100 pF. If the minority carrier lifetime τ_p is 1μs, what is the admittance of the diode at 300 K for a 1 MHz signal?

The diode conductance is given by ($k_B T/e \simeq 0.026$ V)

$$G_s = \frac{eI}{k_B T} = 0.038 \text{ A/V}$$

The diffusion capacitance (using $k = 1/2$) is

$$C_{diff} = \frac{eI\tau_p}{2k_B T} = 1.9 \times 10^{-8} \text{ F}$$

which is much larger than the junction capacitance. Neglecting C_j, we get

$$y = 0.038 + j2\pi \times 10^6 \times 1.9 \times 10^{-8}$$
$$= (0.038 + j\ 0.012) \text{ A/V}$$

5.6.2 Large-Signal Switching of Diodes

An important application of diodes is in large-signal switching of the device where the diode is switched from the conducting state to its non-conducting state. Such large-signal switching finds uses in digital technology, in pulse shaping, and in optoelectronics where the diode is used as a light emitter.

We have seen in our study of the diode that in the forward-biased state, minority charge is injected across the depletion region. In the reverse-bias state, the excess minority charge is below the equilibrium value. To reach these states in diode switching, the minority charge has to be removed and injected. This takes time and the diode temporal response is controlled by the time it takes to introduce and remove the minority charge.

In most practical diodes, one side is doped at a much higher level than the other. For example, we may have $N_a \gg N_d$. In this case the hole minority carrier lifetime will be much larger than the electron lifetime and will dominate the time response of the diode. To understand the time response of the diode, we use the relationship between the excess minority charge and the current in the diode. As noted above, for the asymmetrically doped diode, we will consider only the minority charge injected into the lightly doped region (holes for our case where $N_a \gg N_d$). The total charge Q_p injected into the n-region is (for a long diode)

$$Q_p = eA \int_{W_n}^{\infty} \delta p_n(x)dx \tag{5.78}$$

Using the relation between the charge and the voltage across the diode, we have

$$Q_p = eAL_p p_{no} \left(e^{eV/k_BT} - 1 \right) \tag{5.79}$$

In steady state the current is related to the charge by

$$I = \frac{Q_p}{\tau_p} \tag{5.80}$$

where τ_p is the hole recombination time. For a narrow diode the relevant time is the carrier extraction time from the neutral n-region of width $W_{\ell n} - W_n$, which is given in Chapter 3 (see Example 3.17):

$$\tau_T = \frac{|W_{\ell n} - W_n|^2}{2D_p} \tag{5.81}$$

To obtain a dynamic relation between current and excess charge, we use the equation

$$i(t) = \frac{Q_p}{\tau_p} + \frac{dQ_p}{dt} \tag{5.82}$$

where the first term is due to the recombination of charge and the second is due to the change in the minority charge with time.

Turn-ON

Consider the circuit of Fig. 5.16 where a diode is driven by a square wave pulse with the voltage switching between V_F and V_R. The voltage V_F is much larger than the voltage across the diode under forward-bias conditions. We are interested in finding out how the voltage and current across the diode respond to the external potential. Let us first consider how the diode responds when the voltage pulse switches to V_F. As shown in Fig. 5.16a, the voltage switches at $t = t_1$. Once the diode is forward biased, the current becomes

$$i(t) = \frac{V_F - V_1}{R} \tag{5.83}$$

where V_1 is the voltage across the diode and is related to the minority charge by the relation (see Eqn. 5.35)

$$V_1(t) = k_B T \ell n \left(\frac{p(W_n)}{p_n} \right) \tag{5.84}$$

Since the voltage across the diode is small (~ 0.7 V), we have

$$i(t) \sim \frac{V_F}{R} \tag{5.85}$$

i.e., upon turn-on, the diode current goes to its high value almost instantly, as shown in Fig. 5.16b.

The minority charge in the n-region increases gradually, as shown in Fig. 5.16c. From Eqn. 5.84 we see that the voltage across the diode also increases and saturates at

$$V_1 = k_B T \ell n \left(\frac{I_F}{I_o} \right) \tag{5.86}$$

The time taken for the voltage to saturate to V_1 is approximately $2\tau_p$. The voltage across the diode is shown in Fig. 5.16d.

Turn-OFF

Let us now consider the turn-off behavior of the diode. As shown in Fig. 5.17a, the voltage is switched from V_F to V_R at $t = t_2$. To understand the diode response to this turn-off, we note the relation between the excess hole density on the n-side and the voltage across the diode (see Eqn. 5.39):

$$\delta p(W_n) = p_n \left[\exp \left(\frac{eV}{k_B T} \right) - 1 \right] \tag{5.87}$$

This equation tells us that as long as $\delta p(Wn)$ is positive, the voltage across the diode is essentially the forward bias voltage (~ 0.7 V). The diode current is

$$t < t_2 \quad : \quad i(t) = I_F = \frac{V_F - V_1}{R}$$

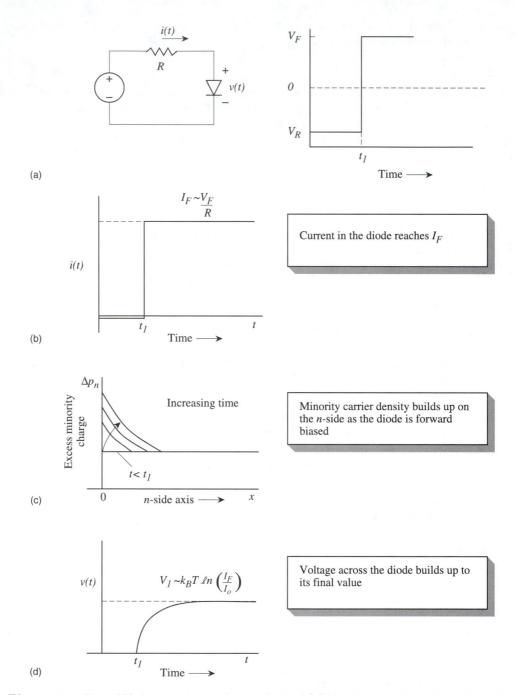

Figure 5.16: Turn-ON characteristics of a *p-n* diode: (a) The voltage switches from V_R to V_F as shown at $t = t_1$. (b) Switching of the current. (c) Change in the minority charge injected into the *n*-side. (d) Voltage across the diode.

$$t = t_2 \quad : \quad i(t) = I_R = \frac{V_R - V_1}{R} \tag{5.88}$$

Before the diode is reverse biased, there is excess minority charge (holes) stored in the n-side. This charge has to be removed. The diode response is controlled by the speed at which this charge is removed. If t_3 is the time at which the excess minority charge is removed, then up to this time, the diode cannot be reverse biased (see Eqn. 5.87). To examine the time response, let us examine the charge control equations

$$t < t_2 \quad : \quad i(t) = I_F = \frac{Q_p}{\tau_p}$$

$$t = t_2 \quad : \quad i(t) = I_R = \frac{Q_p}{\tau_p} + \frac{dQ_p}{dt} \tag{5.89}$$

The general solution of the equation is

$$Q_p(t) = i_R \tau_p + C e^{-t/\tau_p} \tag{5.90}$$

To obtain the constant C, we note that at time just before t_2,

$$Q_p(t) = i_F \tau_p = i_R \tau_p + C e^{-t_2/\tau_p}$$

or

$$C = \tau_p (i_F - i_R) e^{t_2/\tau_p} \tag{5.91}$$

This gives, for the time dependence of the minority charge,

$$t > t_2 \quad : \quad Q_p(t) = \tau_p \left[i_R + (i_F - i_R) e^{(t_2 - t)/\tau_p} \right] \tag{5.92}$$

At $t = t_3$, the entire excess minority charge is removed, i.e., $Q_p(t_3) = 0$. This gives us

$$i_R + (i_F - i_R) e^{-(t_3 - t_2)/\tau_p} = 0$$

or for the long diode, we get

$$t_3 - t_2 = \tau_p \, \ell n \frac{i_F - i_R}{i_R} = \tau_{sd} \tag{5.93}$$

The time $(t_3 - t_2)$ it takes to remove the stored minority charge is called the *storage delay time* τ_{sd}. Until this time, the diode remains forward biased. For the short diode, the time τ_p is replaced by the transit time defined in Eqn. 5.81. We have, for the short diode,

$$t_3 - t_2 = \tau_T \, \ell n \frac{i_F - i_R}{i_R} = \tau_{sd} \tag{5.94}$$

Once the minority charge has been removed, the diode reverse biases in a time controlled by the circuit resistance and the average depletion capacitance of the diode. This time, known as the *transition time*, is

$$\tau_t \sim 2.3 \, RC_j \tag{5.95}$$

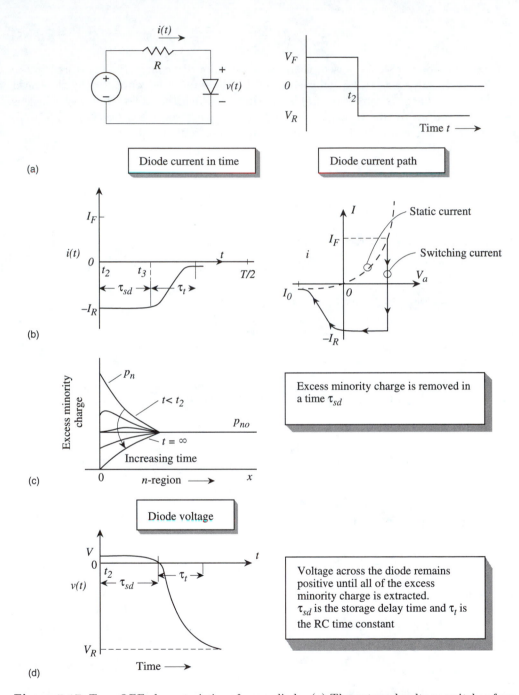

Figure 5.17: Turn-OFF characteristics of a p-n diode: (a) The external voltage switches from V_F to V_R at $t = t_2$. (b) A schematic of the current in the circuit and the current path (right) superimposed on the state I-V of a diode. (c) Minority charge distribution change. (d) Voltage across the diode.

where R is the resistance in the circuit and C_j is the average depletion capacitance.

The discussion of the turn-off process is represented schematically in Fig. 5.17. In Fig. 5.17c we show how the excess minority charge is removed in time. The response of the voltage across the diode is shown in Fig. 5.17d.

EXAMPLE 5.15 A diode is forward biased with $I_F = 3.0$ mA. A reverse bias of 10 V is applied to the diode through a 10 KΩ resistor at $t = 0$. The junction capacitance is 20 pF at zero bias and 10 pF at full reverse bias. Calculate the diode recovery time if $\tau_p = 10^{-7}$ s.

Assuming that there is no voltage drop across the diode, the instant reverse current is

$$|I_R| = \frac{10(V)}{10^4(\Omega)} = 1 \text{ mA}$$

The storage delay time is given by

$$\begin{aligned} \tau_{sd} &= 10^{-7} \times \ell n 2 \\ &= 6.93 \times 10^{-8} \text{ s} \end{aligned}$$

The time τ_t is calculated as (using the average capacitance during switching, 15 pF)

$$\begin{aligned} \tau_t &= \left(2.3 \times 10^4 \times 15 \times 10^{-12}\right) \text{ s} \\ &= 3.45 \times 10^{-7} \text{ s} \end{aligned}$$

The total diode recovery time is

$$\tau = \tau_{sd} + \tau_t = 4.14 \times 10^{-7} \text{ s}$$

5.7 SPICE MODEL FOR A DIODE

In the introduction to this text we have remarked that SPICE has become a standard simulation package for modeling circuits. In addition to modeling resistors, capacitors, and inductors, SPICE has extensive modeling capabilities for devices such as *p-n* diodes, MOSFETs, MESFETs, etc. The input file of the SPICE circuit model contains a variety of parameters that describe the device being used in the circuit. In this section we will discuss the SPICE model for the *p-n* diode and how one can determine the model parameters.

In this chapter we have discussed diode operation based on fundamental physical parameters, such as doping, diffusion coefficient, carrier lifetimes, etc. The SPICE input file does not reach such a basic level. Instead, it uses parameters such as junction capacitance, transit time, built-in voltage, etc.

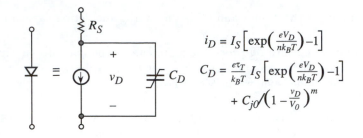

$$i_D = I_S\left[\exp\left(\frac{eV_D}{nk_BT}\right)-1\right]$$

$$C_D = \frac{e\tau_T}{k_BT} I_S\left[\exp\left(\frac{eV_D}{nk_BT}\right)-1\right]$$
$$+ C_{j0}\bigg/\left(1-\frac{v_D}{V_0}\right)^m$$

(a)

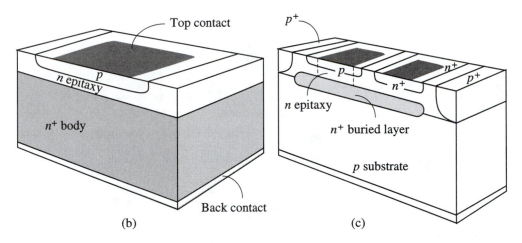

(b)   (c)

Figure 5.18: (a) Model for a *p-n* diode used in SPICE. (b) Cross-section of a discrete *p-n* diode. (c) Cross-section of a discrete *p-n* diode in IC technology.

In Fig. 5.18a we show a model of a *p-n* diode used in SPICE. The model is characterized by a series resistance R_S representing the resistance of the neutral *n* and *p* regions, a current source i_D, and a capacitance C_D. The values of the current and capacitance are shown in Fig. 5.18a.

In Fig. 5.18b and 5.18c we show a cross-section of a discrete *p-n* diode and a *p-n* diode in IC technology. To obtain the SPICE input file parameters, we carry out an exercise in Appendix D where we take material and physical parameters and obtain the SPICE parameters. In Fig. 5.19 we show a set of material and physical parameters and the resulting SPICE input file. Details of the calculations are given in Appendix D. Note that, while obtaining SPICE parameters from the material parameters is a very useful exercise, often these are obtained from experimental measurements. Nevertheless, it is very useful for engineers to understand how parameters like doping, thickness, lifetimes, etc., influence SPICE parameters. Once a SPICE file is known it is straightforward to simulate diode-based circuit performance. Such simulations have now become a standard part of circuit design courses.

p-doping : 10^{16} cm^{-3}

n-doping : 5×10^{15} cm^{-3}

m : 0.5

p-region thickness : 5.0 μm

n-region thickness : 5.0 μm

Area : 100 μm x 100 μm

D_p : 10 cm^2 s

D_n : 30 cm^2 s

$\tau_n = \tau_n = 10^{-7}$s

τ (depletion) $= 10^{-7}$s

Symbol	Name	Parameter name	Units	Default	Example
I	IS	Saturation current	A	1.0E-14	5×10^{-14}
r_s	RS	Ohmic resistance	Ω	0	13.5
n	N	Ideality factor (emission coefficient)		1	1.015
C_{j0}	CJO	Zero-bias depletion capacitance	F	0	2.0×10^{-12}
ϕ_0, V_{bi}	VJ	Built-in potential	V	1	0.68
m	M	Grading coefficient		0.5	
τ_T	TT	Transit time	s	0	12.5×10^{-12}
V_{zk}	BV	Breakdown voltage	V	∞	50
I_{zk}	IBV	Reverse current at V_{zk}	A	10^{-10}	5.35×10^{-10}

Figure 5.19: At the top of this figure are shown basic material parameters for a *p-n* diode. The SPICE parameters calculated using these parameters are shown as well. Details are given in Appendix D.

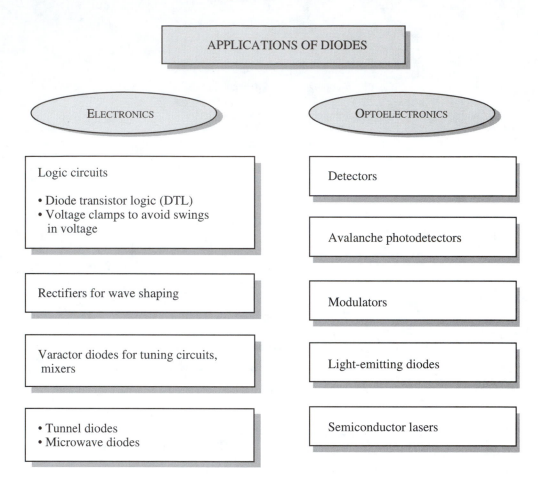

Table 5.1: Some of the important applications of *p-n* semiconductor diodes in electronics and optoelectronics.

5.8 SUMMARY

In this chapter we have addressed one of the most important semiconductor structures—the *p-n* diode. We have observed how the use of the special properties of semiconductors—viz. large change in carrier concentrations controllable by external fields together with the basic transport and recombination phenomena—lead to a highly nonlinear device response. The *p-n* diodes discussed in this chapter find numerous applications such as rectifiers, elements of digital technology, microwave devices, and optoelectronic devices. These applications are listed in Table 5.1. One of the most important applications is in the area of optoelectronic devices. Essentially all the semiconductor devices catering to optoelectronics use the diode concept. These include detectors, avalanche photodetectors, optical modulators, as well as light-emitting diodes and semiconductor lasers. In Chapter 11 we discuss the operation of semiconductor optoelectronic devices.

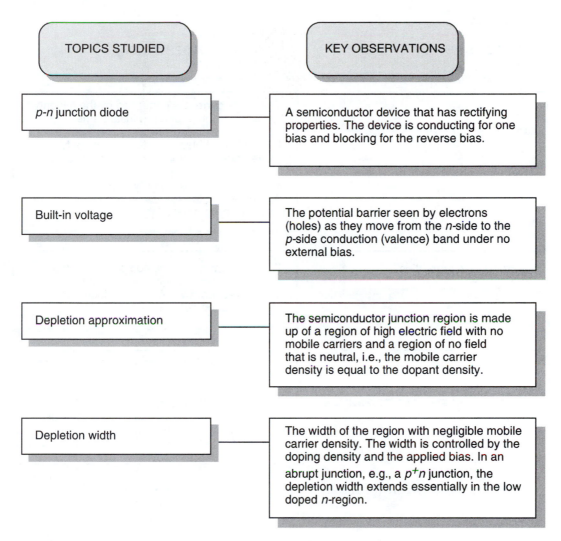

Table 5.2: Summary table.

Summary table 5.2 highlights the key findings of this chapter.

5.9 A BIT OF HISTORY

Before the Second World War, semiconductor materials were of limited importance in the industrial world. The Second World War changed this forever. The key motivation for semiconductor advances came from the radar project. It is interesting to note that more money was spent in perfecting radar than in the Manhattan Project for the atomic

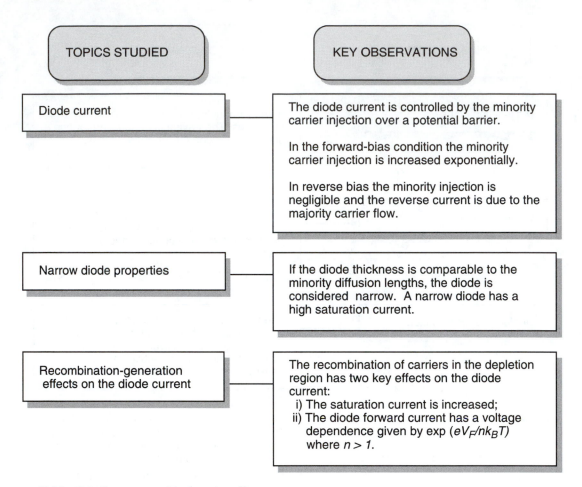

Table 5.2: Summary table (continued).

bomb. Although not as "glamorous" as the atomic bomb project, the radar project gave the Allies a critical edge over the Germans.

The principle of using reflected electromagnetic waves to locate objects was understood around 1920. However, for efficient detection of objects one needed electromagnetic waves with wavelengths of a few centimeters (the microwave range). The British scientists were able to develop the magnetron around 1940 and could generate microwave power. What was missing was a detector that could detect the microwave signals.

The quest for a suitable detector started an enormous research effort into semiconductors and led to the perfection of semiconductor rectification. The fact that a semiconductor crystal could rectify was known and utilized in the early days of radio.

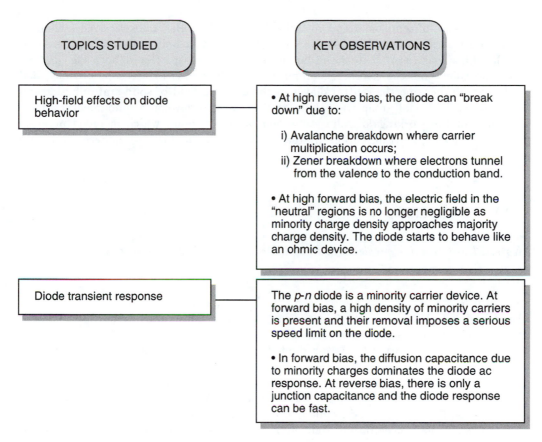

TOPICS STUDIED	KEY OBSERVATIONS
High-field effects on diode behavior	• At high reverse bias, the diode can "break down" due to: i) Avalanche breakdown where carrier multiplication occurs; ii) Zener breakdown where electrons tunnel from the valence to the conduction band. • At high forward bias, the electric field in the "neutral" regions is no longer negligible as minority charge density approaches majority charge density. The diode starts to behave like an ohmic device.
Diode transient response	The *p-n* diode is a minority carrier device. At forward bias, a high density of minority carriers is present and their removal imposes a serious speed limit on the diode. • In forward bias, the diffusion capacitance due to minority charges dominates the diode ac response. At reverse bias, there is only a junction capacitance and the diode response can be fast.

Table 5.2: Summary table (continued).

However, the quality of the semiconductor crystals was rather poor and the rectifiers produced by simply pressing a metal wire to the semiconductor (cat's-whisker detectors) were highly unreliable. In 1942, scientists at Purdue University were awarded a major contract to develop a better understanding of the semiconductor device. The person in charge of this effort was an Austrian-born professor, Karl Lark Horovitz.

The Purdue group initially wanted to consider junctions made in PbS, but soon switched to Ge. At that time work at Bell Labs was focusing on Si as the semiconductor material. The group worked to develop high-purity Ge and studied impurities that could produce better rectification. Eventually this work led to the mass production of Ge diodes by the production arm of Bell Labs and these served as a key component in the radar systems that played a role in the victory over Germany. Ironically, the name germanium was chosen by a German chemist to honor his country!

The success of radar detectors made from germanium was crucial in highlighting the importance of semiconductors in technology. Work at Bell Labs received a tremendous boost as a result of the Second World War. Almost half of the defense research at Bell Labs was devoted to the radar project. Bell Labs perfected both Ge- and Si-based rectifiers and the commercial application of these devices was soon obvious. Bell Labs rapidly became the central player in the exciting field of solid state physics and electronics. A special group was set up at Bell Labs to develop a fundamental understanding of the workings of solid state devices. This group, which included pioneers such as Shockley, Brittain, Moore, Bardeen, etc., led to the birth of modern solid state devices.

5.10 PROBLEMS

Note: Please examine problems in the following Design Problems section as well.

Section 5.2
Problem 5.1 Why does the potential in a *p-n* diode fall mainly across the depletion region and not across the neutral region?

Problem 5.2 An abrupt GaAs *p-n* diode has $N_a = 10^{17}$ cm^{-3} and $N_d = 10^{15}$ cm^{-3}.
(a) Calculate the Fermi level positions at 300 K in the *p* and *n* regions.
(b) Draw the equilibrium band diagram and determine the contact potential V_{bi}.

Problem 5.3 Consider an Si *p-n* diode doped at $N_a = 10^{17}$ cm^{-3}; $N_d = 5 \times 10^{17}$ cm^{-3} at 300 K. Plot the band profile in the neutral and depletion region. Also, plot the electron and hole concentration from the *p-* to the *n*-sides of equilibrium. How good is the depletion approximation?

Problem 5.4 Consider the sample discussed in Problem 5.2. The diode has a diameter of 50 μm. Also calculate the charge in the depletion regions and plot the electric field profile in the diode.

Problem 5.5 An abrupt silicon *p-n* diode at 300 K has a doping of $N_a = 10^{18}$ cm^{-3}, $N_d = 10^{15}$ cm^{-3}. Calculate the built-in potential and the depletion widths in the *n* and *p* regions.

Problem 5.6 A Ge *p-n* diode has $N_a = 5 \times 10^{17}$ cm^{-3} and $N_d = 10^{17}$ cm^{-3}. Calculate the built-in voltage at 300 K. At what temperature does the built-in voltage decrease by 1%?

Section 5.3
Problem 5.7 Explain, using physical arguments, why the reverse current in a *p-n* diode does not change with bias (before breakdown). Would this be the case if the electrons and holes had a constant mobility independent of the electric field?

Problem 5.8 The diode of Problem 5.3 is subjected to bias values of: (a) $V_f = 0.1$ V; (b) $V_f = 0.5$ V; (c) $V_r = 1.0$V; (d) $V_r = 5.0$ V. Calculate the depletion widths and the maximum field F_m under these biases.

Problem 5.9 Consider a p^+n Si diode with $N_a = 10^{18}$ cm^{-3} and $N_d = 10^{16}$ cm^{-3}.

The hole diffusion coefficient in the n-side is 10 cm^2/s and $\tau_p = 10^{-7}$ s. The device area is 10^{-4} cm^2. Calculate the reverse saturation current and the forward current at a forward bias of 0.8 V at 300 K.

Problem 5.10 Consider a p^+n silicon diode with area 10^{-4} cm^2. The doping is given by $N_a = 10^{18}$ cm^{-3} and $N_d = 10^{17}$ cm^{-3}. Plot the 300 K values of the electron and hole currents I_n and I_p at a forward bias of 0.8 V. Assume $\tau_n = \tau_p = 1$ μs and neglect recombination effects. $D_n = 20$ cm^2/s and $D_p = 10$ cm^2/s.

Problem 5.11 A GaAs LED has a doping profile of $N_a = 10^{17}$ cm^{-3}, $N_d = 10^{18}$ cm^{-3} at 300 K. The minority carrier time is $\tau_n = 10^{-8}$ s; $\tau_p = 5 \times 10^{-9}$ s. The electron diffusion coefficient is 100 cm^2 s^{-1} while that of the holes is 20 cm^2 s^{-1}. Calculate the ratio of the electron-injected current (across the junction) to the total current.

Problem 5.12 The diode of Problem 5.11 has an area of 1 mm^2 and is operated at a forward bias of 1.2 V. Assume that 50% of the minority carriers injected recombine with the majority charge to produce photons. Calculate the rate of the photon generation in the n- and p-side of the diode.

Problem 5.13 Consider a GaAs p-n diode with a doping profile of $N_a = 10^{16}$ cm^{-3}, $N_d = 10^{17}$ cm^{-3} at 300 K. The minority carrier lifetimes are $\tau_n = 10^{-7}$ s; $\tau_p = 10^{-8}$ s. The electron and hole diffusion coefficients are 150 cm^2/s and 24 cm^2/s, respectively. Calculate and plot the minority carrier density in the quasi-neutral n and p regions at a forward bias of 1.0 V.

Problem 5.14 Consider a p-n diode made from InAs at 300 K. The doping is $N_a = 10^{16}$ cm$^{-3} = N_d$. Calculate the saturation current density if the electron and hole density of states masses are $0.02m_o$ and $0.4m_o$, respectively. Compare this value with that of a silicon p-n diode doped at the same levels. The diffusion coefficients are $D_n = 800$ cm^2/s; $D_p = 30$ cm^2/s. The carrier lifetimes are $\tau_n = \tau_p = 10^{-8}$s for InAs. For the silicon diode use the values $D_n = 30$ cm^2/s; $D_p = 10$ cm^2/s; $\tau_n = \tau_p = 10^{-7}$s.

Problem 5.15 Consider a p-n diode in which the doping is linearly graded. The doping is given by

$$N_d - N_a = Gx$$

so that the doping is p-type at $x < 0$ and n-type at $x > 0$. Show that the electric field profile is given by

$$F(x) = \frac{e}{2\epsilon} G \left[x^2 - \left(\frac{W}{2}\right)^2 \right]$$

where W is the depletion width, given by

$$W = \left[\frac{12\epsilon \left(V_{bi} - V\right)}{eG} \right]^{1/3}$$

Problem 5.16 A silicon diode is being used as a thermometer by operating it at a fixed forward-bias current. The voltage is then a measure of the temperature. At 300 K, the diode voltage is found to be 0.6 V. How much will the voltage change if the temperature changes by 1 K?

Problem 5.17 Compare the dark currents (i.e., reverse saturation current) in p-n

diodes fabricated from GaAs, Si, Ge, and $In_{0.53}Ga_{0.47}As$. Assume that all the diodes are doped at $N_d = N_a = 10^{18}$ cm^{-3}. The material parameters are (300 K):

$$\text{GaAs} \quad : \quad \tau_n = \tau_p = 10^{-8} \text{ s}; D_n = 100 \text{ cm}^2/\text{s}; D_p = 20 \text{ cm}^2/\text{s}$$
$$\text{Si} \quad : \quad \tau_n = \tau_p = 10^{-7} \text{ s}; D_n = 30 \text{ cm}^2/\text{s}; D_p = 15 \text{ cm}^2/\text{s}$$
$$\text{Ge} \quad : \quad \tau_n = \tau_p = 10^{-7} \text{ s}; D_n = 50 \text{ cm}^2/\text{s}; D_p = 30 \text{ cm}^2/\text{s}$$

When *p-n* diodes are used as light detectors, the dark current is a noise source.

Section 5.4

Problem 5.18 Consider a Si *p-n* diode at 300 K. Plot the I-V characteristics of the diode between a forward bias of 1.0 V and a reverse bias of 5.0 V. Consider the following cases for the impurity-assisted *electron-hole recombination time in the depletion region*: (a) 1.0 μs; (b) 10.0 ns; and (c) 1.0 ns. Use the following parameters:

$$
\begin{aligned}
A &= 10^{-3} \text{ cm}^2 \\
N_a &= N_d = 10^{18} \text{ cm}^{-3} \\
\tau_n &= \tau_p = 10^{-7} \text{ s} \\
D_n &= 25 \text{ cm}^2/\text{s} \\
D_p &= 6 \text{ cm}^2/\text{s}
\end{aligned}
$$

Problem 5.19 Consider a GaAs *p-n* diode with $N_a = 10^{17}$ cm^{-3}, $N_d = 10^{17}$ cm^{-3}. The diode area is 10^{-3} cm^2 and the minority carrier mobilities are (at 300 K) $\mu_n = 3000$ cm^2/V·s; $\mu_p = 200$ cm^2/V·s. The electron-hole recombination times are 10^{-8}s ($\tau_p = \tau_n = \tau$). Calculate the diode current at a reverse bias of 5 V. Plot the diode forward-bias current including generation-recombination current between 0.1 V and 1.0 V.

Problem 5.20 A long base GaAs abrupt *p-n* junction diode has an area of 10^{-3} cm^2, $N_a = 10^{18}$ cm^{-3}, $N_d = 10^{17}$ cm^{-3}, $\tau_p = \tau_n = 10^{-8}$ s, $D_p = 6$ cm^2 s^{-1} and $D_n = 100$ cm^2 s^{-1}. Calculate the 300 K diode current at a forward bias of 0.3 V and a reverse bias of 5 V. The electron-hole recombination time in the depletion regions is 10^{-7}s.

Problem 5.21 Two different processes are used to fabricate a Si *p-n* diode. The first process results in a electron-hole recombination time via impurities in the depletion region of 10^{-7} s while the second one gives a time of 10^{-9} s. Calculate the diode ideality factors for the two cases near a forward bias of 0.9 V. Use the following parameters:

$$
\begin{aligned}
N_a &= N_d = 10^{18} \text{ cm}^{-3} \\
\tau_n &= \tau_p = 10^{-7} \text{ s} \\
D_n &= 25 \text{ cm}^2/\text{s} \\
D_p &= 8 \text{ cm}^2/\text{s}
\end{aligned}
$$

Problem 5.22 Consider a Si diode with the following parameters:

$$
\begin{aligned}
A &= 10^{-3} \text{ cm}^2 \\
N_a &= N_d = 10^{18} \text{ cm}^{-3}
\end{aligned}
$$

$$\tau_n = \tau_p = 10^{-7} \text{ s}$$
$$D_n = 25 \text{ cm}^2/\text{s}$$
$$D_p = 8 \text{ cm}^2/\text{s}$$

The length of the n- and p-sides are 1.0 μm each and the electron-hole impurity-assisted recombination time in the depletion region is 10^{-8} s. Plot the I-V relation of the diode from -5.0 V to 1.0 V. Compare the results for the case where a long diode is made from the same material technology.

Section 5.5

Problem 5.23 The critical field for breakdown of silicon is 4×10^5 V/cm. Calculate the n-side doping of an abrupt p^+n diode that allows one to have a breakdown voltage of 30 V.

Problem 5.24 Consider an abrupt p^+n GaAs diode at 300 K with a doping of $N_d = 10^{16}$ cm^{-3}. Calculate the breakdown voltage. Repeat the calculation for a similarly doped p^+n diode made from diamond. Use Appendix B for the data you may need.

Problem 5.25 What is the width of the potential barrier seen by electrons during band-to-band tunneling in an applied field of 5×10^5 V/cm in GaAs, Si and In$_{0.53}$Ga$_{0.47}$As ($E_g = 0.8$ V)?

Problem 5.26 Consider an Si p-n diode with $N_a = 10^{18}$ cm^{-3}; $N_d = 10^{18}$ cm^{-3}. Assume that the diode will break down by Zener tunneling if the peak field reaches 10^6 V/cm. Calculate the reverse bias at which the diode will break down.

Problem 5.27 Punchthrough diode: For junction diodes that have to operate at high reverse biases, one needs a very thick depletion region. However, in forward-bias conditions this region is undepleted and leads to a series resistance. One uses a p^+-n-n^+ structure in such cases. The width of the n-region is smaller than the depletion region width at breakdown.

Consider two Si p^+-n-n^+ diodes with the n region having a doping of 10^{14} cm^{-3}. In one case the n-region is 150 μm long while in the other case it is 80 μm. What are the reverse-bias voltages that the diodes can tolerate before punchthrough occurs?

Section 5.6

Problem 5.28 A p^+-n silicon diode has an area of 10^{-2} cm^2. The measured junction capacitance is given by (at 300 K)

$$\frac{1}{C^2} = 5 \times 10^8 (2.5 - 4 \text{ V})$$

where C is in units of μF and V is in volts. Calculate the built-in voltage and the depletion width at zero bias. What are the dopant concentrations of the diode?

Problem 5.29 In a long base n^+p diode, the slope of the C_{diff} versus I_F plot is $1.6 \times 10^{-5} F/A$. Calculate the electron lifetime, the stored charge, and the value of the diffusion capacitance at $I_F = 1$ mA.

Problem 5.30 Consider a Si p^+n diode with a long base. The diode is forward-biased (at 300 K) at a current of 2 mA. The hole lifetime in the n-region is 10^{-7} s. Assume that the depletion capacitance is negligible and calculate the diode impedance at the frequency of 100 KHz, 100 MHz, and 500 MHz.

Problem 5.31 Consider a diode with the junction capacitance of 16 pF at zero applied bias and 4 pF at full reverse bias. The minority carrier time is 2×10^{-8} s. If the diode is switched from a state of forward-bias with current of 2.0 mA to a reverse-bias voltage of 10 V applied through a 5kΩ resistance, estimate the response time of the transient.

Problem 5.32 Consider a Si p-n diode at room temperature with following parameters:

$$
\begin{aligned}
N_d &= N_a = 10^{17} \ \text{cm}^{-3} \\
D_n &= 20 \ \text{cm}^2/\text{s} \\
D_p &= 12 \ \text{cm}^2/\text{s} \\
\tau_n &= \tau_p = 10^{-7} \ \text{s}
\end{aligned}
$$

Calculate the reverse saturation current for a long ideal diode. Also estimate the storage delay time for the long diode.

Now consider a narrow diode made from the structure given above. The thickness of the n-side region is 1.0 μm. The thickness of the p-side region is also 1.0 μm. Calculate the reverse saturation current in the narrow diode at a reverse bias of 2.0 volt. Also estimate the storage delay time for this diode.

Problem 5.33 In SPICE—a circuit simulation package—the junction capacitance of a p-n diode is written as

$$
C_{JA} = CJ \left(1 + \frac{V_r}{PB} \right)^{-MJ}
$$

where CJ, PB and MJ are parameters. The value of MJ is 1/2 for abrupt junctions and 1/3 for a linearly graded junction (see Problem 5.15). Use the measured capacitance data shown in Fig. 5.20 to determine CJ, PB, and MJ for the diode.

5.11 DESIGN PROBLEMS

Problem 1 Consider a Si long diode that must be able to operate up to a reverse bias of 10 V. The maximum electric field that the diode can tolerate anywhere within the structure is 5×10^5 V/cm. Design the diode so that the reverse current is *as small as possible within the given specifications*. Assume that $N_a = N_d$. What is the doping density you will use?

Problem 2 Consider a Si short p-n diode with the following parameters:

$$
\begin{aligned}
n\text{-side length} &= 2.0 \times 10^{-4} \ \text{cm} \\
p\text{-side length} &= 2.0 \times 10^{-4} \ \text{cm}
\end{aligned}
$$

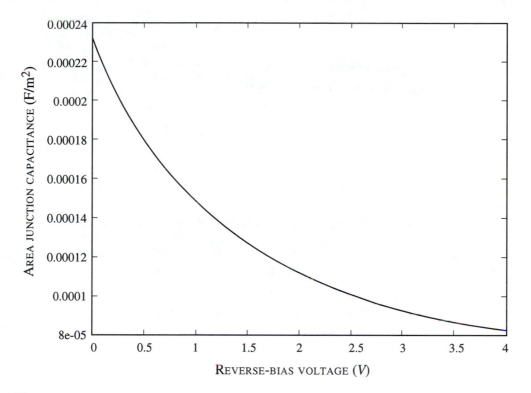

Figure 5.19: Measured C-V data for the diode of Problem 5.33.

$$
\begin{aligned}
n\text{-side doping} &= 10^{17} \text{ cm}^{-3} \\
p\text{-side doping} &= 10^{17} \text{ cm}^{-3} \\
\text{minority carrier lifetime } \tau_n &= \tau_p = 10^{-7} \text{ s} \\
\text{electron diffusion constant} &= 25 \text{ cm}^2/\text{s} \\
\text{hole diffusion constant} &= 10 \text{ cm}^2/\text{s} \\
\text{diode area} &= 10^{-3} \text{ cm}^2
\end{aligned}
$$

Calculate the diode current (assuming that the diode is non-ideal) at a forward bias of 0.1 V and at 0.7 V at 300 K. What are the diode ideality factors *near* the two biasing values?

Problem 3 Give a short discussion on why the reverse-bias current in an ideal *p-n* diode has no voltage dependence. Discuss also the voltage dependence of the reverse-bias current in a non-ideal diode (i.e., a diode with defects).

Problem 4 Consider a Si short (or narrow) *p-n* diode with the following parameters:

$$
\begin{aligned}
n\text{-side thickness} &= 3.0 \ \mu/\text{m} \\
p\text{-side thickness} &= 4.0 \ \mu/\text{m}
\end{aligned}
$$

$$
\begin{aligned}
n\text{-side doping} &= 10^{18} \text{ cm}^{-3} \\
p\text{-side doping} &= 10^{18} \text{ cm}^{-3} \\
\text{minority carrier lifetime } \tau_n &= \tau_p = 10^{-7} \text{ s} \\
\text{electron diffusion constant} &= 30 \text{ cm}^2/\text{s} \\
\text{hole diffusion constant} &= 10 \text{ cm}^2/\text{s} \\
\text{diode area} &= 10^{-4} \text{ cm}^2/\text{s}
\end{aligned}
$$

Calculate the diode current at a forward bias of 0.5 V at 300 K. Also calculate the total excess hole charge (in coulombs) injected into the n-side (from W_n to the diode n-side contact) at this biasing.

Problem 5   Consider a Si long p-n diode with the following parameters:

$$
\begin{aligned}
n\text{-side doping} &= 10^{18} \text{ cm}^{-3} \\
p\text{-side doping} &= 10^{18} \text{ cm}^{-3} \\
\text{minority carrier lifetime } \tau_n &= \tau_p = 10^{-7} \text{ s} \\
\text{electron diffusion constant} &= 30 \text{ cm}^2/\text{s} \\
\text{hole diffusion constant} &= 10 \text{ cm}^2/\text{s} \\
\text{diode area} &= 10^{-4} \text{ cm}^2/\text{s}
\end{aligned}
$$

Calculate the diode current at a forward bias of 1.0 V at 300 K.

An electron comes from the p-side into the depletion region and is swept away by the field to the n-side. *Estimate* the time it takes the electron to cross *the depletion region* at zero applied bias and a reverse bias of 1.0 volt.

Problem 6   Consider a Si long p-n diode with the following parameters:

$$
\begin{aligned}
n\text{-side doping} &= 10^{17} \text{ cm}^{-3} \\
p\text{-side doping} &= 10^{17} \text{ cm}^{-3} \\
\text{minority carrier lifetime } \tau_n &= \tau_p = 10^{-7} \text{ s} \\
\text{electron diffusion constant} &= 30 \text{ cm}^2/\text{s} \\
\text{hole diffusion constant} &= 10 \text{ cm}^2/\text{s} \\
\text{diode area} &= 10^{-4} \text{ cm}^2 \\
\text{carrier lifetime in the depletion region} &= 10^{-8} \text{ s}
\end{aligned}
$$

Calculate the diode current at a forward bias of 0.5 V and 0.6 V at 300 K. What is the ideality factor of the diode in this range?

Problem 7   Consider a narrow diode with the same parameters as given above. Calculate the total electron- and hole-injected charge in the n- and p- sides at a forward bias of 0.4 V. The widths of the n- and p-sides are both 1.0 μm.

Problem 8   Design a p^+n Si diode that can be used in a digital system operating at 1 gigabit per second. Assume that the minority carrier lifetime is 10^7 s. Other parameters can be obtained from the text. Plot the I-V characteristics of this diode. At what applied reverse bias would the entire *n*-region be depleted in this diode?

Problem 9   Discuss how a *p-n* diode can be used as a temperature sensor. Assuming an ideal Si *p-n* diode, calculate the value of x and y where

$$x = \frac{1}{I_o} \frac{dI_o}{dT}, \; y = \frac{1}{I} \frac{dI}{dT}$$

In real diodes the value of x and y is smaller than what is expected for an ideal diode. Discuss the reason for this.

Problem 10   A Si p^+n diode is to be used in the reverse bias state ($V_R = 5V$) as a high-speed detector. Design the diode so it can operate up to a frequency of 5 GHz. Make reasonable assumptions for the material parameters.

Problem 11   Assume that a Si diode suffers Zener breakdown at a field of 2×10^5 V/cm if both *n*- and *p*-sides are doped above 10^{18} cm^{-3}. Design a diode that suffers Zener breakdown at a reverse bias of 5 V. Draw the I-V characteristics for this diode assuming reasonable material parameters.

5.12   READING LIST

- **General**

  - M. S. Tyagi, *Introduction to Semiconductor Materials and Devices* (John Wiley and Sons, New York, 1991).

  - B. G. Streetman and S. Banerjee, *Solid State Electronic Devices* (Prentice-Hall, Englewood Cliffs, NJ, 1999).

  - G. W. Neudeck, "Modular Series on Solid State Devices," Vol. 11, *The P-N Junction Diode*, (Addison-Wesley, Reading, MA, 1983).

  - R. S. Muller and T. I. Kamins, *Device Electronics for Integrated Circuits* (John Wiley and Sons, New York, 1986).

- **Diode Breakdown**

  - M. H. Lee and S. M. Sze, "Orientation Dependence of Breakdown Voltage in GaAs," *Solid State Electronics* **23**, 1007 (1980).

  - S. M. Sze, *Physics of Semiconductor Devices* (John Wiley and Sons, New York, 1981).

  - S. M. Sze and G. Gibbons, *Applied Physics Letters* **8**, 112 (1986).

- **Temporal Response of Diodes**

  - R. H. Kingston, "Switching Time in Junction Diodes and Junction Transistors," *Proc. IRE* **42**, 829 (1954).

– M. S. Tyagi, *Introduction to Semiconductor Materials and Devices* (John Wiley and Sons, New York, 1991).

– D. A. Neamen, *Semiconductor Physics and Devices: Basic Principles* (Irwin, Homewood, IL, 1992).

Chapter

6

SEMICONDUCTOR JUNCTIONS WITH METALS AND INSULATORS

In Chapter 5 we saw how a junction between doped semiconductors can lead to extremely interesting responses in terms of nonlinear I-V characteristics and tunable C-V characteristics. These responses can be used to design devices for a variety of applications. However, semiconductor-semiconductor junctions are not the only kinds of junctions that are important in electronics. Metals are an essential component of electronic and optoelectronic devices. Metals are necessary to connect the semiconductors to the "outside world" of batteries. They are also able, under proper conditions, to produce rectifying junctions that can be used to control the response of semiconductors. Insulators are also an integral part of electronics. These materials provide an isolation between two regions of a device, can be used for bandstructure tailoring, can be used as capacitors, etc. In this chapter we will examine some important properties of the junctions of semiconductors with these materials.

6.1   METALS AS CONDUCTORS: INTERCONNECTS

As we mentioned briefly in Section 1.8, metals are materials in which the highest occupied band is half full of electrons so that the Fermi level lies in the middle of an allowed band. As a result, the electron density at the Fermi level is very high and there are available unoccupied states above the Fermi level that allow a high density of electrons to participate in carrying current. Thus the resistivity of metals is very low. In Table 6.1 we show the resistivities of some important metals used in electronics. In Chapter 3 (see Eqn. 3.8) we related conductivity (the inverse of resistivity) to the electron density and the scattering times. The low resistivity of metals is due to a very high electron concentration.

Metals serve three important roles in electronic circuits: (a) as interconnects, they provide the pathways to pass electrical signals to and from a device; (b) as Schottky barriers, they provide junctions that can provide rectification, have built-in electric fields, and have a variety of uses; (c) as ohmic contacts, they allow electrons or holes to enter and leave the semiconductors with little resistivity.

If one examines an integrated circuit under a microscope, the circuit looks very much like a map of a city. A great deal of the "chip" is covered with metal interconnects that appear as "highways" through which charge travels from one point to another. While these interconnects are obviously passive elements of the circuit, much like highways in a city, they are extremely important. The metal strips making up the interconnect must be able to carry adequate current and make good contact with the devices. Interconnects are deposited on insulators and touch the active devices only through windows that are opened at select points. The region where the metals form a junction with the semiconductor is described by the Schottky barrier or ohmic contact theory discussed in the next sections. Interconnects have linear response described by Ohm's law.

MATERIAL	RESISTIVITY ($\mu\Omega$-cm)
Aluminum (Al)	2.7
Titanium (Ti)	40.0
Tungsten (W)	5.6
Ti-W	15-50
Gold (Au)	2.44
Silver (Ag)	1.59
Copper (Cu)	1.77
Platinum (Pt)	10.0
Silicides	
PtSi	28-35
NiS_2	50

Table 6.1: Resistivities of some metals used in solid state electronics.

Aluminum is a commonly used interconnect material. In bulk, Al is a good conductor, with resistivity of 2.7×10^{-6} Ω-cm. The resistivity in thin-film form can be up to a factor of 20 lower, allowing the thin interconnect film to carry very high current densities. The current-carrying capability of the interconnects is very important since in high-speed devices one has to work with high current densities ($\sim 10^5$ A/cm^2). At such high current densities, a phenomenon known as electromigration occurs that is a major cause of the breakdown of integrated circuits.

In electromigration (also known as the electron wind effect) metal ions are forced to move downstream due to the high electron current densities, mainly along the grain structure. When this happens, the metal film develops voids and hillocks at regions of curves and forks, as shown in Fig. 6.1. The thinning due to voids causes excessive heating and eventually there is a break at this point.

Electromigration is prevented by controlling the grain structure along the microcrystallites that form the metal lines. Larger grains have less surface area and therefore resist electromigration. Some alloying metals, such as copper, accumulate in the grain boundaries, locking the atoms there into place and thus preventing electromigra-

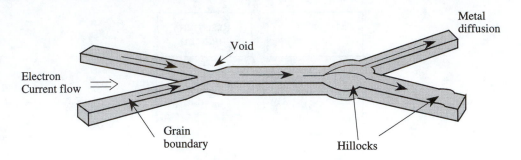

Figure 6.1: The initial steps of failure in metal interconnects. The metal moves under the influence of current along grain boundaries. Voids and hillocks are produced and eventually a break occurs at the voids.

tion. Copper also prevents hillocking of Al films and thus prevents nonuniform thermal effects. It is thus increasingly common to add Cu to Al in small quantities to form interconnects. Since 1998 scientists have been able to make interconnects from copper (see Section 6.5). Since copper has a lower resistivity, interconnect delays in devices have improved.

EXAMPLE 6.1 The resistivity of aluminum at room temperature is 2.7 $\mu\Omega$-cm. Assume that there are 10^{22} electrons per cm^3 participating in the current flow. Calculate the mobility of electrons in aluminum.

The resistivity ρ is related to mobility by the following equations ($J =$ current density; $\sigma = 1/\rho =$ conductivity; $\mu =$ mobility; $F =$ electric field):

$$\begin{aligned} J &= \sigma F \\ &= nev = ne\mu F \end{aligned}$$

so that

$$\rho = \frac{1}{ne\mu}$$

Thus

$$\begin{aligned} \mu = \frac{1}{ne\rho} &= \frac{1}{(10^{22} \text{ cm}^{-3})(1.6 \times 10^{-19} \text{ C})(2.7 \times 10^{-6}\,\Omega\text{-cm})} \\ &= 240.4 \text{ cm}^2/\text{V} \cdot \text{s} \end{aligned}$$

Compare this with a mobility of electrons in GaAs of $\sim$ 8000 cm^2/V·s and in Si of 1500 cm^2/V·s. In spite of the low mobility, the resistivity of metals is low because of the high density of electrons that can carry current.

6.2 SCHOTTKY BARRIER DIODE

The metal-semiconductor junction can, under appropriate conditions, be an active device capable of strongly nonlinear response. The resulting Schottky barrier and the Schottky barrier diode are widely used in important applications. The Schottky diode has characteristics that are essentially similar to those of the p-n diode except that for many applications it has a much faster response.

6.2.1 Schottky Barrier Height

To clarify the properties of the Schottky barrier junction, a metal semiconductor structure is shown in Fig. 6.2a. Figures 6.2b and 6.2c show the band profiles of a metal and a semiconductor. Figure 6.2b shows that the band profile and Fermi level positions between the metal are away from the semiconductor. In Fig. 6.2c the metal and the semiconductor are in contact. The Fermi level E_{Fm} in the metal lies in the band, as shown. Also shown is the work function $e\phi_m$, which is the energy required to take an electron out to the free vacuum state. In the semiconductor, we show the vacuum level along with the position of the Fermi level E_{Fs} in the semiconductor, the electron affinity, and the work function.

We will assume that $\phi_m > \phi_s$ so that the Fermi level in the metal is at a lower position than in the semiconductor. When the junction between the two systems is formed, the Fermi levels should line up at the junction and remain flat in the absence of any current, as shown in Fig. 6.2c. At the junction, the vacuum energy levels of the metal side and semiconductor side must be the same. To ensure the continuity of the vacuum level and keep the Fermi level flat, *the Fermi level must move deeper into the bandgap of the semiconductor at the interface region.* This involves electrons moving out from the semiconductor side to the metal side. *Note that since the metal side has an enormous electron density, the metal Fermi level or the band profile does not change when a small fraction of electrons are added or taken out.* As electrons move to the metal side, they leave behind positively charged fixed dopants, and a dipole region is produced in the same way as for the p-n diode of Chapter 5. The electric field opposes the electron flow and at equilibrium the band bending is such that the Fermi level is flat.

In the ideal Schottky barrier, the height of the barrier at the semiconductor-metal junction (Fig. 6.2c), defined as the difference between the semiconductor conduction band at the junction and the metal Fermi level, is

$$e\phi_b = e\phi_m - e\chi_s \tag{6.1}$$

The electrons coming from the semiconductor into the metal face a barrier denoted by eV_{bi} as shown in Fig. 6.2c. The potential eV_{bi} is called the built-in potential of the junction and is given by

$$eV_{bi} = e\phi_m - e\phi_s \tag{6.2}$$

The height of the potential barrier can be altered by applying an external bias, as in the case in the *p-n* diode, and the junction can be used for rectification, as we shall see below. In Fig. 6.3 we show the case of a metal-*p*-type semiconductor junction where we choose a metal so that $\phi_m < \phi_s$. In this case, at equilibrium the electrons are injected from the metal to the semiconductor, causing a negative charge on the semiconductor side. The bands are bent once again and a barrier is created for hole transport. The height of the barrier seen by the holes is

$$eV_{bi} = e\phi_s - e\phi_m \tag{6.3}$$

According to the discussion so far, the Schottky barrier height for *n*- or *p*-type semiconductors depends upon the metal and the semiconductor properties. However, it is found experimentally that the Schottky barrier height is *usually independent of the metal employed*, as can be seen from Table 6.2. This can be understood qualitatively in terms of a model based upon nonideal surfaces. In this model the metal-semiconductor interface has a distribution of interface states that may arise from the presence of chemical defects (e.g., an oxide film) or broken bonds, etc. The defect region leads to a distribution of electronic levels in the bandgap at the interface, as shown in Fig. 6.4. The distribution may be characterized by a neutral level ϕ_o having the properties that states below it are neutral if filled and above it are neutral if empty. *If the density of bandgap states near ϕ_o is very large, then addition or depletion of electrons to the semiconductor does not alter the Fermi level position at the surface and the Fermi level is said to be pinned.* In this case, as shown in Fig. 6.4, the Schottky barrier height is

$$e\phi_b = E_g - e\phi_o \tag{6.4}$$

and is *almost independent of the metal used*. The model discussed above provides a qualitative understanding of the Schottky barrier heights. However, the detailed mechanism of the interface state formation and Fermi level pinning is quite complex. In Table 6.2 we show Schottky barrier heights for some common metal-semiconductor combinations.

6.2.2   Capacitance Voltage Characteristics

Once the Schottky barrier height is known, the electric field profile, depletion width, depletion capacitance, etc., can be evaluated the same way we obtained the values for the *p-n* junction. The problem for a Schottky barrier on an *n*-type material is identical to that for the abrupt p^+n diode, since there is no depletion on the metal side. One

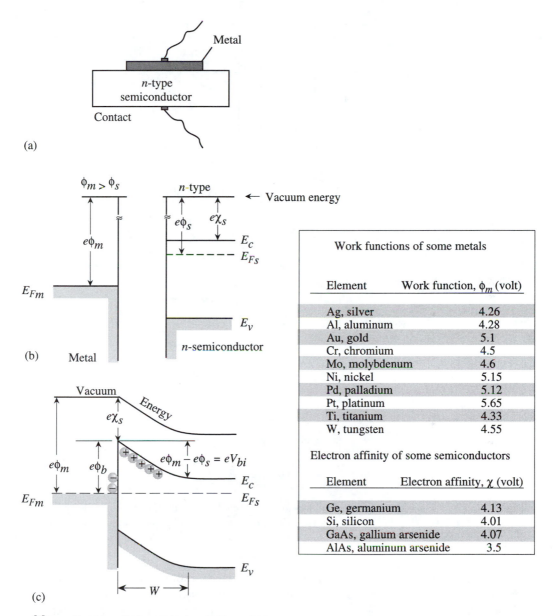

METAL-SEMICONDUCTOR JUNCTION AT EQUILIBRIUM

Figure 6.2: (a) A schematic of a Schottky barrier junction. (b) The various important energy levels in the metal and the semiconductor with respect to the vacuum level. (c) The junction potential produced when the metal and semiconductor are brought together. Due to the built-in potential at the junction, a depletion region of width W is created. Also shown in the inset are the work functions and electron affinities of several systems.

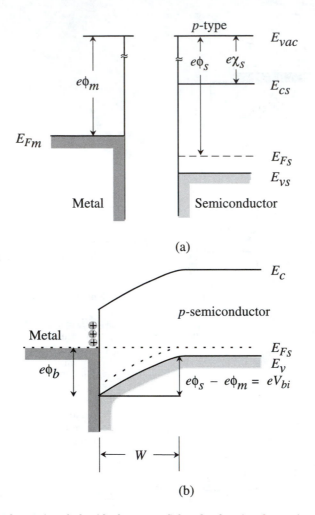

Figure 6.3: A schematic of the ideal p-type Schottky barrier formation. (a) The positions of the energy levels in the metal and the semiconductor; (b) the junction potential and the depletion width.

again makes the depletion approximation; i.e., there is no mobile charge in the depletion region and the semiconductor is neutral outside the depletion region. Then the solution of the Poisson equation (see Chapter 5) gives the depletion width W for an external potential V (see Eqn. 5.29 with $N_a \gg N_d$),

$$W = \left[\frac{2\epsilon(V_{bi} - V)}{eN_d} \right]^{1/2} \tag{6.5}$$

Here N_d is the doping of the n-type semiconductor. Note that there is no depletion on the metal side because of the high electron density there. The potential V is the applied

SCHOTTKY METAL	n Si	p Si	n GaAs
Aluminum, Al	0.7	0.8	
Titanium, Ti	0.5	0.61	
Tungsten, W	0.67		
Gold, Au	0.79	0.25	0.9
Silver, Ag			0.88
Platinum, Pt			0.86
PtSi	0.85	0.2	
NiSi$_2$	0.7	0.45	

Table 6.2: Schottky barrier heights (in volts) for several metals on n- and p-type semiconductors. The barrier height is seen to have a rather weak dependence upon the metal used.

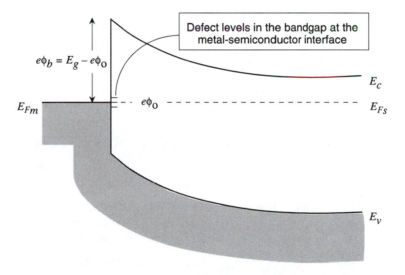

Figure 6.4: At a real metal-semiconductor interface, there are a large number of interface states in the bandgap region. A neutral level ϕ_o is defined so that the interface states above ϕ_o are neutral if they are empty and those below ϕ_o are neutral if they are occupied. If the interface density is high, the Fermi level is "pinned" at ϕ_o and the Schottky barrier is determined by ϕ_o and the material bandgap.

potential, which is positive for forward bias and negative for reverse bias.

The depletion region charge is given from the depletion width as

$$Q_d = eN_dW = [2e\epsilon N_d(V_{bi} - V)]^{1/2} \tag{6.6}$$

The corresponding capacitance for a device with area A is

$$C = A \left| \frac{dQ_d}{dV} \right| = A \left[\frac{e\epsilon N_d}{2(V_{bi} - V)} \right]^{1/2} = \frac{\epsilon A}{W} \tag{6.7}$$

We have, as in the case of the *p-n* diode,

$$\frac{1}{C^2} = \frac{2(V_{bi} - V)}{A^2 e\epsilon N_d} \tag{6.8}$$

Thus the plot of $\frac{1}{C^2}$ and V provides the information on both the built-in potential and the doping density.

EXAMPLE 6.2 A metal-silicon Schottky barrier with a 100 μm diameter has $\frac{1}{C^2}$ vs. V slope of $-3 \times 10^{24} F^{-2}V^{-1}$. Calculate the doping density in the silicon.

The slope of the $\frac{1}{C^2}$ vs. V curve is given by

$$\frac{d\left(\frac{1}{C^2}\right)}{dV} = -\frac{2}{A^2 e\epsilon N_d}$$

Thus

$$
\begin{aligned}
N_d &= \frac{2}{(3 \times 10^{24}\,\text{F}^{-2}\text{V}^{-1}) \times (7.85 \times 10^{-9}\text{m}^2)^2 \times (1.6 \times 10^{-19}\text{C}) \times (8.84 \times 10^{-12} \times 11.9\,\text{Fm}^{-1})} \\
&= 6.42 \times 10^{20}\,\text{m}^{-3} = 6.42 \times 10^{14}\,\text{cm}^{-3}
\end{aligned}
$$

It is important to note that the slope of the $\frac{1}{C^2}$ vs. V relation is constant only for a uniformly doped semiconductor. If, however, the doping is nonuniform, the slope at a particular voltage depends upon the doping at the depletion width edge corresponding to that voltage. Thus, from the relations

$$C = \frac{\epsilon A}{W(V)}$$

and

$$\left. \frac{d\left(\frac{1}{C^2}\right)}{dV} \right|_V = \frac{-2}{A^2 e\epsilon N_d(W)}$$

one can obtain the doping density as a function of depth in the semiconductor.

6.2.3   Current Flow in a Schottky Barrier

Let us consider an *n*-type Schottky barrier, as shown in Fig. 6.5. Electrons can flow from the metal to the semiconductor and from the semiconductor to the metal. When an external bias is applied, current flows in the device. The current flow across a Schottky barrier can involve a number of different mechanisms. The most important and most desirable mechanism is that of thermionic emission, in which electrons with energy greater than the barrier height $e(V_{bi} - V)$ can overcome the barrier and pass across the junction. Note that as the bias changes, the barrier to be overcome by electrons changes and the electron current injected is thus altered. This is shown schematically in Fig. 6.5.

In addition to thermionic emission, electrons can also tunnel through the barrier to generate current. This can be important if the semiconductor is heavily doped so that the depletion width is small. We will examine the dominant thermionic-emission-related current.

Thermionic Emission Current

If we assume that the tunneling current is negligible, the electrons that cross the metal-semiconductor junction must have energies greater than the barrier height at the junction. At the junction the electrons must pass over the barrier and only those electrons that have energies greater than the barrier will pass. The current is thus limited by the junction barrier.

We assume that the electrons in the semiconductor region are distributed according to Boltzmann statistics. Thus the fraction of electrons with energy greater than the barrier of $(V_{bi} - V)$ is (V is positive for the forward-biased diode and negative for the reverse-biased case):

$$n_b = n_o \ \exp\left[-e\frac{(V_{bi} - V)}{k_B T}\right] \tag{6.9}$$

where n_o is the electron density in the neutral region. The density n_o is given in terms of the effective density of electrons N_c as

$$n_o = N_c \ \exp\left[-\frac{(E_c - E_{Fs})}{k_B T}\right] \tag{6.10}$$

Thus (note that the metal side barrier $e\phi_b = eV_{bi} + E_c - E_{Fs}$),

$$n_b = N_c \ \exp\left[-\frac{(e\phi_b - eV)}{k_B T}\right] \tag{6.11}$$

If the electrons are considered to be moving randomly, the average flux of electrons impinging on the metal-semiconductor barrier is approximately $<v> n_b/4$, where

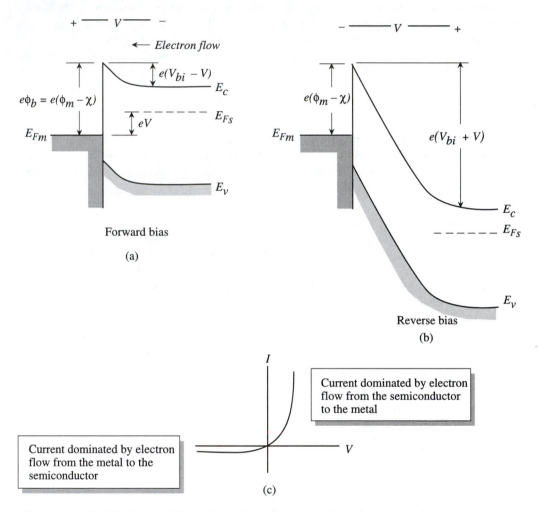

Figure 6.5: (a) The forward bias allows electrons to flow from the semiconductor to the metal side, increasing the current. (b) The reverse bias suppresses the electron flow from the semiconductor side while the flow from the metal side is unaffected. (c) The rectifying characteristics of the Schottky diode.

$< v >$ is the average speed of the electrons. The corresponding current is then (A is the device area)

$$I_{sm} = \frac{eA < v >}{4} N_c \, \exp\left[\frac{-e(\phi_b - V)}{k_B T}\right] \tag{6.12}$$

When the applied bias V is zero, the current flow from the metal into the semiconductor I_{ms} must balance the current flow from the semiconductor to the metal. Thus

$$I_{ms} = I_{sm}(V = 0) = \frac{eA < v >}{4} N_c \, \exp -\left(\frac{e\phi_b}{k_B T}\right) \tag{6.13}$$

When a potential V is applied, the *barrier seen by the electrons coming from the metal side is unchanged* and I_{ms} remains a constant ($= I_s$).

Thus the net current at an applied bias V is, from Eqns. 6.12 and 6.13,

$$\boxed{I = I_{sm} - I_{ms} = I_s \left[\exp\left(\frac{eV}{k_B T}\right) - 1\right]} \tag{6.14}$$

For a Maxwell-Boltzmann distribution of electrons, the average velocity can be shown to be related to the temperature by

$$< v > = \left(\frac{8k_B T}{\pi m^*}\right)^{1/2} \tag{6.15}$$

Substituting for the effective conduction band density N_c (see Eqn. 2.30), we get for $I_s = I_{sm}(V = 0)$ from Eqn. 6.14

$$\boxed{I_s = A \left(\frac{m^* e k_B^2}{2\pi^2 \hbar^3}\right) T^2 \, \exp\left\{-\left(\frac{e\phi_b}{k_B T}\right)\right\}} \tag{6.16}$$

The quantity in parentheses is called the Richardson constant and is denoted by R^*. Its numerical value is given by

$$
\begin{aligned}
R^* &= \frac{m^* e k_B^2}{2\pi^2 \hbar^3} \\
&\cong 120 \left(\frac{m^*}{m_o}\right) A \ \text{cm}^{-2} K^{-2}
\end{aligned}
\tag{6.17}
$$

A detailed formalism shows that the simple theory above gives too large a value of R^*. *Approximate values for the effective Richardson constant are: electrons in Si,*

~ 110 A $cm^{-2}K^{-2}$; electrons in GaAs, 8 A $cm^{-2}K^{-2}$; holes in Si, ~ 32 A $cm^{-2}K^{-2}$; holes in GaAs, ~ 74 A $cm^{-2}K^{-2}$.

The saturation current in the Schottky barrier turns out to be much higher than that in a p-n diode with similar built-in voltage. This results in a turn-on voltage for forward-bias conducting state at a very low applied bias, but also results in a high reverse current. In Fig. 6.5, the band configurations in forward and reverse bias are shown along with the I-V characteristics. It is important to note that the electrons in the semiconductor side see a variable potential barrier as the applied bias changes, while the current from the metal side remains essentially unchanged.

If the barrier height is small (comparable to a few k_BT), one has to worry about a very high saturation current. Since this is to be avoided, one looks for Schottky barriers with a large barrier height. For many narrow bandgap semiconductors, it is difficult to obtain large Schottky barriers and therefore one cannot use this mode of diode fabrication.

In a real device structure the I-V characteristics have the form

$$I = I_s \left[\exp\left(\frac{eV}{nk_BT} \right) - 1 \right] \tag{6.18}$$

as in the case of the real diode. The factor n is the ideality factor and is very close to unity for the Schottky diode, since the device has very little minority carrier participation and recombination current is low.

EXAMPLE 6.3 In a W-n-type Si Schottky barrier the semiconductor has a doping of 10^{16} cm^{-3} and an area of 10^{-3} cm^2.
(a) Calculate the 300 K diode current at a forward bias of 0.3 V.
(b) Consider an Si $p^+ - n$ junction diode with the same area with doping of $N_a = 10^{19}$ cm^{-3} and $N_d = 10^{16}$ cm^{-3}, and $\tau_p = \tau_n = 10^{-6}$ s. At what forward bias will the p-n diode have the same current as the Schottky diode? $D_p = 10.5$ cm^2/s.

From Table 6.2 the Schottky barrier of W on Si is 0.67 V. Using an effective Richardson constant of 110 A $cm^{-2}K^{-1}$, we get for the reverse saturation current

$$I_s = (10^{-3} \ cm^2) \times (110 \ A \ cm^{-2}K^{-2}) \times (300K)^2 \ \exp\left(\frac{-0.67(eV)}{0.026(eV)} \right)$$

$$= 6.37 \times 10^{-8} \ A$$

For a forward bias of 0.3 V, the current becomes (neglecting 1 in comparison to exp (0.3/0.026))

$$I = 6.37 \times 10^{-8} A \ \exp(0.3/0.026)$$

$$= 6.53 \times 10^{-3} \ A$$

In the case of the p-n diode, we need to know the appropriate diffusion coefficients and lengths. The diffusion coefficient is 10.5 cm^2/s, and using a value of $\tau_p = 10^{-6}$s we get $L_p = 3.24 \times$

10^{-3} cm. Using the results for the abrupt $p^+ - n$ junction, we get for the saturation current ($p_n = 2.2 \times 10^4$ cm^{-3}) (note that the saturation current is essentially due to hole injection into the n-side for a p^+-n diode)

$$
\begin{aligned}
I_o &= (10^{-3} \text{ cm}^2) \times (1.6 \times 10^{-19} \text{ C}) \times \frac{(10.5 \text{ cm}^2/\text{s}^{-1})}{(3.24 \times 10^{-3} \text{ cm})} \times (2.25 \times 10^4 \text{ cm}^{-3}) \\
&= 1.17 \times 10^{-14} \text{ A}
\end{aligned}
$$

This is an extremely small value of the current. At 0.3 V, the diode current becomes

$$
I = I_s \exp\left(\frac{eV}{k_B T}\right) = 1.2 \times 10^{-9} \text{ A}
$$

a value which is almost six orders of magnitude smaller than the value in the Schottky diode. For the p-n diode to have the same current that the Schottky diode has at 0.3 V, the voltage required is 0.71 V.

This example highlights the very important differences between Schottky and junction diodes. The Schottky diode turns on (i.e., the current is $\sim$1 mA) at 0.3 V while the p-n diode turns on at closer to 0.7 V.

EXAMPLE 6.4 An important problem in very high-speed transistors (to be discussed in Chapter 8) based on the InAlAs/InGaAs system is the reliability of the Schottky barrier. Consider a Schottky barrier formed on an InAlAs doped n-type at 10^{16} cm^{-3}. Calculate the saturation current density if the Schottky barrier height is (i) 0.7 V; (ii) 0.6 V at 300 K.

The mass of the electrons in InAlAs is 0.08 m_o. The Richardson constant has a value

$$
R^* = 120 \left(\frac{m^*}{m_o}\right) = 9.6 A \text{ cm}^{-2} K^{-2}
$$

The saturation current density then becomes

$$
\begin{aligned}
J_s(\phi_b = 0.7 \text{ V}) &= R^* T^2 \exp\left[-\left(\frac{e\phi_b}{k_B T}\right)\right] \\
&= 1.8 \times 10^{-6} \text{ A/cm}^2 \\
J_s(\phi_b = 0.6 \text{ V}) &= 8.2 \times 10^{-5} \text{ A/cm}^2
\end{aligned}
$$

Thus the current density varies by a very large value depending upon the Schottky barrier value. The Schottky barrier height depends upon the metal-semiconductor interface quality and can be easily affected by fabrication steps. In a p-n diode, on the other hand, the built-in voltage is fixed by doping and is more controllable.

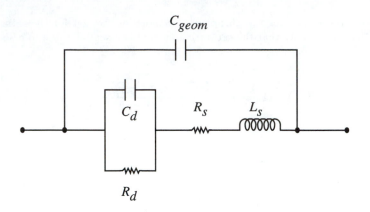

Figure 6.6: Equivalent circuit of a Schottky diode. The Schottky diode, being a majority carrier device, does not have the diffusion capacitance that hampers *p-n* diode performance in the forward-bias mode.

6.2.4   Small-Signal Circuit of a Schottky Diode

The small-signal equivalent circuit of a Schottky diode is shown in Fig. 6.6. One has the parallel combination of the resistance

$$R_d = \frac{dV}{dI} \tag{6.19}$$

and the differential capacitance of the depletion region. The depletion capacitance has the form calculated earlier (see Eqn. 6.7):

$$C_d = A \left[\frac{e N_d \epsilon}{2(V_{bi} - V)} \right]^{1/2} \tag{6.20}$$

These circuit elements are in series with the series resistance R_s (which includes the contact resistance and the resistance of the neutral doped region of the semiconductor) and the parasitic inductance. Finally, one has to include the device geometry capacitance:

$$C_{geom} = \frac{\epsilon A}{L} \tag{6.21}$$

where L is the device length.

There is a very important difference between the equivalent circuit of the Schottky diode and the *p-n* diode discussed in this chapter. This has to do with the *absence of the diffusion capacitance that dominates the forward-bias capacitance of a p-n diode.*

This allows a very fast response of the Schottky diode, which is exploited in a number of applications, as we shall see at the end of this chapter.

EXAMPLE 6.5 Consider the Schottky and *p-n* diodes of Example 6.3. Compare the ac performance of the diodes when the devices are forward biased at 10 mA.

The applied bias for the Schottky diode corresponding to a 10 mA forward current is given by

$$V = \frac{k_B T}{e} \ell n \frac{I}{I_s} = 0.026(V) \times 12$$
$$= 0.31 \text{ V}$$

We need to find the built-in voltage V_{bi} that is equal to the Schottky barrier height minus the value of $\frac{E_c - E_{Fs}}{e}$. The Fermi level position in the neutral semiconductor (E_{Fs}) with respect to the conduction band is given by

$$E_c - E_{Fs} \cong k_B T \ell n \left(\frac{N_c}{N_d} \right)$$
$$\cong 0.026(\text{eV}) \times \ell n \left\{ \frac{2.8 \times 10^{19} (\text{ cm}^{-3})}{10^{16} (\text{ cm}^{-3})} \right\}$$
$$= 0.2 \text{ eV}$$

The built-in potential is

$$V_{bi} = \phi_b - \frac{1}{e}(E_c - E_F) = 0.66 - 0.2$$
$$= 0.46 \text{ V}$$

The diode capacitance is now

$$C_d = 10^{-3}(\text{cm}^2) \sqrt{\frac{(1.6 \times 10^{-19} \text{ C}) \times (8.84 \times 10^{-14} \times 11.9 \text{ F/cm}) \times (10^{16} \text{ cm}^{-3})}{2 \times (0.16(V))}}$$
$$= 72.5 \text{ pF}$$

The small-signal resistance of the Schottky (and *p-n*) diode is given by the derivative of the voltage with respect to current:

$$R = \frac{k_B T}{e} \frac{1}{I} = \frac{0.026(\text{V})}{10^{-2}(\text{A})} = 2.6\Omega$$

In the *p-n* diode, the junction capacitance and the small-signal resistance will be the same as those in the Schottky diode. However, we now have to consider the diffusion capacitance.

The diffusion capacitance is given by

$$C_{diff} = \frac{e}{k_B T} I \tau_p = \frac{10^{-2}(A) \times 10^{-6}(s)}{(0.026 \text{ V})} = 3.8 \times 10^{-7} \text{ F}$$

The diffusion capacitance completely dominates the *p-n* diode response. The RC time constant for the Schottky diode is 1.88×10^{-10} s, while that of the *p-n* diode is 9.9×10^{-7} s. Thus the *p-n* diode is almost 1000 times slower.

6.2.5   Comparison of Schottky and p-n diodes

The Schottky diodes have a number of important advantages over p-n diodes. Some of these are listed in Fig. 6.7. The temperature dependence of the Schottky barrier current is quite weak compared to that of a p-n diode. This is because in a p-n diode, the currents are controlled by the diffusion process of minority carriers, which has a rather strong temperature dependence.

The fact that the Schottky barrier is a majority carrier device gives it a tremendous advantage over p-n diodes in terms of the device speed. Device speed is no longer dependent upon extracting minority charge via diffusion or recombination. By making small devices, the RC time constant of a Schottky barrier can approach a few picoseconds, which is orders of magnitude faster than that of p-n diodes.

Another important advantage of the Schottky diode is the fact that there is essentially no recombination in the depletion region and the ideality factor is very close to unity. In p-n diodes, there is significant recombination in the depletion region and ideality factors range from 1.2 to 2.0.

The thermionic-emission-controlled prefactor gives a current density in the range of $\sim 10^{-7}$ A/cm^2, which is three to four orders of magnitude higher than that of the p-n diode. Thus for a given applied bias, the Schottky barrier has much higher current than the p-n diode. As a result the Schottky diode is preferred as a low-voltage high-current rectifier. However, the reverse current in a Schottky barrier is also quite large, which is a disadvantage for many applications. It should also be noted that since the Schottky barrier quality depends critically on the surface quality, the processing steps are quite critical. For many semiconductors, it is not possible to have a good Schottky contact since the contact is very "leaky" (i.e., there is a very high reverse-bias current). For such materials, the only way to have rectification is by using a p-n junction.

6.3   OHMIC CONTACTS

In order that an electronic device using a semiconductor satisfy the requirements for active devices, some physical property of the device should respond in a controlled manner to an external response. In most devices, the physical property chosen is the current (or conductivity) in the semiconductor. Why current and not other properties like depletion width or capacitance or total charge, etc.? The reason is the ease and speed with which current flow changes can be detected and manipulated. But for current flow to occur, electrons or holes must be able to flow "freely" in and out of the semiconductor. This certainly cannot happen if we try to inject or extract electrons directly through a vacuum (or air)-semiconductor interface. There is a large barrier (the work function)

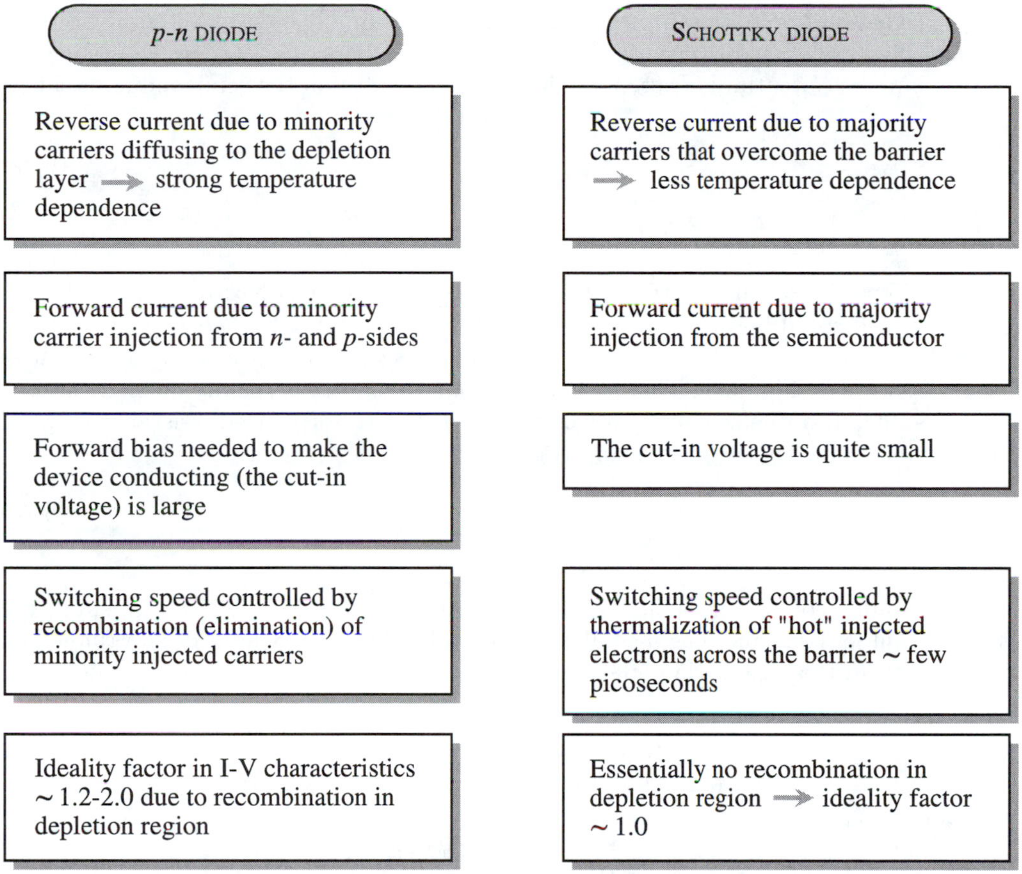

p-n DIODE	SCHOTTKY DIODE
Reverse current due to minority carriers diffusing to the depletion layer ⟶ strong temperature dependence	Reverse current due to majority carriers that overcome the barrier ⟶ less temperature dependence
Forward current due to minority carrier injection from *n*- and *p*-sides	Forward current due to majority injection from the semiconductor
Forward bias needed to make the device conducting (the cut-in voltage) is large	The cut-in voltage is quite small
Switching speed controlled by recombination (elimination) of minority injected carriers	Switching speed controlled by thermalization of "hot" injected electrons across the barrier ~ few picoseconds
Ideality factor in I-V characteristics ~ 1.2-2.0 due to recombination in depletion region	Essentially no recombination in depletion region ⟶ ideality factor ~ 1.0

Figure 6.7: A comparison of some important properties of the *p-n* diode and the Schottky diode.

that restricts the flow of electrons. We have also seen from the previous section that at least in some cases a metal-semiconductor junction also provides a barrier to flow of electrons. However, it is possible to create metal-semiconductor junctions that have a linear non-rectifying I-V characteristic, as shown in Fig. 6.8. Such junctions or contacts are called ohmic contacts.

There are two possibilities for producing ohmic contacts. In the previous section, to produce a Schottky barrier on an *n*-type semiconductor, we needed (for the ideal surface) a metal with a work function larger than that of the semiconductor. Thus, in principle, if we use a metal with a work function smaller than the semiconductor, one should have no built-in barrier. However, this approach is not used in practice because, as we noted in the previous section, the Fermi level at the surface of real semiconductors is pinned because of the high interface density in the gap.

The solution to the ohmic contact problem was briefly mentioned in the previous section when we discussed the depletion width of the Schottky barrier. Let us say we have a built-in potential barrier, V_{bi}. The depletion width on the semiconductor side is

$$W = \left[\frac{2\epsilon V_{bi}}{e N_d}\right]^{1/2} \tag{6.22}$$

Now if near the interface region the semiconductor is heavily doped, the depletion width could be made extremely narrow. In fact, it can be *made so narrow that even though there is a potential barrier, the electrons can tunnel through the barrier with ease*, as shown in Fig. 6.8b. The quality of an ohmic contact is usually defined through the resistance R of the contact over a certain area A. The normalized resistance is called the specific contact resistance r_c and is given by

$$r_c = R \cdot A \tag{6.23}$$

Under conditions of heavy doping where the transport is by tunneling, the specific contact resistance has the following dependence (see Appendix B.2 for tunneling, probability T, through a triangular barrier):

$$\ell n\,(r_c) \propto \frac{1}{\ell n(T)} \propto \frac{(V_{bi})^{3/2}}{F} \tag{6.24}$$

where the field is

$$F = \frac{V_{bi}}{W} \quad \propto \quad (V_{bi})^{1/2}\,(N_d)^{1/2} \tag{6.25}$$

Thus,

$$\begin{aligned} \ell n\,(r_c) \quad &\propto \quad V_{bi} \\ &\propto \quad \frac{1}{\sqrt{N_d}} \end{aligned} \tag{6.26}$$

The resistance can be reduced by using a low Schottky barrier height and doping as heavily as possible. The predicted dependence of the contact resistance on the doping density is, indeed, observed experimentally.

EXAMPLE 6.6 Consider a W-n Si Schottky barrier on silicon doped at 10^{18} and 10^{20} cm^{-3}. Calculate the tunneling probability of electrons for electrons with energies near the conduction band in the two doping cases.

Let us first calculate the depletion width corresponding to a Schottky barrier height of 0.66 V. In such a highly doped semiconductor we can assume that the built-in potential is

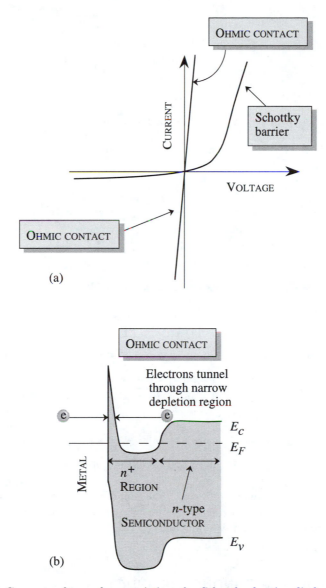

Figure 6.8: (a) Current-voltage characteristics of a Schottky barrier diode and of an ohmic contact. (b) Band diagrams of metal -n^+-n contact. The heavy doping reduces the depletion width to such an extent that the electrons can tunnel through the spiked barrier easily in either direction.

essentially the same as the barrier height. The depletion width is

$$W(n = 10^{18} \text{ cm}^{-3}) = \left[\frac{2 \times (11.9 \times 8.84 \times 10^{-14} \text{ F/cm}) \times (0.66 \text{ V})}{(1.6 \times 10^{-19} \text{ C})(10^{18} \text{ cm}^{-3})}\right]^{1/2}$$

$$= 2.9 \times 10^{-6} \text{ cm} = 290 \text{ Å}$$

$$W(n = 10^{20} \text{ cm}^{-3}) = \frac{W(n = 10^{18} \text{ cm}^{-3})}{\sqrt{100}} = 29 \text{ Å}$$

The electrons with energy just at the conduction bandedge will have to tunnel through an approximately triangular well. We can use the tunneling probability through a triangular well as discussed in Appendix B.2 using an average electric field in the depletion region:

$$\bar{F} = \frac{V_{bi}}{W} \cong \frac{\phi_b}{W}$$

$$\bar{F}(n = 10^{18} \text{ cm}^{-3}) = \frac{0.66 \text{ V}}{2.9 \times 10^{-6} \text{ cm}} = 2.3 \times 10^5 \text{ V/cm}$$

$$\bar{F}(n = 10^{20} \text{ cm}^{-3}) = \frac{0.66 \text{ V}}{2.9 \times 10^{-7} \text{ cm}} = 2.3 \times 10^6 \text{ V/cm}$$

The tunneling probability is now (assuming that the incident electrons impinge with zero energy)

$$T = \exp\left\{-\frac{4(2m^*)^{1/2}(e\phi_b)^{3/2}}{3e\bar{F}\hbar}\right\}$$

Using a density of states mass of $0.34m_o$ for one of silicon's conduction band valleys, we get

$$T(n = 10^{18} \text{ cm}^{-3}) = \exp\left\{-\frac{4 \times (2 \times 0.34 \times 0.91 \times 10^{-30} \text{ kg})^{1/2}(0.66 \times 1.6 \times 10^{-19} \text{ J})^{3/2}}{3 \times (1.6 \times 10^{-19} \text{ C})(2.3 \times 10^7 \text{ V/m})(1.05 \times 10^{-34} \text{ Js})}\right\}$$

$$= \exp(-93) \sim 0$$

In this case the tunneling current is essentially zero and the thermionic current will dominate:

$$T(n = 10^{20} \text{ cm}^{-3}) = \exp(-9.3) = 9.1 \times 10^{-5}$$

We see that at a doping of around 10^{20} cm^{-3}, the tunneling probability starts to become appreciable and the electrons can move across the junction via tunneling. The critical importance of high doping is brought out by this example.

6.4 INSULATOR-SEMICONDUCTOR JUNCTIONS

An important consideration for active devices is the isolation between an input and output signal. Since most electronic devices involve controlling the current flow through a channel, *the point that controls this current must be "insulated" from any changes in current flow.* Insulators are widely used to provide this isolation. Insulators are also used for a wide variety of other applications that we will mention later.

Insulators, as we discussed briefly in Section 1.8, are essentially like semiconductors except that their bandgaps are somewhat larger so that there are negligible electrons or holes available for conduction. The Fermi level lies close to the mid-bandgap. The resistivity of these materials, as well as the breakdown fields for these materials, are extremely high. In addition to these electrical properties, the insulators are very "robust" materials with very strong atomic bonding. Thus they can also be used for protecting or "passivating" the electronic devices.

It is useful to divide insulator-semiconductor combinations into those where there is a good lattice matching between the insulator and semiconductor and those where the matching is not present. Most insulator-semiconductor combinations involve structures that are not lattice-matched. In most cases the insulator and the semiconductor do not even share the same basic lattice type. In this section we will briefly review a few such combinations. Important issues in these junctions are listed in Fig. 6.9. The key issues here revolve around producing an interface with very low density of trapping states. Also, the interface should be as "smooth" as possible, since any roughness can cause scattering of carriers moving near the interface and thus affect mobility.

6.4.1 Silicon Dioxide-Silicon

The SiO_2-Si system is the most important junction in solid state electronics. While there is severe mismatch between SiO_2 structure and Si structure, as shown in Fig. 6.10, the interface quality has improved to such an extent that midgap interface densities as low as 10^{11} eV^{-1} cm^{-2} can be obtained. The ability to produce such high-quality interfaces is responsible for the remarkable successes of the metal-oxide-silicon (MOS) devices. Due to the low interface densities, there is very little trapping of electrons (holes) at the interface so that high-speed phenomena can be predictably used.

No other semiconductor has a natural oxide producing a high-quality interface, giving Si a unique advantage in electronic devices. While attempts have been made to grow GeO_2, success has been poor. The properties of the Si-SiO_2 system will be discussed in more detail in Chapter 9 when we discuss the MOS device.

The SiO_2 film can be produced on Si by simple thermal oxidation using oxygen or steam. The SiO_2 produced is uniform, nonporous, and adheres well to the substrate. The reaction involved is

$$Si \;+\; O_2 \longrightarrow SiO_2$$
$$Si \;+\; 2H_2O \longrightarrow SiO_2 + 2H_2 \tag{6.27}$$

As SiO_2 is deposited on Si, the Si-SiO_2 interface is buried deeper from the SiO_2 surface. An important question is whether the oxygen diffuses into the SiO_2 region to

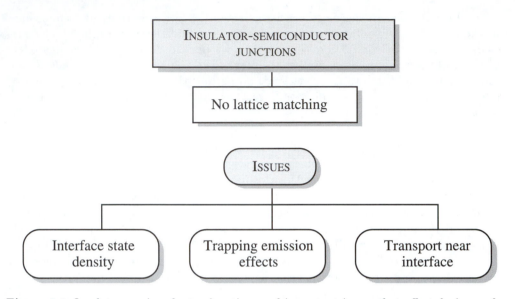

Figure 6.9: Insulator-semiconductor junctions and important issues that affect device performance.

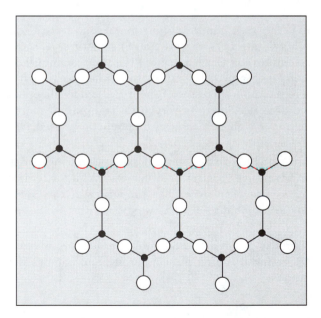

Figure 6.10: In SiO_2, the Si-O distance is 1.62 Å. Each Si atom is surrounded tetrahedrally by four oxygen atoms. Each O atom is bonded by two Si atoms. The O-O distance is $\sim$ 2.65 Å (white circles–O; dark circles–Si). The figure shows a 2D projection of the SiO_2 network. In the Si crystal, the crystal structure is diamond and the interatomic separation is $\sim$ 2.34 Å.

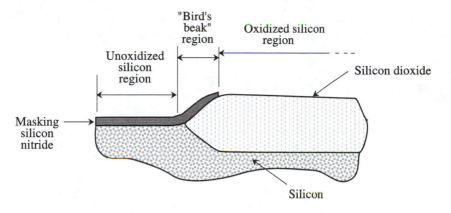

Figure 6.11: When SiO_2 is formed, the surface of the wafer projects out of $\sim 52\%$ of the total SiO_2 thickness. This leads to the "bird's beak" formation during selective oxidation.

react with silicon or whether silicon atoms diffuse outward through SiO_2 to react with oxygen. It is the oxygen atoms that diffuse through SiO_2 to consume silicon atoms. Thus, the Si-SiO_2 interface moves below the original silicon surface. It turns out that for a certain thickness of the SiO_2, 48% of the thickness is due to the Si-SiO_2 interface moving inward from its original position and 52% of the SiO_2 projecting above the original surface, as shown in Fig. 6.11. Thus, if a mask is used (say an Si_3N_4 mask) to prevent oxidation of certain regions, then the oxidation process produces a "bird's beak" near the mask edges as shown.

In addition to being an active region of the MOS devices, SiO_2 also serves as a mask for Si technology. It is used to provide a mask for the diffusion of dopants in select regions of the semiconductor. Important dopants such as B, P, As, and Sb have very low diffusion coefficients in SiO_2, so that it serves as an excellent mask. Other potential donors like Ga, Al, and In have high diffusion coefficients in SiO_2 and are, therefore, not used in Si technology.

6.4.2   Silicon Nitride-Silicon

Silicon nitride (Si_3N_4) is another important film that forms high-quality junctions with Si. It is not easily produced by direct nitridation of silicon and is usually deposited by the reaction of NH_3 with a silicon containing gas, as shown below:

$$3SiH_4 \; + \; 4NH_3 \longrightarrow Si_3N_4 + 12H_2$$
$$3SiCl_4 \; + \; 4NH_3 \longrightarrow Si_3N_4 + 12HCl \tag{6.28}$$

Silicon nitride can be used in a metal-insulator-semiconductor device in Si technology, but its applications are limited. The film is used more as a mask for oxidation of the Si film. It also makes a good material for passivation of finished devices.

6.4.3 Polycrystalline Silicon-Silicon

Although not an insulator or a metal, we include polycrystalline silicon ("poly") in this chapter because of its importance in Si technology. Polysilicon can be deposited by the pyrolysis (heat-induced decomposition) of silane:

$$\text{SiH}_4 \longrightarrow \text{Si} + 2\text{H}_2 \tag{6.29}$$

Depending upon the deposition temperature, microcrystallites of different grain sizes are produced. Typical grain size is ~ 0.1 μm.

Poly films can be doped to low resistivity to produce useful conductors for a number of applications. Poly is often used as a gate of an MOS transistor, as a resistor, or as a link between a metal and the Si substrate to ensure an ohmic contact.

EXAMPLE 6.7 In this example we will study some important concepts in thin-film resistors, which form an important part of semiconductor device technology. The resistors are often made from polysilicon that is appropriately doped. In thin-film technology it is usual to define sheet resistance instead of the resistance of the material. Consider, as shown in Fig. 6.12a, a material of length L, width W, and depth D. The resistance of the material is

$$R = \frac{\rho L}{WD} = \frac{\rho L}{A} \tag{6.30}$$

As we have discussed in Chapter 3, the resistivity ρ is given by

$$\rho = \frac{1}{ne\mu} \tag{6.31}$$

where n is the free carrier density and μ is the mobility of the carriers (the equation can be modified for a p-type material).

The sheet resistance is a measure of the characteristics of a uniform sheet of film. It is defined as ohms per square, as shown in Fig. 6.12b, and is related to the film resistance by

$$R_\square = R\frac{W}{L} \tag{6.32}$$

Consider a film 5000 Å thick doped n-type at 10^{18} cm^{-3}. Assume that the mobility is 100 cm^2/V·s. Calculate the sheet resistance of the film. What is the length of a resistor with resistance of 100 $k\Omega$ if the width of the resistor is 25 μm?

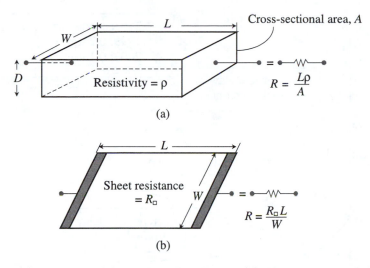

Figure 6.12: (a) A resistor of dimensions $L \times W \times D$. (b) Representation of the resistors in terms of sheet resistance.

From the definition of $R_\square$ and ρ we see that

$$R_\square = \frac{1}{ne\mu D}$$

$$= \frac{1}{\left(10^{18}\ \text{cm}^{-3}\right)\left(1.6 \times 10^{-19}\ \text{C}\right)\left(100\ \text{cm}^2/\text{V}\cdot\text{s}\right)\left(5000 \times 10^{-8}\ \text{cm}\right)}$$

$$= 1250\ \Omega/\square$$

The length of the desired resistor is

$$L = \frac{RW}{R_\square} = 0.2\ \text{cm}$$

This is a fairly large distance on the scale of microelectronics.

6.5 TECHNOLOGY ROADMAP

In this chapter we have discussed the role of metals in interconnects and of insulators as barriers to charge flow. The most important metal used for interconnects is Al (this is expected to change, as we discuss below), while the most important insulator in microelectronic technology is SiO_2. In this chapter we will briefly examine important issues for interconnects and insulators for the next-generation microelectronics.

Interconnects: From Aluminum to Copper to ... ?
Aluminum has been the mainstay of interconnect technology. It has a reasonably high conductivity and can be deposited without problems associated with buckling or peeling

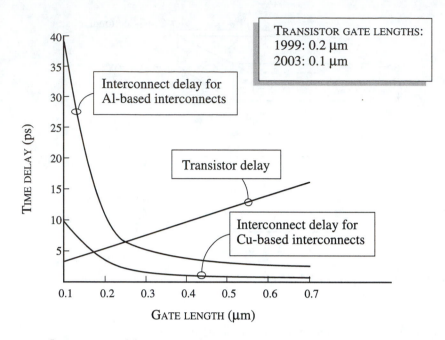

Figure 6.13: Interconnect delay versus transistor gate length for silicon technology. Results are shown for Al and Cu technologies. (Adapted from the website www.sia.com operated by Semiconductor Industry Associates.

off. The key property of interconnects is the resistance associated with them and the maximum current density that can pass through without creating deterioration (structural or electrical). The resistance is controlled by the conductivity, area, and length of the interconnect. It is important to note that the conductivity of a thin film of a material may be quite different from the bulk conductivity due to the importance of surface effects. The ability of the interconnect material to withstand high current densities ($\sim 10^5$ A/cm^2) is an important requirement, since in most microelectronic devices one does need high drive currents.

As the density of devices on a chip becomes higher, the area of a typical interconnect has to become smaller. Thus the RC delay time for signals to propagate from one device to another will increase. Semiconductor Industry Associates (SIA) estimate that interconnect delay will become increasingly dominant and overtake device-switching delay in the first decade of the twenty-first century.

There are two important materials that have promising potential as interconnects. The first one is copper, which, by virtue of its higher conductivity, can give performance improvements over aluminum. The other is high-temperature superconducting materials, which have the property that the conductivity is essentially infinite as long as the temperature is below a certain minimum value T_c.

After decades of research, copper technology finally began to make it into products in 1998. The problems associated with strong adhesion, uniformity, etc., have been overcome and IBM has started offering high-performance microprocessors based on copper interconnects. In Fig. 6.13 we show time delays for switching of MOS transistors in a microprocessor as a function of transistor gate length. Two types of delays are shown. The transistor delay arises from the time it takes electrons to traverse the distance of the transistor (gate length). As gate length decreases, this time decreases, as shown. The interconnect delay is controlled by the RC time constant for the interconnect line. As gate length decreases, interconnect thickness also has to decrease. As a result, the interconnect resistance increases and the delay also increases. We can see that once transistor gate length reaches 0.25 μm, the transistor delay and interconnect delay for Al-based interconnects become equal. As the devices shrink further, the interconnect delay dominates. By going to copper interconnects, the interconnect delay can be dramatically decreased. The ability to use copper in IC technology has been hailed as a major milestone in microelectronics.

The discovery of high-T_c superconductors in the early 1980s is expected to have a major impact on a wide range of technologies. Their impact as interconnects in microelectronic technology has not materialized because of their inability to handle high current densities and because it has been difficult to deposit films with good structure on silicon wafers.

Insulators

Insulators are materials with large bandgap and very low free-carrier densities. Their primary applications in microelectronics are to prevent current flow from one contact to another and to "protect" the structural quality of a device. Silicon-dioxide and silicon-nitride are the most important insulators used in microelectronics. We have already described the benefits of these materials. Here will briefly outline the problems faced by technology over the next decade or so.

The most important challenge to insulators comes from the quantum-mechanical tunneling of electrons through a barrier. The tunneling probability of electrons through a potential barrier depends upon the height and thickness of the barrier. SiO_2 as an insulator offers a high barrier to the flow of electrons or holes. However, as devices shrink, the thickness of the oxide has to shrink. This issue is discussed in Chapter 10. It is expected that by 2001 the oxide thickness will approach 20 Å. At this point, tunneling becomes quite strong and SiO_2 cannot be used as an insulator to further advance Si technology. There is now considerable effort devoted to finding other useful insulators. Silicon nitride is one candidate, although the expected improvements over SiO_2 are small.

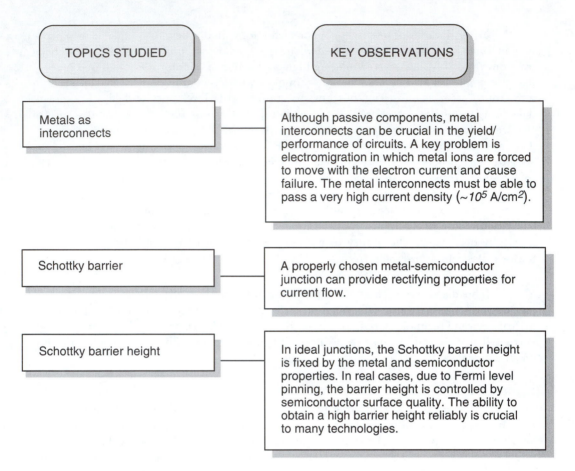

Table 6.3: Summary table.

6.6 SUMMARY

In this chapter we have discussed the properties of the junctions that play an important role in solid state electronics. Metals are an important material system for solid state electronics and, as we have seen, they provide interconnects, Schottky barriers, and ohmic contacts for the technology. In fact, it would be difficult to imagine how solid state devices could be so versatile if there were no metals. The important issues discussed in this chapter are summarized in Table 6.3.

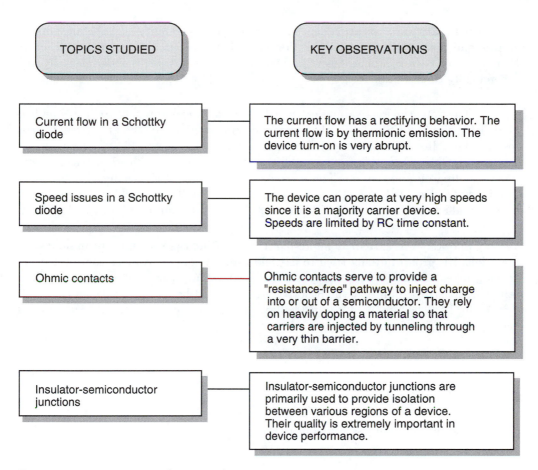

Table 6.3: Summary table (continued).

6.7 A BIT OF HISTORY

The use of solid state components in the electrical industry dates back to the 1910s. It was recognized that certain solid state devices could be used as rectifiers. Selenium was widely used as a rectifier for telegraphy or for photocells. A new advanced rectifier was developed in 1925 based on the junction between copper and copper oxide. The physics of such a device was, however, not understood. A major effort to understand the cuprous oxide rectifier was made at Siemens Laboratory. At this time Walter Schottky was a consultant to Siemens and started to tackle the problem of the junction rectifier. Along with his colleague Deutschmann, Schottky was able to show in 1929 that the rectification occurred in a thin layer at the junction that was only a few hundred angstroms thick. After nearly a decade of work, in 1938 Schottky published his "barrier layer theory" dealing with conduction at a metal-semiconductor junction. While the theory did explain rectification by such junctions, there were many unexplained phenomena. For example, it appeared that the rectification depended strongly on the quality of the semiconductor surface and the metal purity. Indeed, the effect of surface conditions on the "Schottky barrier" properties continues to be an area of active study both experimentally and theoretically to this date.

The potential applications of Schottky's research was apparent to the researchers at Bell Labs in the USA. Bell Labs was fast emerging as one of the premier industrial research labs. However, because of the worldwide recession in the 1930s, a hiring freeze was made. Only when this freeze was lifted in 1936 were capable theoreticians with understanding of quantum mechanics hired to solve many of the challenging problems in solid state physics. Among the first few recruits was William Shockley, who had done his doctoral research in the area of diffusion of electrons in sodium chloride under Slater.

Shockley was extremely interested in duplicating the vacuum tube in solid state materials. His group initially tried to use fine wires in copper oxide as control grids to control current. This, however, failed. When Shockley came across Schottky's research, it immediately struck him that the Schottky barrier could be used to make an electronic device. In 1939 he proposed the field-effect transistor, although the device was not actually developed for some time.

6.8   PROBLEMS

Temperature is 300 K unless stated otherwise.

Section 6.1
Problem 6.1 A 2 μm thick $\times 10$ μm Al interconnect is used in a semiconductor chip. If the length of a particular interconnect is 1 cm, calculate the resistance of the line. Use Table 6.1 for the resistivity data.

Problem 6.2 If a current density of 10^5 A/cm^2 flows in the interconnect of Problem 6.1, calculate the potential drop.

Problem 6.3 Use the resistivity of Al and Cu given in Table 6.1 and use the Drude model (Chapter 1) to calculate the mobility of electrons in these two materials.

Problem 6.4 Discuss why in the analysis of the conductivity of metals we do not consider hole conductivity.

Section 6.2
Problem 6.5 Assume the ideal Schottky barrier model with no interface states for an n-type Si with $N_d = 10^{16}$ cm^{-3}. The metal work function is 4.5 eV and the Si electron affinity is 4 eV. Calculate the Schottky barrier height, built-in voltage, and depletion width at no external bias.

Problem 6.6 A Schottky barrier is formed between Al and n-type silicon with a doping of 10^{16} cm^{-3}. Calculate the theoretical barrier if there are no surface states. Compare this with the actual barrier height. Use the data in the text.

Problem 6.7 Assume that at the surface of GaAs, 50% of all bonds are "defective" and lead to a uniform distribution of states in the bandgap. Each defective bond contributes one bandgap state. What is the two-dimensional density of bandgap states (units of eV^{-1}cm^{-2})? Assume that the neutral level ϕ_o is at midgap. Approximately how much will the Fermi level shift if a total charge density of 10^{12} cm^{-2} is injected into the surface states? This example gives an idea of "Fermi level pinning".

Problem 6.8 The capacitance of a Pt-n-type GaAs Schottky diode is given by

$$\frac{1}{\left(C(\mu F)\right)^2} = 1.0 \times 10^5 - 2.0 \times 10^5 \text{ V}$$

The diode area is 0.1 cm^2. Calculate the built-in voltage V_{bi}, the barrier height, and the doping concentration.

Problem 6.9 Calculate the mean thermal speed of electrons in Si and GaAs at 77 K and 300 K. $m_{Si}^* = 0.3m_o; m_{GaAs}^* = 0.067m_o$.

Problem 6.10 Calculate the saturation current density in an Au Schottky diode made from n-type GaAs at 300 K. Use the Schottky barrier height values given in Table 6.2.

Problem 6.11 Consider an Au n-type GaAs Schottky diode with 50 μm diameter. Plot the current voltage characteristics for the diode between a reverse voltage of 2 V and a forward voltage of 0.5 V.

Problem 6.12 Calculate and plot the I-V characteristics of a Schottky barrier diode between W and n-type Si doped at 5×10^{16} cm^{-3} at 300 K. The junction area is 1 mm^2. Plot the results from a forward current of 0 to 100 mA.

Problem 6.13 In some narrow-bandgap semiconductors, it is difficult to obtain a good Schottky barrier (with low reverse current) due to the very small barrier height. Consider an n-type InGaAs sample. Describe, on physical bases, how the "effective" Schottky barrier height can be increased by incorporating a thin p-type doped region near the surface region.

Problem 6.14 In the text, when we discussed the current flow in a Schottky barrier, we assumed that the current was due to thermionic emission only. This is based on classical physics where it is assumed that only particles with energy greater than a barrier can pass through. Consider a W-n-type GaAs Schottky barrier in which the Schottky barrier triangular potential is described by a field of 10^5 V/cm. The Schottky barrier height is 0.8 V. Calculate the tunneling probability through the triangular barrier as a function of electron energy from $E = 0.4$ eV to $E = 0.8$ eV. The tunneling current increases the Schottky reverse current above the value obtained by thermionic current considerations.

Problem 6.15 In the Schottky barrier, the electrons are injected across the barrier with energies equal to the barrier height. These electrons are very hot. Estimate the "temperature" of these electrons in a typical Si Schottky barrier with a barrier height of $\phi_b = 0.6$ V. (Electron temperature, T_e, is defined by $\frac{3}{2}k_B T_e \sim \langle E_e \rangle$ where $\langle E_e \rangle$ is the average electron energy.)

Problem 6.16 An important consideration in the speed of Schottky barrier diodes is the time it takes hot electrons (see the previous problem) to lose their energy and achieve equilibrium thermal energy. In GaAs, electrons lose excess energy exponentially with a time constant of 1 ps. Consider a W-n-type GaAs Schottky diode with $\phi_b = 0.8$ V. How far will electrons move in the GaAs before they lose 99% of their energy?

Problem 6.17 Consider an Al-n-type Si Schottky diode. The semiconductor is doped at 10^{16} cm^3. Also consider a p-n diode made from Si with the following parameters (the diode is ideal):

$$
\begin{aligned}
N_d &= N_a = 10^{18} \ \text{cm}^{-3} \\
D_n &= 25 \ \text{cm}^2/\text{s} \\
D_p &= 8 \ \text{cm}^2/\text{s} \\
\tau_n &= \tau_p = 10 \ ns
\end{aligned}
$$

Calculate the turn-on voltages for the Schottky and p-n diode. Assume that the current density has to reach 10^5 A/cm^2 for the diode to be turned on.

Section 6.3
Problem 6.18 A gold contact is deposited on GaAs doped at $N_d = 5 \times 20$ cm^{-3}. Calculate the tunneling probability of the electrons to go into the semiconductor.

Problem 6.19 A metal with a work function of 4.2 V is deposited on an n-type silicon semiconductor with an electron affinity of 4.0 V. Assume that there are no interface states. Calculate the doping density for which there is no space charge region at zero applied bias.

Problem 6.20 To fabricate very high-quality ohmic contacts on a large-bandgap material one often deposits a heavily doped low-bandgap material. For example, to make an n-type ohmic contact on GaAs, one may deposit n^+InAs. Discuss why this would help to improve the resistance of the contact.

Section 6.4
Problem 6.21 A (001) Si-SiO$_2$ interface has $\sim 10^{11}$ cm^{-2} interface states. Assume that each state corresponds to one defective bond at the interface. Calculate the fraction of defective bonds for the interface.

Problem 6.22 Calculate the sheet resistance of a 0.5 μm thick poly film doped n-type at 10^{19} cm^{-3}. This film is used to form a resistor of width 20 μm and length 0.1 cm. Calculate the resistance of the contact if the electron mobility is 150 cm^2/V·s.

6.9   READING LIST

  - D. A. Neamen, *Semiconductor Physics and Devices: Basic Principles* (Irwin, Boston, 1997).
  - M. S. Tyagi, *Introduction to Semiconductor Materials and Devices* (John Wiley and Sons, New York, 1991).
  - R. S. Muller and T. I. Kamins, *Device Electronics for Integrated Circuits* (Wiley, New York, 1986).

Chapter

7

BIPOLAR JUNCTION TRANSISTORS

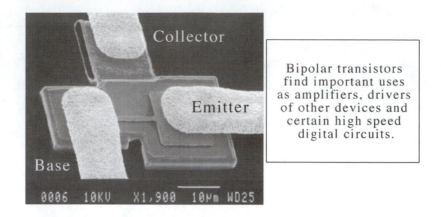

7.1 INTRODUCTION

The bipolar junction transistor was invented in 1948 by John Bardeen and Walter Brittain and its principle of operation was explained by W. Shockley a year later. This was the first three-terminal device in solid state electronics and continues to be a device of choice for many digital and microwave applications. For a decade after its invention, the bipolar device remained the only three-terminal device in commercial applications. This was because field-effect transistors could not be made with reliability at that time. The Si-SiO$_2$ interface discussed in Chapter 6 had serious interface state problems so that the metal oxide semiconductor (MOS) device was not very reliable. However, with the perfection of the Si-SiO$_2$ interface, the MOS device has been able to mount a serious challenge to the bipolar device. A number of applications have been taken over by MOS devices. Also, there is challenge from field-effect devices based on semiconductors such as GaAs, AlGaAs, InGaAs, InP, etc. However, the bipolar and heterojunction bipolar devices have also made steady improvement and continue to be workhorse devices for many applications.

In a transistor, whether a bipolar transistor or a field-effect transistor, the goal is to use a small input to control a large output. The input could be an incoming weak signal to be amplified, or a digital signal. So one has to design a device where an input can effectively control an output. Consider a scenario where you were asked to design a system that would control the flow of water from one point to another. In one case, let's say the water was to flow in a pipe of fixed diameter while in another, it could flow over an open channel. In Fig. 7.1 we show two different ways one could design a system to control the water flow. On the left-hand side sequence of Fig. 7.1 we show how a ground potential "bump" can be used to modify the water flow. Only the fraction of water that is above the bump will flow across the potential profile. The value of the bump could be controlled by an independent control input.

The right-hand sequence of Fig. 7.1 shows a standard faucet in which the faucet

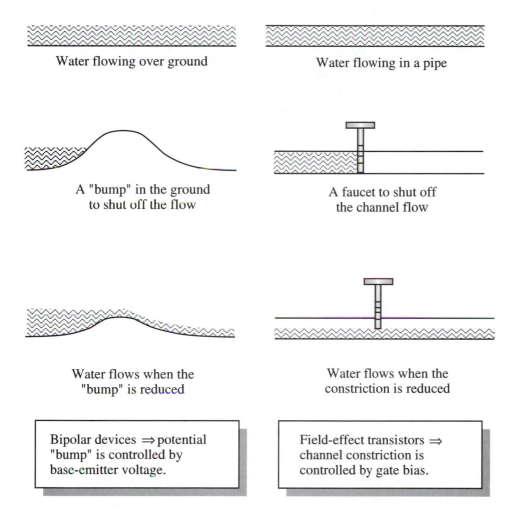

Figure 7.1: Two different ways to control water flow. The BJT and FET use similar approaches to control current flow.

controls the constriction of the pipe and thus the water flow. The bipolar device controls the potential profile in the current flow channel by using the base current as a controlling agent. The FET, on the other hand, is like a tap and controls the channel constriction by applying a gate bias.

As noted earlier, an important requirement for an electronic device is that a small change in the input should cause a large change in the output, i.e., the device should have a high gain. This requirement is essential for amplification of signals, tolerance of high noise margins in digital devices, and the ability to have a large fan-out (i.e., the output can drive several additional devices). Another important requirement is that the input should be isolated from the output. For the faucet example, these two

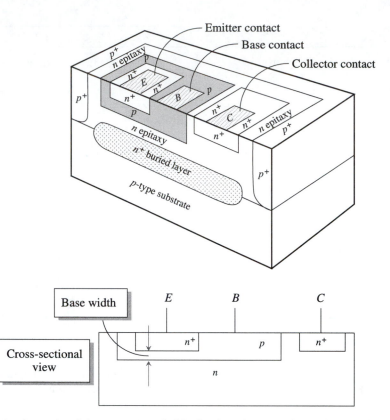

Figure 7.2: A schematic of the structure of a bipolar junction transistor along with a simplified view of the cross-section.

requirements mean we should be able to turn the faucet on and off with little effort and the water should not leak out of the faucet head!

In this chapter we will discuss the bipolar transistor. Sections 7.2 to 7.5 discuss the static operation of the BJT. Sections 7.6 through 7.9 deal with time-dependent issues. The instructor should decide in how much depth these topics should be covered for the class. Section 7.10 deals with heterojunction bipolar transistors.

7.2 BIPOLAR TRANSISTOR: A CONCEPTUAL PICTURE

In Fig. 7.2 we show a schematic of a bipolar junction transistor. The BJT essentially consists of a back-to-back p-n diode. The device could have a doping of the form $n^+ - p$-n or $p^+ - n$-p. We will focus on the $n^+ - p$-n device. The heavily doped region (n^+) is called the emitter, the p-region forms the base, and the lower n region is the collector. The emitter doping N_{de} is much larger than the base doping N_{ab} to ensure that the

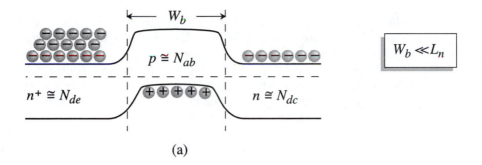

Figure 7.3: (a) Band profile of an unbiased $n^+p\text{-}n$ BJT. (b) Band profile of a BJT biased in the forward active mode.

device has a high current gain, i.e., that a small base current change produces a large collector current charge.

Let us consider a BJT where the emitter base junction (EBJ) is forward biased and the base collector junction (BCJ) is reverse biased. The BJT is said to be biased in the forward active mode. The band profile of the device is shown in Fig. 7.3. Note that the base width W_b is much smaller than the diffusion length of electrons in the p-type base region. Thus, when electrons are injected from the emitter, most of them cross the base without recombining with holes. Due to the strong electric field these electrons see once they reach the collector, they are swept away and form the collector current.

If the diode is n^+-p, the forward bias current is essentially made up of injection of electrons into the p-side (see Example 5.6). This forward-biased current can also be altered by a very small change in the forward bias voltage since the current depends *exponentially on the forward bias value.* The forward-biased n^+ emitter injects electrons into the p-base. Some of the electrons recombine in the base with the holes, but if the base region is less than the diffusion length of the minority carriers, most of them reach the depletion region of the p-n base-collector diode and are swept out to form the collector current. The collector current is proportional to the minority carriers (electrons) that reach the edge of the p-n depletion region, as shown in Fig 7.3b. Since the injected minority carriers are due to the emitter current, we have

$$I_C = BI_{En} \tag{7.1}$$

where I_{En} is the electron part of the emitter current. The factor B is called the base transport factor and its value is less than unity (since some of the electrons recombine in the base and do not reach the collector). The emitter current is made up of electrons injected from the n- to p-sides (I_{En}) and holes injected from p- to n-sides (I_{Ep}). The collector current is related only to the electrons injected and we define the emitter efficiency γ_e as

$$\boxed{\gamma_e = \frac{I_{En}}{I_{En} + I_{Ep}}} \tag{7.2}$$

For optimum devices, γ_e and B should be close to unity. The ratio between the collector and emitter currents is

$$\boxed{\frac{I_C}{I_E} = \frac{BI_{En}}{I_{En} + I_{Ep}} = B\gamma_e = \alpha} \tag{7.3}$$

where α is called the current transfer ratio. This ratio is close to unity in good bipolar devices. In Fig. 7.4 we show a typical circuit for a BJT in the forward-bias active mode. A change in the base current alters the minority carrier density n_p in the base and causes a large change in the collector current. The ratio between the collector current and the *controlling* base current is of great importance since this represents the current amplification. The base current is made up of the hole current injected into the emitter (I_{Ep}) and the hole current due to the recombination in the base with injected electrons from the emitter $(= (1-B)I_{En})$. Thus

$$I_B = I_{Ep} + (1 - B)I_{En} \tag{7.4}$$

where we have assumed there is negligible current associated with the base-collector junction, which is strongly reverse biased. The base-to-collector current amplification factor, denoted by β (also denoted by h_{FE}) is given by

$$\beta = \frac{I_C}{I_B} = \frac{BI_{En}}{I_{Ep} + (1 - B)I_{En}} = \frac{B(I_{En}/I_E)}{1 - B(I_{En}/I_E)}$$

$$= \frac{B\gamma_e}{1 - B\gamma_e} \tag{7.5}$$

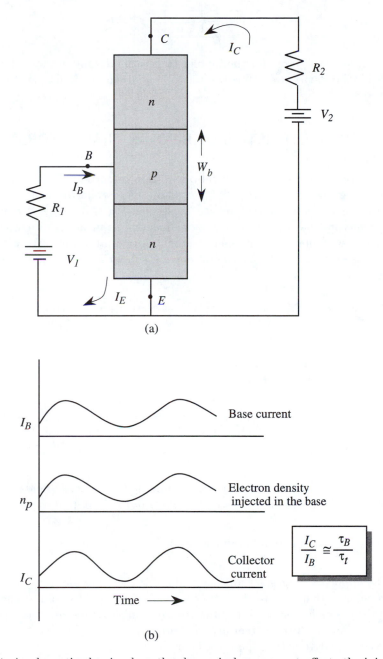

Figure 7.4: A schematic showing how the change in base current affects the injected charge density and the collector current in a bipolar device. (a) A circuit using a bipolar transistor. (b) The effect of base current variation on the injected minority charge and the collector current.

This gives

$$\boxed{\beta = \frac{\alpha}{1 - \alpha}}$$

(7.6)

The factor β can be quite large for the bipolar transistor. Having developed a physical picture of the bipolar device, we will now discuss the mathematical details of the device.

7.3   STATIC CHARACTERISTICS OF BIPOLAR TRANSISTORS

The bipolar device has a complex physical orientation of current flow, as the carriers from the emitter are injected "vertically" across the base while the base charge is injected from the "side" of the device, as can be seen in Fig. 7.5a. However, unless the emitter area is very narrow, the device can be understood using a one-dimensional analysis. Additionally, we will make a number of simplifying assumptions. (We will focus on the *npn* transistors, although the considerations are similar for the *pnp* device.)

1. We will use a BJT with a uniform cross-section and use the one-dimensional analysis.

2. The electrons injected from the emitter diffuse over the base region and the field across the base is small enough that there is no drift.

3. The injected electron (minority) density in the base is much larger than the equilibrium minority carrier density in the base, but is much smaller than the majority (hole) density in the base.

4. As in our *p-n* junction studies, we assume that the electric fields are nonzero only in the depletion regions and are zero in the bulk materials.

5. The collector injection current is negligible.

6. The applied voltages are all steady state. If there are any temporal changes, we assume that the steady state conditions continue to hold at each instant of time. The first subscript of the voltage symbol represents the contact with respect to which the potential is measured. For example, $V_{BE} > 0$ means the base is positive with respect to the emitter.

In real-life devices many of these approximations are not satisfied, and we will examine later what the consequences are for device performance.

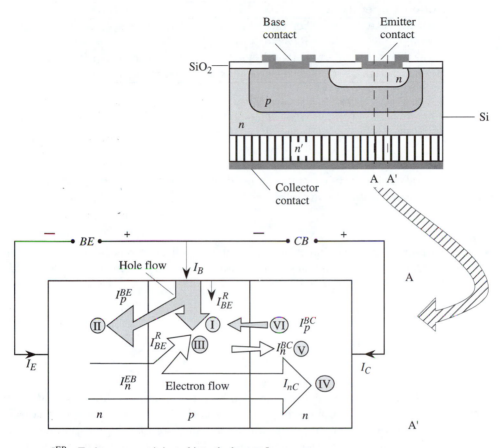

I_n^{EB} = Emitter current injected into the base $\equiv I_{En}$

I_p^{BE} = Base current injected into the emitter $\equiv I_{Ep}$

I_{BE}^{R} = Recombination current in the base region

I_p^{BC} = Hole current injected across reverse-biased base collector junction

I_n^{BC} = Electron current injected across reverse-biased base collector junction

I_{nC} = Electron current coming from the emitter ($\cong I_C$)

Figure 7.5: A schematic of an Si BJT showing the three-dimensional nature of the structure and the current flow. Along the section AA', the current flow can be assumed one-dimensional. The various current components in a BJT are also shown.

Mode of operation	EBJ bias	CBJ bias
Forward active	Forward ($V_{BE} > 0$)	Reverse($V_{CB} > 0$)
Cutoff	Reverse ($V_{BE} < 0$)	Reverse ($V_{CB} > 0$)
Saturation	Forward ($V_{BE} > 0$)	Forward ($V_{CB} < 0$)
Reverse active	Reverse ($V_{BE} < 0$)	Forward ($V_{CB} < 0$)

Table 7.1: Operation modes of the *npn* bipolar transistor. Depending upon the particular application, the transistor may operate in different modes.

In general, a number of currents can be identified in the bipolar device, as shown in Fig. 6.5b. We can identify them as follows:

- Base current: Consists of holes that recombine with electrons injected from the emitter (Component I) and holes that are injected across the emitter-base junction into the emitter (Component II).

- Emitter current: Consists of the current that recombines with the holes in the base region (III) and the part which is injected into the collector (IV).

In addition, there are minority electron (V) and hole (VI) currents that flow in the base-collector junction and are important when the emitter current goes towards zero. In our analysis, we will assume that all the dopants are ionized and the majority carrier density is simply equal to the doping density. The symbols for the doping density are (for the *npn* device): N_{de}—donor density in the emitter; N_{ab}—acceptor density in the base; N_{dc}—donor density in the collector.

7.3.1 Biasing of the BJT

The bipolar device can operate in four distinct biasing modes as shown in Table 7.1. Depending upon the applications, the bipolar device operation may span one or all of these modes. For example, for small-signal applications (microwave devices) one only operates in the forward active mode, while for switching applications the device may have to operate under cutoff and saturation modes and pass through the active mode during the switching. It is important for us to understand the bipolar currents that are generated under all of these operating modes.

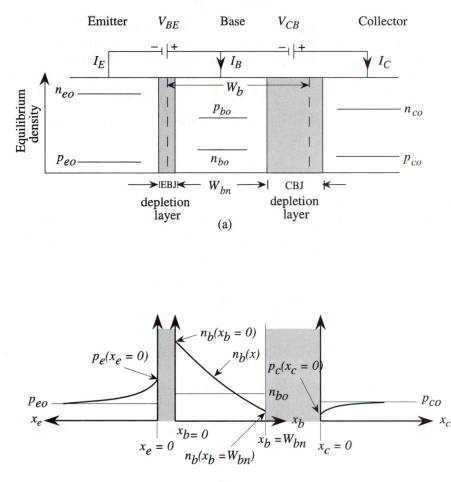

Figure 7.6: (a) The equilibrium carrier concentrations of electrons and holes and positions of the junction depletion regions in the *npn* transistor. (b) Minority carrier distributions in the emitter, base, and collector regions.

7.3.2 Current Flow in a BJT

In order to derive the various current components flowing in the bipolar device, we will make use of our understanding of the *p-n* junction developed in Chapter 5. In that chapter, we made a number of approximations regarding the charge density and current flow across a *p-n* junction. We will retain those approximations for the study of the bipolar device.

For simplicity we will use the different axes and origins shown in Fig. 7.6. The distances are labeled x_e, x_b, and x_c as shown and are measured from the edges of the depletion region. The base width is W_b, but the width of the "neutral" base region is W_{bn} as shown. We assume that W_b and W_{bn} are equal. Later we will study the effect of the two widths being different. Using the p-n diode theory, we have the following relations for the excess carrier densities in the various regions (see Section 5.3.1):

$$
\begin{aligned}
\delta p_e(x_e = 0) &= \text{excess hole density at the emitter side of the EBJ} \\
&\quad \text{(emitter-base junction)} \\
&= p_{eo}\left[\exp\left(eV_{BE}/k_BT\right) - 1\right] & (7.7) \\
\delta n_b(x_b = 0) &= \text{excess electron density on the base side of the EBJ} \\
&= n_{bo}\left[\exp\left(eV_{BE}/k_BT\right) - 1\right] & (7.8) \\
\delta n_b(x_b = W_{bn}) &= \text{excess electron density at the base side of the CBJ} \\
&\quad \text{(collector-base junction)} \\
&= n_{bo}\left[\exp\left(-eV_{CB}/k_BT\right) - 1\right] & (7.9) \\
\delta p_c(x_c = 0) &= \text{excess hole density at the collector side of the CBJ} \\
&= p_{co}\left[\exp\left(-eV_{CB}/k_BT\right) - 1\right] & (7.10)
\end{aligned}
$$

In these expressions the subscripts p_{eo}, n_{bo}, and p_{co} represent the minority carrier equilibrium densities in the emitter, base, and collector, respectively. The total minority carrier concentrations p_e in the emitter, n_b in the base, and p_c in the collector are shown schematically in Fig. 7.6b. The majority carrier densities are n_{eo} $(= N_{de})$, p_{bo} $(= N_{ab})$, and n_{co} $(= N_{dc})$ for the emitter, base, and collector. We will assume that the emitter and collector regions are longer than the hole diffusion lengths L_p, so that the hole densities decrease exponentially away from base regions.

In the base region, the excess electron density is given at the edges of the neutral base region by Eqns. 7.8 and 7.9. To obtain the electron density in the base we must solve the continuity equation using these two boundary conditions, as discussed in Section 3.9. The excess minority carrier density in the base region is given by (see Eqn. 3.56)

$$
\begin{aligned}
\delta n_b(x_b) &= \frac{n_{bo}}{\sinh\left(\frac{W_{bn}}{L_b}\right)}\left\{ \sinh\left(\frac{W_{bn} - x_b}{L_b}\right)\left[\exp\left(\frac{eV_{BE}}{k_BT}\right) - 1\right]\right. \\
&\quad \left. + \sinh\left(\frac{x_b}{L_b}\right)\left[\exp\left(-\frac{eV_{CB}}{k_BT}\right) - 1\right]\right\} & (7.11)
\end{aligned}
$$

The form of the total minority carrier densities (i.e., background and excess) is shown in Fig. 7.6b. The electron distribution in the base is almost linear, as can be seen, and is assumed to be so for some simple applications. Once the excess carrier spatial distributions are known we can calculate the currents as we did for the p-n diode. We assume that the emitter-base currents are due to carrier diffusion once the device is biased. We have, for a device of area A and diffusion coefficients D_b and D_e in the base and emitter, respectively,

$$I_{En} = I_n^{EB} = eAD_b \left.\frac{d\delta n_b(x)}{dx_b}\right|_{x_b=0} \tag{7.12}$$

$$I_{Ep} = I_p^{BE} = -eAD_e \left.\frac{d\delta p(x)}{dx_e}\right|_{x_e=0} \tag{7.13}$$

These are the current components shown in Fig. 7.5b and represent the emitter current components III and IV. Assuming an exponentially decaying hole density into the emitter, we have, as in the case of a *p-n* diode (see Eqn. 5.42),

$$I_{Ep} = -A \left(\frac{eD_e p_{eo}}{L_e}\right) \left[\exp\left(\frac{eV_{BE}}{k_B T}\right) - 1\right] \tag{7.14}$$

Using the electron distribution derived earlier (Eqn. 7.11), we have for the electron part of the emitter current

$$
\begin{aligned}
I_{En} &= -\frac{eAD_b n_{bo}}{L_b \sinh\left(\frac{W_{bn}}{L_b}\right)} \left\{ \cosh\left(\frac{W_{bn}-x_b}{L_b}\right) \left[\exp\left(\frac{eV_{BE}}{k_B T}\right) - 1\right] \right. \\
&\quad \left. \left. - \cosh\left(\frac{x_b}{L_b}\right) \left[\exp\left(-\frac{eV_{CB}}{k_B T}\right) - 1\right] \right\} \right|_{at\ x_b=0} \\
&= -\frac{eAD_b n_{bo}}{L_b \sinh\left(\frac{W_{bn}}{L_b}\right)} \left\{ \cosh\left(\frac{W_{bn}}{L_b}\right) \left[\exp\left(\frac{eV_{BE}}{k_B T}\right) - 1\right] \right. \\
&\quad \left. - \left[\exp\left(-\frac{eV_{CB}}{k_B T}\right) - 1\right] \right\}
\end{aligned}
\tag{7.15}
$$

Note that for high emitter efficiency and high current gain the hole current into the emitter I_{Ep} *can be reduced if the emitter is very heavily doped so that p_{eo} is very small.* We will return to this issue later. The total emitter current is now

$$
\begin{aligned}
I_E = I_{En} + I_{Ep} &= -\left\{ \frac{eAD_b n_{bo}}{L_b} \coth\left(\frac{W_{bn}}{L_b}\right) + \frac{eAD_e p_{eo}}{L_e} \right\} \\
&\quad \left[\exp\left(\frac{eV_{BE}}{k_B T}\right) - 1\right] + \frac{eAD_b n_{bo}}{L_b \sinh\left(\frac{W_{bn}}{L_b}\right)} \left[\exp\left(-\frac{eV_{CB}}{k_B T}\right) - 1\right]
\end{aligned}
\tag{7.16}
$$

The collector current components can be obtained by using the same approach. Thus we have

$$I_n^{BC} = eAD_b \left.\frac{d\delta n_b(x_b)}{dx_b}\right|_{x_b=W_{bn}} \tag{7.17}$$

$$I_p^{BC} = eAD_p \left.\frac{d\delta p(x_c)}{dx_c}\right|_{x_c=0} \tag{7.18}$$

Using the results shown in the first part of Eqn. 7.15 at $x_b = W_{bn}$, we have

$$
I_n^{BC} = - \frac{eAD_b n_{bo}}{L_b \sinh(W_{bn}/L_b)} \left[\exp\left(\frac{eV_{BE}}{k_B T} \right) - 1 \right]
$$
$$
+ \frac{eAD_b n_{bo}}{L_b} \coth\left(\frac{W_{bn}}{L_b} \right) \left[\exp\left(-\frac{eV_{CB}}{k_B T} \right) - 1 \right] \qquad (7.19)
$$

The hole current on the collector side is the same as for a reverse-biased p-n junction:

$$
I_p^{BC} = - \frac{eAD_c p_{co}}{L_c} \left[\exp\left(-\frac{eV_{CB}}{k_B T} \right) - 1 \right] \qquad (7.20)
$$

From the way we have defined the currents, the two current components flow along $+x$ direction. If we define I_C as the total current flowing from the collector into the base, we have

$$
-I_C = \left[\frac{eAD_c p_{co}}{L_c} + \frac{eAD_b n_{bo}}{L_b} \coth\left(\frac{W_{bn}}{L_b} \right) \right] \left[\exp\left(-\frac{eV_{CB}}{k_B T} \right) - 1 \right]
$$
$$
- \frac{eAD_b n_{bo}}{L_b \sinh\left(\frac{W_{bn}}{L_b} \right)} \left[\exp\left(\frac{eV_{BE}}{k_B T} \right) - 1 \right] \qquad (7.21)
$$

We now have all the various current components in the bipolar device ($I_B = I_E - |I_C|$). It is interesting to point out that if the base region W_{bn} is much smaller than the diffusion length, the electron gradient in the base region is simply given by the linear value

$$
\frac{d\delta n_b(x_b)}{dx_b} \longrightarrow \frac{\delta n_b(x_b) - \delta n_b(x_b = 0)}{W_{bn}}
$$

This can be seen by using the approximations

$$
\sinh(\alpha) = \frac{e^\alpha - e^{-\alpha}}{2} = \alpha + \frac{\alpha^3}{3!} + \frac{\alpha^5}{5!} + \cdots
$$
$$
\cosh(\alpha) = \frac{e^\alpha + e^{-\alpha}}{2} = 1 + \frac{\alpha^2}{2!} + \frac{\alpha^4}{4!} \cdots
$$

and retaining only first-order terms for $\alpha \left(= \frac{W_b}{L_b} \right) \ll 1$.

For the *forward active mode if we ignore the current flow in the reverse-biased BCJ and assume that the base width is small so that the electron density in the base is linear, we get*

$$
I_E = \frac{-eAD_b n_{b0}}{L_b} \coth\left(\frac{W_{bn}}{L_b} \right) \left[\exp\left(\frac{eV_{BE}}{k_B T} \right) - 1 \right]
$$
$$
- \frac{eAD_e p_{e0}}{L_e} \left[\exp\left(\frac{eV_{BE}}{k_B T} \right) - 1 \right] \qquad (7.22)
$$

Here the first part is due to electron injection from the emitter into the base and the second part is due to the hole injection from the base into the emitter. The collector current is

$$I_C = \frac{eAD_b n_{b0}}{L_b \sinh\left(\frac{W_{bn}}{L_b}\right)} \left[\exp\left(\frac{eV_{BE}}{k_B T}\right) - 1\right] \tag{7.23}$$

Assuming that $W_{bn} \ll L_b$, we can expand the hyperbolic functions. The base current is the difference between the emitter and collector current. We find that

$$I_B = \frac{eAD_e p_{e0}}{L_e}\left[\exp\left(\frac{eV_{BE}}{k_B T}\right) - 1\right] + \frac{eAD_b n_{b0} W_{bn}}{2L_b^2}\left[\exp\left(\frac{eV_{BE}}{k_B T}\right) - 1\right] \tag{7.24}$$

The first part represents the hole current injected from the base into the emitter and the second part represents the hole current recombining with electrons injected from the emitter.

It is important to examine each of the current components and look at the dependence not only on the applied bias, but also on the material parameters such as doping densities, diffusion coefficient, diffusion lengths, and base dimensions. We will exploit this knowledge to design high-performance devices. The key feature to appreciate is that a very small base current can control a large collector current.

It is useful to recast the prefactor of the first term in the emitter current (Eqn. 7.22) in a different form. The prefactor, which we will denote by I_S (we assume that $W_{bn} \ll L_b$ so that $\coth\alpha = 1/\alpha$), is

$$I_S = \frac{eAD_b n_{bo}}{W_{bn}} = \frac{e^2 A^2 D_b n_i^2}{eAN_{ab} W_{bn}} = \frac{e^2 A^2 D_b n_i^2}{Q_G}$$

where Q_G is called the Gummel number for the transistor. It has a value

$$Q_G = eAN_{ab} W_{bn} \tag{7.25}$$

and denotes the charge in the base region of the device. For high current gain we want to have a small Gummel number. However, if the Gummel number is too small, the base resistance becomes too large and this degrades the device performance.

To understand the operation of a BJT as an amplifier or a switching device it is useful to examine the device under conditions of saturation, forward active (or reverse active), and cutoff. In Fig. 7.7 we show the band profile and the minority carrier distribution for each of these modes. Note that in saturation where both EBJ and BJT are forward biased, a large minority carrier density (electrons for the *npn* device) is injected into the base region. This plays an important role in device switching, as will be discussed later. In the cutoff mode there is essentially no minority charge in the base, since the EBJ and BCJ are both reverse biased. In the forward active mode, the mode used for amplifiers, the EBJ is forward biased while the BCJ is reverse biased. Under this mode $I_C \gg I_B$, providing current gain.

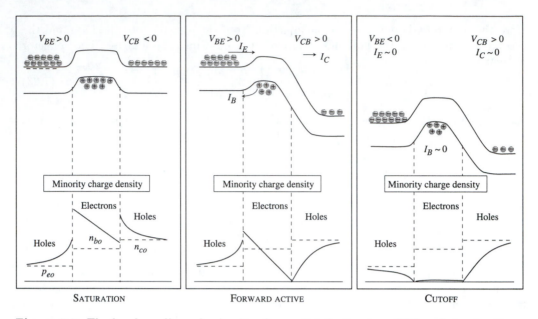

Figure 7.7: The band profile and minority charge distribution in a BJT under saturation, forward active, and cutoff modes.

7.3.3   Operating Configuration of a BJT

The bipolar transistor can be biased in one of three different configurations shown in Fig. 7.8a. Each configuration has its own benefits and the configuration chosen depends upon the applications. The full I-V characteristics of a BJT in the common-base and the common-emitter configuration are shown in Fig. 7.8b. In the common-base configuration the cutoff mode occurs when the emitter current is zero. Once the emitter current is finite, the collector current does not go to zero at $V_{CB} = 0$. The BCJ has to be forward biased (~ 0.7 V for Si devices) to balance the injected emitter current.

In the common-emitter mode, the cutoff mode occurs when the base current is zero and indicates the region where the EBJ is no longer forward biased. The saturation region is represented by the region where $V_{CE} = V_{BE}$ and both EBJ and BCJ are forward biased.

As noted earlier, when the transistor is used in a small-signal amplification, the device operates in the active mode where there is a high current or power gain. However, in the switching mode, the device moves from a cutoff (non-conducting) towards a saturation (conducting) mode. This will be discussed in Section 7.6 where we will examine these modes further.

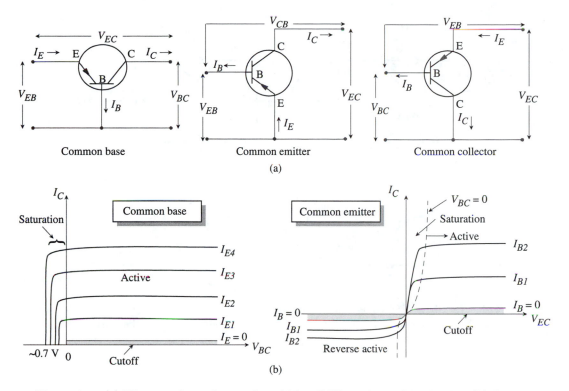

Figure 7.8: (a) Three configurations under which a BJT can be used in circuits. (b) Current-voltage characteristics of a BJT in the common-base and common-emitter configuration.

7.3.4 The Ebers-Moll Model

The current-voltage relations we have obtained above can be rewritten in a form that is extremely suitable for developing a simple model for the BJT. We will discuss this model next.

We rewrite the current-voltage relations obtained in the previous section so that the transistor may be represented by its equivalent circuit. We can write the currents given by Eqns. 7.16 and 7.21 as

$$I_E = -I_{ES} \left[\exp\left(\frac{eV_{BE}}{k_B T} \right) - 1 \right] + \alpha_R I_{CS} \left[\exp\left(-\frac{eV_{CB}}{k_B T} \right) - 1 \right] \quad (7.26)$$

$$I_C = \alpha_F I_{ES} \left[\exp\left(\frac{eV_{BE}}{k_B T} \right) - 1 \right] - I_{CS} \left[\exp\left(-\frac{eV_{CB}}{k_B T} \right) - 1 \right] \quad (7.27)$$

where

$$I_{ES} = \frac{eAD_b n_{bo}}{L_b} \coth\left(\frac{W_{bn}}{L_b} \right) + \frac{eAD_e p_{eo}}{L_e} \quad (7.28)$$

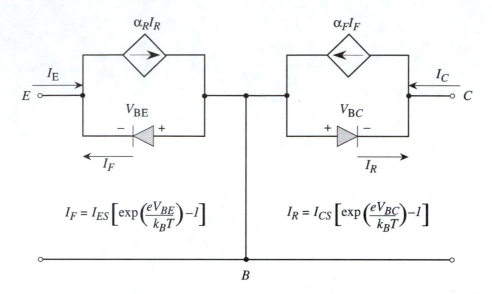

Figure 7.9: The Ebers-Moll equivalent circuit of a bipolar transistor looks at the device as made up of two coupled diodes. Current injected in one diode influences the current in the other diode and vice versa.

$$I_{CS} = \frac{eAD_c p_{co}}{L_c} + \frac{eAD_b n_{bo}}{L_b} \coth\left(\frac{W_{bn}}{L_b}\right) \tag{7.29}$$

$$\alpha_F I_{ES} = \alpha_R I_{CS} = \frac{eAD_b n_{bo}}{L_b \sinh\left(\frac{W_{bn}}{L_b}\right)} \tag{7.30}$$

In these equations, the parameter α_F represents the common-base current gain in the forward active mode, I_{CS} gives the reverse-bias BCJ current, α_R is the common-base current gain for the inverse active mode (i.e., EBJ is reverse biased and CBJ is forward biased) and I_{ES} gives the reverse-bias EBJ current. These equations represent two diodes that are coupled to each other. These equations are the basis of the Ebers-Moll model, which has the four parameters mentioned above. The equations lead to an equivalent circuit given in Fig. 7.9. The Ebers-Moll model is not used for device and circuit design, but provides a useful physical description of the bipolar device.

An important application of the Ebers-Moll model is to find the conditions for the saturation mode. In the common-emitter mode, the saturation condition is given by

$$V_{CE}(sub) = V_{BE} + V_{CB} = V_{BE} - V_{BC} \tag{7.31}$$

In this mode both V_{BE} and V_{BC} $(= -V_{CB})$ are positive. Note that we have an additional constraint on the currents, i.e.,

$$I_E + I_B + I_C = 0 \tag{7.32}$$

Using this equation to eliminate I_E from Eqn. 7.26, we can obtain the values of V_{BE} and V_{BC} in terms of I_C, I_B, and the parameters I_{ES}, I_{CS}, α_R, and α_F. This gives for $V_{CE(sat)}$

$$V_{CE(sat)} = V_{BE} - V_{BC} = \frac{k_B T}{e} \ell n \left[\frac{I_C(1 - \alpha_R) + I_B}{\alpha_F I_B - (1 - \alpha_F)I_C} \cdot \frac{I_{CS}}{I_{ES}} \right] \tag{7.33}$$

Substituting for I_{CS}/I_{ES} from Eqn. 7.30, we get

$$V_{CE(sat)} = \frac{k_B T}{e} \ell n \left[\frac{I_C(1 - \alpha_R) + I_B}{\alpha_F I_B - (1 - \alpha_F)I_C} \cdot \frac{\alpha_F}{\alpha_R} \right] \tag{7.34}$$

Typical values of $V_{CE(sat)}$ are 0.1 to 0.2 V, as can be seen in Example 7.2. In Section 7.6 we will discuss the BJT saturation voltages in greater detail. The Ebers-Moll model is useful in understanding the large signal parameters of the transistor, but for small-signal high-frequency behavior the model has many deficiencies. An important advance in that direction is the hybrid-pi model discussed in Section 7.9.

EXAMPLE 7.1 A silicon *pnp* transistor at 300 K has a device area of 10^{-3} cm^2, base width of 1 μm, and minority carrier diffusion length (for both electrons and holes) of 10 μm. The doping profiles are $N_{ae} = 2 \times 10^{18}$ cm^{-3}, $N_{db} = 10^{17}$ cm^{-3} and $N_{ac} = 10^{16}$ cm^{-3}. Calculate the collector current for the forward active mode cases where (a) $V_{EB} = 0.6$ V; (b) $I_B = -2.0$ μA. Assume that $D_b = 6$ cm^{2}s^{-1} and $D_e = 20$ cm^{2}s^{-1}. Using $n_i^2 = 2.25 \times 10^{20}$ cm^{-3} for Si, we get

$$p_{bo} = 2.25 \times 10^3 \text{ cm}^{-3}; \ n_{eo} = 112.5 \text{ cm}^{-3}; \ n_{co} = 2.25 \times 10^4 \text{ cm}^{-3}$$

Let us use the results given by Eqn. 7.23, (applied to a *pnp* device),

$$I_C \simeq \frac{e A D_b p_{bo}}{L_b \sinh\left(\frac{W_b}{L_b}\right)} \left[\exp\left(\frac{e V_{EB}}{k_B T}\right) - 1 \right]$$

$$I_C = \frac{(1.6 \times 10^{-19} \text{ C}) \times (10^{-3} \text{ cm}^2) \times (6 \text{ cm}^2 \text{ s}^{-1}) \times (2.25 \times 10^3 \text{ cm}^{-3})}{(10 \times 10^{-4} \text{ cm}) \sinh\left(\frac{1.0}{10}\right)} \left[\exp\left(\frac{0.6}{0.026}\right) - 1 \right]$$

$$= 0.2268 \text{ mA}$$

For part (b), given the value of the base current, we can evaluate the base-emitter bias from Eqn. 7.24:

$$I_B = \left(3.6 \times 10^{-16} + 1.08 \times 10^{-16}\right) \left[\exp\left(\frac{e V_{BE}}{k_B T}\right) - 1 \right] = 2.0 \ \mu A$$

This gives

$$V_{BE} = 0.576 \text{ V}$$

EXAMPLE 7.2 An *npn* silicon transistor has the following parameters in the Ebers-Moll model at 300 K:

$$\alpha_F = 0.99; \ \alpha_R = 0.25$$

Calculate the saturation voltage V_{CE} when (a) $I_C = 1$ mA and $I_B = 20$ μA; (b) $I_C = 5$ mA and $I_B = 75$ μA.

In the first case we have

$$V_{CE(sat)} = 0.026 \, \ell n \left[\frac{1(1 - 0.25) + (0.02)}{(0.99)(0.02) - (1 - 0.99)(1)} \left(\frac{0.99}{0.25} \right) \right]$$
$$= 0.149 \text{ V}$$

In the second case we get

$$V_{CE(sat)} = 0.167 \text{ V}$$

We see that as I_B increases the value of $V_{CE(sat)}$ increases, but the change is rather small.

7.4 BJT STATIC PERFORMANCE PARAMETERS

Now that we have derived the current characteristics of the BJT, it is useful to examine the important performance parameters for the device and see their dependence on material parameters. It is important to note at this point that the physical dimensions of the device appear in our analysis so far only via area A and the base width W_b (the base width is larger than W_{bn} by the depletion distances as shown in Fig. 7.6). We will focus on the forward active mode of the device so that we have the conditions

$$eV_{BE} \gg k_BT \tag{7.35}$$

$$eV_{CB} \gg k_BT \tag{7.36}$$

In a well-designed bipolar transistor we always have $W_b \ll L_b$.

7.4.1 Emitter Injection Efficiency, γ_e

The emitter injection efficiency is the ratio of the electron current (in the *npn* BJT) due to the electron injection from the emitter to the total emitter current. Thus,

$$\gamma_e = \frac{I_{En}}{I_{En} + I_{Ep}} \tag{7.37}$$

Obviously, for high emitter efficiency, the back-injected current from the base I_{Ep} should be minimal. Under the voltage approximations made we have from Eqns. 7.22 and 7.24,

$$I_{Ep} = -\frac{eAD_e p_{eo}}{L_e} \exp \left(\frac{eV_{BE}}{k_BT} \right) \tag{7.38}$$

$$I_{En} \cong -\frac{eAD_b n_{bo}}{L_b \tanh \left(\frac{W_{bn}}{L_b} \right)} \exp \left(\frac{eV_{BE}}{k_BT} \right) \tag{7.39}$$

Thus the emitter efficiency becomes

$$\gamma_e = \frac{1}{1 + (p_{eo}D_eL_b/n_{bo}D_bL_e)\tanh(W_{bn}/L_b)} \tag{7.40}$$

If the base width is small compared to the electron diffusion length, the $\tanh(W_{bn}/L_b)$ can be replaced by (W_{bn}/L_b) and we have

$$\gamma_e \cong \frac{1}{1 + (p_{eo}D_eW_{bn}/n_{bo}D_bL_e)} \sim 1 - \frac{p_{eo}D_eW_{bn}}{n_{bo}D_bL_e} \tag{7.41}$$

Thus for γ_e to be close to unity, we should design the device so that $W_{bn} \ll L_e$ and $p_{eo} \ll n_{bo}$. Thus a small base width and a heavy emitter doping compared to the base doping are essential. Of course, the base width cannot be arbitrarily reduced because of the problems of punchthrough and high base resistance, to be discussed later.

7.4.2 Base Transport Factor B

The base transport factor is the ratio of the electron current reaching the base-collector junction to the current injected at the emitter-base junction. As the electrons travel through the base, they recombine with the holes so that the base transport factor is less than unity. We have from Eqns. 7.22 and 7.23 (in the forward active mode, the collector current is essentially due to electron injection from the emitter)

$$B = \frac{I_C}{I_{En}} \cong \frac{1}{\cosh\left(\frac{W_{bn}}{L_b}\right)} \tag{7.42}$$

Again for small base width we have

$$B \cong 1 - \frac{W_{bn}^2}{2L_b^2} \tag{7.43}$$

This result is intuitively expected since a small base width means less recombination and all the injected current will reach the collector.

7.4.3 Collector Efficiency γ_c

The collector efficiency is the ratio of the electron current that reaches the collector to the base-collector current. Due to the high reverse bias at the base-collector junction, essentially all the electrons are swept into the collector so that the collector efficiency can be taken to be unity.

7.4.4   Current Gain

As discussed in Section 7.2, there are two important gain expressions. The parameter α defined as the ratio of the collector current to the emitter current is given by

$$
\alpha = \frac{I_C}{I_E} = \frac{BI_{En}}{I_{En} + I_{Ep}} = \gamma_e B
$$

$$
= \left[1 - \frac{p_{eo}D_e W_{bn}}{n_{bo}D_b L_e}\right]\left[1 - \frac{W_{bn}^2}{2L_b^2}\right] \tag{7.44}
$$

Obviously there is no current gain (α is slightly less than unity) since the output collector current can at most be equal to the emitter current.

As discussed in Section 7.2, the ratio of the collector current to the base current is extremely important since it is the base current that is used to control the device state. This is given by

$$
\beta = \frac{\alpha}{1 - \alpha} \tag{7.45}
$$

An important parameter characterizing the device performance is the transconductance, which describes the control of the output current (I_C) with the input bias (V_{BE}). The transconductance is ($I_C \propto \exp(eV_{BE}/k_BT)$)

$$
g_m = \frac{\partial I_C}{\partial V_{BE}} = \frac{eI_C}{k_BT} = \frac{e\beta I_B}{k_BT} \tag{7.46}
$$

The transconductance of bipolar devices is extremely high compared to that of field-effect transistors of similar dimensions. This is because of the exponential dependence of I_C on V_{BE} in contrast to a weaker dependence of current on "gate bias" for field effect transistors.

EXAMPLE 7.3 Consider an npn Si bipolar transistor that is to be designed with an emitter injection efficiency of $\gamma_e = 0.995$. To maintain a reasonable base resistance, the base is doped p-type with $N_a = 10^{16} \text{cm}^{-3}$. Calculate the n-type doping needed in the emitter. Assume that $L_e = 10W_{bn}$ and $D_e \sim D_b$.

The emitter efficiency is given by Eqn. 7.41 (for the assumptions in this example)

$$
\gamma_e \cong 1 - \frac{p_{eo}}{10n_{bo}} = 0.995
$$

If we use the relation (law of mass action for a homogeneous semiconductor)

$$
p_{eo}n_{eo} = n_{bo}p_{bo} = n_i^2
$$

we have

$$
\frac{n_{eo}}{p_{bo}} = \frac{\text{Emitter doping}}{\text{Base doping}} = 20
$$

Thus the emitter doping required is 2×10^{17} cm^{-3}.

EXAMPLE 7.4 In a bipolar transistor, the base width plays a key role in the design considerations. If the width is too small, the device can suffer punchthrough and may have too large base resistance (which is inversely proportional to the base width). If the base width is too large, the base transport factor becomes small.

Consider an *npn* Si bipolar transistor with base doping of 10^{16} cm^{-3}. If the electron diffusion coefficient in the base is 10 cm^2 s^{-1} and the electron lifetime is 10^{-6} s, calculate the base width required to have a base transport factor of 0.997. Assume that the neutral basewidth W_{bn} is equal to the actual basewidth W_b. The base transport factor is given by

$$B = \frac{1}{\cosh \frac{W_{bn}}{L_b}} \sim 1 - \frac{W_{bn}^2}{2L_b^2} = 0.997$$

This gives

$$W_{bn} = 0.08 L_b$$

The electron diffusion length is

$$L_b = \sqrt{D_b \tau_B} = 3.16 \times 10^{-3} \text{ cm}$$

The base width then has to be 2.5 μm.

Note that if the semiconductor was a direct gap material like GaAs with a minority carrier recombination time of 10^{-9} s, the base width would have to be ~ 800 Å.

EXAMPLE 7.5 Consider the transistor of Example 7.1. Use the results of the example to calculate the current gain and the transconductance of the device.

In Example 7.1 we saw that in part (a) the collector current is 0.2268 mA when $V_{EB} = 0.6$ V. The base current at this biasing is, from Eqn. 7.23, $I_B = 4.92$ μA. The current gain is thus

$$\beta = \frac{0.2268 \times 10^{-3}}{4.92 \times 10^{-6}} = 46$$

The transconductance of the device is

$$g_m = \frac{e I_C}{k_B T}$$

The transconductance depends upon the biasing and for case (a) of Example 7.1 it becomes

$$g_m = \frac{0.2268 \times 10^{-3}}{0.026} = 8.72 \text{ mA/V} = 8.72 \text{ mS}$$

EXAMPLE 7.6 This example will examine the dependence of current gain on the minority carrier recombination in the base. Consider a silicon *npn* transistor with the following parameters at 300 K:

$$
\begin{aligned}
\text{Emitter doping,} \quad & N_{de} && = && 5 \times 10^{17} \text{ cm}^{-3} \\
\text{Base doping,} \quad & N_{ab} && = && 10^{17} \text{ cm}^{-3} \\
\text{Base width,} \quad & W_b && = && 1.0 \ \mu\text{m} \\
\text{Diffusion coefficient,} \quad & D_b = D_e && = && 20 \text{ cm}^2/\text{s}
\end{aligned}
$$

Calculate the current gain for the two cases: (a) minority carrier lifetime for the electrons and holes $= 10^{-6}$ s; (b) minority carrier lifetime $= 10^{-8}$ s. Such a reduction in lifetime can be obtained in silicon by introducing defects.

In the first case, the diffusion lengths are

$$
\begin{aligned}
L_e &= \sqrt{D_e \tau_e} = (20 \times 10^{-6})^{1/2} = 44.7 \ \mu\text{m} \\
L_b &= 44.7 \ \mu\text{m}
\end{aligned}
$$

Also we have

$$
\begin{aligned}
p_{eo} &= \frac{n_i^2}{N_{de}} = \frac{(1.5 \times 10^{10})^2}{5 \times 10^{17}} = 4.5 \times 10^2 \text{ cm}^{-3} \\
n_{bo} &= \frac{n_i^2}{N_{ab}} = \frac{(1.5 \times 10^{10})^2}{10^{17}} = 2.25 \times 10^3 \text{ cm}^{-3}
\end{aligned}
$$

The current gain α is

$$
\begin{aligned}
\alpha &= \left[1 - \frac{p_{eo} D_e W_{bn}}{n_{bo} D_b L_e}\right]\left[1 - \frac{W_{bn}^2}{2L_b^2}\right] \\
&= \left[1 - \frac{(4.5 \times 10^2)(20)(1.0 \times 10^{-4})}{(2.25 \times 10^3)(20)(44.7 \times 10^{-4})}\right]\left[1 - \frac{(1.0 \times 10^{-4})^2}{2(44.7 \times 10^{-4})^2}\right] \\
&= (1 - 4.47 \times 10^{-3})(1 - 2.5 \times 10^{-4}) = (0.9955)(0.9998) \\
&= 0.9953
\end{aligned}
$$

The current gain is dominated in this case by the emitter efficiency since the base transport factor (second term) is close to unity. The current gain β is

$$
\beta = \frac{\alpha}{1 - \alpha} = 210.7
$$

In the second case, we have

$$
\begin{aligned}
L_e &= \sqrt{D_e \tau_e} = (20 \times 10^{-8})^{1/2} = 4.47 \ \mu\text{m} \\
L_b &= 4.47 \ \mu\text{m}
\end{aligned}
$$

The current gain α is

$$
\begin{aligned}
\alpha &= \left[1 - \frac{(4.5 \times 10^2)(20)(1.0 \times 10^{-4})}{(2.25 \times 10^3)(20)(4.47 \times 10^{-4})}\right]\left[1 - \frac{(1 \times 10^{-4})^2}{2(4.47 \times 10^{-4})^2}\right] \\
&= \left[1 - 4.47 \times 10^{-2}\right]\left[1 - 2.5 \times 10^{-2}\right] = (0.9553)(0.974) \\
&= 0.931
\end{aligned}
$$

The gain β is

$$
\beta = 13.6
$$

One sees a big loss in β as the minority carrier time decreases. Once again it is worth pointing out that the minority carrier times in direct gap semiconductors like GaAs are $\sim 10^{-9}$ s compared to $\sim 10^{-6}$ s in defect-free but doped silicon. This gives GaAs devices a poor gain. However, by using heterostructure concepts (Section 7.10) it is possible to have excellent gain in GaAs-based devices by using very short base widths. However, for high speed the minority carrier lifetime must be small. Thus one has to optimize gain and high-speed performance.

EXAMPLE 7.7 To make a high-speed BJT, the emitter thickness should be as small as possible, so that the emitter resistance is low. Once the emitter thickness (let us discuss an *npn* device) is smaller than the hole diffusion length L_e, the considerations of a narrow diode apply. This affects the emitter injection efficiency, since the hole current injected into the emitter from the base increases. Consider the following device parameters for an Si BJT :

$$
\begin{array}{rl}
\text{Base doping} & N_{ab} = 10^{16} \text{ cm}^{-3} \\
\text{Base width} & W_b = 0.5 \ \mu\text{m} \\
\text{Base electron diffusion length} & L_b = 10 \ \mu\text{m} \\
\text{Emitter diffusion length} & L_e = 4.0 \ \mu\text{m} \\
\text{Emitter thickness} & W_e = 10 \ \mu\text{m} \\
\text{Base diffusion coefficient} & D_b = 30 \text{ cm}^2/\text{s} \\
\text{Emitter diffusion coefficient} & D_e = 10 \text{ cm}^2/\text{s} \\
\text{Emitter doping} & N_{de} = 10^{18} \text{ cm}^{-3}
\end{array}
$$

Calculate the emitter efficiency for the forward active device. How does the emitter efficiency change when the emitter thickness becomes 1 μm? (Use the discussion in Section 5.3.3 on narrow diodes.)

For a finite emitter, the hole current injected from the base to the emitter has a similar expression as the electron-emitter current into the finite base. We have

$$
I_{Ep} = \frac{eAD_e p_{eo}}{L_e} \ \coth \ \left(\frac{W_e}{L_e} \right) \left\{ \exp \ \left(\frac{eV}{k_B T} \right) - 1 \right\}
$$

If $W_e \gg L_e$ the coth term becomes unity and we get back the results discussed in the text. The emitter efficiency is now

$$
\gamma_e \cong \frac{\frac{D_b n_{bo}}{L_b} \ \coth \ \left(\frac{W_b}{L_b} \right)}{\frac{D_b n_{bo}}{L_b} \ \coth \ \left(\frac{W_b}{L_b} \right) + \frac{D_e p_{eo}}{L_e} \ \coth \ \left(\frac{W_e}{L_e} \right)}
$$

If the emitter thickness is 10 μm we get

$$
\coth \ \left(\frac{W_b}{L_b} \right) \ = \ 20
$$

$$
\coth \ \left(\frac{W_e}{L_e} \right) \ = \ 1.014
$$

$$
\gamma_e \ = \ 0.9996
$$

If the emitter thickness is 1 μm, we have

$$
\coth \ \left(\frac{W_e}{L_e} \right) \ = \ 4
$$

$$
\gamma_e \ = \ 0.997
$$

Thus a shorter emitter width reduces the injection efficiency. However, a shorter emitter improves the device speed by reducing emitter resistance.

7.5 SECONDARY EFFECTS IN REAL DEVICES

In the derivations of the bipolar device characteristics, we have made a number of simplifying assumptions. There are important secondary effects that make the device characteristics deviate from those derived so far. These deviations have important effects on circuit design as well as on the limits of device performance.

7.5.1 Base Width Modulation: Early Effect and Punchthrough

The quantity W_{bn} that has appeared in our derivation for the current-voltage characteristics is not the metallurgical base width W_b, but the distance between the two depletion regions of the transistor. For $W_{bn} \ll L_b$ we can see from Eqn. 7.23 that the collector current is

$$I_C \propto \frac{1}{W_{bn}}$$

The depletion region changes with the bias conditions, so that W_{bn} and consequently the current will have an additional bias dependence. This base width modulation and its effect on the current flow are known as the Early effect, after J. M. Early who first explained the effect.

Under normal operation, the base-emitter junction is forward biased so that the depletion width is very small. However, the depletion width is quite large for the reverse-biased collector-base junction. The depletion width is proportional to $V_{CB}^{1/2}$ so that W_{bn} will decrease as the base-collector voltage increases. As W_{bn} decreases, γ, the emitter efficiency, and B, the base transport factor, approach unity. The gain of the device consequently increases. This leads to the I-V characteristics shown schematically in Fig. 7.10, where the current does not saturate as V_{CE} is increased. This is not desirable in a device, since one would prefer the output to depend only on the input and not on the output voltage. The $I_C - V_{CE}$ curves intersect the voltage axis at a voltage V_A called the Early voltage. One would like to have $|V_A|$ as large as possible for good input-output isolation. The Early voltage can be determined from the I-V characteristics of the device, as discussed in Example 7.8.

It must also be noted that if the reverse bias is increased too much the width W_{bn} can go to zero, leading to the punchthrough effect.

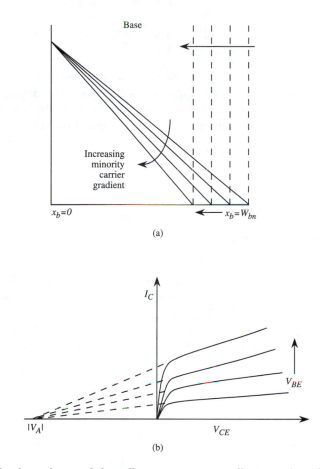

Figure 7.10: The dependence of the collector current on collector emitter bias. (a) The edge of the base side of the EBJ depletion region moves into the metallurgical base as V_{CE} increases; (b) the shortened base causes the collector current to increase with V_{CE}. The voltage V_A is called the Early voltage.

The depletion width on the base side at punchthrough is given by (V_{pt} is the punchthrough voltage; depletion width is equal to the base width)

$$W_b = \left\{ \frac{2\epsilon(V_{bi} + V_{pt})N_{dc}}{eN_{ab}(N_{dc} + N_{ab})} \right\}^{1/2} \tag{7.47}$$

where N_{dc} and N_{ab} are the collector and base doping. Neglecting V_{bi} (which is ~ 0.7 V for Si) compared to V_{pt}, we get for the punchthrough voltage

$$V_{pt} = \frac{eW_b^2 N_{ab}(N_{dc} + N_{ab})}{2\epsilon N_{dc}} \tag{7.48}$$

To avoid punchthrough, the base doping should be high while the collector doping should be low. However, a high base doping reduces the emitter efficiency, as discussed above.

7.5.2   Drift Effects in the Base: Nonuniform Doping

We have assumed so far that the base doping is uniform and consequently there is no built-in electric field in the base region. In real devices the doping can be quite nonuniform, especially if the doping is done by ion implantation. The nonuniform doping causes a built-in field that can help or hinder the carriers injected into the base from the emitter. Of course, if the doping can be made non-uniform in a controlled manner, it can be exploited to shorten the base transit time.

7.5.3   Avalanche Breakdown

Just as in the case of the *p-n* diode, the avalanche process limits the collector-base voltage that the transistor can sustain. This then sets the limit on the power that can be obtained by the transistor. The breakdown due to the impact ionization (avalanching) is reflected in the I-V characteristics of the transistor in a manner shown in Fig. 7.11. In the common-base configuration, the breakdown occurs at a well defined collector-base voltage BV_{CBO}. On the other hand, for the common-emitter configuration, the breakdown is not as sharply reflected in the device output characteristics. The breakdown in the common-emitter configuration also occurs at a lower value of V_{CE} than it does in the common-base configuration.

In the common-base configuration, as V_{CB} is increased, the breakdown is essentially similar to that of a single *p-n* junction discussed in Chapter 5. The current coming from the emitter has little effect on the breakdown. However, in the common-emitter configuration, as soon as the impact ionization process starts, say, in an *npn* BJT, the secondary holes are injected into the base and act as a base current, leading to increased emitter current and eventual current runaway as the process snowballs.

7.5.4   High Injection: Thermal Effects

At high injection levels there is thermal heating of the bipolar device since a power level of $I_C V_{BC}$ is dissipated. As the temperature of the device changes the current usually increases further. This is due to the exponential dependence of the injection current and the increase in the recombination time (which increases faster than the increase in the base transit time), which increases the base transport factor. As the current increases, a further increase in heat dissipation occurs until the device can be burnt out if proper design considerations are not met.

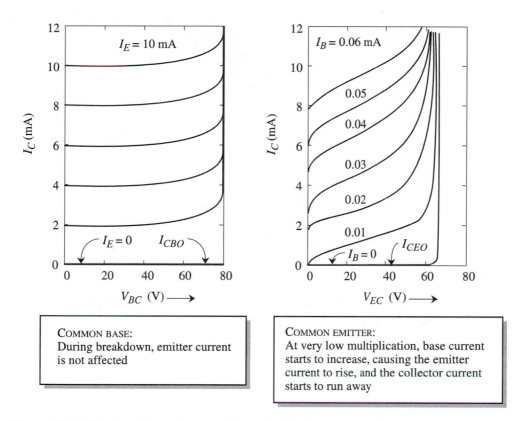

Figure 7.11: The breakdown characteristics of a bipolar transistor in the common-base and common-emitter configurations.

7.5.5 High Injection: Base Pushout Effect

An important limit on the collector current density also arises from the base pushout or Kirk effect. When the electron density injected into the collector starts approaching the background donor density, the field on the depletion region starts to decrease. The electron density of the injected carriers is

$$n = \frac{J_C}{ev_s} \tag{7.49}$$

where J_C is the collector current density and v_s is the saturated velocity of the electrons. As n increases, the space charge $(N_{dc} - n)$ decreases, leading to an increase in the depletion width. Eventually the depletion width covers the entire collector region. This occurs at a critical injected density n_c with a value

$$n_c - N_{dc} = \frac{2(V_{CB} + V_{bi})\epsilon}{eW_c^2} \tag{7.50}$$

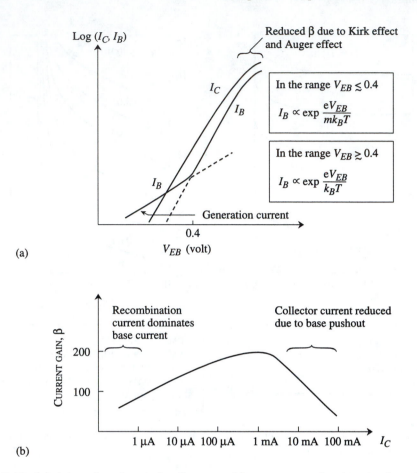

(a)

(b)

Figure 7.12: (a) A log plot of typical collector and base currents showing the degradation of the current gain at low injection values and at high injection values. (b) A schematic plot of current gain β in the forward active mode as a function of collector current.

where V_{bi} is the built-in voltage at the base-collector junction and W_c is the collector width. Further injection allows the flat band region in the base to move out towards the collector, thus effectively increasing the base width. To avoid this effect, especially in small devices, one has to keep a certain minimum collector doping.

An important consequence of the Kirk effect is that the collector current decreases and, as a result, the current gain decreases. An additional effect that reduces current gain at high injection is the Auger effect. In this process electrons and holes recombine. The process is proportional to n^2p and becomes important at high injection. In Fig. 7.12a and 7.12b we show the influence of the Kirk effect and of the Auger effect on the device currents and on the current gain.

7.5.6 Low Injection Effects and Current Gain

In our calculations for the BJT junction currents we have assumed that the junctions are "ideal," i.e., there is no current flow due to recombination-generation effects. In Chapter 5, Section 5, we discussed how non-ideal effects arising from recombination generation in the depletion region alter the current flowing in the junction. This effect is particularly important under low injection (i.e., low values of V_{EB}) conditions.

If we examine the forward-biased EBJ for a device operating in the forward active mode, the base current will have an "ideal" current component and a "non-ideal" current component arising from generation-recombination. Referring to Eqn. 5.55, we can write

$$I_B = \frac{eAD_e n_{oe}}{L} \exp\left(\frac{eV_{EB}}{k_B T}\right) + \frac{eAn_i W_{EBJ}}{2\tau} \exp\left(\frac{eV_{EB}}{2k_B T}\right)$$

where the second term is due to recombination in the emitter-base junction depletion region (W_{EBJ}). The recombination time is τ. The base current may be written as

$$I_B = I_S \exp\frac{eV}{mk_B T}$$

where m is the junction ideality factor.

The collector current is not greatly influenced by the recombination-generation process. At low injection the recombination-generation part of the base current dominates and as a result, the current gain β is reduced. As the injection (V_{EB} value) is increased, the recombination part becomes negligible and the value of β reaches its ideal value calculated earlier. In Fig. 7.12a we show typical values of the base current and collector current as a function of V_{EB}.

In Section 7.3.4 we discussed the Ebers-Moll model for bipolar transistors. This model does not account for some of the issues discussed in this section. A more advanced model that includes more realistic effects is the Gummel-Poon model. Three important effects are incorporated in the Gummel-Poon model:

- Recombination current in the emitter depletion region under low injection levels.
- Reduction of current gain at high injection levels.
- Finite output conductance in terms of an Early voltage, V_A.

7.5.7 Current Crowding Effect

The picture we have developed for the BJT is a one-dimensional picture. In reality, the base current flows along the directions perpendicular to the emitter, as can be seen from Fig. 7.5. There is a voltage drop IR across the base cross-section that becomes increasingly important at high injections and high frequencies. As a result of this potential drop, the edge of the emitter may be forward biased but the "core" of the emitter region may not be forward biased. Higher current densities would thus flow along the edges of the emitter. This effect is called emitter crowding and, because of it, the high injection effects discussed above can be important even at low total current values.

Emitter crowding has an adverse effect on power transistors where high current values are required. It is essential for these transistors that the emitter be properly designed. Computer simulation techniques are used to study the current flow so that an optimum emitter can be used. The emitter crowding effects can be suppressed by increasing the perimeter-to-area ratio of the emitter. This is often done by using long fingers for the emitter and base contacts in the "interdigitated" approach shown in Fig. 7.13.

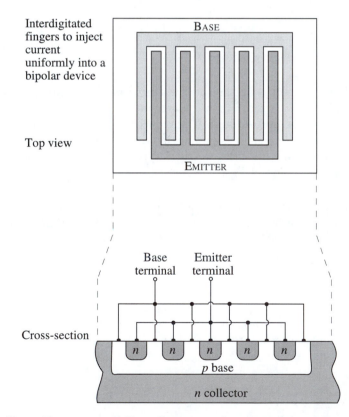

Figure 7.13: To avoid current crowding effects near the edges of the emitter contact, a large number of finger contacts are used in an interdigitated approach. Shown are the top view and the cross-section of a typical device.

EXAMPLE 7.8 Consider an *npn* silicon transistor at 300 K with a base doping of 5 × 10^{16} cm^{-3} and a collector doping of 5 × 10^{15} cm^{-3}. The width of the base region is 1.0 μm. Calculate the change in the base width as V_{CB} changes from 1.0 to 5.0 V. Also calculate how the collector current changes and determine the Early voltage. Assume that $D_b = 20$ cm^2/s, $V_{BE} = 0.7$ V, and $W_b \ll L_b$.

The depletion width at the base-collector junction is shared between the base and the collector region. The extent of depletion into the base side is given by

$$\Delta W_b = \left\{ \frac{2\epsilon_s(V_{bi} + V_{CB})}{e} \left(\frac{N_{dc}}{N_{ab}(N_{ab} + N_{dc})} \right) \right\}^{1/2}$$

The built-in voltage V_{bi} is given by

$$V_{bi} = \frac{k_B T}{e} ln \frac{N_{ab} N_{dc}}{n_i^2} = (0.026(V)) \times ln(1.1 \times 10^{12})$$

$$= 0.721 \text{ V}$$

For an applied bias of 1 V, we get

$$\Delta W_b = \left\{ \frac{2 \times (8.84 \times 10^{-14} \times 11.9 \text{ F/cm})(1.72 \text{ V})}{(1.6 \times 10^{-19} \text{ C})} \times \frac{1}{5.5 \times 10^{17} \text{ cm}^{-3}} \right\}^{1/2}$$

$$= 6.413 \times 10^{-6} \text{ cm} = 0.064 \ \mu\text{m}$$

The neutral base width is thus

$$W_{bn} = 0.936 \ \mu\text{m}$$

When the collector-base voltage increases to 5 V, we get

$$\Delta W_b = 0.117 \ \mu\text{m}$$

The neutral base width is

$$W_{bn} = 0.883 \ \mu\text{m}$$

In the limit of $W_{bn} \gg L_b$ we have for the collector current density (using Eqn. 7.23 with sinh $(W_{bn}/L_b) \sim W_{bn}/L_b$)

$$J_C = \frac{e D_b n_{bo}}{W_{bn}} \exp \left(\frac{e V_{BE}}{k_B T} \right)$$

where

$$n_{bo} = \frac{n_i^2}{N_{ab}} = \frac{2.25 \times 10^{20}}{5 \times 10^{16}} = 4.5 \times 10^3 \text{ cm}^{-3}$$

For the base-collector bias of 1 V, we have $W_{bn} = 0.936 \ \mu$m. The collector current density is then

$$J_C = \frac{(1.6 \times 10^{-19} \text{ C}) \times (20 \text{ cm}^2 \text{ s}^{-1}) \times (4.5 \times 10^3 \text{ cm}^{-3})(4.93 \times 10^{11})}{(0.936 \times 10^{-4} \text{ cm})}$$

$$= 75.8 \text{ A/cm}^2$$

When the collector-base bias changes to 5 V, the current density becomes

$$J_C = 80.35 \text{ A/cm}^2$$

The slope of the J_C vs. V_{CE} curve is then

$$\frac{dJ_C}{dV_{CE}} \cong \frac{\Delta J_C}{\Delta V_{CE}} \cong \frac{80.35 - 75.8}{4} \ \text{Acm}^{-2} \ \text{V}^{-1}$$

We define the Early voltage V_A through the relation

$$\frac{dJ_C}{dV_{CE}} = \frac{J_C}{V_{CE} + V_A}$$

Equating the two relations, we get

$$V_A \cong 64.9 \ \text{V}$$

EXAMPLE 7.9 Consider an *npn* silicon bipolar transistor with a base width of 0.2 μm, a base doping of 5×10^{16} cm^{-3}, and a collector doping of 10^{16} cm^{-3}. Calculate the punchthrough voltage of the transistor.

The punchthrough voltage is simply given by (Eqn. 7.48)

$$V_{pt} = \frac{(1.6 \times 10^{-19} \ \text{C})(0.2 \times 10^{-4} \ \text{cm})^2 (5 \times 10^{16} \ \text{cm}^{-3})(6 \times 10^{16} \ \text{cm}^{-3})}{2(11.9 \times 8.85 \times 10^{-14} \ \text{F/cm})(10^{16} \ \text{cm}^{-3})} = 9.27 \ \text{V}$$

Note that the depletion width on the collector side is (N_{ab}/N_{dc}) times the base width at punchthrough. Thus the total depletion width is 1.2 μm. The average field at punchthrough is therefore

$$F = \frac{9.27}{1.2 \times 10^{-4}} = 7.7 \times 10^4 \ \text{V/cm}$$

This field is much lower than the field needed for avalanche breakdown (which is $\sim 4 \times 10^5$ V/cm for Si). Thus, for this particular transistor, the punchthrough mechanism will be the dominant one for device breakdown.

EXAMPLE 7.10 Consider a *npn* Si-BJT at 300 K with the following parameters:

$$
\begin{aligned}
N_{de} &= 10^{18} \ \text{cm}^{-3} \\
N_{ab} &= 10^{17} \ \text{cm}^{-3} \\
N_{dc} &= 5 \times 10^{16} \ \text{cm}^{-3} \\
D_b &= 30.0 \ \text{cm}^2/\text{s} \\
L_b &= 15.0 \ \mu\text{m} \\
D_e &= 10.0 \ \text{cm}^2/\text{s} \\
L_e &= 5.0 \ \mu\text{m}
\end{aligned}
$$

(i) Calculate the maximum base width, W_b, that will allow a current gain β of 100 when the EBJ is forward biased at 1.0 V and the BCJ is reverse biased at 5.0 V.

(ii) Describe two advantages and two disadvantages of making the base smaller.

Since the base width is much smaller than the base diffusion length, we have for the emitter and collector current

$$I_E = \left[\frac{eAD_b n_{b0}}{W_{bn}} + \frac{eAD_e p_{e0}}{L_e} \right] \left[\exp\left(\frac{eV_{BE}}{k_B T} \right) - 1 \right]$$

$$I_C = \left[\frac{eAD_b n_{b0}}{W_{bn}} \right] \left[\exp(\frac{eV_{BE}}{k_B T}) - 1 \right]$$

The base current is the difference of these and the current gain β is

$$\beta = \frac{I_C}{I_B} = \frac{D_b n_{b0} L_e}{D_e p_{e0} W_{bn}} = 100$$

This gives for W_{bn}

$$W_{bn} = 1.5 \times 10^{-4} \text{ cm}$$

This is the neutral base width. The actual base width will be larger and we need to calculate the depletion on the base side at the BCJ due to the biasing of the device. Since the EBJ is strongly forward biased, there is essentially no depletion of the base at this junction.

The built-in voltage on the BCJ is

$$V_{bi} = \frac{k_B T}{e} ln \left(\frac{N_{ab} N_{dc}}{n_i^2} \right) = 0.8 \text{ V}$$

Using the V_{bi} value we find that the depletion width on the base side of the EBJ for a 5 volt bias at the base collector junction is

$$\Delta W(V = 5 \text{ V}) = 1.59 \times 10^{-5} \text{ cm}$$

and the base width becomes

$$W_b = W_{bn} + 1.59 \times 10^{-5} = 1.659 \times 10^{-4} \text{ cm}$$

(ii) Two disadvantages of a shorter base:
• The output conductance will suffer and the collector current will have a stronger dependence on V_{CB}.
• The device may suffer punchthrough at a lower bias.

Two advantages:
• The current gain will be higher.
• The device speed will be faster.

EXAMPLE 7.11   Consider a *npn* Si-BJT at 300 K with the following parameters:

$$
\begin{aligned}
N_{de} &= 10^{18} \text{ cm}^{-3} \\
N_{ab} &= 10^{17} \text{ cm}^{-3} \\
N_{dc} &= 10^{16} \text{ cm}^{-3} \\
D_b &= 30.0 \text{ cm}^2/\text{s} \\
L_b &= 10.0 \ \mu\text{m} \\
W_b &= 1.0 \ \mu\text{m} \\
D_e &= 10 \text{ cm}^2/\text{s} \\
L_e &= 10.0 \ \mu\text{m} \\
\text{Emitter thickness} &= 1.0 \ \mu\text{m} \\
\text{Device area} &= 4.0 \times 10^{-6} \text{ cm}^2
\end{aligned}
$$

Calculate the emitter efficiency and gain β when the EBJ is forward biased at 1.0 V and the BCJ is reverse biased at 5.0 V.

Calculate the output conductance of the device defined by

$$g_o = \frac{\Delta I_C}{\Delta V_{CB}}$$

To solve this problem we need to calculate the neutral base width in the device. Also note that since the emitter thickness is small compared to the carrier diffusion length in the emitter, we will use the narrow diode theory to calculate the emitter efficiency, as discussed in Example 7.7.

Using the parameters given, the built-in voltage in the BCJ is

$$V_{bi} = \frac{k_B T}{e} ln \left(\frac{10^{17}.10^{16}}{2.25 \times 10^{20}} \right) = 0.757 \text{ V}$$

The depletion width on the base side of the BCJ is found to be

$$\delta W(5.0 \text{ V}) = 8.296 \times 10^{-6} \text{ cm}$$

and

$$\delta W(6.0 \text{ V}) = 8.981 \times 10^{-6} \text{ cm}$$

Thus the neutral base width is

$$W_{bn}(5.0 \text{ V}) = 9.17 \times 10^{-5} \text{ cm}$$

The emitter efficiency is (for a narrow emitter of width W_e)

$$\gamma_e = 1 - \frac{p_{e0} D_e W_{bn}}{n_{b0} D_b W_e} = 0.969$$

We find that the base transport factor is

$$B = 1 - \frac{W_{bn}^2}{2L_b^2} = 0.996$$

This gives

$$\alpha = \gamma_e B = 0.9656$$

and the current gain is

$$\beta = \frac{\alpha}{1 - \alpha} = 28$$

The collector current is

$$I_C = \frac{e A D_b n_{b0}}{W_{bn}} \left[\exp\left(\frac{e V_{BE}}{k_B T} \right) - 1 \right] - \frac{e A D_b n_{b0} W_{bn}}{2 L_b^2} \left[\exp\left(\frac{e V_{BE}}{k_B T} \right) - 1 \right]$$

with the second part being negligible.

We find that

$$I_C(5.0 \text{ V}) = 23.79 \text{ A}$$

We now calculate the neutral base width when the BCJ is reverse biased at 6.0 V. This is

$$W_{bn}(6.0 \ V) = 9.1 \times 10^{-5} \ \text{cm}$$

This gives

$$I_C(6.0 \ V) = 23.973 \ A$$

The output conductance is now

$$g_o = 0.183 \ \Omega^{-1}$$

7.6 A CHARGE-CONTROL ANALYSIS

This and the next section provide a detailed discussion of how bipolar junction transistors behave in switching applications. Instructors should decide how much detail they want to cover and may choose to qualitatively discuss Section 7.7 if there are severe time constraints in the course.

The two junctions of the BJT can be biased in several ways to produce four operating modes for the transistor. It is important to examine these modes and the simplification that arises in the expressions for the transistor currents. In this section we will describe the transistor in terms of the charge in the device. The charge-control description describes the behavior of the device in terms of the charges in different regions of the device and various time constants related to charge flow. This approach is quite useful in understanding the switching behavior of the device.

Forward Active Mode
In this mode the emitter-base junction (EBJ) is forward biased, while the base-collector junction (BCJ) is reverse biased. We will use the subscript F to denote various terms in the forward active mode. The currents are given by the Ebers-Moll model discussed in Section 7.3.4 ($eV_{CB} \gg k_B T$ in this mode):

$$
\begin{aligned}
I_E &= I_{ES} \exp\left(\frac{eV_{BE}}{k_B T}\right) + \alpha_R I_{CS} \\
I_C &= \alpha_F I_{ES} \exp\left(\frac{eV_{BE}}{k_B T}\right) + I_{CS}
\end{aligned}
\tag{7.51}
$$

Here we assume that the emitter and collector current have the same direction. If we express $I_{ES} \exp(eV_{BE}/k_B T)$ in the second equation using the first equation, we can write

$$
\begin{aligned}
I_C &= \alpha_F (I_E - \alpha_R I_{CS}) + I_{CS} \\
&= \alpha_F I_E + I_{CS}(1 - \alpha_F \alpha_R)
\end{aligned}
\tag{7.52}
$$

Using $I_E = I_B + I_C$, we have

$$I_C = \alpha_F I_B + \alpha_F I_C + I_{CS} \left(1 - \alpha_F \alpha_R\right) \tag{7.53}$$

or

$$
\begin{aligned}
I_C &= \frac{\alpha_F}{1 - \alpha_F} I_B + \frac{I_{CS} \left(1 - \alpha_F \alpha_R\right)}{1 - \alpha_F} \\
&= \beta_F I_B + \left(\beta_F + 1\right) I_{CS} \left(1 - \alpha_F \alpha_R\right)
\end{aligned} \tag{7.54}
$$

where

$$\beta_F = \frac{\alpha_F}{1 - \alpha_F} \tag{7.55}$$

β_F represents the forward active current gain I_C/I_B for the transistor.

It is useful to examine the charge in the device in the forward active mode. In Fig. 7.7 we show the minority charge injected in the emitter base and collector. The excess minority charge injected into the base region is given by

$$Q_F = \frac{e A W_{bn} n_{b0}}{2} \left(\exp \frac{e V_{BE}}{k_B T} - 1\right) \tag{7.56}$$

We can define the collector current in terms of the excess charge by defining a time constant τ_F which is the forward transit time of minority carriers through the base. We have

$$Q_F = \tau_F I_C \tag{7.57}$$

As discussed in Chapter 3 (see Example 3.17), the forward transit time is

$$\tau_F = \frac{W_{bn}^2}{2 D_b} \tag{7.58}$$

The base current I_B is due to recombination in the neutral base with the minority charge and hole injection into the emitter. These two effects can be summarized by a time constant τ_{BF} and we can write

$$I_B = \frac{Q_F}{\tau_{BF}} \tag{7.59}$$

The current gain is then

$$\beta_F = \frac{I_C}{I_B} = \frac{\tau_{BF}}{\tau_F} \tag{7.60}$$

It is useful to relate the collector and base current to the charge Q_F in case the charge is changing in time. This is useful in understanding the switching behavior of the device. The collector current is simply given by

$$i_C(t) = \frac{Q_F(t)}{\tau_F} \tag{7.61}$$

The base current must include a time-dependent change in Q_F:

$$i_B(t) = \frac{Q_F(t)}{\tau_{BF}} + \frac{dQ_F}{dt} \qquad (7.62)$$

The time constants τ_F and τ_{BF} play very important roles in the switching properties of the bipolar device, as we shall see later.

In addition to the minority charge in the base, a change in biasing influences the charge in the depletion regions associated with the EBJ and BCJ. These regions of charge are shown in Fig. 7.14. We define a quantity $Q_V(V)$ that is the difference in the depletion region charge at $V = 0$ and at an applied bias of V:

$$Q_V(V) = Q_j(0) - Q_j(V) \qquad (7.63)$$

If C_{j0} is the depletion capacitance at $V = 0$, we have, from Chapter 5,

$$Q_V(V) = \frac{3}{2} C_{j0} V_{bi}^{1/3} \left[V_{bi}^{2/3} - (V_{bi} - V)^{2/3} \right] \qquad (7.64)$$

There is this additional charge at the EBJ, $Q_{VE}(V)$, and at the BCJ, $Q_{VC}(V)$, which has to be modulated when the device conditions change. Including the depletion region current, the current-charge equations can be written as

$$
\begin{aligned}
i_C &= \frac{Q_F}{\tau_F} - \frac{dQ_{VC}}{dt} \\
i_B &= \frac{Q_F}{\tau_{BF}} + \frac{dQ_F}{dt} + \frac{dQ_{VC}}{dt} + \frac{dQ_{VE}}{dt} \\
i_E &= Q_F \left(\frac{1}{\tau_F} + \frac{1}{\tau_{BF}} \right) + \frac{dQ_F}{dt} + \frac{dQ_{VE}}{dt}
\end{aligned}
\qquad (7.65)
$$

Reverse Active Mode

In the reverse active mode the EBJ is reverse biased while the BCJ is forward biased. Note that bipolar devices are asymmetrically doped, i.e., $N_{de} \gg N_{dc}$, and the reverse active mode has a poor current gain. The current and excess minority charge can be written, in analogy with the forward active mode case (note that the collector is now acting as the emitter):

$$
\begin{aligned}
Q_R &= \frac{e A W_{bn} n_{b0}}{2} \left(\exp \frac{e V_{BC}}{k_B T} - 1 \right) \\
i_E &= \frac{-Q_R}{\tau_R} + \frac{dQ_{VE}}{dt} \\
i_B &= \frac{Q_R}{\tau_{BR}} + \frac{dQ_R}{dt} + \frac{dQ_{VC}}{dt} + \frac{dQ_{VE}}{dt} \\
i_C &= -Q_R \left(\frac{1}{\tau_R} + \frac{1}{\tau_{BR}} \right) - \frac{dQ_R}{dt} - \frac{dQ_{VC}}{dt}
\end{aligned}
\qquad (7.66)
$$

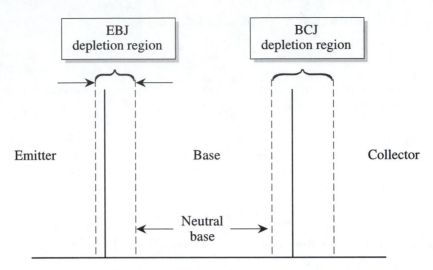

Figure 7.14: A schematic of the depletion regions of the emitter-base and base-collector junctions.

Cutoff Mode

In the cutoff mode, both junctions are reverse biased and we may write

$$
\begin{aligned}
I_E &= -I_{ES} + \alpha_R I_{CS} \\
I_C &= -\alpha_F I_{ES} + I_{CS}
\end{aligned}
\tag{7.67}
$$

In the cutoff mode the terminal currents are extremely small and there is an effective open circuit at the terminals.

Saturation Mode

In the saturation mode the EBJ and the BCJ are both forward biased. In this case a good approximation to the current-voltage equations is

$$
\begin{aligned}
I_E &= I_{ES} \exp\left(\frac{eV_{BE}}{k_B T}\right) - \alpha_R I_{CS} \exp\left(\frac{eV_{BC}}{k_B T}\right) \\
I_C &= \alpha_F I_{FS} \exp\left(\frac{eV_{BE}}{k_B T}\right) - I_{CS} \exp\left(\frac{eV_{BC}}{k_B T}\right)
\end{aligned}
\tag{7.68}
$$

In saturation there is charge injected into the base from the emitter (Q_F) and the collector (Q_R). The charge in the depletion region charge is negligible, since the junction voltages do not change much once the device is in saturation. The current-charge relations can be written as

$$
\begin{aligned}
i_C &= \frac{Q_F}{\tau_F} - Q_R\left(\frac{1}{\tau_R} + \frac{1}{\tau_{BR}}\right) - \frac{dQ_R}{dt} \\
i_B &= \frac{Q_F}{\tau_{BF}} + \frac{Q_R}{\tau_{BR}} + \frac{d}{dt}\left(Q_F + Q_R\right)
\end{aligned}
\tag{7.69}
$$

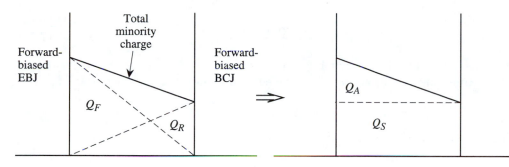

Figure 7.15: Minority charge in the base of a BJT in saturation mode. Charge is injected from the emitter and the collector into the base. The figure on the right shows a representation of the charge in terms of a uniform charge Q_S and charge Q_A.

The charges Q_F and Q_R are shown in Fig. 7.15a. The total base charge may be written as shown on the right-hand side of Fig. 7.15b:

$$
\begin{aligned}
Q_B &= Q_F + Q_R \\
&= Q_A + Q_S
\end{aligned}
\tag{7.70}
$$

where Q_A represents the charge at the edge of saturation (EOS) and Q_S is the overdrive charge that drives the device into saturation. The charge Q_A can be written as

$$
\begin{aligned}
Q_A &= \tau_F I_{C(EOS)} = \tau_{BF} I_{B(EOS)} \\
\frac{I_{C(EOS)}}{I_{B(EOS)}} &= \beta_F
\end{aligned}
\tag{7.71}
$$

The overdrive charge Q_S can be written as

$$
Q_S = \tau_S I_{BS}
\tag{7.72}
$$

where I_{BS} is the base current over and above $I_{B(EOS)}$ that brings the device to the edge of saturation. The time τ_S is the weighted mean of τ_{BF} and τ_{BR}.

The static base current in saturation is

$$
I_B = I_{B(EOS)} + I_{BS}
\tag{7.73}
$$

The instantaneous value of the base current is

$$
i_B(t) = \frac{Q_A}{\tau_{BF}} + \frac{Q_S}{\tau_S} + \frac{dQ_S}{dt}
\tag{7.74}
$$

We can use the relation

$$
\frac{Q_A}{\tau_{BF}} = \frac{\tau_F I_{C(EOS)}}{\tau_{BF}} = \frac{I_{C(EOS)}}{\beta_F}
\tag{7.75}
$$

so that we have

$$i_B(t) - \frac{I_{C(EOS)}}{\beta_F} = \frac{Q_S}{\tau_S} + \frac{dQ_S}{dt} \tag{7.76}$$

We will see later that in the switching characteristics of the BJT, the overdrive charge and the time constant τ_S appearing in the equation above play a critical role.

7.6.1   Junction Voltages at Saturation

Having discussed the various operating modes of the BJT, we will now obtain expressions for the junction voltages as the device goes into the saturation mode. These voltages are useful in studying the behavior of a BJT as a logic element. In Fig. 7.16 we show a simple model of the BJT in the saturation mode. Let us apply Kirchhoff's voltage law (KVL) to the voltage values:

$$\begin{aligned} V_{CE} &= V_{CB} + V_{BE} \\ &= -V_{CB} + V_{BE} \end{aligned} \tag{7.77}$$

Thus

$$V_{CE(sat)} = V_{BE(sat)} - V_{CB} \tag{7.78}$$

To obtain $V_{BE(sat)}$ we multiply the second of Eqns. 7.68 by α_R and subtract the resulting equation from the first of Eqns. 7.68. This gives

$$I_E - \alpha_R I_C = I_{ES}(1 - \alpha_F \alpha_R)e^{eV_{BE}/k_BT} \tag{7.79}$$

Using $I_E = I_B + I_C$, we find

$$I_B + I_C(1 - \alpha_R) = I_{ES}(1 - \alpha_F \alpha_R)e^{eV_{BE}/k_BT} \tag{7.80}$$

This gives for $V_{BE(sat)}$

$$V_{BE(sat)} = \frac{k_BT}{e}\ell n\frac{I_B + I_C(1 - \alpha_R)}{I_{EO}} \tag{7.81}$$

where

$$I_{EO} = I_{ES}(1 - \alpha_F \alpha_R) \tag{7.82}$$

In a similar manner, the value of $V_{BC(sat)}$ is

$$V_{BC(sat)} = \frac{k_BT}{e}\ell n\frac{\alpha_F I_B - I_C(1 - \alpha_F)}{I_{CO}} \tag{7.83}$$

with

$$I_{CO} = I_{CS}(1 - \alpha_F \alpha_R) \tag{7.84}$$

From these values of $V_{BE(sat)}$ and $V_{BC(sat)}$ we have

$$V_{CE(sat)} = \frac{k_BT}{e}\ell n\left[\frac{I_B + I_C(1 - \alpha_R)}{\alpha_F I_B - I_C(1 - \alpha_F)} \cdot \frac{I_{CO}}{I_{EO}}\right] \tag{7.85}$$

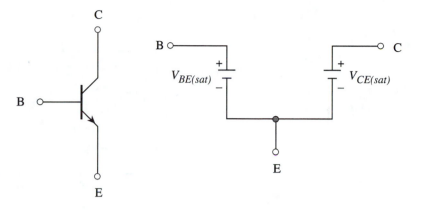

Figure 7.16: The BJT and a simple model for the behavior in the saturation mode.

Note that

$$\frac{I_{CO}}{I_{EO}} = \frac{I_{CS}}{I_{ES}} = \frac{\alpha_F}{\alpha_R} \tag{7.86}$$

The equation for $V_{CE(sat)}$, after some simple manipulation, can be written as

$$V_{CE(sat)} = \frac{k_B T}{e} \ell n \frac{\dfrac{1}{\alpha_R} + \dfrac{I_C}{I_B}\dfrac{1-\alpha_R}{\alpha_R}}{1 - \dfrac{I_C}{I_B}\dfrac{1-\alpha_F}{\alpha_F}} \tag{7.87}$$

We finally substitute for the current gains $\beta_R = \alpha_R/(1-\alpha_R)$, $\beta_F = \alpha_F/(1-\alpha_F)$ to get

$$V_{CE(sat)} = \frac{k_B T}{e} \ell n \frac{\dfrac{1}{\alpha_R} + \dfrac{I_C}{I_B}\dfrac{1}{\beta_R}}{1 - \dfrac{I_C}{I_B}\dfrac{1}{\beta_F}} \tag{7.88}$$

Typical values of $V_{CE(sat)}$ from the expression derived here are ~ 50 mV. If one adds to this value the voltage drop across the neutral regions of the emitter and the collector, we find that $V_{CE(sat)}$ is ~ 0.1 V. Typical values for the various junction voltages, including the bulk resistance (i.e., terminal voltages), are

$$
\begin{aligned}
V_{BE(sat)} &\sim 0.8 \text{ V} \\
V_{CE(sat)} &\sim 0.1 \text{ V}
\end{aligned} \tag{7.89}
$$

7.7 BIPOLAR TRANSISTOR AS AN INVERTER

The inverter forms the basic building block of digital technology. Combined with other circuits, the inverter leads to logic gates, such as AND, OR, etc. These gates, in turn, can be used to implement the full range of Boolean algebra. While much of the digital

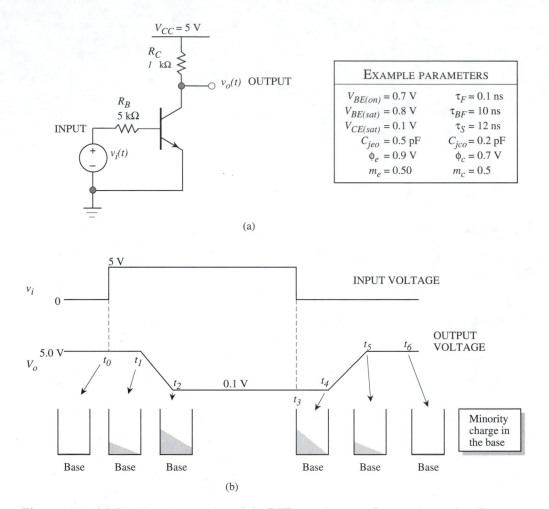

Figure 7.17: (a) Circuit representation of the BJT as an inverter. Parameters used to illustrate the switching phenomenon are given. (b) Input-output for an inverter. The charge in the base at each point in time is shown.

logic is now based on the MOSFET, a device to be discussed later in this text, bipolar devices continue to be used in applications where high speed or high current drive is of importance. In this section we will examine some important aspects of the bipolar inverter.

The basic inverter circuit is shown in Fig. 7.17. The circuit is a simple common-emitter switch. The input voltage form v_i is used to control the state of the switch between the collector and the emitter. The form of the input applied bias is also shown in Fig. 7.17b. We see that at time $t = t_0$, the pulse is switched on from a value 0 to V_2 (chosen as 5.0 V in our study) while at time t_3, it is switched back.

The response of the output voltage v_0 is not instantaneous but shows a time delay, as shown schematically in Fig. 7.17b. The various time delays associated with the response are described below.

Turn ON:

• Cutoff to Active Region: The time delay $t_d = t_1 - t_0$ is associated with going from the cutoff state where both the EBJ and BCJ are reverse biased to the onset of the active region where the EBJ is turned ON. The time delay is due to the charging effects of the base and the depletion region associated with the EBJ and BCJ.
• Active Region to Saturation Region: The time $t_f = t_2 - t_1$ (called fall time) is associated with taking the transistor from the onset of the active region to the saturation region. The times t_d and t_f are the time it takes the transistor to switch on and the output voltage to go from V_{CC} to 0.

In Fig. 7.17b we show the minority charge in the base at times t_0, t_1, and t_2 as the device turns on.

Turn OFF:

• Saturation Region to Edge of Active Region: The time $t_s = t_4 - t_3$ is the time it takes for the excess charge in the base to be removed so that the device reaches the active region.
• Active Region to the Cutoff Region: The time $t_r = t_5 - t_4$ (the rise time) is the time it takes for the transistor to go from the active region to the edge of the cutoff region.
• Finally, for the transistor to completely recover to the OFF state the depletion charge has to be removed. This is the recovery time $t_6 - t_5$. This time is not apparent from the output voltage.

In Fig. 7.17b we show the minority charge in the base as we go from the ON state to the OFF state. The time $t_s + t_r$ is the time it takes for the inverter output voltage to go from its low value to high value.

The calculation of the time constants discussed above requires one to find the time it takes to introduce and remove charges in the base and the depletion regions of the device. *To understand the various time constants involved, let us consider an inverter with the specific device parameters and resistances shown in Fig. 7.17a.*

From Cutoff to Active Region ($t_0 \rightarrow t_1$):
The initial conditions for these states are:

$$t = t_0 \quad :$$
$$V_{BE}(t_0) = 0$$

$$V_{BC}(t_0) = -V_{CC} = -5 \text{ V}$$

$$i_B(t_0) = \frac{v_i(t_0) - V_{BE}(t_0)}{R_B} = 1.0 \text{ mA}$$

$t = t_1$:

$$V_{BE}(t_1) = V_{BE(ON)} = 0.7 \text{ V}$$

$$V_{BC}(t_1) = V_{BE(ON)} - V_{CC} = -4.3 \text{ V}$$

$$i_B(t_1) = \frac{v_i - V_{BE(ON)}}{R_B} = 0.86 \text{ mA} \tag{7.90}$$

The average base current during the time is 0.93 mA. To find the time $t_1 - t_0$ we use the relation

$$\int_{t_0}^{t_1} i_B(t)dt = [Q_{VE}(t_1) - Q_{VE}(t_0)] + [Q_{VC}(t_1) - Q_{VC}(t_0)]$$

where, as discussed in Section 7.6, Q_{VE} and Q_{VC} are the excess charges in the EBJ and BCJ. The emitter junction charge is

$$\Delta Q_E = Q_{VE}(t_1) - Q_{VE}(t_0) = C_{eq}\Delta V_{BE}$$

$$C_{eq} = -C_{ejo}\frac{\phi_e}{(\Delta V_{BE})(1 - m_e)}\left\{\left[1 - \frac{V_{BE}(t_1)}{\phi_e}\right]^{1-m_e} - \left[1 - \frac{V_{BE}(t_0)}{\phi_e}\right]^{1-m_e}\right\}$$

where m_e is the grading parameter for the junction, as discussed in Eqn. 5.68, and ϕ_e is the built-in voltage. Using the parameters for the device given in Fig. 7.17a and the values of V_{BE}, we find

$$C_{eq} = 0.68 \text{ pF}$$

$$\Delta Q_E = (0.68 \text{ pF})(0.7 \text{ V}) = 0.476 \text{ pC}$$

The collector junction has a capacitance charge

$$\Delta Q_C = Q_{VC}(t_1) - Q_{VC}(t_0) = C_{eq}\Delta V_{BC}$$

with

$$C_{eq} = \frac{-(0.2)(0.7)}{(-4.3 + 5.0)(0.5)}\left[\left(1 + \frac{4.3}{0.7}\right)^{1/2} - \left(1 + \frac{5.0}{0.7}\right)^{1/2}\right] = 0.072 \text{ pF}$$

Thus

$$\Delta Q_C = (0.072 \text{ pF})(0.7 \text{ V}) = 0.051 \text{ pC}$$

We now have

$$i_B(\text{average})(t_1 - t_0) = 0.476 + 0.051 = 0.527 \text{ pC}$$

$$t_1 - t_0 = \frac{(0.527 \times 10^{-12} \text{ C})}{0.93 \times 10^{-3} \text{ A}} = 0.57 \text{ ns} \tag{7.91}$$

Active Region to Edge of Saturation (EOS) Region $(t_1 \rightarrow t_2)$:
In the time period $(t_2 - t_1)$ the device moves from the active region to the edge of saturation. The various voltages in the device have the following values at t_2:

$$
\begin{aligned}
V_{BE}(t_2) &= 0.8 \text{ V} \\
V_{BC}(t_2) &= 0.8 - 0.1 = 0.7 \text{ V} \\
i_B(t_2) &= \frac{v_i - V_{BE(sat)}}{R_B} = 0.84 \text{ mA}
\end{aligned}
\tag{7.92}
$$

Using the base current-charge control equation, we get

$$
\int_{t_1}^{t_2} i_B(t)dt = \frac{1}{\tau_{BF}} \int_{t_1}^{t_2} Q_F(t)dt + \Delta Q_F + \Delta Q_{VE} + \Delta Q_{VC}
$$

where $Q_F(t)$ is the minority charge injected in the base and

$$
\Delta Q_F = Q_F(t_2) - Q_F(t_1)
$$

The minority charge injected is

$$
Q_F(t_2) = I_{C(EOS)}\tau_F
$$

where

$$
I_{C(EOS)} = \frac{V_{CC} - V_{CE}(t_2)}{R_C} = \frac{4.9 \text{ V}}{10^3 \ \Omega} = 4.9 \text{ mA}
$$

This gives

$$
\begin{aligned}
Q_F = (4.9 \text{ mA})(0.2 \text{ ns}) &= 0.98 \text{ pC} \\
\Delta Q_F(t_2) = 0.98 - 0.0 &= 0.98 \text{ pC}
\end{aligned}
$$

As in the previous discussion on $(t_1 - t_0)$, we find that

$$
\begin{aligned}
\Delta Q_E &= 0.125 \text{ pC} \\
\Delta Q_C &= 0.47 \text{ pC}
\end{aligned}
$$

We now assume that the charge $Q_F(t)$ rises linearly from 0 to $Q_F(t_2)$ as shown in Fig. 7.18. This then gives

$$
\begin{aligned}
i_B(\text{average})(t_2 - t_1) &= \frac{1}{\tau_{BF}} \frac{(0.98 \text{ pC})}{2}(t_2 - t_1) + \Delta Q_F + \Delta Q_E + \Delta Q_C \\
(0.85 \times 10^{-3})(t_2 - t_1) &= 4.9 \times 10^{-5}(t_2 - t_1) + 1.575 \times 10^{-12} \text{ C}
\end{aligned}
$$

This gives

$$
(t_2 - t_1) = \frac{1.575 \times 10^{-12}}{(0.85 \times 10^{-3} - 4.9 \times 10^{-5})} = 1.97 \text{ ns}
\tag{7.93}
$$

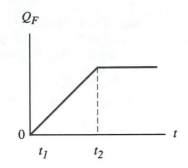

Figure 7.18: Model for buildup of the base charge as the device goes from turn ON to edge of saturation.

The total time required to turn the device ON and for the output to go from a high value to a low value is therefore $(0.57 + 1.97)$ ns $= 2.54$ ns. Let us now consider the device turn-OFF time.

Deep Saturation to Active Region $(t_3 \rightarrow t_2)$:
During the switching ON of the transistor, the device goes well into saturation. This adds a large amount of excess charge in the base. This charge is extracted as the BCJ starts to reverse bias. The time it takes for the excess charge to be extracted is approximately τ_S, which is the mean of τ_{BF} and τ_{BR}, discussed in the previous section. Thus, to a first approximation,

$$t_s = t_4 - t_3 \sim \tau_S$$

An accurate solution of the charge-control equation shows that

$$t_s = \tau_S \ell n \frac{i_{BF} - i_{BR}}{(I_{C(EOS)}/\beta_F - i_{BR})}$$

Here i_{BF} is the current before switching and i_{BR} is the base current between t_3 and t_4:

$$i_{BF} = \frac{v_i - V_{BE(sat)}}{R_B} = \frac{5 - 0.8}{5 \times 10^3} = 0.84 \text{ mA}$$

$$i_{BR} = \frac{0 - 0.8}{R_B} = -0.16 \text{ mA}$$

This, for our case, gives

$$t_s = 12 \, \ell n \frac{0.84 - 0.16}{(4.9/100) + 0.16} = 14.16 \text{ ns} \tag{7.94}$$

Active Region to Edge of Cutoff $(t_4 \rightarrow t_5)$:
In this case the excess charge from the base and depletion regions has to be removed.

The problem is similar to what we encounter in calculating $t_2 - t_1$. The charge terms have the same value as calculated earlier, except the sign is reversed:

$$\int_{t_4}^{t_5} i_B(t)dt = \frac{1}{\tau_{BF}} \int_{t_4}^{t_5} Q_F(t)dt + \Delta Q_F + \Delta Q_E + \Delta Q_C$$

$$i_B(t_4) = \frac{0 - 0.8}{5 \times 10^3} = -0.16 \text{ mA}$$

$$i_B(t_5) = \frac{0 - 0.7}{5 \times 10^3} = -0.14 \text{ mA}$$

Using the average value of 0.15 mA, we get

$$-(0.15 \times 10^{-3})(t_5 - t_4) = \frac{1}{10 \times 10^{-9}} \left(\frac{0.98 \text{ pC}}{2} \right)(t_5 - t_4) - 1.575 \times 10^{-12} \text{ C}$$

$$t_5 - t_4 = \frac{1.575 \times 10^{-12} \text{ C}}{(0.15 \times 10^{-3} - 4.9 \times 10^{-5}) \text{ A}} = 15.6 \text{ ns} \qquad (7.95)$$

Recovery Region $(t_5 \rightarrow t_6)$:
As far as the output voltage is concerned, the output has reached its final value at $t = t_5$. However, the transistor has not recovered completely, since there is a depletion charge that needs to be removed. The base currents at t_5 and t_6 are

$$i_B(t_5) = \frac{0 - 0.7}{5 \times 10^3} = -0.14 \text{ mA}$$

$$i_B(t_6) = 0$$

The mean base current is -0.035 mA. We have already calculated the depletion charge in the cutoff region discussion:

$$\Delta Q_V = -0.527 \text{ pC}$$

Thus the recovery time is

$$(t_6 - t_5) = \frac{-0.527 \times 10^{-12} \text{ C}}{-0.035 \times 10^{-3} \text{ A}} = 15.0 \text{ ns} \qquad (7.96)$$

The discussion in this section shows how various device parameters influence the switching of the inverter. We see that the turn-OFF time for the device is very long because of the time taken to remove all the minority charge injected in the base region.

7.7.1 Schottky Transistor for Faster Switching

In our discussions of the Schottky diode in Chapter 6, we mentioned that since the Schottky diode is a majority carrier device, its transient response is much faster than

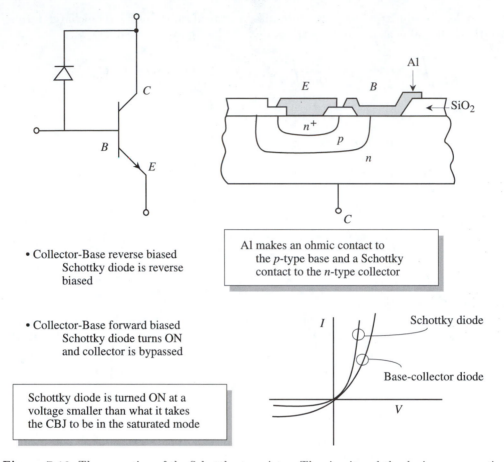

- Collector-Base reverse biased Schottky diode is reverse biased

- Collector-Base forward biased Schottky diode turns ON and collector is bypassed

Al makes an ohmic contact to the *p*-type base and a Schottky contact to the *n*-type collector

Schottky diode is turned ON at a voltage smaller than what it takes the CBJ to be in the saturated mode

Figure 7.19: The operation of the Schottky transistor. The circuit and the device cross-section are shown. The I-V characteristics of the base-collector diode and the Schottky diode are also shown.

that of the bipolar devices. Also, the turn-ON voltage of the Schottky diode is much smaller than that of a *p-n* diode. The properties of the Schottky diode are used to speed up the response of the BJT. The Schottky transistor is shown schematically in Fig. 7.19 along with a simple cross-section of the device structure. As shown in the figure, the metal makes an ohmic contact to the base, but forms a Schottky barrier on the collector. When the transistor is in the cutoff or active mode, the base-collector and the Schottky diode are reverse biased. The Schottky diode thus has no influence on the device. However, when the transistor starts to go into saturation, the diode becomes forward biased and the voltage across the base-collector junction is clamped to the forward turn-ON bias of the diode. The turn-ON voltage of the Schottky diode is much smaller than that of the base-collector junction (for the same current value) and, therefore, the diode allows the excess base current to pass through it. Thus the device does not go into saturation mode and the extraction of the excess base charge becomes

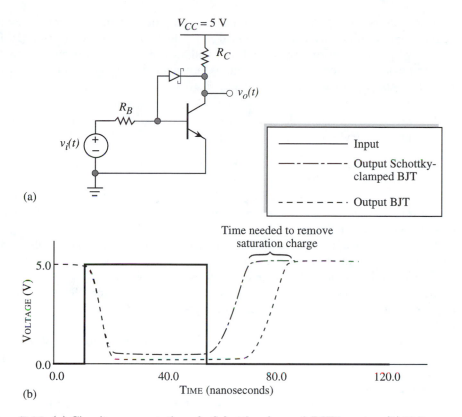

Figure 7.20: (a) Circuit representation of a Schottky-clamped BJT inverter. (b) Voltage versus time for a Schottky-clamped BJT and a conventional BJT.

fast. The device can now be switched in a much shorter time. In Fig. 7.20 we show the switching of a conventional BJT and a Schottky-clamped BJT. We can see that the faster switching of the Schottky-clamped device arises from the time needed to remove saturation charge during device turn-OFF. The Schottky transistor is an important component of the non-saturated bipolar logic and is used in applications where speed is an important concern.

7.8 HIGH-FREQUENCY BEHAVIOR OF BJT

In order to develop a full understanding of the high-frequency operation of a BJT, we need to go beyond the Ebers-Moll model, which neglected a number of important issues.

Hybrid-Pi Model
An important application of bipolar transistors is in the amplification of high-frequency

signals. For understanding the transistor performance in such applications, a small-signal equivalent circuit is needed. The hybrid-pi model is one such description. In Fig. 7.21a we show the cross-section of a typical BJT and also describe the various parameters that need to be incorporated into a small-signal equivalent circuit. The equivalent circuit known as the hybrid-pi model is shown in Fig. 7.21b.

We will briefly discuss the elements of the equivalent circuit. It will be useful for the student to review the small-signal properties of the *p-n* diode (Section 5.6.1):

- r_e: This is the resistance associated with the emitter and depends upon the doping of the emitter and the mobility of the carriers.
- C_π: Since the EBJ is forward biased, the diffusion capacitance associated with the injection of minority electrons into the base can be quite important. By use of a narrow base this capacitance is reduced.
- r_π: This is the resistance associated with the forward-biased EBJ and is dependent upon the value of the injected emitter current.
- C_{je}: This is the junction capacitance of the forward-biased EBJ.
- r_b: This is the base resistance and depends upon the base doping. To reduce r_b, the base doping must be high, but this adversely affects the emitter efficiency.
- $g_m V_{b'e'}$: This represents the current source for the transistor.
- r_o: This is the output resistance and is due to the finite slope of the $I_C - V_{CE}$ curves. It should be as high as possible to ensure that the device current is controlled only by the input signal.
- C_s: This represents the capacitance between the doped collector and the substrate.
- r_μ: This is the resistance of the reverse-biased BCJ and should be very large.
- C_μ: This is the junction capacitance associated with the BCJ.
- r_c: This is the resistance associated with the collector. Usually the collector is doped lightly near the base and heavily further down to ensure a low value of r_c.

A simplified version of the equivalent circuit model results if only the diffusion capacitance and resistance of the forward-biased EBJ are included. The simple model is shown in Fig. 7.21c.

An important parameter representing the frequency response is the cutoff frequency f_T related to the total delay time τ_{ec} representing the propagation of carriers from the emitter to the collector. The cutoff frequency is

$$f_T = \frac{1}{2\pi\tau_{ec}} \tag{7.97}$$

The total delay time τ_{ec} is made up of a number of delay components. We may, in general, write

$$\tau_{ec} = \tau_e + \tau_t + \tau_d + \tau_c \tag{7.98}$$

where τ_{ec} is the emitter to collector delay, τ_e is the EBJ capacitance charging time, τ_t is the base transit time, τ_d is the transit time through the collector depletion region, and

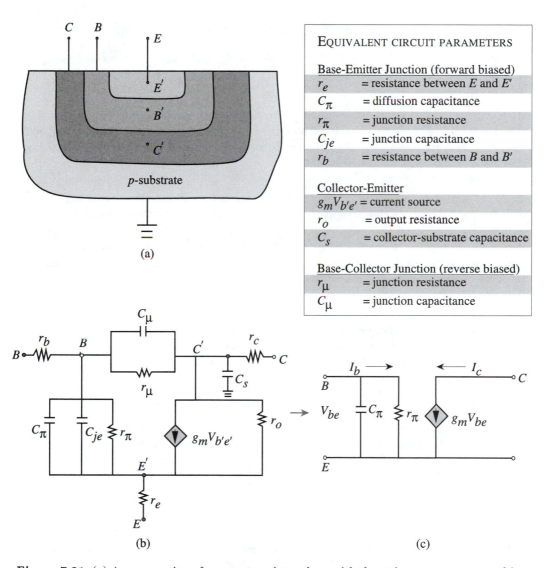

(a)

EQUIVALENT CIRCUIT PARAMETERS	
Base-Emitter Junction (forward biased)	
r_e	= resistance between E and E'
C_π	= diffusion capacitance
r_π	= junction resistance
C_{je}	= junction capacitance
r_b	= resistance between B and B'
Collector-Emitter	
$g_m V_{b'e'}$	= current source
r_o	= output resistance
C_s	= collector-substrate capacitance
Base-Collector Junction (reverse biased)	
r_μ	= junction resistance
C_μ	= junction capacitance

(b)

(c)

Figure 7.21: (a) A cross-section of an *npn* transistor along with the various parameters used in the equivalent circuit; (b) the hybrid-pi model for the bipolar device; (c) a simplified hybrid-pi model in which several parameters are ignored.

τ_c is the collector capacitance charging time. These times are given by

$$\tau_e = r_e' C_{je}$$

$$\tau_t = \frac{W_b^2}{2D_b}$$

$$\tau_d = \frac{W_{dc}}{v_s}$$

$$\tau_c = r_c(C_\mu + C_s) \tag{7.99}$$

where r_e' is the emitter resistance obtained from the slope of the I_E vs. V_{BE} curves, C_{je} is defined in Fig. 7.21, W_{dc} is the depletion width of the base-collector junction, and v_s is the saturation velocity. The other resistances and capacitances are defined in Fig. 7.21.

To achieve a high cutoff frequency one needs narrow emitter stripes (i.e., small area devices), large emitter current (to reduce r_e'), very thin base regions (to reduce τ_t), and low parasitic capacitances. Of course, one also needs to use a material with superior transport properties. The value of f_T initially increases with I_C due to a decrease in τ_e, but it then starts to decrease because of the Kirk effect discussed earlier, which causes the base region to extend out into the collector so that τ_t starts to increase.

Another important figure of merit of a transistor is the highest frequency at which the device can deliver a power gain of unity into a properly matched load. This frequency is called the maximum frequency of oscillation, f_{max}, and is obviously the upper limit beyond which the device is useless as an amplifier.

EXAMPLE 7.12 Consider an *npn* transistor with the following properties at 300 K:

Emitter current,	I_E	=	1.5 mA
EBJ capacitance,	C_{je}	=	$2pF$
Base width,	W_b	=	0.4 μm
Diffusion coefficient,	D_b	=	60 cm^2/s
Width of collector depletion region,	W_{dc}	=	2.0 μm
Collector resistance,	r_C	=	30Ω
Total collector capacitance,	$(C_s + C_\mu)$	=	$0.4pF$

Calculate the cutoff frequency of this transistor. How will the cutoff frequency change (i) if the emitter current level is doubled? (ii) if the base thickness is halved?

The emitter resistance r_e' is given by (see Eqn. 5.72 for the resistance of a forward-biased diode)

$$r_e' = \frac{dI_E}{dV_{BE}} \cong \frac{k_B T}{e I_E} = \frac{0.026}{1.5 \times 10^{-3}} = 17.3 \ \Omega$$

This gives

$$\tau_e = r_e' C_{je} = (17.3)(2 \times 10^{-12}) = 34.6 \ \text{ps}$$

The base transit time is

$$\tau_t = \frac{W_b^2}{2D_b} = \frac{(0.4 \times 10^{-4})^2}{2 \times 60} = 13.3 \text{ ps}$$

The collector transit time is

$$\tau_d = \frac{W_{dc}}{v_s} = \frac{(2.0 \times 10^{-4})}{1 \times 10^7} = 20 \text{ ps}$$

The collector charging time is

$$\tau_c = r_c(C_\mu + C_s) = 30(0.4 \times 10^{-12}) = 12 \text{ ps}$$

The total time is

$$\tau_{ec} = 34.6 + 13.3 + 20 + 12 = 79.9 \text{ ps}$$

The cutoff frequency is

$$f_T = \frac{1}{2\pi\tau_{ec}} = \frac{1}{2\pi(79.9 \times 10^{-12} \text{ s})} = 1.99 \text{ GHz}$$

If the emitter current is doubled (assuming no other change occurs), the time τ_e is reduced by half. This gives a cutoff frequency of 2.54 GHz. Similarly, if the base width is reduced by half, the base transit time becomes 3.3 ps and the cutoff frequency becomes 2.08 GHz. In this problem the dominant source of delay is the emitter junction.

7.9   SPICE MODEL FOR A BIPOLAR TRANSISTOR

To understand how bipolar transistors behave in a circuit, mathematical models have been developed. We will discuss a simple model used in SPICE simulations of the transistor. It is important to note that a number of more advanced-level models are needed for state-of-the-art devices.

In Fig. 7.22 we show the Ebers-Moll model and the model used in SPICE. The model is represented by nonlinear current sources I_E, I_C and the charge stored in the base and the depletion regions of the device. The currents may be written as

$$\begin{aligned} I_E &= \frac{I_S}{\alpha_F}\left(e^{eV_{BE}/k_BT} - 1\right) - I_S\left(e^{eV_{BC}/k_BT} - 1\right) \\ I_C &= I_S\left(e^{eV_{BE}/k_BT} - 1\right) - \frac{I_S}{\alpha_R}\left(e^{eV_{BC}/k_BT} - 1\right) \end{aligned} \quad (7.100)$$

with

$$I_S = \alpha_F I_{ES} = \alpha_R I_{CS} \quad (7.101)$$

The ohmic resistances of the neutral base, collector, and emitter are represented by the resistors r_b, r_c, and r_e shown in Fig. 7.22a.

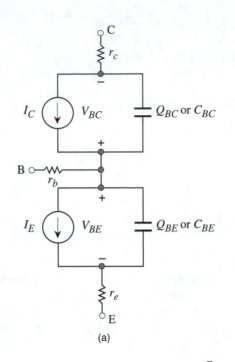

(a)

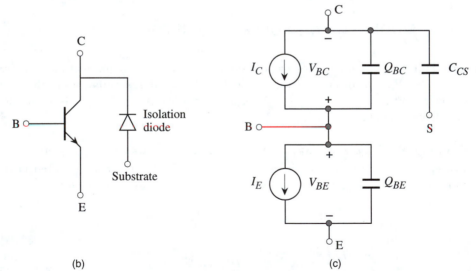

(b)   (c)

Figure 7.22: (a) A SPICE BJT model for a discrete device. (b) A representation of a BJT in an integrated circuit. (c) A SPICE BJT model, including the isolation diode.

The charge stored in the BJT is modeled by Q_{BE} and Q_{BC} and includes the neutral base charge and the charge in the depletion region of the EBJ and BCJ. The charges are given by (see Eqn. 5.68 for junction capacitance)

$$Q_{BE} = \tau_F I_S \left(e^{V_{BE}/k_B T} - 1 \right) + C_{jeo} \int_0^{V_{BE}} \left(1 - \frac{V}{\phi_e} \right)^{-m_e} dV$$

$$Q_{BC} = \tau_R I_S \left(e^{V_{BC}/k_B T} - 1 \right) + C_{jco} \int_0^{V_{BC}} \left(1 - \frac{V}{\phi_c} \right)^{-m_c} dV \qquad (7.102)$$

where C_{jeo} and C_{jco} are the zero-bias junction capacitances for the EBJ and BCJ. ϕ_e and ϕ_c are the built-in voltages for the EBJ and BCJ, respectively. The parameters m_e and m_c are the grading factors for the EBC and BCJ. For abrupt junctions, these are 0.5.

It is possible to use capacitance instead of charge in the SPICE model. The capacitances are

$$C_{BE} = \frac{dQ_{BE}}{dV_{BE}} = \frac{e\tau_F I_S}{k_B T} e^{V_{BE}/k_B T} + \frac{C_{jeo}}{[1 - (V_{BE}/\phi_e)]^{m_e}}$$

$$C_{BC} = \frac{dQ_{BC}}{dV_{BC}} = \frac{e\tau_R I_S}{k_B T} e^{V_{BC}/k_B T} + \frac{C_{jco}}{[1 - (V_{BC}/\phi_c)]^{m_c}} \qquad (7.103)$$

There is an additional element that is needed to describe a BJT in an integrated circuit. In addition to the model described above (valid for a discrete device) one has to recognize that in an IC, the substrate of a device is shared by all devices. Thus, each device has to be isolated from others. This is done by using a reverse-biased diode. This diode is modeled by a current source (diode leakage current) in parallel with a capacitance (depletion region capacitance).

In Fig. 7.22b we show the model diagram for a *npn* transistor in an integrated circuit. The capacitance C_{CS} between the collector and substrate is

$$C_{CS} = \frac{C_{jso}}{(1 - V_{SC}/\phi_s)^{m_s}} \qquad (7.104)$$

The parameters used in the model discussed above form the input file of the SPICE modeling for a circuit. In Table 7.2 we show a typical file.

7.10 BJT DESIGN LIMITATIONS: NEED FOR BAND TAILORING AND HBTs

In the BJT, once a material system is chosen (say Si, Ge, GaAs, etc.), the only flexibility one has in the device design is the doping levels and the device dimensions. We will

Symbol	Name	Parameter	Units	Example
I_S	IS	Saturation current	A	2.0E-15
β_F	BF	Forward current gain		100
β_R	BR	Reverse current gain		1
r_b	RB	Base resistance	Ω	100
r_c	RC	Collector resistance	Ω	50
r_e	RE	Emitter resistance	Ω	1
C_{je0}	CJE	B-E zero-bias depletion capacitance	F	1.0E-12
ϕ_e	VJE	B-E built-in potential	V	0.8
m_e	MJE	B-E junction grading factor		0.5
C_{JC0}	CJC	B-C zero-bias depletion capacitance	F	0.5E-12
ϕ_c	VJC	B-C built-in potential	V	0.7
m_c	MJC	B-C junction grading factor		0.5
C_{JS0}	CJS	Zero-bias collector-substrate capacitance	F	3.0E-21
ϕ_s	VJS	Substrate junction built-in potential	V	0.6
m_s	MJS	Substrate junction grading factor		0.5
τ_F	TF	Forward transit time	s	2.0E-10
τ_R	TR	Reverse transit time	s	1.0E-8

Table 7.2: Parameters of a typical SPICE simulation program for a BJT.

show in this section why this is a serious handicap for high-performance devices. Let us examine the material parameters controlling the device performance parameters. We have (see Eqn. 7.44)

$$\alpha = \left[1 - \frac{p_{eo}D_e W_{bn}}{n_{bo}D_b L_e}\right]\left[1 - \frac{W_{bn}^2}{2L_b^2}\right] \tag{7.105}$$

and the current gain β is

$$\beta = \frac{\alpha}{1 - \alpha} \tag{7.106}$$

For α to be close to unity and β to be high, it is essential that: (i) the emitter doping be much higher than the base doping, i.e., for an *npn* device ($n_{eo} \gg p_{bo}$); and (ii) the base width be as small as possible. In fact, the product $p_{bo}W_b$, called the Gummel number, should be as small as possible. However, a small base with relatively low doping (usually in BJTs $n_{eo} \sim 10^2$-$10^3 p_{bo}$) introduces a large base resistance, which adversely affects the device performance. From this point of view, the Grummel number should be as high as possible.

One may argue that the emitter should be doped as much as possible maintaining $n_{eo} \gg p_{bo}$ and yet having a high enough base doping to ensure low base resistance. However, a serious problem arises from the bandgap shrinking of the emitter region that is very heavily doped. This issue was discussed in Section 2.7.3. If the emitter has a different bandgap (due to heavy doping) from the base, the equations for the excess hole and electrons injected across the EBJ change and affect the transistor performance.

If we assume that hole injection across the EBJ is a dominant factor, the current gain of the device becomes

$$\beta = \frac{\alpha}{1-\alpha} \simeq \frac{n_{bo} D_b L_e}{p_{eo} D_e W_{bn}} \tag{7.107}$$

If the emitter bandgap shrinks by ΔE_g due to doping, the hole density for the same doping changes by an amount that can be evaluated using the change in the intrinsic carrier concentration,

$$n_{ie} \left(E_g - \Delta E_g \right) = n_{ie} \left(E_g \right) \exp \left(\frac{\Delta E_g}{2 k_B T} \right) \tag{7.108}$$

where ΔE_g is positive in our case. Thus the value of p_{eo} changes as

$$p_{eo} \left(E_g - \Delta E_g \right) \quad \propto \quad n_{ie}^2 \left(E_g - \Delta E_g \right) \tag{7.109}$$

$$= \quad p_{eo} \left(E_g \right) \exp \left(\frac{\Delta E_g}{k_B T} \right) \tag{7.110}$$

The bandgap decrease with doping is given for Si by (N_d is in units of cm^{-3})

$$\Delta E_g = 22.5 \left(\frac{N_d}{10^{18}} \cdot \frac{300}{T(K)} \right)^{1/2} \text{ meV} \tag{7.111}$$

The expression is reasonable up to a doping of 10^{19} cm^{-3}. At higher doping levels, the bandgap shrinkage is not so large. For example, at a doping of 10^{20} cm^{-3}, the shrinkage is ~ 160 meV and not 225 meV as given by the equation above. As a result of the bandgap decrease, the gain of the device decreases as (for the case where $L_b \gg W_{bn}$)

$$\beta = \frac{D_b N_{de} L_e}{D_e N_{ab} W_{bn}} \exp \left(-\frac{\Delta E_g}{k_B T} \right) \tag{7.112}$$

where we have used the fact that

$$\frac{p_{eo}}{n_{bo}} = \frac{N_{ab}}{N_{de}} \tag{7.113}$$

where N_{ab} and N_{de} are the acceptor and donor levels in the base and emitter, respectively. As a result of this for a fixed base doping, as the emitter doping is increased,

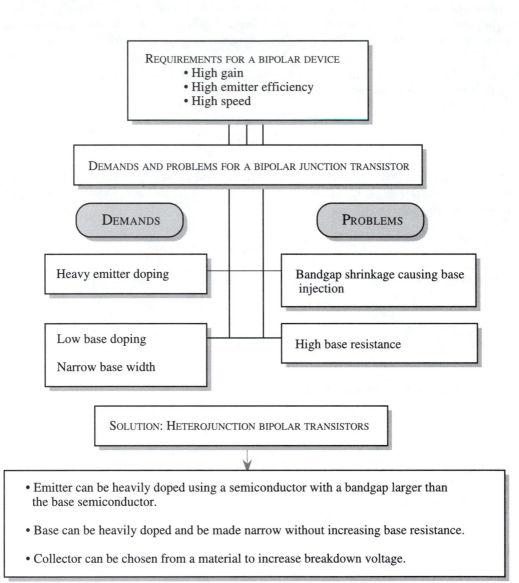

Figure 7.23: Conflicting requirements for high-performance BJTs. Band tailoring offers reconciliation of all these requirements.

initially the current gain increases, but then as bandgap shrinkage increases, the current gain starts to decrease.

From the discussion above, it is clear that the conflicting requirements of heavy emitter doping, low base doping, small base width, etc., as shown in Fig. 7.23, cannot be properly met by a BJT in which the same bandgap semiconductor is used for the emitter and the base, i.e., by a homojunction transistor. This led Shockley and Kroemer in the '50s to conceive of bipolar devices in which the bandgap could change from one region to another. Typically, in these heterojunction bipolar transistors (HBJT or HBT), the *emitter is made from a wide-gap material.* This would dramatically suppress the injection of holes from the base to the emitter. In a typical HBT the emitter is made from a material that has a bandgap that is, say, > 0.2 eV larger than the bandgap of the material used in the base. Near the base side, the emitter material composition is graded so that there is a smooth transition in the bandgap from the emitter side to the base side. A typical example of an HBT structural layout is shown in Fig. 7.24a. In the case shown, the emitter material is AlGaAs, which has a larger bandgap than GaAs, used for the base and the collector.

In Fig. 7.24b we show the band profile for the emitter and the base region. We can see that if ΔE_g is the bandgap difference between the emitter material bandgap and the base material bandgap, this difference appears across the valence band potential barrier, seen by holes. Thus, holes in the base see an increased barrier for injection into the emitter. As a result, the emitter efficiency dramatically increases. The suppression in hole injection current is

$$\frac{I_{Ep}(HBT)}{I_{Ep}(BJT)} = \exp\left(\frac{-\Delta E_g}{k_B T}\right)$$

The gain β in the device correspondingly increases.

For emitter injection limited gain (for $L_b \gg W_{bn}$),

$$\beta_{max} = \frac{D_b N_{de} L_e}{D_e N_{ab} W_{bn}} \exp\left(\frac{\Delta E_g}{k_B T}\right) \tag{7.114}$$

Typically $\Delta E_g/k_B T$ is chosen to be ~ 10, so that an improvement of 10^4 occurs. Now one can dope the base heavily without paying the penalty of low gain.

Due to the heavy doping now allowed in wide emitter HBTs, the base can be made narrow without too large a base resistance or the danger of punchthrough. This also avoids the Kirk effect to some extent. Also, the Early effect is greatly reduced since the base charge is essentially independent of V_{CB}. The emitter doping can also be reduced now so that the emitter depletion region can increase, thus reducing the emitter capacitance. Thus the wide-gap emitter HBT allows one to overcome essentially all the problems associated with optimizing the BJT.

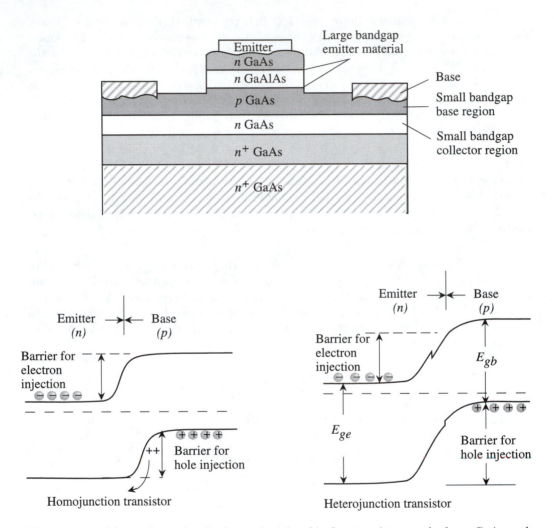

Figure 7.24: (a) A schematic of a heterojunction bipolar transistor made from GaAs and AlGaAs. (b) The band profile for a homojunction and heterojunction transistor. In the homojunction device, the barrier seen by the electrons injected from the emitter into the base and holes injected from the base to the emitter is the same. In the HBT these barriers are different. The spike shown in the HBT conduction band is smoothed out by using a grading in the bandgap between the emitter and the base.

EXAMPLE 7.13 In this example we will examine the effects of bandgap narrowing due to doping on transistor characteristics. Consider a silicon *npn* transistor in which the emitter doping changes from 10^{18} to 10^{20} cm^{-3}. Calculate the change of the hole concentration in the emitter at 300 K.

In silicon the bandgap narrowing is (Eqn. 7.111)

$$\Delta E_g \left(10^{18} \text{ cm}^{-3}\right) \cong 0.022 \text{ eV}$$

$$\Delta E_g \left(10^{20} \text{ cm}^{-3}\right) \cong 0.225 \text{ eV}$$

It is important to note from the discussion following Eqn. 7.111 that the actual bandgap decrease at a doping of 10^{20} cm^{-3} is actually only ~ 160 meV. The student can re-work this problem with this more realistic value.

In the absence of any bandgap shrinkage we note that n_i^2 for Si is 2.25×10^{20} cm^{-3}. This value decreases due to the shrinkage. We get

$$p_{eo} = \frac{n_i^2}{n_{eo}} \exp \left(\frac{\Delta E_g}{k_B T}\right)$$

When n_{eo} is 10^{18} cm^{-3}

$$p_{eo} = 5.24 \times 10^2 \text{ cm}^{-2}$$

when

$$n_{eo} = 10^{20} \text{ cm}^{-3}$$

$$p_{eo} = 1.26 \times 10^4 \text{ cm}^{-3}$$

Thus for the higher doped emitter, the hole density in the emitter *increases rather than decreases*. As discussed in the text, this creates a problem for emitter efficiency. The equation used above for bandgap shrinkage overestimates the reduction in the gap, especially at high doping. The example, however, does point towards the problem of heavily doped emitters.

EXAMPLE 7.14 Consider an *npn* GaAs BJT that has a doping of $N_{de} = 5 \times 10^{17}$ cm^{-3}, $N_{ab} = 10^{17}$ cm^{-3}. Compare the emitter efficiency of this device with that of a similarly doped HBT where the emitter is Al$_{0.3}$Ga$_{0.7}$As and the base is GaAs. The following parameters characterize the devices at 300 K:

Electron diffusion constant in the base,	D_b	$= 100 \text{ cm}^2/\text{s}$
Hole diffusion constant in the emitter,	D_e	$= 15 \text{ cm}^2/\text{s}$
Base width,	W_b	$= 0.5 \text{ } \mu\text{m}$
Bandgap discontinuity,	ΔE_g	$= 0.36 \text{ eV}$
Minority carrier length for holes,	L_e	$= 1.5 \text{ } \mu\text{m}$

For GaAs we have the emitter and base minority carrier concentrations

$$p_{eo} = \frac{n_i^2}{N_{de}} = \frac{(2.2 \times 10^6)^2}{5 \times 10^{17}} = 9.7 \times 10^{-6} \text{ cm}^{-3}$$

$$n_{bo} = \frac{n_i^2}{N_{ab}} = \frac{(2.2 \times 10^6)^2}{10^{17}} = 4.84 \times 10^{-5} \text{ cm}^{-3}$$

The emitter efficiency is

$$\gamma_e = 1 - \frac{p_{eo} D_e W_b}{n_{bo} D_b L_e} \quad = \quad 1 - \frac{(9.7 \times 10^{-6})(15)(0.5 \times 10^{-4})}{(4.84 \times 10^{-5})(100)(1.5 \times 10^{-4})}$$

$$= \quad 0.99$$

In the HBT, the value of p_{eo} is greatly suppressed. The new value is approximately

$$p_{eo}(Al_{0.3}Ga_{0.7}As) \quad = \quad \frac{n_i^2(GaAs)}{N_{de}} \exp \left(-\frac{\Delta E_g}{k_B T} \right)$$

$$= \quad p_{eo}(GaAs) \exp \left(-\frac{\Delta E_g}{k_B T} \right) = 9.4 \times 10^{-12} \ cm^{-3}$$

In this case the emitter efficiency is essentially unity.

7.11   BIPOLAR TRANSISTORS: A TECHNOLOGY ROADMAP

In this section we will discuss some of the important design considerations in the performance of bipolar devices. Bipolar devices must compete with the field effect transistor (FETs) and in many respects the two classes of families carry out similar functions. This puts a tremendous pressure on the BJT and HBT device designers to design the best devices in a given material system.

Bipolar devices are exploiting both fabrication techniques and new material systems to produce superior devices. A survey of the development of advanced devices was given in Fig. 7.25. We will now give a brief overview of these developments.

7.11.1   Si Bipolar Technology

In spite of the superior performance of HBTs, the Si bipolars continue to be the workhorse devices for both digital and some microwave applications. The advances in Si technology have come from two directions. The first direction relates to advanced fabrication technology and the second one relates to the use of polysilicon as a contact for the emitters.

The fabrication-technology-related advances in Si bipolars have resulted from: (i) self-aligned emitter and base contacts, which allow extremely precise placement of the base contact next to the emitter contact and thus reduce parasitic resistances; (ii) trench isolation, which allows very dense packing of the transistors without cross-talk. This involves etching narrow grooves around the transistor down to the substrate, lining them with SiO_2, and filling them with polysilicon. This greatly reduces the isolation capacitance; (iii) sidewall contact process, which dramatically reduces the extrinsic base

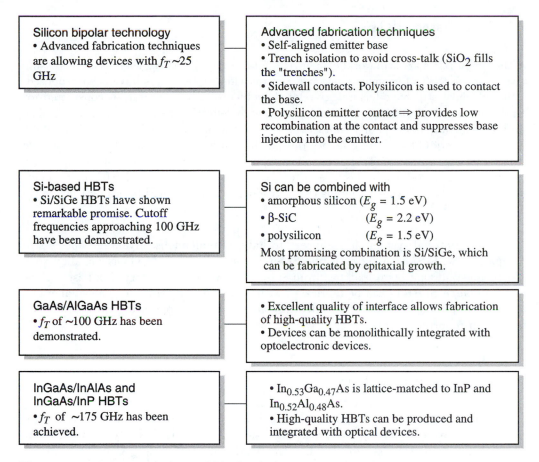

Figure 7.25: A survey of advanced bipolar devices.

collector capacitance. In this process polysilicon is used to contact the base and is isolated from the collector by a thick oxide. The device becomes essentially one-dimensional as a result and also becomes quite symmetric between the emitter and the collector.

The second source of improvements in Si bipolar devices is the use of polysilicon to contact the emitter. The advantages of polysilicon over metal contacts arise from the boundary conditions the contact places on the hole density injected into the emitter from the base. The boundary condition is very important for the thin emitters needed for high-frequency applications. The hole density goes to zero at a normal ohmic contact due to the very large recombination rate with the electrons. In the case of polysilicon the hole density goes to zero gradually so that the hole injection current is similar to that of a thick emitter. Due to this, the base injection into the emitter is strongly suppressed.

With advanced technology in use, Si BJTs have reached f_T values of ~25 GHz.

7.11.2   Si-Based HBTs

Although Si BJTs are still workhorse devices for most applications, there is an increasing interest in Si HBTs for obvious reasons. Several wide-gap emitters have been proposed, although most still have technology-related problems. Among materials considered for emitters are: (i) amorphous Si, which has a large "effective bandgap" ($\sim$1.5 eV). The problems include poor-quality contacts to amorphous Si; (ii) β-SiC with bandgap of 2.2 eV. The material has a strong lattice mismatch with Si and it is not clear how reliable the technology will be; (iii) semi-insulating polycrystalline Si, which has a gap of 1.5 eV. High current gains have been reported for this system; (iv) use of III-V compounds like GaP. The main problem here is the cross-doping issue since Si dopes GaP while Ga and P dope Si.

A material system that appears to have a tremendous advantage and is still compatible with Si technology is the Si-SiGe system. The $Si_{1-x}Ge_x$ is an alloy with lattice constant that is mismatched from Si by $4x\%$. However, for very thin base regions n-Si/p-SiGe/n-Si HBTs can be fabricated with very high performance. The smaller gap of SiGe suppresses hole injection into the emitter. Devices operating up to 100 GHz have been reported in this material system.

7.11.3   GaAs/AlGaAs HBTs

In Chapter 3 we discussed the bandstructure of GaAs and AlAs systems. The two semiconductors have excellent lattice matching ($\sim$0.14%) and high-quality GaAs/AlGaAs heterostructures can be grown. The bandgap of the alloy $Al_xGa_{1-x}As$ up to compositions of $x \sim 0.45$ is given by

$$E_g(x) = 1.42 + 1.247x$$

Above $x \sim 0.45$, the material becomes indirect and is usually not used for most device applications because of poor transport and optical properties.

GaAs material has a high bandgap and thus the intrinsic carrier concentration is quite low ($\sim 2.2 \times 10^6$ cm^{-3}) at room temperature. Thus the semi-insulating GaAs can have a very high resistivity ($\sim 5 \times 10^8$ Ω-cm), with the result that there is essentially negligible capacitance between the substrate and the interconnects or the collector. This is a serious problem for Si at high frequencies.

An important advantage of GaAs technology is that the electronic devices can be monolithically integrated with optoelectronic devices, leading to optoelectronic integrated circuits (OEICs), which are certainly not possible for Si technology (so far).

Another important advantage of GaAs technology is the ability to fabricate millimeter microwave integrated circuits (MMICs) in which the active and passive elements of the circuit are all made on the same chip. MMIC technology is quite advanced in GaAs while it is still primitive in Si.

In the GaAs/AlGaAs system, HBTs with f_T values around 100 GHz have been achieved, making this material system an important player in microwave technology.

7.11.4   InGaAs/InAlAs and InGaAs/InP HBTs

An important consideration in the development of any material technology is the substrate availability. One must have a high-quality substrate that is lattice-matched to the material and has very few defects. There are three main substrates that have reached a very high quality level: Si, GaAs, and InP. The material systems $In_{0.53}Ga_{0.47}As$ ($E_g \sim 0.75$ eV) and $In_{0.52}Al_{0.48}As$($E_g \cong 1.4$ eV) are lattice-matched to InP. Thus the $In_{0.53}Ga_{0.47}As/In_{0.52}Al_{0.48}As$ and InGaAs/InP both can be exploited for high-performance HBTs. InGaAs has extremely attractive electronic properties and is therefore the material of choice for all high-speed/high-frequency applications. The InGaAs/InP HBTs have achieved f_T values of over 250 GHz.

7.12   SUMMARY

In this chapter we have studied the bipolar device, which is an important three-terminal device with high gain and excellent input-output isolation. The key issues covered in this chapter are given in the summary table (Table 7.3).

7.13   A BIT OF HISTORY

As the reader may be well aware, vacuum tubes were the predecessor of semiconductor three-terminal devices. The vacuum tube with a cathode, anode, and a control grid that formed the third terminal provided all the desirable properties of an active device. It had excellent input-output isolation, high amplification, and in the early days of semiconductor research appeared much more reliable than the semiconductor devices. Of course, the biggest problem with the vacuum tube was its size, and this was, perhaps, one of the most important reasons why the vacuum tube gave way to the semiconductor transistor.

The idea of duplicating the vacuum tube concept in a semiconductor occurred to Pohl and his co-worker Hilsch around 1938. Professor Pohl had a very healthy interaction with the budding German electronics industries and as a result was in an ideal position

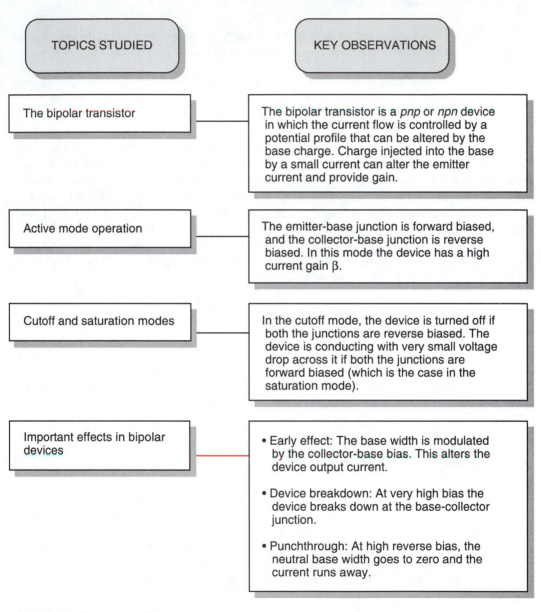

TOPICS STUDIED

KEY OBSERVATIONS

The bipolar transistor

The bipolar transistor is a *pnp* or *npn* device in which the current flow is controlled by a potential profile that can be altered by the base charge. Charge injected into the base by a small current can alter the emitter current and provide gain.

Active mode operation

The emitter-base junction is forward biased, and the collector-base junction is reverse biased. In this mode the device has a high current gain β.

Cutoff and saturation modes

In the cutoff mode, the device is turned off if both the junctions are reverse biased. The device is conducting with very small voltage drop across it if both the junctions are forward biased (which is the case in the saturation mode).

Important effects in bipolar devices

- Early effect: The base width is modulated by the collector-base bias. This alters the device output current.

- Device breakdown: At very high bias the device breaks down at the base-collector junction.

- Punchthrough: At high reverse bias, the neutral base width goes to zero and the current runs away.

Table 7.3: Summary table.

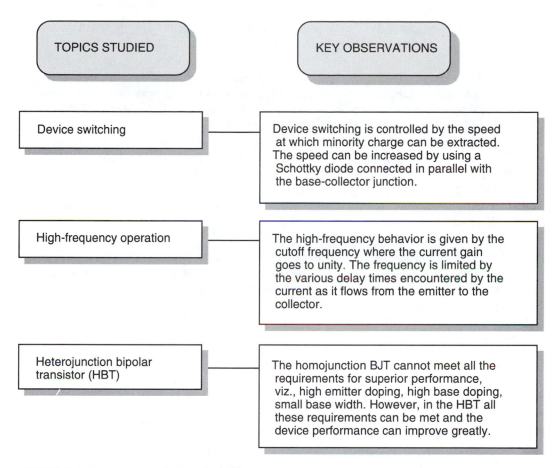

Table 7.3: Summary table (continued).

to sense the need for new physical phenomena for devices. In 1938 Pohl and Hilsch presented results on "Control of Electric Currents with Three Electrode Crystal and a Model of a Barrier Layer." In this work they were able to duplicate the features of a vacuum tube in a semiconductor. The semiconductor used was potassium bromide that was connected to two different metal electrodes, only one of which could inject electrons into the semiconductor. A grid was built close to the cathode and the injected charge was controlled by the grid bias, just as in a vacuum tube. An amplification as high as 100 was achieved in their experiment, but the device was not suitable because of lack of adequate technology at the time.

The real glory of inventing a working three-terminal transistor in semiconductors goes to the Bell Labs group involving Bardeen, Brittain, and Shockley. As we discussed in the history section of Chapter 5, the radar work during the Second World War was crucial in developing the semiconductor diode. High-purity Ge was now available to researchers. Since Bell Labs was intimately involved in diode research, it was natural that after the war the labs would try to exploit semiconductors for other devices.

While the diode research was largely carried out without fully understanding the basic physics of electrons in semiconductors and metals, it was clear that future work would require detailed physical understanding. The research director of Bell Labs, Kelly, recognized the need to attract bright minds to his lab so that the fundamental areas could be addressed. In 1945, Shockley took charge of the semiconductor group at Bell Labs as a theorist; with him were experimentalists Brittain and Pearson, circuit specialist H. Moore, physical chemist R. Gibney, and John Bardeen. While the group was investigating a variety of fundamental phenomena in semiconductors, they wished to create a semiconductor amplifier that would open new vistas in the fields of microwaves and radios.

Shockley initially decided to focus on the field-effect transistor in which a control electrode would control the current flow by changing the number of carriers flowing. This idea had already been patented by Julius Lilienfield in 1926, although it had not been realized in practice. Shockley and his co-workers could not succeed in the experiment, although the failure led to development of the surface state model for the Schottky barrier discussed in Chapter 6.

To overcome the surface states, an electrolyte was deposited on the semiconductor and this led to a thin insulator film on which a metal electrode was deposited. In one such experiment a gold film was deposited to control the number of carriers in a Ge crystal. However (as was realized later), the insulator film was somehow washed away! As a result, this control electrode, instead of changing the flowing carriers, injected holes into the semiconductor. These injected holes influenced the current flow in the Ge device. This new concept was discovered while the group was looking for the field-effect device and was a result of an inadvertent removal of the insulator film!

Further experiments led to the point contact transistor, which led to bipolar technology. On December 23, 1947, the transistor was used to demonstrate amplification of voice signals. The work was, however, not disclosed to the outside world until June 22, 1948, when it was shown to the Defense Department. On July 30, 1948 it was introduced to the press.

The initial reaction to the transistor was rather lukewarm. Although they were used in some applications in the telephone system and hearing aids, it was felt the technology was unreliable. Also, vacuum tubes had been miniaturized to a considerable extent. The U.S. military came to the rescue of this new technology when it decided to bear the cost of research and development of the solid state transistor. The military's interest stemmed from the realization that if these new devices were reproducible they could greatly improve the new weapons systems that were already requiring several thousand tubes around 1952.

The point contact transistor (in whose invention Shockley was not very actively involved) was a forerunner of the modern junction transistor. Shockley realized that it was very difficult to control the spacing of the electrodes in a point contact transistor and came up with the idea of having a buried junction with a different doping. The junction transistor was demonstrated at Bell Labs in 1950. This junction transistor could be produced with greater reliability, and Bell Labs exchanged the technology of this new device with a number of new companies, including Texas Instruments, which was later to become a key player in solid state electronics. Work at Texas Instruments led to the first silicon bipolar transistor and also the first integrated circuit.

The bipolar transistor and its nurturing during the initial period by the U.S. Defense Department was a key step in the explosion in solid state electronics. It was relatively inexpensive to set up device fabrication facilities and, in the 1950s, about two dozen companies were involved in transistor fabrication. Many of these companies were run by entrepreneurs who were dissatisfied with the conditions in big companies. It was a start of a new era in science and technology.

Silicon bipolar devices continue to play a key role in modern solid state technology, even half a century after their invention. As noted in this chapter, the key contribution to high-performance bipolar technology has been the concept of HBT, formulated by Herbert Kroemer. BJTs and HBTs appear well poised to serve the scientific community for a long time to come.

7.14 PROBLEMS

Note: Please examine problems in the following Design Problems section as well.

Temperature is 300 K unless otherwise specified.

Section 7.2
Problem 7.1. Sketch the energy bands and electric field profile in a n^+pn and a p^+np bipolar transistor (a) in equilibrium; (b) biased in the forward active mode.

Sections 7.3–7.4
Problem 7.2 An *npn* silicon bipolar transistor at 300 K has the dopings $N_{de} = 8 \times 10^{19}$ cm^{-3}, $N_{ab} = 10^{17}$ cm^{-3}, and $N_{dc} = 10^{16}$ cm^{-3}. The transistor is operating in the active mode with $V_{BE} = 0.5$ V and $V_{CB} = 2.0$ V. Plot the minority carrier profile in the device. Assume the following parameters: $W_b = 10$ μm; $D_b = 30$ cm^2 s^{-1}; $\tau_B = 10^{-7}$ s $= \tau_E = \tau_C$; $D_e = 10$ cm^2 s^{-1}; $D_c = 10$ cm^2 s^{-1}.
Problem 7.3 Show that for a narrow base *npn* transistor of area A the excess charge

injected into the base is given by

$$\bar{Q} = -\frac{en_{bo}AW}{2}\left\{\left[\exp\left(\frac{eV_{BE}}{k_BT}\right) - 1\right] + \left[\exp\left(\frac{-eV_{CB}}{k_BT}\right) - 1\right]\right\}$$

Problem 7.4 In a *pnp* silicon transistor at 300 K, the base doping is 5×10^{16} cm^{-3}. The base width is 1.0 μm and $L_b = 10.0$ μm. What is the minority carrier charge in the base (a) at $V_{EB} = 0.5$ V; $V_{BC} = 1.0$ V, (b) when both emitter base and base collector junctions are reverse biased at -2.0 V. The area of the device is 10^{-2} cm^2.

Problem 7.5 Consider an *npn* transistor with the following parameters:

$$D_b = 20 \text{ cm}^2 \text{ s}^{-1} \qquad\qquad D_e = 10 \text{ cm}^2 \text{ s}^{-1}$$
$$N_{de} = 5 \times 10^{18} \text{ cm}^{-3} \qquad\qquad N_{ab} = 5 \times 10^{16} \text{ cm}^{-3}$$
$$N_{dc} = 5 \times 10^5 \text{ cm}^{-3} \qquad\qquad W_b = 1.0 \text{ } \mu\text{m}$$
$$\tau_B = \tau_E = 10^{-7} \text{ s} \qquad\qquad n_i^2 = 2.25 \times 10^{20} \text{ cm}^{-6}$$
$$A = 10^{-2} \text{ cm}^2$$

Calculate the collector current in the active mode with an applied emitter base bias of 0.5 V. What is the collector current when the base current is now increased by 20%?

Problem 7.6 An Si *npn* transistor at 300 K has an area of 1 mm^2, base width of 1.0 μm, and dopings of $N_{de} = 10^{18}$ cm^{-3}, $N_{ab} = 10^{17}$ cm^{-3}, $N_{dc} = 10^{16}$ cm^{-3}. The minority carrier lifetimes are $\tau_E = 10^{-7} = \tau_B$; $\tau_C = 10^{-6}$ s. Calculate the collector current in the active mode for (a) $V_{BE} = 0.5$ V, (b) $I_E = 2.5$ mA, and (c) $I_B = 5$ μA. The base diffusion coefficient is $D_b = 20$ cm^{2}s^{-1}.

Problem 7.7 An *npn* silicon transistor is operated in the inverse active mode (i.e., collector-base is forward biased and emitter base is reverse biased). The doping concentrations are $N_{de} = 10^{18}$ cm^{-3}; $N_{ab} = 10^{17}$ cm^{-3}, and $N_{dc} = 10^{16}$ cm^{-3}. The voltages are $V_{BE} = -2$ V, $V_{BC} = 0.6$ V. Calculate and plot the minority carrier distribution in the device. Also calculate the current in the collector and the emitter. The device parameters are: $W_b = 1.0$ μm, $\tau_E = \tau_B = \tau_C = 10^{-7}$s, $D_b = 20$ cm^{2}s^{-1}, $D_e = 10$ cm^{2}s^{-1}, $D_c = 25$ cm^{2}s^{-1}, A = 1 mm^2.

Problem 7.8 Calculate the error made in the emitter efficiency expression (i.e., Eqn. 7.41 versus 7.40) when one makes the approximation given in the text for tanh. Obtain the error as a function of the ratio of L_b to W_{bn}.

Problem 7.9 Plot the dependence of the base transport factor in a bipolar transistor as a function of W_b/L_b over the range $10^{-2} \le W_b/L_b \le 10$. Assume that the emitter efficiency is unity. How does the common-emitter current gain vary over the same range of W_b/L_b?

Problem 7.10 In an *npn* bipolar transistor, calculate and plot the dependence of the emitter efficiency on the ratio of N_{ab}/N_{de} in the range $10^{-2} \le N_{ab}/N_{de} \le 1$. Calculate the results for the cases: (a) $D_e = D_b, L_e = L_b, W_b = L_b$, and (b) $D_e = 0.2D_b, L_e = 0.2L_b, W_b = 0.1L_b$.

Problem 7.11 In a uniformly doped *npn* bipolar transistor, the following current values are measured (see Fig. 7.5 for the current definitions):

$$I_{En} = 1.2 \text{ mA} \qquad\qquad I_{Ep} = 0.1 \text{ mA}$$

$$I_C = I_{nC} = 1.19 \text{ mA} \qquad I_{BE}^R = 0.1 \text{ mA}$$

Determine the parameters α, β, γ_e for the transistor.

Problem 7.12 Consider an *npn* bipolar transistor at 300 K with the following parameters:

$$
\begin{array}{lll}
N_{de} = 5 \times 10^{18} \text{ s}; & N_{ab} = 5 \times 10^{16} \text{ cm}^{-3}; & N_{dc} = 10^{15} \text{ cm}^{-3} \\
D_e = 10 \text{ cm}^2 \text{ s}; & D_b = 15 \text{ cm}^2 \text{ s}; & D_c = 20 \text{ cm}^2 \text{ s} \\
\tau_E = 10^{-8} \text{ s}; & \tau_B = 10^{-7} \text{s}; & \tau_C = 10^{-6} \text{ s} \\
W_b = 1.0 \text{ } \mu\text{m}; & A = 0.1 \text{ mm}^2 &
\end{array}
$$

Calculate the emitter current and the collector current as well as the values of α, γ_e, β for $V_{BE} = 0.6$ V; $V_{CE} = 5$ V.

Problem 7.13 The mobility of holes in silicon is 100 cm^2/V·s. It is required that a BJT be made with a base width of 0.5 μm and base resistivity of no more than 1.0 Ω-cm. It is also desired that the emitter injection efficiency be at least 0.999. Calculate the emitter doping required. The various device parameters are

$$
\begin{array}{lcl}
L_b & = & 10 \text{ } \mu\text{m} \\
L_e & = & 10 \text{ } \mu\text{m} \\
D_e & = & 10 \text{ cm}^2/\text{s} \\
D_b & = & 20 \text{ cm}^2/\text{s}
\end{array}
$$

What is the current gain β of the device? Assume $W_{bn} = W_b$.

Problem 7.14 Consider a *npn* Si-BJT at 300 K with the following parameters:

$$
\begin{array}{lcl}
N_{de} & = & 10^{18} \text{ cm}^{-3} \\
N_{ab} & = & 10^{17} \text{ cm}^{-3} \\
N_{dc} & = & 5 \times 10^{16} \text{ cm}^{-3} \\
D_b & = & 30.0 \text{ cm}^2/\text{s} \\
L_b & = & 15.0 \text{ } \mu\text{m} \\
D_e & = & 10.0 \text{ cm}^2/\text{s} \\
L_e & = & 5.0 \text{ } \mu\text{m}
\end{array}
$$

(i) Calculate the maximum base width, W_b, that will allow a current gain β of 100 when the EBJ is forward biased at 1.0 V and the BCJ is reverse biased at 5.0 V.

(ii) Describe two advantages and two disadvantages of making the base smaller.

Problem 7.15 The $V_{CE}(sat)$ of an *npn* transistor decreases as the base current increases for a fixed collector current. In the Ebers-Moll model, assume $\alpha_F = 0.995$, $\alpha_R = 0.1$, and $I_C = 1.0$ mA. At 300 K, at what base current is the $V_{CE}(sat)$ value equal to (a) 0.2 V, (b) 0.1 V?

Problem 7.16 Consider an *npn* bipolar device in the active mode. Express the base current in terms of $\alpha_F, \alpha_R, I_{ES}, I_{CS}$, and V_{BE}, using the Ebers-Moll model.

Problem 7.17 Derive the expressions for the emitter and collector current for a *pnp* transistor in analogy with the equations derived in the text for the *npn* transistor.

Problem 7.18 An *npn* silicon bipolar device has the following parameters in the Ebers-Moll model at 300 K:

$$\alpha_F = 0.99; \ \alpha_R = 0.2$$

Calculate the saturation voltage V_{CE} for $I_C = 5$ mA and $I_B = 0.2$ mA. Why is I_C/I_B not equal to $\alpha_F/(1 - \alpha_F)$?

Section 7.5

Problem 7.19 A silicon *pnp* transistor at 300 K has a doping of $N_{ae} = 10^{18}$ cm^{-3}, $N_{db} = 5 \times 10^{16}$ cm^{-3}, $N_{ac} = 10^{15}$ cm^{-3}. The base width is 1.0 μm. The value of D_b is 10 cm^2/s and $\tau_B = 10^{-7}$ s. The emitter base junction is forward biased at 0.7 V. Using the approximation that the minority carrier distribution in the base can be represented by a linear decay, calculate the hole diffusion current density in the base at (a) $V_{CB} = 5$ V (reverse bias), (b) $V_{CB} = 15$ V.

Problem 7.20 A uniformly doped *npn* bipolar transistor is fabricated to within $N_{de} = 10^{19}$ cm^{-3} and $N_{dc} = 10^{16}$ cm^{-3}. The base width is 0.5 μm. Design the base doping so that the punchthrough voltage is at least 25 V in the forward active mode.

Problem 7.21 An *npn* silicon bipolar transistor has a base doping of 10^{16} cm^{-3} and a heavily doped collector region. The neutral base width is 1.0 μm. What is the base collector reverse bias when punchthrough occurs?

Problem 7.22 The punchthrough voltage of a Ge *pnp* bipolar transistor is 20 V. The base doping is 10^{16} cm^{-3}, and the emitter and collector dopings are 10^{18} cm^{-3}. Calculate the zero bias base width. If $\tau_B = 10^{-6}$ s, what is the α of the transistor at a 10 V reverse bias across the collector-base junction at 300 K? The hole diffusion coefficient in the base is 40 cm^{2}s^{-1}.

Problem 7.23 In a silicon *npn* transistor, the doping concentrations in the emitter and collector are $N_{de} = 10^{18}$ cm^{-3} and $N_{dc} = 5 \times 10^{15}$ cm^{-3}, respectively. The neutral base width is 0.6 μm at $V_{BE} = 0.7$ V and $V_{CB} = 5$ V. When V_{CB} is increased to 10 V, the minority carrier diffusion current in the base increases by 5%. Calculate the base doping and the Early voltage if $D_b = 20$ cm^2/s and $\tau_B = 5 \times 10^{-7}$s.

Problem 7.24 Consider a *npn* Si-BJT at 300 K with the following parameters:

$$
\begin{aligned}
N_{de} &= 10^{18} \text{ cm}^{-3} \\
N_{ab} &= 10^{17} \text{ cm}^{-3} \\
N_{dc} &= 10^{16} \text{ cm}^{-3} \\
D_b &= 30.0 \text{ cm}^2/\text{s} \\
L_b &= 10.0 \ \mu\text{m} \\
W_b &= 1.0 \ \mu\text{m} \\
D_e &= 10 \text{ cm}^2/\text{s} \\
L_e &= 10.0 \ \mu\text{m} \\
\text{Emitter thickness} &= 1.0 \ \mu\text{m}
\end{aligned}
$$

$$\text{Device area} \quad = \quad 4.0 \times 10^{-6} \quad \text{cm}^2$$

Calculate the emitter efficiency and gain β when the EBJ is forward biased at 1.0 V and the BCJ is reverse biased at 5.0 V. Calculate the output conductance of the device defined by

$$g_o = \frac{\Delta I_C}{\Delta V_{CB}}$$

Problem 7.25 An important advance in Si bipolar transitors is the use of polysilicon emitters. If a normal ohmic contact is made to an emitter, the injected minority density goes to zero at the ohmic contact boundary. In polysilicon emitters, heavily doped polysilicon forms the contact to the emitter. The minority density does not go to zero at the polysilicon contact, but decreases to zero well inside it. This allows one to have very thin emitter contacts for high-speed operation. Discuss the disadvantage of such a contact over a normal ohmic contact in a thin emitter. (Consider the emitter efficiency and how it is affected by a thin emitter by using the discussions in Chapter 5 on the narrow p-n diode.)

Problem 7.26 Consider an npn BJT with a base width of 0.5 μm and base doping of 10^{17} cm^{-3}. The hole mobility is 200 cm^2/V·s. An emitter stripe of 25 μm ×100 μm is placed to form the EBJ. If a base current of 100 μA passes in the device and the EBJ is forward biased at 0.7 V at the edge of the emitter, *estimate the value of the forward bias of the EBJ at the middle of the emitter.* Discuss the possible problems that the biasing difference could cause. (Assume that the base current is flowing through an area 100 μm ×0.5 μm.)

Problem 7.27 From our discussions of narrow p-n diodes, the importance of the boundary conditions imposed on the injected minority charge at the contact is quite obvious. We have used the condition that the minority charge density goes to zero at the contact. This is a reasonable approximation for the metal contact. One approach to defining the boundary conditions at any interface is through the concept of a recombination velocity. The recombination velocity v_{recom} is defined via the relation (say, for holes as minority charge)

$$J_p |_{boundary} = e \; v_{recom} \; \delta p|_{boundary}$$

where the current and the excess charge are evaluated at the boundary. Using the expression for minority current in terms of the diffusion coefficient and the charge density gradient, we have

$$D_p \frac{d(\delta p)}{dx} |_{boundary} = v_{recom} \; \delta p|_{boundary}$$

Consider the case where an excess hole density of $\delta p(x_1)$ is injected across a depletion region into an n-side. The boundary of the contact is at a position x_2. The distance $(x_2 - x_1) \ll L_p$ so that the hole concentration can be assumed to decrease linearly. Express $\delta p(x_2)$ in terms of the surface recombination velocity, $\delta p(x_1)$, and the diffusion coefficient D_p and $(x_2 - x_1)$.

Section 7.8

Problem 7.28 In a particular BJT, the base transit time forms 20% of the total delay time of the charge transport. The base width is 0.5 μm and the diffusion constant is $D_b = 25$ cm^{2}s. Calculate the cutoff frequency for the device.

Problem 7.29 A silicon *npn* bipolar transistor has a cutoff frequency at 300 K limited by base transit time. The cutoff frequency is 1 GHz. Estimate the base width if the base doping is 10^{16} cm^{-3}. The minority carrier mobility in the base is 500 cm^2/V·s.

Section 7.10

Problem 7.30 Consider an *npn* silicon bipolar transistor in which $W_b = 2.0$ μm, $L_e = L_b = 10.0$ μm, and $D_e = D_b = 10$ cm^{2}s^{-1}. Assume that $N_{ab} = 10^{16}$ cm^{-3}. What is the emitter injection efficiency for $N_{de} = 10^{18}, 10^{19}$ and 10^{20} cm^{-3} when (a) bandgap narrowing is neglected, (b) bandgap narrowing is included?

Problem 7.31 A silicon *npn* bipolar transistor is to be designed so that the emitter injection efficiency at 300 K is $\gamma_e = 0.995$. The base width is 0.5 μm and $L_e = 10W_b, D_e = D_b$, and $N_{de} = 10^{19}$ cm^{-3}. Calculate the base doping required with and without bandgap narrowing effects.

Problem 7.32 Consider a GaAs/AlGaAs HBT in which an injector efficiency of 0.999 is required at 300 K. The emitter and base dopings are both 10^{18} cm^{-3}. The base width is 0.1 μm. The carrier diffusion coefficients are $D_b = 60$ cm^{2}s, and $D_e = 20$ cm^{2}s. The carrier lifetimes are $\tau_B = \tau_E = 10^{-8}$s. Calculate the Al fraction needed in the emitter of the HBT.

Problem 7.33 Due to the high base doping possible, the base of an HBT can be very narrow. Consider a GaAs/AlGaAs HBT where the GaAs base is 500 Å. The minority charge diffusion coefficient is 100 cm^2/V·s in the base. Calculate the base transit time limited cutoff frequency of this device.

7.15   DESIGN PROBLEMS

Problem 1   Consider a *npn* Si-BJT at 300 K with the following parameters:

$$
\begin{aligned}
N_{de} &= 10^{18} \text{ cm}^{-3} \\
N_{ab} &= 10^{17} \text{ cm}^{-3} \\
N_{dc} &= 10^{16} \text{ cm}^{-3} \\
D_b &= 30.0 \text{ cm}^2/\text{s} \\
L_b &= 10.0 \ \mu\text{m} \\
W_b &= 1.0 \ \mu\text{m} \\
D_e &= 10 \text{ cm}^2/\text{s} \\
L_e &= 5.0 \ \mu\text{m} \\
\text{electron mobility in the emitter} &= 500 \text{ cm}^2 \text{ V}^{-1} \text{ s}^{-1}
\end{aligned}
$$

$$\text{area} = 5.0 \times 10^{-7} \text{ cm}^2$$

Calculate the emitter efficiency and gain β when the EBJ is forward biased at 1.0 V and the BCJ is reverse biased at (a): 5.0 V and (b) 10.0 V.

For high-speed operation, it is found that the BJT discussed above has too large an emitter resistance. The device designer wants to limit the emitter resistance (keeping the area unchanged) to 2.0 Ω. Calculate the emitter efficiency and β for the new device using the case (a) given above.

Problem 2 Consider a *npn* Si-BJT at 300 K with the following parameters:

$$
\begin{aligned}
N_{de} &= 10^{18} \text{ cm}^{-3} \\
N_{ab} &= 10^{17} \text{ cm}^{-3} \\
N_{dc} &= 5 \times 10^{16} \text{ cm}^{-3} \\
D_b &= 30.0 \text{ cm}^2/\text{s} \\
L_b &= 15.0 \ \mu\text{m} \\
D_e &= 10.0 \text{ cm}^2/\text{s} \\
L_e &= 5.0 \ \mu\text{m}
\end{aligned}
$$

Design the maximum base width, W_b, that will allow a current gain β of 100 when the EBJ is forward biased at 1.0 V and the BCJ is reverse biased at 5.0 V. You may make the following approximations:

- The reverse bias collector current is zero.
- W_b is much smaller than L_b.

Problem 3 Consider a *npn* Si-BJT at 300 K with the following parameters:

$$
\begin{aligned}
N_{de} &= 10^{18} \text{ cm}^{-3} \\
N_{ab} &= 10^{17} \text{ cm}^{-3} \\
N_{dc} &= 10^{16} \text{ cm}^{-3} \\
D_b &= 30.0 \text{ cm}^2/\text{s} \\
L_b &= 10.0 \ \mu\text{m} \\
W_b &= 1.0 \ \mu\text{m} \\
D_e &= 10 \text{ cm}^2/\text{s} \\
L_e &= 10.0 \ \mu\text{m} \\
\text{emitter thickness} &= 1.0 \ \mu\text{m} \\
\text{device area} &= 4.0 \times 10^{-6} \text{ cm}
\end{aligned}
$$

(a) Calculate the emitter efficiency and gain β when the EBJ is forward biased at 1.0 V and the BCJ is reverse biased at 5.0 V.

(b) Calculate the output conductance of the device defined by

$$g_o = \frac{\Delta I_C}{\Delta V_{CB}}$$

Problem 4   Consider an *npn* Si-BJT at 300 K with the following parameters:

$$
\begin{aligned}
N_{de} &= 10^{18} \text{ cm}^{-3} \\
N_{ab} &= 10^{17} \text{ cm}^{-3} \\
N_{dc} &= 5 \times 10^{16} \text{ cm}^{-3} \\
D_b &= 20.0 \text{ cm}^2/\text{s} \\
L_b &= 15.0 \ \mu\text{m} \\
D_e &= 10.0 \text{ cm}^2/\text{s} \\
L_e &= 5.0 \ \mu\text{m} \\
\text{emitter dimensions} &= 100 \ \mu\text{m} \times 100 \ \mu\text{m}
\end{aligned}
$$

(a) Calculate the base width, W_b, that will allow a current gain β of 200 when the EBJ is forward biased at 0.8 V and the BCJ is reverse biased at 5.0 V. Design the base width so that the gain goal is achieved and the base resistance is minimum.

(b) Estimate the base resistance. Note that the base hole current flows *sideways* into the device (Fig. 7.5). The hole mobility in the base is 300 cm^2/V·s.

You may make the following approximations :

- The reverse bias collector current is zero.
- W_b is much smaller than L_b.

7.16   READING LIST

- **General**

    - D. A. Neaman, *Semiconductor Physics and Devices, Basic Principles* (Irwin, Boston, 1997).

    - G. W. Neudeck, *The Bipolar Junction Transistor* (Vol. 3 of the Modular Series on Solid State Devices, Addison-Wesley, Reading, MA, 1989).

    - D. J. Roulston, *Bipolar Semiconductor Devices* (McGraw-Hill, New York, 1990).

    - B. G. Streetman and S. Bannerjee, *Solid State Electronic Devices* (Prentice-Hall, Englewood Cliffs, NJ, 2000).

    - D. A. Hodges and H. G. Jackson, *Analysis and Design of Digital Integrated Circuits* (McGraw-Hill, New York, 1988).

Chapter

8

FIELD EFFECT TRANSISTORS: JFET AND MESFET

Chapter at a Glance

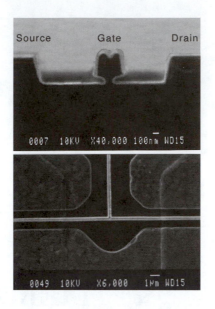

Figure 8.1: MESFETs and JFETs are important devices for high-speed, low-noise amplifiers, D/A and A/D converters, and much "front-end" processing where high speed is critical. These devices exploit materials, like GaAs, InP, and InGaAs, that have transport properties that are superior to Si. (Top) A cutoff cross-section of a 0.1 μm MESFET. (Bottom) Top view of the MESFET.

8.1 INTRODUCTION

The field-effect transistor (FET) is, perhaps, the most important electronic device in modern solid state technology. The tremendous versatility of this device combined with high yield and reliability have allowed it to become a workhorse for virtually every application. As noted in the previous chapter, the first transistors were based on the bipolar concept and because of the poor quality of Si-SiO$_2$ interfaces, FETs could not compete with BJTs. However, as this interface quality improved, many of the BJT functions were replaced by the metal-oxide semiconductor FET (MOSFET). In addition, new kinds of FETs based on materials like GaAs and InP have now been fabricated. These metal semiconductor FETs are called MESFETs. Figure 8.1 shows a GaAs-based MESFET.

The basic concept behind the FET is quite simple and is illustrated in Fig. 8.2. The device consists of an active channel through which electrons flow from the source to the drain. The source and drain contacts are ohmic contacts. The conductivity of the channel is modulated by a potential applied to the gate. This results in the modulation of the current flowing in the channel. *An important consideration in this process is to isolate the gate from the channel current flow. If the gate is not well isolated from the*

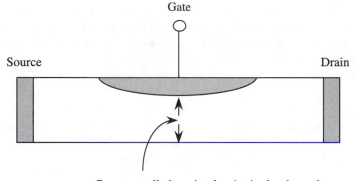

Gate controlled carrier density in the channel

Figure 8.2: The physical principle behind the FET involves the use of a gate to alter the charge in a channel. The gate potential changes thus result in corresponding changes in the current flow.

channel, it draws a lot of current, leading to a device that has a poor gain (i.e., the ratio of output power (or current) to input power (or current). The gate isolation is done in a variety of ways, leading to a number of different devices. In the MOSFET, the gate is isolated from the channel by an oxide. This is the basis of the silicon devices. In the metal-semiconductor FET or MESFET, the gate forms a Schottky barrier with the semiconductor and the gate current is small in the useful range of gate voltages. In the junction FET or JFET, a *p-n* junction is used in reverse bias to isolate the gate. In general, for the purposes of our discussion, the FETs can be divided into two main groups: (i) devices where the gate isolation is achieved by the use of an insulator between the gate and the active channel where the electron (or hole) flow occurs. If the insulator bandgap is large, electrons (or holes) can be "induced" into the channel *without* any doping, by using strong bandbending; (ii) devices in which gate isolation is achieved by using a Schottky barrier or a *p-n* junction. Here dopants are used to provide free carriers and the gate serves to alter the channel conductivity by changing the depletion width. The gate isolation technology is a key ingredient of the FET in all these configurations.

The Si-based FETs use the MOSFET concept while most III-V compound semiconductors are based on the MESFET or JFET concepts. In the MOSFET free carriers are produced in a channel by the process of "inversion" where a gate bias pulls the bandedges to a point lower than the Fermi level and thus creates free carriers. *However, to use this approach, there should be excellent insulation between the gate and the channel. In other words, there should be a high bandgap material separating the gate and the channel, otherwise there will be a large current between the gate and the channel, thus defeating the role of the gate.* Only for Si have scientists been able to find such a desirable insulator (although devices known as HIGFETs—heterostructure insulated gate field-effect transistors—based on GaAs/AlGaAs technology use principles similar to MOSFETs). This insulator is simply the oxide SiO_2, which is produced with a very high-quality interface with Si. The high quality of the Si-SiO$_2$ is primarily responsible for the pre-eminence of silicon in the electronics arena and the pre-eminence of MOS-FET in the transistor arena. The inability to develop high-quality insulators on other semiconductors has been rather frustrating for electrical engineers, and with each passing year it seems more and more that nature has, indeed, some special closeness to silicon!

For most semiconductors it is not possible to find a high-quality insulator that can be placed between the gate and the semiconductor channel to prevent current leakage into the gate. To isolate the gate from the channel we exploit the fact that in a reverse-biased *p-n* junction or a Schottky barrier, there is very little current flow. Thus if the gate is formed from such junctions *and the device operation conditions are such that the gate-semiconductor junction is always reverse biased* then there will be a very small gate current. The gate can then modulate the depletion region under it and thus control the conductivity of the semiconductor channel. The JFET and the MESFET are based on these principles.

8.2   JFET AND MESFET

The simplest field-effect transistor is the junction field-effect transistor or JFET. This particular form of the FET is not used as widely as the MESFET, but it does find some niche applications. The main difference between the JFET and the MESFET is that the JFET is based on a p^+-n (or n^+-p) junction while the MESFET is based on the Schottky barrier junction. An important difference arises from the choice of the two junctions. We have seen from Chapters 5 and 6 that the reverse current in a Schottky junction is much larger than that in a *p-n* junction. *As a result the JFET is especially used in materials for which it is difficult to produce a large Schottky barrier.*

In Fig. 8.3a we show a simple JFET or MESFET structure. The device is based on a low-conductivity substrate on which an *n*-type region is grown to form an "active" conducting channel of thickness h. The gate is formed by a p^+ region (n^+ region for a *p*-type FET) or a Schottky barrier. The source and drain are ohmic contacts. In Fig. 8.3b we show a case where the active channel is partially depleted (say at zero gate bias). A negative bias on the gate reverse biases the gate-ohmic conductor junction (for an *n*-type device) and alters the width of the depletion region. This allows the gate to modulate the conductance of the device. In Fig. 8.3c we show a case where the channel is completely depleted.

To conceptually understand the operation of the JFET or the MESFET, consider the cases shown in Fig. 8.4. In Fig. 8.4a we show the device with a small source-drain bias V_{DS} and no gate bias. As the gate bias is increased and the gate semiconductor junction is reverse biased, the current through the channel decreases until eventually

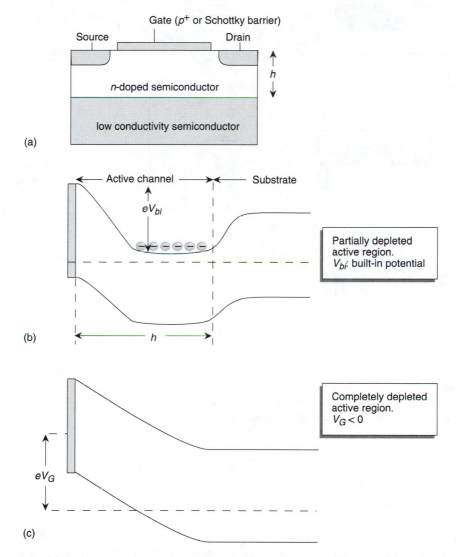

Figure 8.3: (a) A schematic of a JFET or MESFET showing the source, drain, and gate. The channel width of this device is h; (b) the band profile when the applied gate bias is zero as is the source-drain bias; (c) the band profile with a negative gate bias so that the channel is depleted.

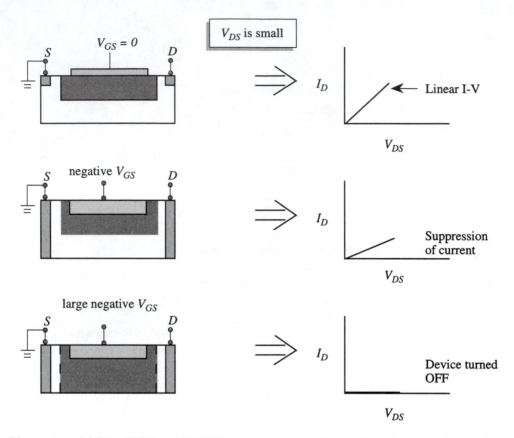

Figure 8.4: (a) The JFET or MESFET under zero gate bias so that the channel has a large opening. Such a device is called a depletion-mode device; (b) the device with a negative gate bias showing reduction in the channel opening and current; (c) the gate bias is large and negative and the channel is pinched off with current in the channel zero.

the channel is "pinched off" and there are no free carriers in it.

Consider now the case where the gate bias is fixed and the drain bias is increased, as shown in Fig. 8.5. As the drain bias becomes more and more positive, the gate semiconductor junction near the drain becomes more reverse biased. Eventually, the channel is pinched off near the drain side. At this point the current cannot increase even if the drain voltage is increased. This is called the saturation region. Once the device reaches saturation, the current in the channel remains more or less unchanged.

If h is the width as shown in Fig. 8.3 and V_{bi} is the built-in voltage, the gate bias required to pinch off the channel is simply given by the depletion approximation

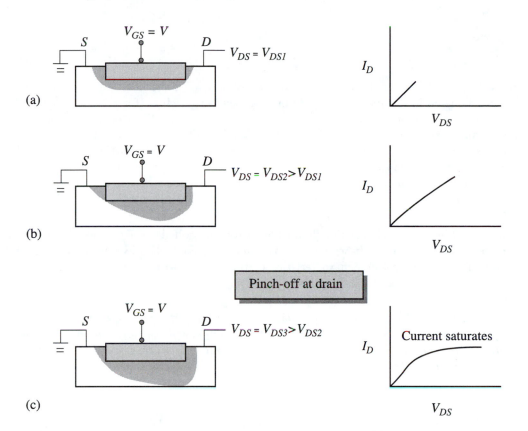

Figure 8.5: The effect of increased drain bias at a fixed gate bias. (a) the drain bias is small; (b) the drain bias is increased and the channel is constricted near the drain; (c) the drain bias is increased to the point that the channel is pinched off at the drain side. The drain current saturates as shown.

(see Section 5.2)

$$h = \left\{ \frac{2\epsilon(V_{bi} - V_{GS})}{eN_d} \right\}^{1/2} \tag{8.1}$$

Note that in an n-channel device, if the device is not pinched off by V_{bi}, then a negative gate bias will cause pinch-off. *In a p-channel device, a positive bias is needed for pinch-off.*

The gate bias needed at pinch-off will be denoted by V_T, the threshold bias. The pinch-off voltage V_p (called the intrinsic pinch-off voltage) is defined by

$$V_p = \frac{eN_d h^2}{2\epsilon} \tag{8.2}$$

and the gate bias needed for pinch-off for the n-channel device is

$$V_T = V_{bi} - V_p \tag{8.3}$$

If the voltage V_p is smaller than the built-in potential V_{bi}, the device channel is completely depleted in the absence of a gate bias. A positive gate bias (for n-channel devices) can allow the channel to have free charge and be conducting. Such devices are said to be enhancement-mode devices. On the other hand, if V_p is larger than V_{bi}, the device has free charge in the channel at $V_G = 0$ since the channel is only partially depleted. A negative gate bias will then turn the device off, i.e., deplete the channel. Such devices are said to operate in the depletion mode. Electronic circuits may use enhancement- or depletion-mode devices or even combinations of them, depending upon the application.

While in the JFET, the channel width is controlled by a reverse-biased p^+n junction, in the MESFET discussed next, the channel is controlled by a Schottky barrier metal. The current-voltage characteristics of the two devices are based upon similar physics and will be discussed in the next section.

The JFET approach of gate isolation is used for materials in which it is difficult to get a large Schottky barrier height. In Chapters 5 and 6 we saw that for the same material technology, the p-n diode has a lower reverse current than a Schottky barrier. Thus the Schottky barrier is quite "leaky" and is not used unless the barrier height is large.

EXAMPLE 8.1 Consider a p^+n JFET of n-channel Si. The p^+ doping is 10^{18} cm^{-3} and the n channel doping is 10^{17} cm^{-3}. The width of the channel is 0.25 μm. Calculate the built-in voltage and the bias required to pinch off the channel at 300 K.

The built-in voltage of a p^+n diode is given by

$$
\begin{aligned}
V_{bi} &= \frac{k_B T}{e} \ell n \left(\frac{N_a N_d}{n_i^2} \right) = (0.026 \text{ V}) \ell n \frac{10^{18} \times 10^{17}}{(1.5 \times 10^{10})^2} \\
&= 0.877 \text{ V}
\end{aligned}
$$

The total voltage drop required to pinch the channel is

$$
\begin{aligned}
V_{bi} - V_G &= \frac{e h^2 N_d}{2\epsilon} = \frac{\left(1.6 \times 10^{-19} C\right)\left(0.25 \times 10^{-4} \text{ cm}\right)^2 \left(10^{17} \text{ cm}^{-3}\right)}{2(11.9)\left(8.85 \times 10^{-14} \text{ F/cm}\right)} \\
&= 4.83 \text{ V}
\end{aligned}
$$

Thus at the pinch-off the gate bias is

$$V_G = 0.877 - 4.83 = -3.95 \text{ V}$$

EXAMPLE 8.2 Consider an n-MESFET made from GaAs doped at 10^{17} cm^{-3}. Calculate the gate current density under normal operation if:

(i) the gate is made from a Schottky metal with a barrier $\phi_b = 0.8$ V;

(ii) the gate is made from a heavily doped p^+ GaAs.

You may use the following parameters:

$$
\begin{aligned}
D_p &= 20 \text{ cm}^2/\text{s} \\
L_p &= 1.0 \text{ } \mu\text{m} \\
R^* &= 8 \text{ Acm}^{-2}K^{-2}
\end{aligned}
$$

The gate current under normal operation is just the reverse-bias current of the junction between the gate and the semiconductor. For the Schottky case we have (see Chapter 6)

$$
J_s = R^* T^2 \exp\left(-\frac{e\phi_b}{k_B T}\right)
$$

This gives

$$
J_s = 8 \times (300)^2 \times 4.34 \times 10^{-14} = 3.125 \times 10^{-8} \text{ A/cm}^2
$$

For the p^+-gate we have from p-n diode theory (see Chapter 5)

$$
J_0 = \frac{e D_p p_n}{L_p}
$$

This gives

$$
J_0 = \frac{1.6 \times 10^{-19} \times 20 \times 3.38 \times 10^{-5}}{10^{-4}} = 1.08 \times 10^{-18} \text{ A/cm}^2
$$

We see that the gate current is much smaller for the JFET case. However for the GaAs case considered here, the MESFET gate current is small enough for most applications.

8.3 CURRENT-VOLTAGE CHARACTERISTICS

The JFET and MESFET form an important class of transistors whose use in electronic technologies is increasing rapidly. It has already found important applications in high-speed digital processing as well as in microwave circuits. The devices can be reliably fabricated using GaAs and InP semiconductors.

The MESFET is a relatively simple device to fabricate and has a configuration shown in Fig. 8.6. Commonly used materials for MESFETs are GaAs and InP. The substrate material for the GaAs MESFET is high-resistivity semi-insulating GaAs, which is produced by carefully introducing impurities that have levels near the midgap of GaAs.

The active region of the device is made by an n-channel that is produced by either in-situ doping in epitaxial processes or ion implantation. The source and drain contacts are ohmic contacts and the gate is formed by a Schottky barrier. The operation

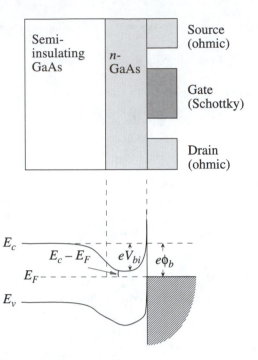

Figure 8.6: A schematic of a GaAs MESFET. Also shown are the energy band profile under the gate region and some important device parameters.

of this device is governed by fairly complicated transport theory, especially in materials like GaAs where the velocity-field relations are quite complex. We will present a rather simplified model that provides insight into the device operation. Before presenting the model calculations, let us briefly review the physics behind the device operation.

In Fig. 8.7 we show the MESFET cross-section along with the depletion width under the gate region. In the absence of any bias, there is a uniform depletion region under the gate region, as shown in Fig. 8.7a. If the gate bias is made more negative, the depletion width spreads further into the active region until eventually the channel is completely depleted. Thus, as the gate bias is increased (to negative values), the total charge available for conduction decreases until the channel is pinched off. This gate control is similar to that of the JFET shown in Fig. 8.4. If the drain bias is increased, the depletion region becomes larger near the drain side, as shown in Fig. 8.7b.

If the gate bias is fixed and the drain voltage is increased towards positive values, current starts to flow in the channel. Initially, the device behaves as an ohmic resistor. However, as the drain voltage is increased, the depletion width towards the drain end starts to increase, since the potential difference between the gate and the drain end of the channel increases. The channel then starts to pinch off at the drain end. As this happens, the current starts to saturate. Eventually, at very large drain bias

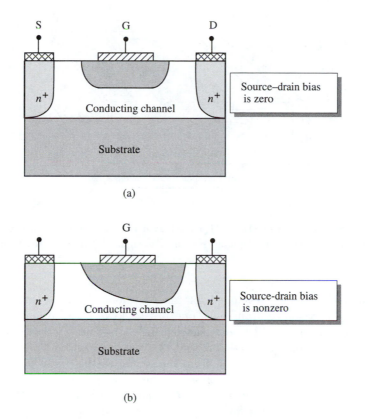

Figure 8.7: A schematic of a MESFET showing the depletion width under the gate. (a) In the absence of a source-drain bias, the depletion width is uniform and is controlled by the gate bias. (b) In the presence of a source-drain bias, the depletion width is greater in the drain side.

values, the device "breaks down" and the current shows a runaway behavior.

A full analysis of the current is rather complex even in the one-dimensional problem because one needs to solve the current continuity and Poisson equations self-consistently. We will simply use the approximations below:

- The mobility of the electrons is constant and independent of the electric field. We know that this is true only at low electric fields. At high fields the velocity of the electrons saturates. The analysis is, therefore, valid only if the field in the channel ($\sim$ drain bias divided by the channel length) is less than 2-3 kV/cm. Since this is usually not true in most modern devices, the analysis is only semi-quantitative and is useful more to convey the physics of the device operation.

- We assume the gradual channel approximation introduced by Shockley. In the absence of any source-drain bias, the depletion width is simply given by the one-

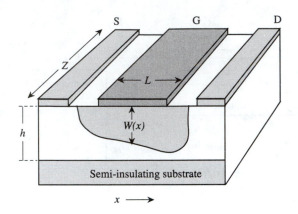

Figure 8.8: A schematic of the MESFET model used to obtain the I-V characteristics. The current flow occurs only in the undepleted region. The channel potential at any point x in the channel is $V(x)$.

dimensional model we developed for the *p-n* diode. However, strictly speaking, when there is a source-drain bias, one has to solve a two-dimensional problem to find the depletion width and, subsequently, the current flow. In the gradual channel approximation, we assume that field in the direction from the gate to the substrate is much stronger than from the source to the drain, i.e., the potential varies "slowly" along the channel as compared to the potential variation in the direction from the gate to the substrate. Thus the depletion width, at a point x along the channel, is given by the potential at that point using the simple one-dimensional results. This approximation is good if the gate length L is greater than the channel depth, which is typically a few hundred angstroms.

Under these approximations the current in the drain is given by (field $= -dV/dx$)

$$I_D = \text{channel area} \times (\text{charge density}) \times (\text{mobility}) \times (\text{field})$$

$$= Z[h - W(x)]eN_d\,\mu_n\,\frac{dV}{dx} \tag{8.4}$$

where $W(x)$ is the depletion width as shown in Fig. 8.8 and h is the channel thickness. Thus $h - W(x)$ is the channel opening. The depletion width at a point x is given in terms of the gate voltage V_{GS}, the built-in voltage V_{bi}, and the channel voltage $V(x)$ by the depletion equation

$$W(x) = \left[\frac{2\epsilon\,[V(x) + V_{bi} - V_{GS}]}{eN_d}\right]^{1/2} \tag{8.5}$$

We substitute for $W(x)$ in Eqn. 8.4 and integrate (I_D is constant throughout the channel) to get

$$I_D \int_0^L dx = e\mu_n N_d Z \int_0^{V_{DS}} \left[h - \left\{ \frac{2\epsilon\left[V(x) + V_{bi} - V_{GS}\right]}{eN_d} \right\}^{1/2} \right] dV \qquad (8.6)$$

which gives (after dividing by L)

$$I_D = \frac{e\mu_n N_d Z h}{L} \left\{ V_{DS} - \frac{2\left[(V_{DS} + V_{bi} - V_{GS})^{3/2} - (V_{bi} - V_{GS})^{3/2}\right]}{3(eN_d h^2/2\epsilon)^{1/2}} \right\} \qquad (8.7)$$

We denote by g_o the channel conductance when the channel is completely open,

$$\boxed{g_o = \frac{e\mu_n N_d Z h}{L}} \qquad (8.8)$$

Using the definition of the pinch-off voltage V_p, we have

$$V_p = \frac{eN_d h^2}{2\epsilon} \qquad (8.9)$$

The drain current versus drain voltage characteristics can be written as

$$I_D = g_o \left\{ V_{DS} - \frac{2\left[(V_{DS} + V_{bi} - V_{GS})^{3/2} - (V_{bi} - V_{GS})^{3/2}\right]}{3V_p^{1/2}} \right\} \qquad (8.10)$$

It must be remembered that this equation was derived under the condition that the gate and drain voltages are such that there is no pinch-off near the drain region, i.e.,

$$W(L) = \left\{ 2\epsilon \frac{V_{DS} + V_{bi} - V_{GS}}{eN_d} \right\}^{1/2} < h \qquad (8.11)$$

We assume that to the first approximation, when pinch-off occurs, the drain current saturates. What happens once saturation occurs will be discussed at the end of this section. The drain voltage at which saturation occurs is

$$V_{DS}(sat) = V_p - V_{bi} + V_{GS} \qquad (8.12)$$

and the saturated channel current becomes, from Eqn. 8.10 (the drain current does not change in our simple model once $V_{DS} \geq V_{DS}(sat)$),

$$\boxed{I_D(sat) = g_o \left[\frac{V_p}{3} - V_{bi} + V_{GS} + \frac{2(V_{bi} - V_{GS})^{3/2}}{3V_p^{1/2}} \right]} \qquad (8.13)$$

An important parameter of the device is the transconductance g_m, which defines the control of the gate on the drain current. From Eqn. 8.10, the transconductance becomes

$$g_m = \left. \frac{dI_D}{dV_{GS}} \right|_{V_{DS}=constant} = g_o \frac{(V_{DS} + V_{bi} - V_{GS})^{1/2} - (V_{bi} - V_{GS})^{1/2}}{V_p^{1/2}} \qquad (8.14)$$

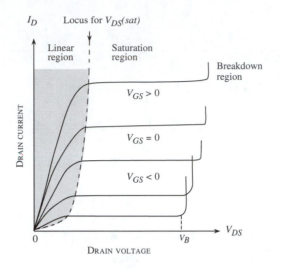

Figure 8.9: Typical I-V characteristics of an *n*-MESFET. In the Shockley model discussed in the text, it is assumed that once pinch-off of the channel occurs, the current saturates. In the figure, V_B is the breakdown voltage.

From Eqns. 8.8 and 8.14 we can see that the transconductance is improved by using a higher-mobility material as well as a shorter channel length. An improved transconductance means the gate has a greater control over the channel. This results in higher gain and high-frequency capabilities, as will be discussed later.

When the source-drain voltage is small, the expression for the current can be simplified by using

$$V_{DS} \ll V_{bi} - V_{GS} \tag{8.15}$$

Using the Taylor series, we then get from Eqn. 8.10,

$$I_D = g_o \left[1 - \left(\frac{V_{bi} - V_{GS}}{V_p} \right)^{1/2} \right] V_{DS} \tag{8.16}$$

The device is ohmic in this regime, as shown in Fig. 8.9, with a transconductance

$$g_m = \frac{g_o V_{DS}}{2 V_p^{1/2} (V_{bi} - V_{GS})^{1/2}} \tag{8.17}$$

In the saturation regime the transconductance is, from Eqn. 8.13,

$$g_m(sat) = g_o \left[1 - \left(\frac{V_{bi} - V_{GS}}{V_p} \right)^{1/2} \right] \tag{8.18}$$

In the model discussed here (known as the Shockley model) the current cannot be calculated beyond pinch-off. At pinch-off, the channel width becomes zero so that the electron velocity must, in principle, go to infinity to maintain constant current. This, of course, does not happen. We will now discuss, using physical arguments, what happens in the saturation region.

8.3.1 The Saturation Regime

In the simple model used so far, when the drain voltage exceeds $V_D(sat)$ for a given gate bias, the channel pinches off at the drain side and the model is unable to predict what happens to the current in this saturation regime. According to the simple depletion region picture, once the channel pinches off, there are no mobile carriers and the current should go to zero. In reality, the simple depletion picture breaks down and the current tends to saturate. We can explain this feature by the simple physical argument given below.

Consider the two generic materials Si and GaAs. In materials like silicon the velocity-field relations are such that the velocity increases monotonically with the applied field and eventually saturates. In GaAs, the velocity peaks at a field $F_p(\sim 3 \text{ kV/cm})$ and then decreases and gradually saturates. As shown schematically in Figs. 8.5b and 8.5c, as the drain bias is increased, the device starts to pinch off near the drain end of the gate. Just before pinch-off starts, the channel opening starts to become very small and since the current given by $en(x)[h - W(x)]v(x)$ must remain constant, either the carrier concentration, $n(x)$, must increase or the velocity, $v(x)$, must increase near the drain end of the gate. *However, since $v(x)$ can increase only to the saturation velocity, $n(x)$ must increase to maintain a constant current flow. Thus an accumulation region is created just under the drain end of the gate.* Also, as can be seen from Figs. 8.5b and 8.5c, the channel again "opens up" as one moves from the drain end of the gate towards the drain. In this region, $n(x)$ becomes less than the background doping level, creating a depletion region.

The presence of an accumulation layer under the drain end of the gate and a depletion region between the gate and the drain results in a nearly constant current flow even after the channel is pinched off.

EXAMPLE 8.3 Consider a GaAs MESFET with a gold Schottky barrier of barrier height

0.8 V. The n-channel doping is 10^{17} cm^{-3} and the channel thickness is 0.25 μm. Calculate the 300 K threshold voltage for the MESFET.

The total voltage at the surface of the MESFET to cause depletion of the channel is given by

$$V_p = \frac{eN_d h^2}{2\epsilon} = \frac{\left(1.6 \times 10^{-19} \text{ C}\right)\left(0.25 \times 10^{-4} \text{ cm}\right)^2 \left(10^{17} \text{ cm}^{-3}\right)}{2(13.2)\left(8.85 \times 10^{-14} \text{F/cm}\right)} = 4.31 \text{ V}$$

To find the built-in potential at the surface we must find the difference between the conduction band and Fermi level in the channel. This is given by the Boltzmann approximation

$$\begin{aligned}
E_c - E_F &= \frac{k_B T}{e}\ell n\left(\frac{N_d}{N_c}\right) = -(0.026)\ell n\left(\frac{10^{17}}{4.45 \times 10^{17}}\right) \\
&= 0.04 \text{ V}
\end{aligned}$$

Here N_c used above is the effective density of states for GaAs at 300 K. The built-in potential is now

$$V_{bi} = 0.8 - 0.04 = 0.76 \text{ V}$$

The gate bias required for pinch-off is now

$$V_T = V_{GS}(\text{pinch-off}) = 0.76 - 4.31 = -3.55 \text{ V}$$

A negative bias means the device is ON in the absence of any bias. Such a device is called a depletion-mode device.

EXAMPLE 8.4 Consider the device in Example 8.3. Calculate the maximum channel thickness at which the device is OFF when no gate bias is applied, i.e., the device is an enhancement MESFET.

We need to find the thickness h at which the value of V_p is the same at V_{bi}. This gives

$$\begin{aligned}
h &= \sqrt{\frac{2\epsilon V_{bi}}{eN_d}} = \left[\frac{2(13.1)\left(8.85 \times 10^{-14}\right)(0.76)}{\left(1.6 \times 10^{-19}\right)\left(10^{17}\right)}\right]^{1/2} \\
&= 0.1 \ \mu\text{m}
\end{aligned}$$

Thus, by tailoring the channel depth, the device can be made into an enhancement-mode or depletion-mode MESFET.

EXAMPLE 8.5 Consider a GaAs MESFET with the following parameters:

$$\begin{aligned}
\text{Schottky barrier height} &= 0.8 \text{ V} \\
\text{Channel doping} &= 10^{17} \text{ cm}^{-3} \\
\text{Channel depth} &= 0.06 \ \mu\text{m}
\end{aligned}$$

Calculate the gate bias needed to open up the MESFET channel.

The built-in voltage of this device is (from Example 8.2)

$$V_{bi} = 0.76 \text{ V}$$

The total pinch-off voltage is

$$V_p = \frac{eN_d h^2}{2\epsilon} = \frac{\left(1.6 \times 10^{-19} \text{ C}\right)\left(0.06 \times 10^{-4} \text{ cm}\right)^2 \left(10^{17} \text{ cm}^{-3}\right)}{2(13.2)(8.85 \times 10^{-14} \text{ F/cm})}$$

$$= 0.25 \text{ V}$$

The gate bias required to open the channel is then

$$V_{GS} = V_{bi} - V_p = 0.76 - 0.25 = 0.51 \text{ V}$$

In this case a positive gate bias is needed to turn on the transistor and the device is called an enhancement-mode device.

EXAMPLE 8.6 Consider a GaAs MESFET with the following parameters:

Channel mobility,	μ_n	=	$6000 \text{ cm}^2/\text{V}\cdot\text{s}$
Schottky barrier height,	ϕ_b	=	0.8 V
Channel depth,	h	=	$0.25 \text{ } \mu\text{m}$
Channel doping,	N_d	=	$5 \times 10^{16} \text{ cm}^{-3}$
Channel length,	L	=	$2.0 \text{ } \mu\text{m}$
Gate width,	Z	=	$25 \text{ } \mu\text{m}$

Calculate the 300 K saturation current when a gate bias of 0.0 V and -1.0 V is applied to the MESFET. Also calculate the transconductance of the device at these biases.

We need to calculate the built-in voltage and the internal pinch-off voltage V_p. The built-in voltage is

$$V_{bi} = \phi_b - \frac{k_B T}{e} \ell n \left(\frac{N_c}{N_d}\right) = 0.8 - 0.026 \ell n \left(\frac{4.45 \times 10^{17}}{5 \times 10^{16}}\right)$$

$$= 0.74 \text{ V}$$

The internal pinch-off potential V_p is

$$V_p = \frac{eN_d h^2}{2\epsilon} = \frac{\left(1.6 \times 10^{-19} X\right)\left(5 \times 10^{16} \text{ cm}^{-3}\right)\left(0.25 \times 10^{-4} \text{ cm}\right)^2}{2(13.2)(8.85 \times 10^{-14} \text{ F/cm})}$$

$$= 2.16 \text{ V}$$

The value of g_o of the channel is

$$g_o = \frac{e\mu_n N_d Z h}{L}$$

$$= \frac{\left(1.6 \times 10^{-19} \text{ C}\right)\left(6000 \text{ cm}^2/\text{V}\cdot\text{s}\right)\left(5 \times 10^{16} \text{ cm}^{-3}\right)\left(25 \times 10^{-4} \text{ cm}\right)\left(0.25 \times 10^{-4} \text{ cm}\right)}{2 \times 10^{-4} \text{ cm}}$$

$$= 0.015 \text{ } \Omega^{-1}$$

The current is (from Eqn. 8.13)

$$V_{GS} = 0 : \ V_D(sat) \ = \ 0.015 \left[\frac{2.16}{3} - 0.74 + \frac{2(0.74)^{3/2}}{3(2.16)^{1/2}} \right]$$

$$= \ 4.03 \text{ mA}$$

$$V_{GS} = -1.0 \ V : \ I_D(sat) \ = \ 0.015 \left[\frac{2.16}{3} - 0.74 - 1.0 + \frac{2(1.74)^{3/2}}{3(2.16)^{1/2}} \right]$$

$$= \ 0.32 \text{ mA}$$

The transconductance in the saturated regime is given by Eqn. 8.18. We have

$$V_{GS} = 0 : \ g_m(sat) \ = \ 0.015 \left[1 - \left(\frac{0.74}{2.16} \right)^{1/2} \right]$$

$$= \ 6.2 \text{ mS}$$

If the transconductance is normalized in the commonly used units of millisiemens per millimeter (mS/mm) we get

$$\frac{g_m(sat)}{Z} = 248 \text{ mS/mm}$$

At the gate bias of -1.0 V, we have

$$g_m(sat) \ = \ 0.015 \left[1 - \left(\frac{1.74}{2.16} \right)^{1/2} \right]$$

$$= \ 1.54 \text{ mS (or 61.6 mS/mm)}$$

8.4   EFFECTS IN REAL DEVICES

The simple analysis discussed above is not very accurate in describing real MESFET devices, although it gives a reasonable description. We will now discuss some of the important issues in real devices and point out the shortcomings of our model.

8.4.1   Velocity-Field Relations in Real Devices

In our simple analysis we have assumed a constant-mobility description of the carrier velocity dependence on its applied electric field. In this model, the carrier velocity keeps on increasing with the applied field, while, in real semiconductors, the velocity saturates. The result is that the I-V characteristics calculated in our simple model give values too high for the current at a given applied bias. In Appendix C we show the velocity-field (v-F) relations for several important semiconductors. To account for realistic v-F relations, several approaches are taken. One approach is to assume a velocity of the form

$$v(F) = \frac{\mu_n F}{1 + \frac{\mu_n F}{v_s}} \tag{8.19}$$

where v_s is the saturation velocity. It can be shown that it results in a reduction of the drain current obtained earlier by a factor of approximately $(1 + \mu_n V_{DS}/v_s L)$.

Another approach is based upon the two-region model where the mobility is constant in one region up to a field of F_p, and for fields beyond this the velocity is constant $(= v_s)$. For GaAs the value of F_p is ~ 3.5 kV/cm.

A third model applicable to small devices ($\lesssim 1.0$ μm) assumes that the velocity of the electrons is saturated, regardless of the field. These models give current values lower than obtained from the constant mobility model.

The models discussed above lead to I-V relations that can be put in simple analytic forms. However, more accurate analysis of the devices requires two-dimensional models that are purely numerical. These models are quite computer-intensive, but are critical in device design. Sophisticated computer programs are developed to analyze devices. These programs provide important insight into the development of advanced device structures. This insight is becoming increasingly crucial as devices become more complex and costly to fabricate, so that a "hit-or-miss" approach to device evolution becomes prohibitively expensive.

8.4.2 Channel Length Modulation

In our simple model for the current in the JFET or the MESFET, the current is inversely proportional to the channel length L. The channel length, however, is not a constant as far as the effective behavior of the device is concerned. The effective channel length is the length of the neutral region, which decreases as V_{DS} increases beyond the value of $V_{DS}(sat)$. As a result, the drain current does not remain constant beyond $V_{DS}(sat)$, but increases so that the output conductance of the device is not zero but finite.

When the drain voltage reaches $V_{DS}(sat)$, the channel pinches off at the drain end. If V_{DS} is increased, the pinch-off point moves toward the source, and a potential $V_{DS}(sat)$ is supported in a length L' (the effective channel length) while a voltage $(V_{DS} - V_{DS}(sat))$ is supported across the space charge of length $\Delta L = L - L'$. This effect is shown in Fig. 8.10.

The length of the depletion region is

$$\Delta L = \left[\frac{2\epsilon(V_{DS} - V_{DS}(sat))}{e N_d} \right]^{1/2} \tag{8.20}$$

We can now see how the drain current changes once $V_{DS} > V_{DS}(sat)$. We assume that the depletion length ΔL extends equally into the channel region and the drain region,

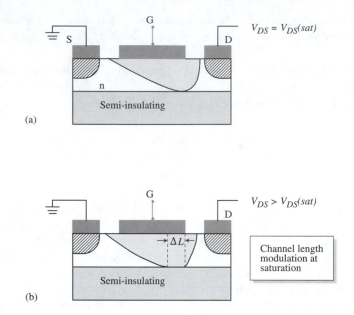

Figure 8.10: The modulation of channel length in MESFETs in the saturation region. (a) At $V_{DS} = V_{DS}(sat)$, the channel is pinched off near the drain. (b) In the case of $V_{DS} > V_{DS}(sat)$, the effective channel length shrinks as the pinch-off point moves into the channel.

so that the actual decrease in the channel length is only $\sim \frac{1}{2}\Delta L$. Using the relation (using a Taylor series expansion of the denominator)

$$\frac{1}{L - \frac{1}{2}\Delta L} = \frac{1}{L}\left(1 + \frac{\Delta L}{2L}\right),\tag{8.21}$$

we get

$$I'_D\left(V_{DS} > V_{DS}(sat)\right) = I_D(sat)\left(1 + \frac{\Delta L}{2L}\right)\tag{8.22}$$

Since ΔL is a function of V_{DS}, the output conductance

$$g_D = \left.\frac{\partial I_D}{\partial V_{DS}}\right|_{V_{GS}=constant}\tag{8.23}$$

is finite.

EXAMPLE 8.7 Consider the n-channel GaAs MESFET discussed in Example 8.6. If the gate bias $V_{GS} = 0$ V, how much does the output current change as a result of channel length modulation when the drain bias changes from $V_{DS}(sat) + 1.0$ V to $V_{DS}(sat) + 1.5$ V? What is the corresponding channel output resistance?

The change in the channel length is given by Eqn. 8.20 as

$$\Delta L \left(V_{DS}(sat) + 1.0 \text{ V} \right) = \left[\frac{(2 \times 13.2) \left(8.85 \times 10^{-14} \text{ F/cm} \right) (1.0 \text{ V})}{(1.6 \times 10^{-19} \text{ C}) (5 \times 10^{16} \text{ cm}^{-3})} \right]^{1/2}$$

$$= 0.17 \ \mu m$$

$$\Delta L \left(V_{DS}(sat) + 1.5 \text{ V} \right) = 0.21 \ \mu m$$

The currents at these biases are given by (using $I_D(sat)$ from Example 8.5)

$$I_D' \left(V_{DS}(sat) + 1.0 \text{ V} \right) = I_D(sat) \left(1 + \frac{\Delta L}{2L} \right)$$

$$= 4.03 \left(1 + \frac{0.17}{4.0} \right) = 4.20 \text{ mA}$$

$$I_D' \left(V_{DS}(sat) + 1.5 \text{ V} \right) = 4.03 \left(1 + \frac{0.21}{4.0} \right) = 4.24 \text{ mA}$$

The output resistance of the source-drain channel at saturation is

$$r_{DS} = \frac{0.5 \text{ V}}{4.0 \times 10^{-5} A} = 1.25 \times 10^4 \ \Omega$$

This is to be contrasted with the simple model where the channel modulation is ignored and the output resistance is infinity. Note that if the doping density in the channel is low, the value of r_{DS} will decrease.

8.5 HIGH-FREQUENCY, HIGH-SPEED ISSUES

Like the bipolar transistors, the FETs are important three-terminal devices for high-frequency microwave applications as well as high-speed digital applications. The FET has excellent input-output isolation since the gate is insulated from the channel where the current flows. It has a high gain since with proper design a small gate bias can produce a large change in the output bias. It is a device of choice for many applications.

8.5.1 Small-Signal Characteristics

To develop a simple picture for high-frequency response of the FET, we consider a typical MESFET where the gate controls the channel current. The equivalent circuit of a MESFET and the source of the various terms are shown in Fig. 8.11. A change of charge ΔQ on the gate produces the change ΔQ in the channel (assuming charge neutrality). If Δt is the time taken by the device to respond to this change, the change in the current in the channel is

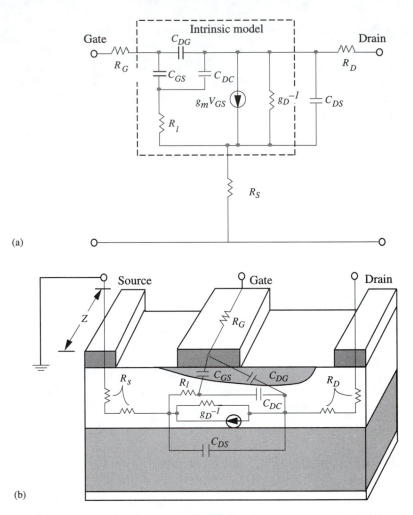

Figure 8.11: (a) Equivalent circuit of a MESFET. (b) Cross-section of a MESFET indicating the origins of the elements.

$$\delta I_D = \frac{\delta Q}{\Delta t} \tag{8.24}$$

where I_D is the current flowing between the source and the drain. The time Δt can be interpreted as the average transit time t_{tr} for the electrons to move through the device. The transistor transconductance can be related to the transit time. The transistor intrinsic transconductance is given by

$$g_m = \left. \frac{\partial I_D}{\partial V_G} \right|_{V_D}$$

$$= \left. \frac{\partial I_D}{\partial Q} \right|_{V_D} \left. \frac{\partial Q}{\partial V_G} \right|_{V_D}$$

$$= \frac{C_G}{\Delta t} = \frac{C_G}{t_{tr}} \tag{8.25}$$

where C_G is the gate-to-channel capacitance and describes the relationship between the gate voltage and the gate charge. The intrinsic transconductance is thus inversely proportional to the carrier transit time. The gate capacitance can be characterized by the gate-source capacitance C_{GS} and the gate-drain capacitance C_{DG} shown in Fig. 8.11b.

We can also define the output conductance g_D, which describes the effect of the drain bias on the drain current as

$$g_D = \left. \frac{\partial I_D}{\partial V_{DS}} \right|_{V_{GS}} \tag{8.26}$$

In addition to the intrinsic circuit elements discussed above, important extrinsic parasitic elements are the gate resistance R_G, the drain resistance R_D, and the source resistance R_S, which represents the series resistance of the ohmic contact and the channel region between the source and the gate. Also, we have the drain-to-substrate and drain-to-channel capacitances C_{DS} and C_{DC} respectively. These parameters lead to a simplified circuit model for the FET device shown in Fig. 8.11. This figure shows the equivalent circuit and the physical origin of the circuit elements discussed above.

An important characterization parameter is the forward current gain cutoff frequency f_T, which is measured with the output short-circuited. The parameter f_T defines the maximum frequency at which the current gain becomes unity.

If the capacitance charging time is the limiting factor, *at the cutoff frequency the gate current I_{in} is equal to the magnitude of the output channel current $g_m V_{GS}$.* The input current is the current due to the gate capacitor, and for a small-signal sinusoidal signal we have

$$I_{in} = j\omega C_G V_{GS} \tag{8.27}$$

Equating this to $g_m V_{GS}$ at $\omega = 2\pi f_T$, we get for the cutoff frequency

$$\boxed{f_T = \frac{g_m}{2\pi C_G} = \frac{1}{2\pi t_{tr}}} \tag{8.28}$$

where t_{tr} represents the transit time of the electrons through the channel. The frequency response is therefore improved by using materials with better transport properties and shorter channel lengths. In our constant-mobility model, the maximum value of the transconductance is, from Eqn. 8.18 (with $V_{bi} = V_{GS}$),

$$g_m(sat) = g_o = \frac{e\mu_n N_d Z h}{L} \tag{8.29}$$

The gate capacitance is

$$C_G = \frac{\epsilon Z L}{h} \tag{8.30}$$

The maximum cutoff frequency is then obtained by using the values of $g_m(sat)$ and C_G in Eqn. 8.28,

$$\boxed{f_T(max) = \frac{e\mu_n N_d h^2}{2\pi\epsilon L^2}} \tag{8.31}$$

In reality, this expression gives an overestimation of the cutoff frequency be-cause of the limitations of the constant-mobility model. If we depart from the constant-mobility model and instead assume that the carriers are moving at a saturated velocity, the transit time t_{tr} is simply

$$t_{tr} = \Delta t = \frac{L}{v_s} \tag{8.32}$$

and the cutoff frequency becomes (from Eqn. 8.28)

$$\boxed{f_T = \frac{v_s}{2\pi L}} \tag{8.33}$$

It may be noted that the source resistance R_S has an effect of reducing the effective transconductance of the device. In the presence of a source resistance the gate bias is V_{GS}' since a part of the input voltage drops across the resistance R_S. The drain current is

$$I_D = g_m V_{GS}' \tag{8.34}$$

Also we have

$$V_{GS} = V_{GS}' + \left(g_m V_{GS}'\right) R_S = (1 + g_m R_S) V_{GS}' \tag{8.35}$$

The drain current now becomes

$$I_D = \frac{g_m V_{GS}}{1 + g_m R_S} = g_m' V_{GS}' \tag{8.36}$$

where g_m' is the extrinsic transconductance and is smaller than g_m.

Another important characterization parameter for the FET is the maximum available gain (MAG) and the frequency at which the MAG becomes unity. This max-imum frequency f_{max} represents the maximum frequency at which the device can still amplify a signal. The limits on f_{max} are governed by both voltage and current amplifi-cation, while f_T is governed by current amplification alone. Thus depending upon the particular device, f_{max} can be larger or smaller than f_T.

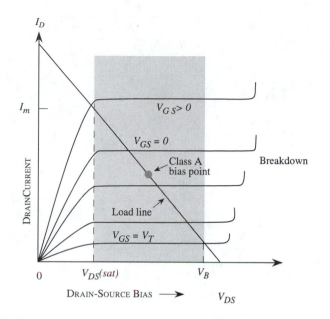

Figure 8.12: I-V characteristics for an n-MESFET showing the load line for large-signal switching. The maximum current that can be drawn is I_m, beyond which the gate starts to conduct significantly. The maximum voltage is V_B, the breakdown voltage.

8.5.2 Large-Signal Analog Applications

One of the most important uses of the FET devices is the large-signal amplification for power amplifiers. As a power amplifier, the device is operated in the saturated region of the I-V characteristics (class A operation). The gate bias swing causes a swing in the drain bias from the breakdown voltage, V_B to $V_{DS}(sat)$, as shown in Fig. 8.12. The maximum power output is given by (this is a time-averaged output, hence the factor 8 in the denominator)

$$P_m = \frac{I_m(V_B - V_{DS}(sat))}{8} \tag{8.37}$$

assuming that the drain current swings from 0 to I_m.

For high-power applications, one would like $V_{DS}(sat)$, the point at which the load line intersects the I-V curve's linear region, to be as small as possible. This requires a high-mobility material with low R_S and R_D. One also wants V_B to be as high as possible. V_B is governed by the breakdown voltage so that a high bandgap material is needed for high V_B. Also, a tradeoff exists between high V_B and the f_T of the device. For example, we have the approximate relation (using Eqn. 8.33 for f_T)

$$V_B f_T \sim V_B \frac{v_s}{2\pi L} \sim \frac{F_{crit} v_s}{2\pi} \tag{8.38}$$

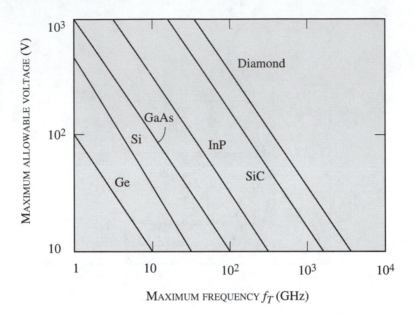

Figure 8.13: Plot of the maximum operating voltage for transistors made of selected semiconductors as a function of estimated f_T. The f_T estimates are based on the steady-state velocity-field curve for each material. (After M. W. Geis, N. N. Efremow, and D. D. Rathman, "Summary Abstract: Device Applications of Diamond," *J. Vac. Sci. Technol. A6*, 1953 (1988).)

where $F_{crit} \sim V_B/L$ is the field in the channel with length L. The maximum field that the channel will tolerate is determined by the breakdown field F_{crit}. Thus one has the limiting value of the $V_B f_T$ product as shown in Fig. 8.13 for several semiconductors. It is clear that if one wants higher power from a given semiconductor-based device, the high-frequency response has to be sacrificed. The maximum power is (assuming $V_D(sat) \sim 0$ and the signals are sinusoidal)

$$P_{max} = \frac{I_m V_B}{8} \tag{8.39}$$

For GaAs, the maximum power below f_T is ~ 1.4 W/mm of the channel. Work on large bandgap semiconductors such as GaN, SiC, and C is motivated by high-power applications, which can allow a larger V_B. In this context it is worth noting that in high-power microwave applications (satellite communication transmission, radio station transmission, radars, etc.), the tubes still dominate the market. Since neither Si nor GaAs can reach the high power levels needed for many applications, materials like SiC and C are being vigorously studied in a number of laboratories.

EXAMPLE 8.8 Consider an Si JFET with the following parameters:

$$
\begin{aligned}
\text{Channel mobility,} \quad & \mu_n & = \quad & 1000 \text{ cm}^2/\text{ V} \cdot \text{s} \\
\text{Channel depth,} \quad & h & = \quad & 0.5 \ \mu\text{m} \\
\text{Doping density,} \quad & N_d & = \quad & 10^{16} \text{ cm}^{-3} \\
\text{Channel length,} \quad & L & = \quad & 2.0 \ \mu\text{m}
\end{aligned}
$$

Calculate the maximum cutoff frequency of the device.

The maximum cutoff frequency in the constant-mobility model is given by

$$
\begin{aligned}
f_T(max) \quad = \quad & \frac{e \mu_n N_d h^2}{2\pi \epsilon L^2} \\
= \quad & \frac{\left(1.6 \times 10^{-19} C\right)\left(1000 \text{ cm}^2/\text{ V} \cdot \text{s}\right)\left(10^{16} \text{ cm}^{-3}\right)\left(0.5 \times 10^{-4} \text{ cm}\right)^2}{2(3.1416)(11.9)\left(8.85 \times 10^{-14} \text{ F/cm}\right)\left(2.0 \times 10^{-4} \text{ cm}\right)^2} = 15.37 \text{ GHz}
\end{aligned}
$$

If we use the saturation velocity model ($V_S = 10^7$ cm/s) we get from Eqn. 8.33, $f_T = 7.9$ GHz. This is a more reasonable value.

EXAMPLE 8.9 A 1.0 μm MESFET is fabricated from Si and GaAs and a source-drain bias of 0.5 V is used. Assuming that the electric field is constant in the channel, calculate the transit time for the electrons in the channel. What are the corresponding values of f_T?

The electric field in the channel is

$$
F = \frac{0.5 \text{ V}}{10^{-4} \text{ cm}} = 5 \times 10^3 \text{ V/cm}^{-1}
$$

The velocity of electrons at this field in Si and GaAs is $\sim 5 \times 10^6$ cm/s and 1.0×10^7 cm/s respectively, from the velocity-field relations for the two materials. The transit times are therefore

$$
\begin{aligned}
t_{tr}(\text{Si}) \tau_{Si} \quad = \quad & \frac{10^{-4} \text{ cm}}{5 \times 10^6 \text{ cm/s}} = 20 \text{ ps} \\
t_{tr}(\text{GaAs}) \quad = \quad & \frac{10^{-4}}{10^7} = 10 \text{ ps}
\end{aligned}
$$

The corresponding frequencies are

$$
\begin{aligned}
f_T(\text{Si}) \quad = \quad & \frac{1}{2\pi t_{tr}} = 7.96 \text{ GHz} \\
f_T(\text{GaAs}) \quad = \quad & 15.9 \text{ GHz}
\end{aligned}
$$

Note that if a higher source-drain bias is allowed, the field would increase and the velocities of the electron in GaAs and Si would be closer. Thus GaAs would produce faster devices if the applied bias is small, i.e., for low-power dissipation devices.

EXAMPLE 8.10 For the design of a particular high-power circuit, it is desired to have the maximum source-drain voltage to be 100 V. Calculate the maximum frequency of operation if one were to use MESFETs based on Si, GaAs, and SiC. Use the following material parameters:

Property	Si	GaAs	SiC
Saturation velocity	10^7 cm/s	10^7 cm/s	2×10^7 cm/s
Breakdown field	3×10^5 V/cm	4×10^5 V/cm	3×10^6 V/cm

The minimum channel length associated with the applied bias of 100 V at which the material will break down is

$$L_B = \frac{V_B}{F_{crit}}$$

and the corresponding cutoff frequency is

$$f_T = \frac{v_s}{2\pi L_B}$$

This gives the following results:

$$\text{Si} : L_B = \frac{100\text{ V}}{3 \times 10^5\text{ V/cm}} = 3.3\ \mu\text{m}$$

$$f_T = \frac{10^7\text{ cm/s}}{2\pi(3.3 \times 10^{-4}\text{ cm})} = 4.8\text{ GHz}$$

$$\text{GaAs} : L_B = \frac{100}{4 \times 10^5} = 2.5\ \mu\text{m}$$

$$f_T = \frac{10^7}{2\pi(2.5 \times 10^{-4})} = 6.4\text{ GHz}$$

$$\text{SiC} : L_B = \frac{100}{3 \times 10^6} = 0.33\ \mu\text{m}$$

$$f_T = \frac{2 \times 10^7}{2\pi(0.33 \times 10^{-4})} = 96\text{ GHz}$$

Clearly the advantage of the large bandgap SiC is evident from this example.

8.6 SUMMARY

In this chapter we have studied the JFET and MESFET devices. These devices are playing an increasingly important role in modern electronics. Although not as dominant as the MOSFETs that will be discussed in the next chapter, these devices are finding important uses in high-performance applications. The summary table (Table 8.1) provides an overview of the various concepts developed and discussed in this chapter.

8.7 PROBLEMS

Note: Please examine problems in the following Design Problems section as well.

Section 8.2
Problem 8.1 Discuss the reasons why one needs a large Schottky barrier value for the gate in a MESFET.

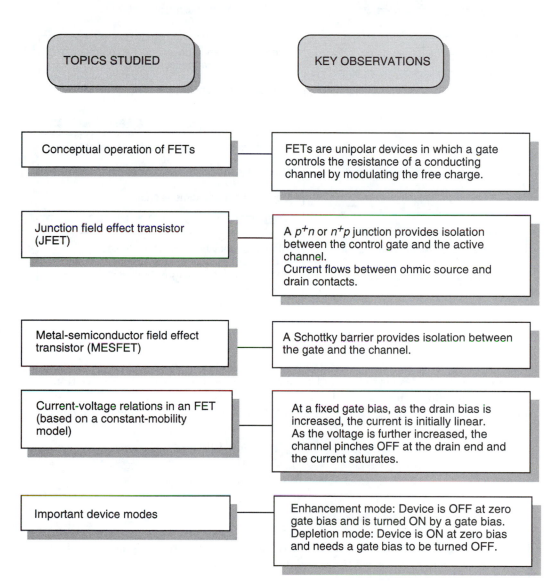

Table 8.1: Summary table.

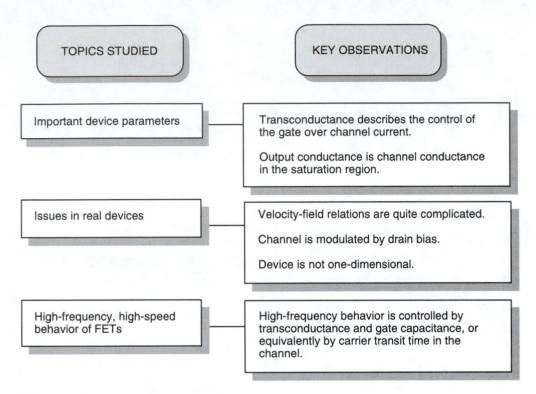

Important device parameters

Transconductance describes the control of the gate over channel current.

Output conductance is channel conductance in the saturation region.

Issues in real devices

Velocity-field relations are quite complicated.

Channel is modulated by drain bias.

Device is not one-dimensional.

High-frequency, high-speed behavior of FETs

High-frequency behavior is controlled by transconductance and gate capacitance, or equivalently by carrier transit time in the channel.

Table 8.1: Summary table (continued).

Problem 8.2 By drawing the band profile of a MESFET, discuss the restrictions on the gate bias values that can be allowed.

Problem 8.3 Consider an n-channel Si JFET at 300 K with the following parameters:

$$
\begin{aligned}
p^+\text{-doping,} \quad & N_a &=& \quad 5 \times 10^{18} \text{ cm}^{-3} \\
n\text{-doping,} \quad & N_d &=& \quad 10^{17} \text{ cm}^{-3} \\
\text{Channel thickness,} \quad & h &=& \quad 0.5 \ \mu\text{m}
\end{aligned}
$$

(a) Calculate the internal pinch-off for the device. (b) Calculate the gate bias required to make the width of the undepleted channel 0.25 μm.

Problem 8.4 Consider a GaAs JFET with the same characteristics as those of the Si device in Problem 8.3. Repeat the calculations for this GaAs device.

Problem 8.5 An n-type In$_{0.53}$Ga$_{0.47}$As epitaxial layer doped at 10^{16} cm^{-3} is to be used as a channel in a FET. A decision is to be made whether the JFET or MESFET technology is to be used for the device. In the JFET technology a p^+ region can be made with a doping of 5×10^{17} cm^{-3}. In the MESFET technology a Schottky barrier with a height of 0.4 V is available. Which technology will you use? Give reasons considering gate isolation issues. ($R^* = 5$ Acm^{-2}K^{-2}; $D_p = 20$ cm^2/s; $D_n = 50$ cm^2/s; $L_n = 5$ μm; $L_p = 5$ μm.)

Problem 8.6 Consider a p-channel Si JFET with the following parameters:

$$
\begin{aligned}
p\text{-doping}, \qquad N_a &= 5 \times 10^{16} \text{ cm}^{-3} \\
n^{+}\text{-doping}, \qquad N_d &= 5 \times 10^{18} \text{ cm}^{-3} \\
\text{Channel depth}, \quad h &= 0.25 \ \mu m
\end{aligned}
$$

(a) Calculate the internal pinch off for the device as well as the gate bias needed for pinch off.

(b) Calculate the width of the undepleted channel for gate biases of $V_{GS} = 1$ V and $V_{GS} = 2$ V for $V_{DS} = 0$.

Section 8.3

Problem 8.7 Consider an n-channel GaAs MESFET with the following parameters:

$$
\begin{aligned}
\text{Schottky barrier height}, \quad \phi_b &= 0.8 \text{ V} \\
\text{Channel doping}, \qquad N_d &= 5 \times 10^{16} \text{ cm}^{-3} \\
\text{Channel width}, \qquad h &= 0.8 \ \mu m
\end{aligned}
$$

Calculate the minimum width of the undepleted channel (near the drain side) with $V_{GS} = 0.5$ V when (a) $V_{DS} = 0.0$ V; (b) $V_{DS} = 1.0$ V; (c) $V_{DS} = 2.0$ V; (d) $V_{DS} = 10$ V.

Problem 8.8 (a) An n-type GaAs MESFET is to be designed so that the device is just turned off at a gate voltage of $V_{GS} = 0$ V. The Schottky barrier height ϕ_b is 0.8 V and the channel thickness is 0.2 μm. Calculate the channel doping required. To calculate the depletion region thickness (only) you may assume that $V_{bi} \cong \phi_b$ - 0.1 V.

(b) If a gate bias of 0.2 V is applied, calculate the *gate current*.

(c) What is the saturation drain current when the gate bias is 0.2 V? Compare the gate current with the drain current.

$$
\begin{aligned}
\text{Mobility}, \qquad \mu_n &= 5000 \text{ cm}^2 / \text{V} \cdot \text{s} \\
\text{Gate length}, \qquad L &= 2.0 \ \mu m \\
\text{Gate width}, \qquad Z &= 20.0 \ \mu m \\
\text{Channel width}, \quad h &= 0.2 \ \mu m
\end{aligned}
$$

Problem 8.9 In the text we used the constant-mobility model to obtain the relation between the drain current and the gate and drain voltages below pinch-off. Obtain a result for the depletion region $h - h(x)$ as a function of x (distance from source to drain) for a gate bias V_{GS} and a drain bias V_{DS}. Use the symbols used in the text for other device parameters.

Problem 8.10 Consider the design of two n-channel GaAs MESFETs with the following parameters:

$$
\begin{aligned}
\text{Schottky barrier height}, \quad \phi_b &= 0.8 \text{ V}; \qquad\qquad 0.8 \text{ V} \\
\text{Channel doping}, \qquad N_d &= 2 \times 10^{16} \text{ cm}^{-3}; \quad 2 \times 10^{17} \text{ cm}^{-3}
\end{aligned}
$$

The first sequence belongs to one device and the second sequence to the other. Calculate the depths of the channel needed for each device so that the devices are just turned off in the absence of any gate bias.

Problem 8.11 Consider an n-channel GaAs MESFET at 300 K with the following parameters:

$$\text{Schottky barrier height,} \quad \phi_b \ = \ 0.8 \text{ V}$$
$$\text{Channel thickness,} \quad h \ = \ 0.25 \ \mu m$$

Calculate the channel doping needed so that the device turns off at a gate bias of $V_{GS} = V_T = 0.5$ V.

Problem 8.12 Consider an n-channel Si MESFET at 300 K with the following known parameters:

$$\text{Barrier height,} \quad \phi_b \ = \ 0.7 \text{ V}$$
$$\text{Channel doping,} \quad N_d \ = \ 10^{16} \text{ cm}^{-3}$$

It is found that when a gate bias of $V_{GS} = -0.3$ V is applied ($V_{DS} = 0$), the channel is just fully depleted. Calculate the channel depth h for the device.

Problem 8.13 Consider a GaAs n-channel MESFET at 300 K with the following parameters:

$$\text{Schottky barrier height,} \quad \phi_b \ = \ 0.8 \text{ V}$$
$$\text{Electron mobility,} \quad \mu_n \ = \ 6000 \text{ cm}^2/ \text{ V} \cdot \text{s}$$
$$\text{Channel width,} \quad Z \ = \ 25 \ \mu m$$
$$\text{Channel length,} \quad L \ = \ 1.0 \ \mu m$$
$$\text{Channel depth,} \quad h \ = \ 0.25 \ \mu m$$
$$\text{Channel doping,} \quad N_d \ = \ 1.0 \times 10^{17} \text{ cm}^{-3}$$

(a) Calculate the gate bias $V_{GS} = V_T$ needed for the device to just turn off.
(b) Calculate $V_D(sat)$ for gate biases of $V_{GS} = -1.5$ V and $V_{GS} = -3.0$ V.
(c) Calculate the saturation drain current for the cases considered in part b.

Problem 8.14 Consider an n-channel GaAs MESFET at 300 K with the parameters of Problem 8.13. Calculate the transconductance of the device in the saturation region for the gate biases $V_{GS} = -1.5$ V and $V_{GS} = -2.0$ V. Express the results in terms of mS/mm.

Problem 8.15 Consider an n-channel Si MESFET at 300 K. The following parameters define the MESFET:

$$\text{Schottky barrier height,} \quad \phi_b \ = \ 0.8 \text{ V}$$
$$\text{Channel mobility,} \quad \mu_n \ = \ 1000 \text{ cm}^2/ \text{ V} \cdot \text{s}$$
$$\text{Channel doping,} \quad N_d \ = \ 5 \times 10^{16} \text{ cm}^{-3}$$
$$\text{Channel length,} \quad L \ = \ 1.5 \ \mu m$$
$$\text{Channel depth,} \quad h \ = \ 0.25 \ \mu m$$
$$\text{Gate width,} \quad Z \ = \ 25 \ \mu m$$

(a) Calculate the turn-on voltage V_T for the structure.
(b) Calculate $V_{DS}(sat)$ at a gate bias of $V_{GS} = 0$. Also calculate the device transconductance.
(c) If the device turn-on voltage is to be $V_T = -2.0$ V, calculate the additional doping

needed for the channel.

Problem 8.16 In a MESFET, as the gate length shrinks, the channel doping has to be increased. Discuss the reasons for this.

Section 8.4

Problem 8.17 Consider a GaAs n-channel MESFET operating under conditions such that one can assume that the field in the channel has a constant value of 5.0 kV/ cm^{-1}. The channel length is 2.0 μm. Calculate the transit time for an electron to traverse the channel if one assumes a constant mobility of 7500 cm^2/V·s. What would the time be if the correct velocity-field relations plotted in Appendix C were used?

Problem 8.18 Consider a 1.0 μm channel length n-channel Si MESFET operating under the condition that the average field in the channel is 15 kV/cm. Assume the electric field in the channel is constant at this value. Calculate the electron transit time assuming a constant mobility of 1000 cm^2/V·s and using the velocity-field relations for Si given in the text.

Problem 8.19 Consider two n-channel GaAs MESFETs operating at a source-drain bias of 2.0 V. Assume that the electric field in the channel is constant and has a value of V_{DS}/L where $L = 1.0$ μm for one device and 5 μm for the second. Calculate the transit time for electrons in the two devices using two models for transit: (a) constant-mobility model with $\mu = 6000$ cm^2/V·s; (b) correct velocity-field relations for the velocity. Use the curves given in Appendix B for the velocity field. Note that the discrepancy in the two models is larger for the shorter channel device.

Problem 8.20 Consider an n-channel GaAs MESFET with the following parameters:

Schottky barrier height,	ϕ_b	=	0.8 V
Channel doping,	N_d	=	5×10^{16} cm^{-3}
Channel depth,	h	=	0.5 μm
Channel mobility,	μ_n	=	5000 cm^2/ V · s
Channel length,	L	=	1.5 μm
Channel width,	Z	=	20.0 μm

Calculate the value of $V_{DS}(sat)$ at $V_{GS} = 0$. Also calculate the output resistance of the channel at $V_{DS} = V_{DS}(sat) + 2.0$ V.

Problem 8.21 Consider an n-channel GaAs with the same parameters as the device in Problem 8.20 except for the channel length. A maximum value of V_{DS} is 10.0 V for the device, and it is required that the effective channel length L' at $V_{GS} = 0$ and the maximum drain voltage should be no less than 90% of the actual channel length L. What is the smallest channel length L that satisfies this requirement?

Section 8.5

Problem 8.22 In this problem we will consider the effect of the source resistance on the device transconductance. Consider an n-channel GaAs MESFET with the following parameters:

$$\begin{aligned}
\text{Schottky barrier height,} \quad & \phi_b & = & \quad 0.8 \text{ V} \\
\text{Gate length,} \quad & L & = & \quad 3.0 \ \mu\text{m} \\
\text{Channel mobility,} \quad & \mu_n & = & \quad 6000 \text{ cm}^2/\text{ V} \cdot \text{s} \\
\text{Channel doping,} \quad & N_d & = & \quad 5 \times 10^{16} \text{ cm}^{-3} \\
\text{Channel depth,} \quad & h & = & \quad 0.5 \ \mu\text{m} \\
\text{Gate width,} \quad & Z & = & \quad 25 \ \mu\text{m}
\end{aligned}$$

Calculate the intrinsic transconductance of the device. If the source-to-gate separation is 0.5 μm, calculate the value of the extrinsic transconductance.

Problem 8.23 Calculate the maximum cutoff frequency for the ideal device of Problem 8.19 (with the source resistance assumed equal to zero). Calculate the degradation in the cutoff frequency due to the effect of the source series resistance.

Problem 8.24 Consider an n-type GaAs MESFET at 300 K with the following parameters:

$$\begin{aligned}
\text{Schottky barrier height,} \quad & \phi_b & = & \quad 0.8 \text{ V} \\
\text{Channel doping,} \quad & N_d & = & \quad 10^{17} \text{ cm}^{-3} \\
\text{Channel mobility,} \quad & \mu_n & = & \quad 6000 \text{ cm}^2/\text{ V} \cdot \text{s} \\
\text{Channel depth,} \quad & h & = & \quad 0.2 \ \mu\text{m} \\
\text{Channel width,} \quad & Z & = & \quad 2.0 \ \mu\text{m} \\
\text{Channel length,} \quad & L & = & \quad 1.0 \ \mu\text{m}
\end{aligned}$$

Calculate the maximum cutoff frequency using the constant-mobility model and the saturation velocity model.

Problem 8.25 An important effect in short-channel FETs made from high-mobility materials like GaAs and InGaAs is the "velocity overshoot effect." The average time for scattering τ_{sc} in such materials is $\sim$1.0 ps. If the electron transit time is less than 1 ps, the electron moves "ballistically," i.e., without scattering. Consider a FET in which the average electric field is 20 kV/cm. Electrons are injected at the source with thermal velocities and move in the average electric field towards the drain. Estimate the gate length at which the velocity overshoot effect will become important for Si, GaAs, and InAs. Assume that the average scattering time is 1 ps for all three materials. Assume electron effective masses of 0.26 m_0, 0.067 m_0, and 0.02 m_0, respectively.

8.8 DESIGN PROBLEMS

Problem 1 Consider an n-MESFET made from GaAs operating in an ON state. Sketch schematically (i.e., only semi-quantitatively) the electric field in the channel below the gate going from the source to the drain for the following cases:
(a) the device is in the linear regime, i.e., the drain bias is very small.
(b) the device is under a high drain bias (i.e., $V_D \sim V_D(sat)$).
Give reasons for your results.

Problem 2 A field-effect transistor is to be made from the high-speed material n-InGaAs. The doping is 10^{17} cm^{-3}. The bandgap of the material is 0.8 eV and the maximum Schottky barrier height possible is 0.4 eV. In the device the maximum gate leakage current density allowed is 10^{-2} Acm^{-2}. Discuss how you would design the FET using the MESFET and JFET approach.

$$
\begin{aligned}
R^* &= 4.7 \text{ Acm}^{-2}\text{K}^{-2} \\
D_p &= 25 \text{ cm}^2/\text{s} \\
L_p &= 1.5 \ \mu\text{m} \\
n_i &= 2 \times 10^{11} \text{ cm}^{-3}
\end{aligned}
$$

Discuss the limitations on the gate bias for the MESFET and the JFET.

Problem 3 An n-MESFET is made from GaAs doped at 10^{17} cm^{-3}. The gate width Z is 50.0 μm and the gate length is 2.0 μm and the channel thickness h is 0.25 μm.

To characterize the gate properties, the gate semiconductor current is measured and is found to have the value (at 300 K)

$$I_G = 3.12 \times 10^{-14}[\exp(eV/K_B T) - 1] \text{ A}$$

where V is the bias between the gate and the semiconductor. The mobility in the semiconductor is measured to be 4000 cm^2/V·s.
(a) Calculate the threshold voltage V_T for the device.
(b) Calculate the transconductance at saturation when the gate bias is $V_{GS} = -2.0$ V.

8.9   READING LIST

- **General**

    - D. A. Neaman, *Semiconductor Physics and Devices* (Irwin, Boston, MA, 1997).

    - R. F. Pierret, *Field Effect Devices* (Vol. 4 of the Modular Series on Solid State Devices, Addison-Wesley, Reading, MA, 1990).

    - M. Shur, *Physics of Semiconductor Devices* (Prentice-Hall, Englewood Cliffs, NJ, 1990).

    - S. M. Sze, *Physics of Semiconductor Devices* (Wiley, New York, 1981).

Chapter

9

FIELD EFFECT TRANSISTORS: MOSFET

Chapter at a Glance

9.1 INTRODUCTION

High yield, low cost, and dense packing are considerations that have pushed the MOS-FET to the status of the most widely used device in information technology hardware. As noted in Section 8.1, silicon is the only semiconductor on which a high-quality insulator (SiO_2) can be grown with ease. The quality of the Si/SiO_2 interface is good enough to allow electrons and holes moving near the interface to have reasonable mobilities.

A major advantage of the MOSFET technology is that both *n*-type and *p*-type devices can be made on the same substrate. The resulting technology, called *complementary MOSFET* or CMOS, has become increasingly critical, since it consumes very little power. With increased demands for mobile communication and computation, power consumption is of great value.

In this chapter we will examine the operation of the MOSFET. We will first discuss the MOS capacitor (without worrying about the source and drain) and examine how charge is induced in the MOS channel by "inversion." It is important to note that in a MOSFET, unlike the MESFET or JFET, charge can be induced in the MOS *without doping*. With doping in the silicon channel, one has additional control on the charge. Once we discuss the MOS capacitor we will examine the operation of the MOSFET. For completeness, we will start with some simple structural and fabrication issues.

9.2 MOSFET: STRUCTURE AND FABRICATION

MOSFETs can be made so that the current from the source to drain is carried by electrons (NMOS), by holes (PMOS), or in the case of complementary MOSFET (CMOS), by electrons and by holes in two devices. In Fig. 9.1 we show a schematic of an NMOS device. The structure starts with a *p*-type substrate. A voltage applied to the gate "inverts" the polarity of the carriers and produces electrons near the oxide-semiconductor interface. This will be discussed in the next section.

The body or substrate of a MOSFET can also be contacted and a bias applied to it. This proves a useful facility to alter device operation, as will be discussed later. In Fig. 9.1 the gate has a length L and width Z. (It is useful to note that in most circuit-related books, the width of gate is denoted by W. Here we use the symbol W for depletion width in the semiconductor.) The gate length is one of the most critical parameters controlling device performance.

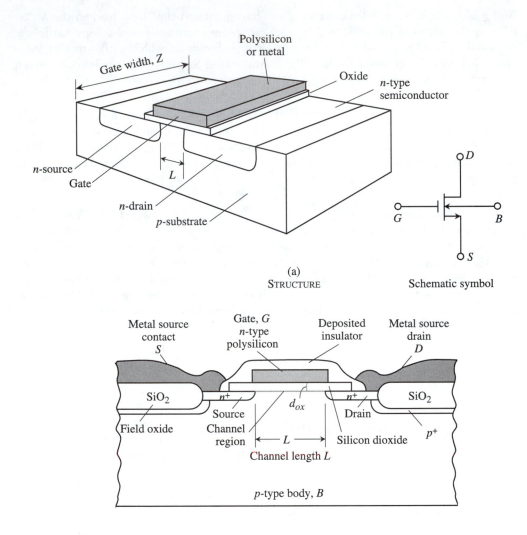

(a)
STRUCTURE   Schematic symbol

(b)
CROSS-SECTIONAL VIEW

Figure 9.1: (a) A schematic of an NMOS device. Also shown is a symbol for the device. The contact B denotes the body or substrate of the device. (b) A cross-section of the NMOS.

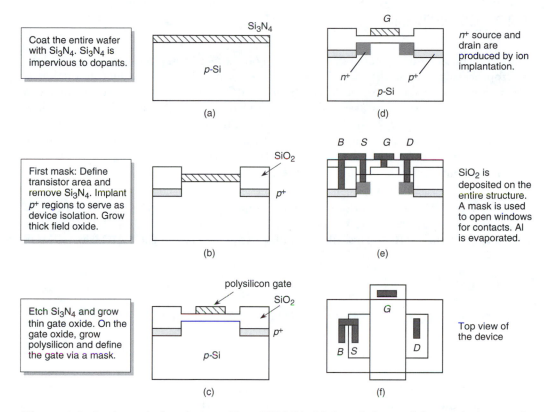

Coat the entire wafer with Si₃N₄. Si₃N₄ is impervious to dopants.

(a)

First mask: Define transistor area and remove Si₃N₄. Implant p^+ regions to serve as device isolation. Grow thick field oxide.

(b)

Etch Si₃N₄ and grow thin gate oxide. On the gate oxide, grow polysilicon and define the gate via a mask.

(c)

n^+ source and drain are produced by ion implantation.

(d)

SiO₂ is deposited on the entire structure. A mask is used to open windows for contacts. Al is evaporated.

(e)

Top view of the device

(f)

Figure 9.2: A schematic description of how NMOS is fabricated. State-of-the-art devices need more steps, since several additional implants are required to optimize the device performance.

9.2.1 NMOS Fabrication

In Chapter 4 we have discussed the important fabrication steps in semiconductor technology. In Fig. 9.2 we show the basic processes used in the fabrication of a NMOS. A brief discussion of the steps used in the fabrication is also given with the drawings. Note in Fig. 9.2 that:

• A polysilicon gate is often used instead of a metal gate. This is to reduce the threshold voltage value.

• A p^+ implant is done in regions outside the transistor. This serves to isolate devices. As shown in Fig. 9.2e, the body and source contacts are tied together. This ensures that there is no current in the source-substrate diode. Since the drain is at a positive bias with respect to the source, the drain-substrate diode is also cut off. This ease of isolation of the devices is a key reason why very high-density MOS circuitry can be produced.

9.2.2   CMOS Technology

CMOS technology has become the most widely used technology, finding use in wireless, microprocessors, memories, and a host of other applications. The chief attraction is low power dissipation. Since both NMOS and PMOS are to be fabricated on the same substrate, additional steps are needed compared to the NMOS case discussed earlier. The cross-section of a typical CMOS is shown in Fig. 9.3a. As can be seen, the PMOS transistor is fabricated within an n-type well that is implanted or diffused into the p-substrate. The n-well acts as the body or substrate for the PMOS. In addition to creating the n-well, one needs to do a p^+ implant for the source and drain of the transistor.

The PMOS transistor occupies more chip area than the NMOS transistor. This is to compensate for the lower mobility of holes. The Z/L value for PMOS is about two times that of the NMOS. This allows similar current values in the NMOS and PMOS devices.

9.3   METAL-OXIDE-SEMICONDUCTOR CAPACITOR

A tremendous advantage that Si technology has over any other semiconductor is the availability of a high-quality oxide SiO_2 that can be formed on Si. We discussed some of the properties of the Si-SiO_2 interface in Section 6.4. It is the high degree of perfection of this interface that has allowed Si to overtake Ge, which was the initial material of choice for transistors. Also, the Si-SiO_2 interface perfection has been the reason why bipolar devices have been replaced by field-effect devices in many applications.

Before addressing the MOSFET device we first examine the MOS capacitor, which has the configuration shown in Fig. 9.4. An oxide layer is grown on top of a p-type semiconductor and a metal contact is placed on the oxide. In general, the oxide could be any large bandgap insulator. The main purpose of the oxide layer is to provide an isolation between the metal and the semiconductor so that there is essentially no current flow between the top metal contact and the metal contact at the bottom of the semiconductor.

Let us first consider the ideal structure where there are no states in the bandgap regions at the interface between the oxide and the semiconductor. In reality, because of the mismatch between the oxide lattice and the Si lattice, there are some defect states at the interface, which will be discussed later. In Fig. 9.4b we show the energy levels associated with the metal, the oxide, and the semiconductor. When the capacitor is formed, the Fermi levels align themselves so that there is no gradient. The conduction band of the oxide is at a higher position than the conduction band of the semiconductor. In general, if the work functions in the metal and the semiconductors are different, the band alignments are as shown in Fig. 9.4c. There is a fairly high potential barrier

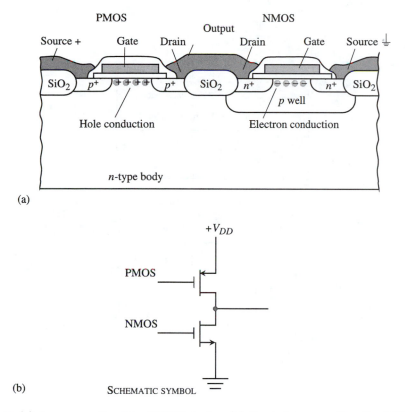

Figure 9.3: (a) A cross-section of a CMOS device. (b) Symbol representing the CMOS.

between the metal and the semiconductor. At zero applied bias, the bandbending at the semiconductor is determined by the difference in the work functions of the metal and the semiconductor. If one applies an external bias, the bandbending can be removed (thus reaching the flat band condition) by applying a compensating bias V_{fb} given by

$$
\begin{aligned}
eV_{fb} &= e\phi_m - e\chi_s - (E_c - E_F) \\
&= e\phi_m - e\phi_s = e\phi_{ms}
\end{aligned}
\tag{9.1}
$$

where ϕ_m and ϕ_s are the metal and semiconductor work functions, ϕ_{ms} is their difference, χ_s is the semiconductor electron affinity, E_c the semiconductor conduction bandedge, and E_F the Fermi level. In Fig. 9.5 we show the values of ϕ_{ms} for several different metals as a function of doping density. *Note that if there are defect charges in the oxide, the flat band voltage is modified due to an additional potential drop in the oxide.* Starting from the flat band position, there are three important regimes of biasing in the MOS capacitor, as shown in Fig. 9.6.

(i) *Hole Accumulation*: If a negative bias is applied between the metal and

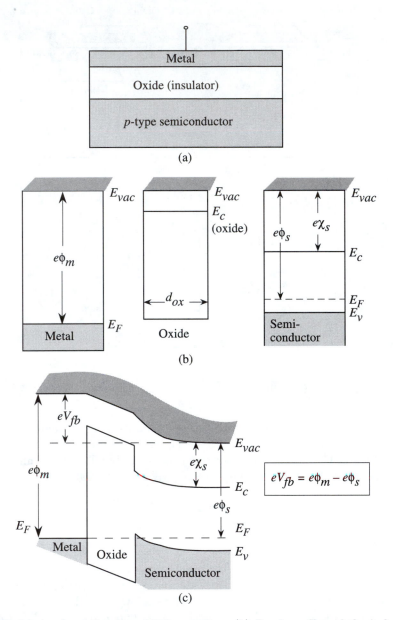

Figure 9.4: (a) A schematic of an MOS capacitor. (b) Band profiles of the isolated metal, oxide, and semiconductor. Shown are the metal work function, semiconductor work function, and electron affinity. (c) Band profile of the MOS junction.

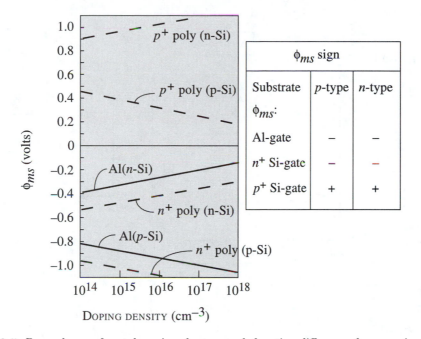

Figure 9.5: Dependence of metal-semiconductor work function difference for some important gate metals used in MOS devices. Also shown are the signs of ϕ_{ms} for three different gate types for NMOS and PMOS.

the semiconductor, a negative charge is deposited on the metal and an equal positive charge is accumulated at the semiconductor-oxide interface. This occurs due to the bandbending shown in Fig. 9.6a, where the valence bands are bent to come closer to the Fermi level, causing an accumulation of holes at the interface. The difference between the Fermi level in the metal and the semiconductor is the applied bias.

(ii) *Depletion*: If a positive bias is applied to the metal with respect to the semiconductor, the bands bend as shown in Fig. 9.6b. The Fermi level in the metal is lowered by an amount eV with respect to the semiconductor, causing the valence band to move away from the semiconductor Fermi level near the interface. As a result the hole density near the interface falls below the bulk value in the p-type semiconductor.

(iii) *Inversion*: If the positive bias on the metal side is increased further, eventually the conduction band at the oxide-semiconductor region comes close to the Fermi level in the semiconductor. As a result, the *electron* density at the interface starts to increase. If the positive bias is increased until E_c comes quite close to the semiconductor Fermi level near the interface, the electron density becomes very high and the semiconductor near the interface has electrical properties of an n-type semiconductor. This is shown in Fig. 9.6c. However, it is important to note that the n-type property is produced not by doping the semiconductor but by "inverting" the bands by an external

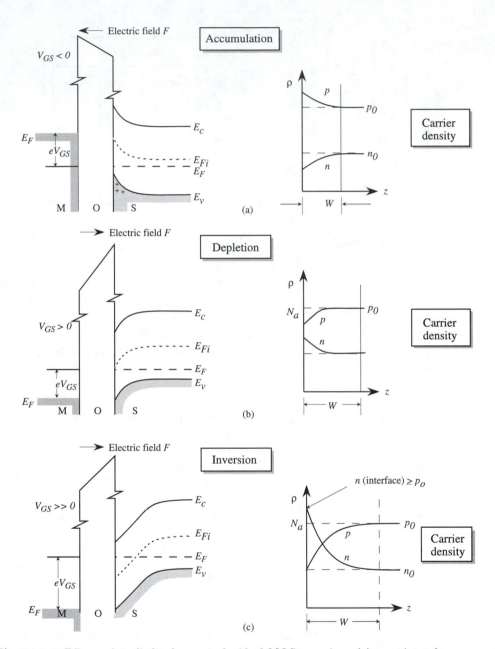

Figure 9.6: Effects of applied voltage on the ideal MOS capacitor: (a) negative voltage causes hole accumulation in the p-type semiconductor; (b) positive voltage depletes holes from the semiconductor surface; and (c) a larger positive voltage causes inversion—an "n-type" layer at the semiconductor surface. The figures also show the electron and hole distributions in each case. W represents the depletion width and N_a is the background acceptor concentration.

bias. This approach of producing n-type behavior without doping has many advantages over the usual doping. In particular, one does not have scattering due to dopants, and the carrier freezeout effect can be suppressed. A very high carrier density can be induced if the insulator has a large bandgap and has a good-quality interface with the semiconductor.

Due to the importance of the inversion regime in the MOSFET, let us examine it in quantitative detail. In Fig. 9.7 we show the bandbending of the semiconductor in the inversion. The bandbending is described by the quantity $e\phi$, *which measures the position of the intrinsic Fermi level with respect to the bulk intrinsic Fermi level.* The surface bandbending at the oxide-semiconductor interface is described in terms of the potential eV_s as shown in Fig. 9.7. When eV_s is zero, the bands are flat. As eV_s becomes positive we have depletion, and when eV_s is positive and larger than $e\phi_F$, the difference between the intrinsic Fermi level and the Fermi level, the inversion occurs.

Let us develop a more quantitative criterion for the inversion when the material properties become n-type. There is no precise way to define the point where the system is strongly n-type at the surface. One definition uses the criterion that inversion occurs when the n-type concentration at the interface is equal to the bulk p-type concentration. *Thus the intrinsic level E_{Fi} should be at a position $e\phi_F$ below the Fermi level at the interface.* Thus the surface bandbending is given by

$$\boxed{V_s(inv) = -2\phi_F} \tag{9.2}$$

Note that for an NMOS, the substrate is p-type and ϕ_F is negative and a positive bias V_s is needed to cause inversion. For a PMOS the substrate is n-type and ϕ_F is positive. A negative bias is needed to cause inversion. From Chapter 1 (see Eqn. 1.46), using Boltzmann statistics,

$$\phi_F = -\frac{k_B T}{e}\, \ell n\, \frac{p}{n_i} \sim -\frac{k_B T}{e}\, \ell n\, \frac{N_a}{n_i} \tag{9.3}$$

where N_a is the acceptor density and n_i is the intrinsic carrier concentration. The inversion criterion then becomes

$$V_s(inv) = -2\frac{k_B T}{e}\ell n\frac{N_a}{n_i} \tag{9.4}$$

This definition is, however, not unique or universally applicable, and other definitions for inversion have also been suggested in the literature. The important consideration is that at the onset of inversion there is an electron charge density of $\sim 10^{11}$ cm^{-2} at the surface so that the interface region's conductivity is high. Let us now evaluate the charge in the semiconductor channel. The electron concentration is approximately given

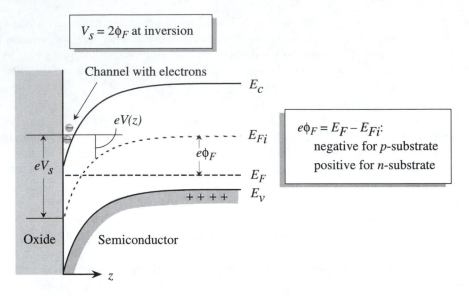

Figure 9.7: Bandbending of the semiconductor in the inversion mode. The interface potential is V_s. A simple criterion for inversion is that $V_s = 2\phi_F$.

by the Boltzmann distribution. In the bulk region, this concentration is

$$n_o = n_i \, \exp \, (E_F - E_{Fi})/k_B T = n_i \, \exp \, \left(\frac{e\phi_F}{k_B T} \right) \tag{9.5}$$

We are interested in calculating the carrier concentration in the MOS near the Si-SiO$_2$ interface.

We will now solve the Poisson equation using these charge-potential relations so that we can relate the charge to externally applied potential. In Fig. 9.5 we showed a schematic of the charge distribution of the electrons and holes in the semiconductor channel under various conditions. A more detailed overview of the charge, electric field, and potential in the inversion regime is shown in Fig. 9.8. The areal charge density on the metal Q_m is balanced by the channel depletion charge Q_d and the inversion charge Q_n. We are interested in calculating the gate voltage needed to cause inversion in the channel. This voltage is called the threshold voltage and is a very critical parameter of the MOS capacitor and the MOSFET.

The total surface charge density is related to the surface field by Gauss' law and is

$$|Q_s| = \epsilon_s \, |F_s| \tag{9.6}$$

This charge Q_s is the total surface charge density at the semiconductor-oxide interface region and includes the induced free charge (in inversion) and the background charge. The charge Q_s goes to zero when the bands are flat.

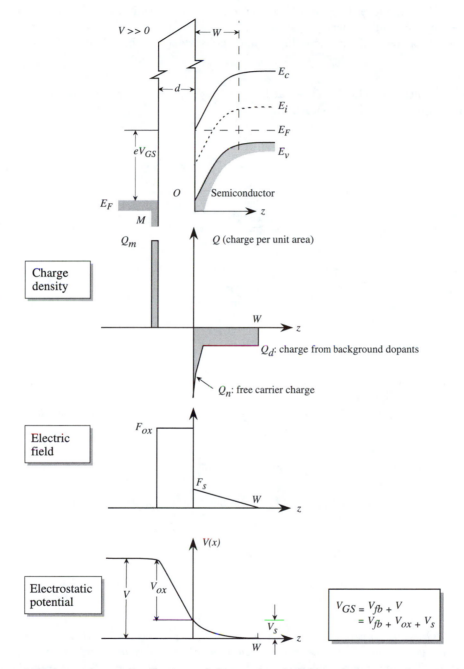

Figure 9.8: Approximate distributions of charge, electric field, and electrostatic potential in the ideal MOS capacitor in inversion. Once inversion begins, the depletion width W does not increase further because of the high mobile electron density at the interface region.

We can relate the gate voltage to the surface potential V_s by using the continuity of the electric displacement across the oxide-semiconductor interface (F_s and F_{ox} are the electric fields in the semiconductor and the oxide at the interface):

$$\epsilon_s F_s = \epsilon_{ox} F_{ox} \tag{9.7}$$

The gate voltage is related to the flat band voltage V_{fb} and voltage across the oxide and the semiconductor (V_{ox}) as shown in Fig. 9.8:

$$V_{GS} = V_{fb} + V_{ox} + V_s \tag{9.8}$$

We have

$$\begin{aligned} V_{ox} &= F_{ox} \cdot d_{ox} = \frac{\epsilon_s F_s d_{ox}}{\epsilon_{ox}} \\ &= \frac{\epsilon_s F_s}{C_{ox}} \end{aligned} \tag{9.9}$$

where C_{ox} is the oxide capacitance *per unit area* ($= \epsilon_{ox}/d_{ox}$). Thus

$$V_{GS} = V_{fb} + V_s + \frac{\epsilon_s F_s}{C_{ox}} = V_{fb} + V_s + \frac{Q_s}{C_{ox}} \tag{9.10}$$

This equation provides the link between the gate bias and the channel charge and will be used to study the capacitance of the MOS structure. Let us evaluate the threshold voltage V_T applied to the gate at which inversion starts in the channel. When inversion just occurs, the charge in the channel is *essentially due to the depletion charge* ($= N_a W$) *since the total free charge is still small.* Using the relation between the depletion width W and the potential V_s,

$$W = \left(\frac{2\epsilon_s |V_s|}{e N_a} \right)^{1/2} \tag{9.11}$$

the areal charge density ($Q_s = e N_a W$) becomes (using $V_S(inv) = 2\phi_F$)

$$Q_s = (2\epsilon_s e N_a |V_s|)^{1/2} = (4\epsilon_s e N_a |\phi_F|)^{1/2} \tag{9.12}$$

so that the threshold voltage is, from Eqn. 9.10,

$$V_T = V_{GS} \left(V_s = -2\phi_F \right) = V_{fb} - 2\phi_F + (4e\epsilon_s N_a |\phi_F|)^{1/2} \frac{1}{C_{ox}} \tag{9.13}$$

Once the inversion condition is satisfied, the depletion width does not change since the large density of free carriers induced after inversion starts prevent further depletion. The maximum depletion width is given by using $V_s = -2\phi_F$ in Eqn. 9.11 as

$$\boxed{W_{max} = \left(\frac{4\epsilon_s |\phi_F|}{e N_a} \right)^{1/2}} \tag{9.14}$$

Using Eqns. 9.6 and 9.13, the field at the surface at the inversion condition is

$$F_s = \left(\frac{4eN_a |\phi_F|}{\epsilon_s} \right)^{1/2} \tag{9.15}$$

The threshold voltage is extremely important in determining the performance of MOS-based devices. Variations in V_T can cause serious problems in the yield of the devices. It is useful to point out that if the substrate or body is biased at a value V_{sub}, the threshold voltage needed at the gate would simply be modified as

$$V_{GS} = V_T + V_{sub} \tag{9.16}$$

where V_T is given by Eqn. 9.13.

If the body is at a bias V_{SB} with respect to the inversion region, then the surface potential needed to cause inversion becomes $-2\phi_F + V_{SB}$. Replacing this value for V_s in Eqns. 9.12 and 9.14, we get, for the threshold voltage,

$$V_T = V_{fb} - 2\phi_F + [2e\epsilon_s N_a + |-2\phi_F + V_{SB}|]^{1/2} \frac{1}{C_{ox}} \tag{9.17}$$

We finally discuss the effect of interface traps due to defects on the threshold voltage. Let $N_t(x)$ be the position-dependent trap density in the MOS device in the oxide region. The traps will have additional charge, which will cause a voltage drop across the insulator. The voltage drop will cause a shift in the threshold voltage that is given by Gauss' law and the superposition principle (see Problem 9.9) as

$$\Delta V_T = \frac{-e}{C_{ox}} \int_o^{d_{ox}} \frac{z N_t(z)}{d_{ox}} dz \tag{9.18}$$

Since the variations in V_T can have serious consequences in the device turn-on (as we will see in the MOSFET discussions), the control of trap states is crucial. In the earlier stages of MOSFET development, this was the most important issue (the traps were created by Na ions) in device reliability.

From Eqn. 9.18, it can be seen that the effect of the interface trap charge on the threshold voltage depends upon where the charge is. It has the least effect if it is near the gate ($z = 0$), and has the maximum effect if it is at the Si-SiO$_2$ interface ($z = d_{ox}$). If Q_{ss} is the effective charge density per unit area at the oxide-semiconductor interface, the potential drop will occur across the oxide with a value

$$\Delta V_{ox} = \frac{-Q_{ss}}{C_{ox}}$$

Parameter	NMOS	PMOS
Substrate	*p*-type	*n*-type
ϕ_{ms}		
Al-gate	−	−
n^+ Si-gate	−	−
p^+ Si-gate	+	+
ϕ_F	−	+
Q_{ox}	+	+
γ	+	−
C_{ox}	+	+
Source-to-body voltage V_{SB}	+	−

Table 9.1: Signs for various terms in the threshold voltage equation for a MOSFET.

Adding the voltage shift due to interface charge, the threshold voltage expression becomes

$$V_T = V_{fb} - 2\phi_F + \left[2e\epsilon_s N_a \left| -2\phi_F + V_{SB} \right|^{1/2} \right] \frac{1}{C_{ox}} - \frac{Q_{ss}}{C_{ox}} \tag{9.19}$$

Defining a parameter, γ, known as the body factor as

$$\gamma = \frac{1}{C_{ox}} \sqrt{2e\epsilon_s N_a} \tag{9.20}$$

we can write the equation for the threshold voltage as

$$V_T = V_{TO} + \gamma \left(\sqrt{ \left| -2\phi_F + V_{SB} \right| } - \sqrt{ 2 \left| \phi_F \right| } \right) \tag{9.21}$$

where V_{TO} is the threshold voltage when $V_{SB} = 0$. The expressions given above are valid for NMOS or PMOS. Of course, N_a has to be replaced by substrate doping N_d in the case of a PMOS. The signs for various terms in the threshold voltage equation for NMOS and PMOS are given in Table 9.1.

EXAMPLE 9.1 Assume that the inversion in an MOS capacitor occurs when the surface potential is twice the value of $e\phi_F$. What is the maximum depletion width at room temperature of a structure where the *p*-type silicon is doped at $N_a = 10^{16} \text{cm}^{-3}$?

At room temperature, the intrinsic carrier concentration is $n_i = 1.5 \times 10^{10} \text{cm}^{-3}$ for Si. Thus, we have for the potential ϕ_F,

$$\phi_F = -\frac{k_B T}{e} \, \ell n \, \frac{N_a}{n_i} = (0.026 eV) \ell n \left(\frac{10^{16}}{1.5 \times 10^{10}} \right)$$

$$= -0.347 \text{ V}$$

The corresponding space charge width is

$$W = \left[\frac{4\epsilon_s |\phi_F|}{e N_a} \right]^{1/2} = \left[\frac{4 \times 11.9 \times (8.85 \times 10^{-14})(0.347)}{1.6 \times 10^{-19} \times 10^{16}} \right]^{1/2}$$

$$= 0.30 \ \mu\text{m}$$

EXAMPLE 9.2 Consider an aluminum-SiO$_2$-Si MOS device. The work function of Al is 4.1 eV, the electron affinity for SiO$_2$ is 0.9 eV, and that of Si is 4.15 eV. Calculate the potential V_{fb} if the Si doping is $N_a = 10^{14} \text{cm}^{-3}$.

The potential V_{fb} is given by

$$eV_{fb} = e\phi_m - (e\chi_s + (E_c - E_F))$$

The position of the Fermi level is

$$E_F = E_{Fi} + k_B T \ \ell n \ \frac{N_a}{n_i}$$

below the conduction band. Also, $E_{Fi} = E_g/2$ where for Si, $E_g = 1.11$ eV. Using $T = 300$ K, we get

$$E_F = 0.555 + 0.026 \ \ell n \ \left(\frac{10^{14}}{1.5 \times 10^{10}} \right) = 0.783 \text{ eV}$$

below the conduction band. Thus

$$V_{fb} = 4.1 - (4.15 + 0.783) = -0.833 \text{ eV}$$

EXAMPLE 9.3 Consider a p-type silicon doped to $3 \times 10^{16} \text{cm}^{-3}$. The SiO$_2$ has a thickness of 500 Å. An n^+ polysilicon gate is deposited to form the MOS capacitor. The work function difference $V_{fb} = -1.13$ eV for the system; temperature = 300 K. Calculate the threshold voltage if there is no oxide charge and if there is an oxide charge of 10^{11}cm^{-2}.

The position of the Fermi level is given by (measured from the intrinsic Fermi level)

$$\phi_F = -0.026 \ \ell n \ \left(\frac{3 \times 10^{16}}{1.5 \times 10^{10}} \right) = -0.376 \text{ V}$$

Under the assumption that the charge Q_s is simple $N_a W$ where W is the maximum depletion width, we get

$$
\begin{aligned}
Q_s &= (4\epsilon_s e N_a |\phi_F|)^{1/2} \\
&= \left(4 \times (11.9) \times (8.85 \times 10^{-14} \text{ F/cm}) \ (1.6 \times 10^{-19} \text{ C})(3 \times 10^{16} \text{ cm}^{-3})(0.376 \text{ V}) \right)^{1/2} \\
&= 8.64 \times 10^{-8} \text{ C cm}^{-2}
\end{aligned}
$$

In the absence of any oxide charge, the threshold voltage is

$$V_T = -1.13 + 2(0.376) + \left(8.64 \times 10^{-8}\right)\left(\frac{500 \times 10^{-8}}{3.9(8.85 \times 10^{-14})}\right)$$

$$= 0.874 \text{ V}$$

In the case where the oxide has trap charges, the threshold voltage is shifted by

$$\Delta V_T = \left(10^{11}\right)\left(1.6 \times 10^{-19}\right)\left(\frac{500 \times 10^{-8}}{3.9 \times 8.85 \times 10^{-14}}\right)$$

$$= -0.23 \text{ V}$$

It can be seen from this example that oxide charge can cause a significant shift in the threshold voltage of an MOS device.

EXAMPLE 9.4 Consider an n-MOSFET made from Si-doped p-type at $N_a = 5 \times 10^{16}$ cm^{-3} at 300 K. The other parameters for the device are the following:

$$V_{fb} = -0.5 \text{ V}$$
$$\mu_n = 600 \text{ cm}^2 \text{ V}^{-1} \text{ s}^{-1}$$
$$\mu_p = 200 \text{ cm}^2 \text{ V}^{-1} \text{ s}^{-1}$$

The inversion condition is $V_s = -2\phi_F$. Assume that the electrons induced under inversion are in a region 200 Å wide near the Si/SiO$_2$ interface.

(i) Calculate the channel conductivity near the Si-SiO$_2$ interface under flat band condition and at inversion.

(ii) Calculate the threshold voltage.

(i) Assuming that all of the donors are ionized, we have at flat band

$$p = N_a = 5 \times 10^{16} \text{ cm}^{-3}$$

This gives

$$\sigma(fb) = (5 \times 10^{16} \text{ cm}^{-3})(1.6 \times 10^{-19} \text{ C})(200 \text{ cm}^2/ \text{ V} \cdot \text{s}) = 1.6 \text{ } (\Omega \text{ cm})^{-1}$$

At inversion with $V_s = -2\phi_F$ we have

$$n(\text{interface}) = p(\text{bulk}) = 5 \times 10^{16} \text{ cm}^{-3}$$

This gives (near the interface)

$$\sigma(inv) = (5 \times 10^{16} \text{ cm}^{-3})(1.6 \times 10^{-19} C)(600 \text{ cm}^2/ \text{ V} \cdot \text{s}) = 4.8 \text{ } (\Omega \text{ cm})^{-1}$$

(ii) To calculate the threshold voltage we need ϕ_F. This is given by

$$\phi_F = -\frac{k_B T}{e} \ell n\left(\frac{p}{p_i}\right) = -0.39 \text{ V}$$

Using the parameters given and the equation for the threshold voltage we get

$$V_T = -0.5 + 0.78 + 1.637 \text{ V} = 1.93 \text{ V}$$

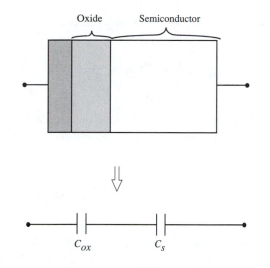

Figure 9.9: A simple equivalent circuit of the MOS capacitor.

9.4 CAPACITANCE-VOLTAGE CHARACTERISTICS OF THE MOS STRUCTURE

An important consideration of the MOS structure is the capacitance as a function of applied voltage. In the previous section we have already derived the basic relations between the charge and the voltage. We will now use them to derive the C-V characteristics, which provide an important characterization tool to check on the quality of the structure. In the C-V measurements, a dc bias V is applied to the gate, and a small ac signal ($\sim$ 5-10 mV) is applied to obtain the capacitance at the bias applied.

The capacitance of the MOS structure is the series combination of the oxide capacitance C_{ox} and the semiconductor capacitance C_s, as shown in Fig. 9.9. The semiconductor areal capacitance is

$$C_s = \frac{dQ_s}{dV_s} \qquad (9.22)$$

and the total areal capacitance is

$$C_{mos} = \frac{C_{ox}C_s}{C_{ox} + C_s} \qquad (9.23)$$

We remind the reader that the capacitance values discussed are capacitance per unit area.

The three important regimes of accumulation, depletion, and inversion that we have discussed are reflected in the C-V characteristics. In the accumulation region (negative V_{GS}), the holes accumulate at the surface and C_s is much larger than C_{ox}.

This is because a small change in bias causes a large change in Q_s in the accumulation regime. The MOS capacitance is then

$$C_{mos} \cong C_{ox} = \frac{\epsilon_{ox}}{d_{ox}} \tag{9.24}$$

As the gate voltage becomes positive and the channel is depleted of holes, the depletion capacitance becomes important. The depletion capacitance is simply given by ϵ_s/W, and the total capacitance becomes

$$C_{mos} = \frac{C_{ox}}{1 + \frac{C_{ox}}{C_s}} = \frac{\epsilon_{ox}}{d_{ox} + \frac{\epsilon_{ox}W}{\epsilon_s}} \tag{9.25}$$

As the device gets more and more depleted, the value of C_{mos} decreases, as shown in Fig. 9.10. At the inversion condition, the depletion width reaches its maximum value W_{max}. At this point there is essentially no free carrier density. The minimum capacitance is

$$C_{mos}(min) = \frac{\epsilon_{ox}}{d_{ox} + \frac{\epsilon_{ox}W_{max}}{\epsilon_s}} \tag{9.26}$$

where W_{max} is defined by Eqn. 9.14.

If the bias is further increased, the free electrons start to collect in the inversion regime and the depletion width remains unchanged with bias. The capacitance of the semiconductor again increases since a small change in V_s causes a large change in Q_s. The capacitance of the MOS device thus returns towards the value of C_{ox}:

$$C_{mos}(inv) = C_{ox} = \frac{\epsilon_{ox}}{d_{ox}} \tag{9.27}$$

Another important point on the C-V characteristics is the point where the bands become flat. Near the flat band position, a small change $\delta V(z)$ of potential produces a charge density given by

$$\delta \rho(z) = p_o \left[\exp \left(\frac{e\delta V(z)}{k_B T} \right) - 1 \right] \sim \frac{e\delta V(z) p_o}{k_B T} \tag{9.28}$$

where p_o is the hole charge density at the flat band. Upon solving the Poisson equation, we get a capacitance of (see Example 9.5 for the solution)

$$C_s = \frac{\epsilon_s}{\sqrt{\frac{k_B T}{e} \frac{\epsilon_s}{eN_a}}} \tag{9.29}$$

The flat band capacitance of the MOS device is then

$$C_{mos}(fb) = \frac{\epsilon_{ox}}{d_{ox} + \frac{\epsilon_{ox}}{\epsilon_s} \sqrt{\frac{k_B T}{e} \frac{\epsilon_s}{eN_a}}} \tag{9.30}$$

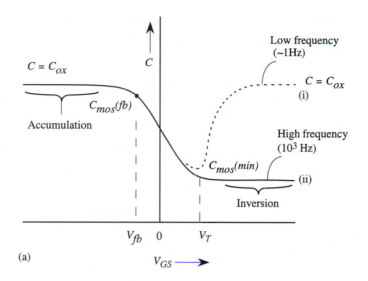

(a)

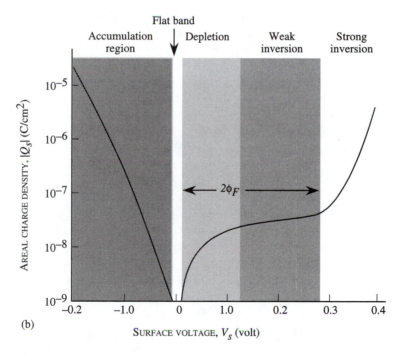

(b)

Figure 9.10: (a) The dependence of MOS capacitance on voltage. Curve (i) is for low frequencies while curve (ii) is for high frequencies. Also shown are the various important regions in the capacitance-voltage relations. (b) The charge density $|Q_s|$ is shown schematically as a function of the surface potential V_s. The dependence of the charge in various regimes is shown.

The capacitance voltage curve for a typical MOS capacitor is shown in Fig. 9.10a. Also shown are the important regions and capacitance points discussed above. A schematic description of how the surface charge density behaves as a function of the surface potential is shown in Fig. 9.10b.

In the discussion above, we have assumed that the electrons needed in the inversion regime can be supplied instantaneously as the gate bias is charged. This is not the case in the *p*-type semiconductor. The excess electrons needed are introduced into the channel by *e-h* generation, by thermal generation processes, or by diffusion of the minority carriers from the *p*-type substrate. *Since the generation process takes a finite time, the inversion sheet charge can follow the voltage only if the voltage variations are slow.* If the variations are fast, the capacitance due to the free electrons makes no contribution and the capacitance is dominated by the depletion capacitance. Thus, under high-frequency measurements, the capacitance does not show a turnaround and remains at the value $C_{mos}(min)$, as shown in Fig. 9.10. The capacitance in the inversion regime starts to decrease even at frequencies of 10 Hz and at 10^4 Hz it reaches the low value of $C_{mos}(min)$.

In the above discussion we have assumed that there are no fixed charges in the oxide region. In reality, there may be two kinds of charges associated with the oxide. The first are charges in the oxide near the Si/SiO_2 interface and are due to impurity ions that may be incorporated during the oxide fabrication process. In addition, there may be interface charges that are due to the non-ideal nature of the Si-SiO_2 interface. The fixed charge density is not dependent upon the gate bias, but the interface charge depends upon the applied bias. The two thus have different effects on the C-V characteristics.

The presence of the fixed charge simply causes a voltage drop across the oxide given by

$$\Delta V = \frac{-Q_{ss}}{C_{ox}} \tag{9.31}$$

where Q_{ss} is the fixed charge density (cm^{-2}) in the oxide. As a result, if Q_{ss} is positive the entire C-V curve shifts to a more negative value. Since the charge Q_{ss} is independent of the gate bias, the entire C-V curve shifts as shown schematically in Fig. 9.11a. The value of Q_{ss} can be obtained by measuring the shift as compared with the calculated ideal curve. Such measurements are very important for characterizing the quality of MOS devices.

The interface charge has a somewhat different effect on the C-V characteristics. In an ideal system, there are no allowed electron states in the bandgap of a semiconductor. However, since the Si-SiO_2 interface is not ideal, a certain density of interface states are produced that lie in the bandgap region.

In contrast to the fixed charge, electrons can flow into and out of these interface states depending upon the position of the Fermi level. The character of the interface

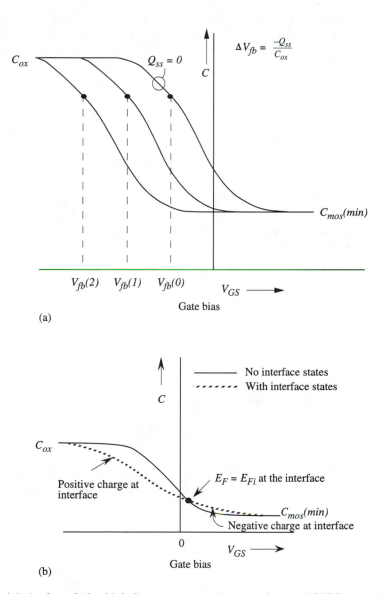

Figure 9.11: (a) A plot of the high-frequency capacitance voltage of MOS capacitors with different values of the fixed oxide charge. The shift of the C-V curves from an ideal calculated result allows one to calculate the fixed charge value. (b) The effect of the interface states is to smear out the C-V curves.

states is defined as "acceptor-like" and "donor-like." An acceptor state is neutral if the Fermi level is below the state (i.e., the state is unoccupied) and becomes negatively charged if the Fermi level is above it (i.e., the state is occupied). The donor state is neutral if the Fermi level is above it (i.e., the state is occupied) and positively charged when it is empty. As a result, when the position of the Fermi level is altered, the charge at the interface changes.

When the interface charge is positive, the C-V curve shifts towards negative voltages, while when it is negative, the curve shifts towards positive voltages. This is shown schematically in Fig. 9.11b. The C-V curve is thus "smeared out" due to the presence of interface states. In modern high-quality MOS structures, the interface state density is maintained below 10^{10} cm^{-2}, so that the effect is negligible.

The MOS structure by itself is usually not used as an electronic device, but the MOSFET device utilizes the concepts we have developed in these sections. As we have pointed out a number of times, the MOSFET is perhaps the most important electronic device; it will be discussed next.

EXAMPLE 9.5 Derive the relation for the capacitance per unit area of the MOS capacitor at the flat band condition.

We start from Eqn. 9.28, according to which the charge density near the flat band is

$$\delta\rho(z) = \frac{e\delta V(z)p_o}{k_BT} = \frac{eN_a\delta V(z)}{k_BT}$$

The Poisson equation then gives us

$$\frac{d^2\delta V(z)}{dz^2} = -\frac{eN_a\delta V(z)}{k_BT}$$

The solution is a simple exponentially decaying function:

$$\delta V(z) = \delta V_s \ \exp\left(-bz\right)$$

where

$$b = \sqrt{\frac{eN_a}{k_BT}}$$

The charge density is now

$$\delta\rho(z) = \frac{eN_a}{k_BT}$$

The areal charge density is obtained by integrating this from the interface into the bulk, with the result

$$\mid \delta Q_s \mid = \int_o^\infty \rho(z)dz = \frac{eN_a}{b\ k_BT}\delta V_s$$

The capacitance is now

$$C_s = \frac{\delta Q_s}{\delta V_s} = \frac{eN_a}{b\ k_BT}$$

which gives the result given in Eqn. 9.30 when the value of b is used.

EXAMPLE 9.6 Consider a MOS capacitor made on a p-type substrate with doping of $10^{16} \, \text{cm}^{-3}$. The SiO_2 thickness is 500 Å and the metal gate is made from aluminum. Calculate the oxide capacitance, the capacitance at the flat band, and the minimum capacitance at threshold.

The oxide capacitance is simply given by

$$C_{ox} = \frac{\epsilon_{ox}}{d_{ox}} = \frac{3.9 \times 8.85 \times 10^{-14}}{500 \times 10^{-8}} = 6.9 \times 10^{-8} \, \text{F/cm}^2$$

To find the minimum capacitance, we need to find the maximum depletion width at the threshold voltage. The value of ϕ_F is given by

$$\phi_F = -0.026 \, \text{V} \, \ell n \left(\frac{N_a}{n_i} \right) = 0.026 \, \ell n \left(\frac{10^{16}}{1.5 \times 10^{10}} \right)$$

$$= -0.347 \, \text{V}$$

The maximum depletion width (assuming $V_s = -2\phi_F$) is

$$W_{max} = \left(\frac{4\epsilon |\phi_F|}{e N_a} \right)^{1/2} = \left\{ \frac{4(11.9 \times 8.85 \times 10^{-14})(0.347)}{1.6 \times 10^{-19} \times 10^{16}} \right\} = 0.3 \times 10^{-4} \, \text{cm}$$

The minimum capacitance is now

$$C_{min} = \frac{C_{ox} C_s}{C_{ox} + C_s} = \left(\frac{\epsilon_{ox}}{d_{ox} + \frac{\epsilon_{ox}}{\epsilon_s} W_{max}} \right)$$

$$= 2.3 \times 10^{-8} \, \text{F/cm}^2$$

The capacitance under flat band conditions is

$$C_{fb} = \frac{\epsilon_{ox}}{d_{ox} + \frac{\epsilon_{ox}}{\epsilon_s} \sqrt{\left(\frac{k_B T}{e} \right) \left(\frac{\epsilon_s}{e N_a} \right)}}$$

$$= \frac{3.9 \times (8.85 \times 10^{-14})}{(500 \times 10^{-8}) + \frac{3.9}{11.9} \sqrt{\frac{0.026 \times 11.7 \times 8.85 \times 10^{-14}}{1.6 \times 10^{-19} \times 10^{16}}}}$$

$$= 5.42 \times 10^{-8} \, \text{F/cm}^2$$

It is interesting to note that C_{fb} is $\sim 80\%$ of C_{ox} and C_{min} is $\sim 33\%$ of C_{ox}.

9.5 METAL-OXIDE-SEMICONDUCTOR FIELD-EFFECT TRANSISTOR

In the previous sections we have seen how a gate voltage on an MOS structure can change the charge density of free electrons at the oxide-semiconductor interface. The free

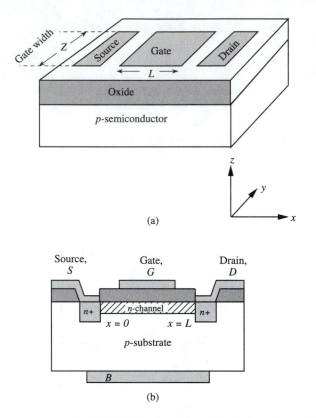

Figure 9.12: A schematic of the MOSFET structure. (a) A top view of the device; (b) a cross-section of the device. The source and drain contacts are n-type ohmic contacts. The gate is isolated from the channel by the oxide.

electrons can obviously carry a current in the direction parallel to the interface. Thus, in principle, the gate can modulate the conductivity of the MOS channel if additional contacts are placed across the channel. This leads to the MOSFET structure shown in Fig. 9.12.

In addition to the gate metal, which is isolated from the conducting channel, we have source and drain ohmic contacts. Thus, if a bias is applied between the source and drain, a current will flow in the channel. The current will be determined by the free charge in the channel, which is controlled by the gate voltage. Thus the device has the essential features required of an electronic device. We will now study the device properties quantitatively.

9.5.1 Current-Voltage Characteristics

The MOSFET is a complex device for which a proper analysis requires the study of charge flow in three dimensions. As can be seen from Fig. 9.12, the geometry of the MOSFET is different in the x, y, and z directions. The full three-dimensional analysis of the MOSFET requires complex numerical techniques. However, we will develop a simplified approach that gives a good semi-quantitative understanding of the current-voltage characteristics of the device.

Qualitatively, we can see how the MOSFET I-V characteristics behave. When a bias is applied between the source and the drain, current flows in the channel near the Si-SiO$_2$ interface. The charge density in the channel is controlled by the gate bias as well as the source-drain bias. The gate bias can thus modulate the current flow in the channel. As discussed for the MESFET or JFET case (Section 8.3), we make the following approximations for our model:

- The mobility of the electrons is constant and independent of the electric field. We know that this is true only at low electric fields. At high fields the velocity of the electrons saturate. The analysis is, therefore, valid only if the field in the channel ($\sim$ drain bias divided by the channel length) is less than 2-3 kV/cm. Since this is usually not true in modern devices, the analysis is only semi-quantitative and is useful primarily to convey the physics of the device operation.

- We assume the gradual channel approximation introduced by Shockley. In the absence of any source-drain bias, the depletion width is simply given by the one-dimensional model we developed for the p-n diode. However, strictly speaking, when there is a source-drain bias, one has to solve a two-dimensional problem to find the depletion width and, subsequently, the current flow. In the gradual channel approximation, we assume that field in the direction from the gate to the substrate is much stronger than from the source to the drain, i.e., the potential varies "slowly" along the channel as compared to the potential variation in the direction from the gate to the substrate. Thus the depletion width, at a point x along the channel, is given by the potential at that point using the simple one-dimensional results. This approximation is good if the gate length L is larger than the channel depth, which is typically a few hundred angstroms.

In the analysis discussed here we will assume that the source is grounded and all voltages are referred to the source. The first step in obtaining the I-V curves is to establish the relation between the channel charge at any point x along the channel and the gate voltage in the MOSFET. This is given by Eqns. 9.6 and 9.10. Using the gradual channel approximation for the induced charge in the channel, we can treat the charge-voltage problem as a one-dimensional problem. The induced charge per unit area, once we are in the inversion region, is

$$Q_s = C_{ox}\left[V_{GS} - V_T - V_c(x)\right] \qquad (9.32)$$

We know that

$$
\begin{aligned}
V_c(x) &= 0 & \text{at the source} \\
&= V_{DS} & \text{at the drain}
\end{aligned}
\qquad (9.33)
$$

We also assume that the body bias is zero. The case of finite body bias will be discussed later. We will now make a constant-mobility approximation to find the current in the device. The current will be assumed to be due to carrier drift and is given by (current = surface charge density × mobility × electric field × gate width)

$$I_D = Q_s \mu_n \frac{dV_c(x)}{dx} Z \qquad (9.34)$$

where Z is the width of the device. The current I_D is constant through any cross-section of the channel. We rewrite Eqn. 9.34 as

$$I_D dx = Q_s \, \mu_n \, dV_c(x) Z \qquad (9.35)$$

The integration of Eqn. 9.35 from the source ($x = 0$) to the drain ($x = L$) after using the value of Q_s gives ($V_c(L) = V_{DS}$)

$$I_D = \frac{\mu_n Z C_{ox}}{L}\left[\left\{V_{GS} - V_T - \frac{V_{DS}}{2}\right\}\right]V_{DS} \qquad (9.36)$$

Here L is the length of the device.

Let us define parameters k and k' to define the prefactor in the equation above:

$$k = \frac{\mu Z C_{ox}}{L} = \frac{k' Z}{L} \qquad (9.37)$$

From Eqn. 9.32 we see that for a sufficiently high drain bias, the channel mobile charge becomes zero (the channel is said to have *pinched off*) at the drain side. This defines the saturation drain voltage $V_{DS}(sat)$, i.e.,

$$Q_s(V_{DS}) = Q_s(V_{DS}(sat)) = 0$$

The pinch-off occurs at the drain end of the channel.

$$V_{DS}\left(Q_s(x = L) = 0\right) = V_{DS}(sat) = V_{GS} - V_T \qquad (9.38)$$

Our derivation of the current is valid only up to pinch-off. Beyond pinch-off the current problem is quite complex. The electron density is no longer given by the simple

Boltzmann distribution function, since the electrons are traveling at high speeds and high energies. The electron kinetic energy is no longer $\sim 3/2k_BT$, but is quite large. *If the non-equilibrium nature of electron transport is considered, it turns out that beyond pinch-off, the current is essentially constant.*

Let us now consider some of the limiting cases of the I-V characteristics.

Linear or Ohmic Region

In the case where the drain bias V_{DS} is less than $V_{DS}(sat)$

$$V_{DS} < V_{DS}(sat) = V_{GS} - V_T \qquad (9.39)$$

For very small drain bias values, the current increases linearly with the drain bias, since the quadratic term in V_{DS} in Eqn. 9.37 can be ignored. The current in this linear regime is

$$\boxed{I_D = k\left[(V_{GS} - V_T)V_{DS}\right]} \qquad (9.40)$$

where V_T is the gate voltage required to "turn on" the transistor by starting the inversion.

Saturation Region

The analysis discussed above is valid up to the point where the drain bias causes the channel to pinch off at the drain end. The saturation current now becomes, after substituting for $V_{DS}(sat)$ in Eqn. 9.36,

$$\boxed{\begin{aligned} I_D(sat) &= k\left\{(V_{GS} - V_T)^2 - \frac{(V_{GS} - V_T)^2}{2}\right\} \\ &= \frac{k}{2}(V_{GS} - V_T)^2 \end{aligned}} \qquad (9.41)$$

Thus once saturation starts, the gate current has a square-law dependence upon the gate bias.

Material and Device Parameters

Important material and device parameters can be extracted from the I-V characteristics of the MOSFET. At low drain bias we can ignore the quadratic term in V_{DS}. The drain current is given by

$$I_D = \frac{Z\mu_n C_{ox}}{L}(V_{GS} - V_T)V_{DS} \qquad (9.42)$$

so that the extrapolation of the low drain bias current points gives the threshold voltage V_T. This is shown schematically in Fig. 9.14. Also, if the drain current is measured at two different values of V_{GS} while keeping V_{DS} fixed, the mobility in the channel can be determined, since

$$I_{D2} - I_{D1} = \frac{Z\mu_n C_{ox}}{L}(V_{GS2} - V_{GS1})V_{DS} \qquad (9.43)$$

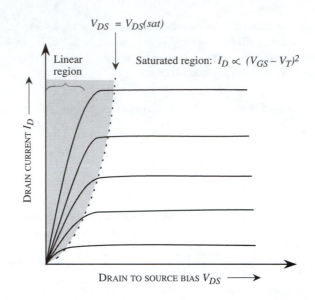

Figure 9.13: A schematic of the I-V characteristics of a MOSFET. In the ohmic region the current increases linearly with the drain bias for a fixed gate bias. In the simple model considered, once $V_{DS} > V_{DS}(sat)$, the current is independent of V_{DS}.

where I_{D1} and I_{D2} are the currents at gate biases of V_{GS1} and V_{GS2}. Since Z, L and C_{ox} are known, the inversion channel mobility can be obtained. *It is worth noting that the mobility in a MOSFET channel is usually much smaller than the mobility in bulk silicon. This is because of the strong scattering that occurs due to the roughness of the Si-SiO$_2$ interface. Typical MOSFET mobilities are $\sim$ 600 cm^2/V·s while typical mobilities in bulk silicon are $\sim$ 1000 cm^2/V·s.*

The performance of the MOSFET as a device is defined via two important parameters, the drain conductance (output conductance) and the transconductance.

The drain conductance is defined as

$$g_D = \left. \frac{\partial I_D}{\partial V_{DS}} \right|_{V_{GS}=\text{constant}} \tag{9.44}$$

At low drain biases we get from Eqn. 9.40 for the ohmic region

$$g_D = \frac{Z \mu_n C_{ox}}{L}(V_{GS} - V_T) \tag{9.45}$$

In the saturation region in our simple model, the drain conductance is zero. In real devices g_D is not zero at saturation, as discussed in Section 9.6 below. The FET is usually used with other loads so that during the circuit operation the voltage V_{DS} may change. If g_D is not zero, the current through the FET will change as V_{DS} changes,

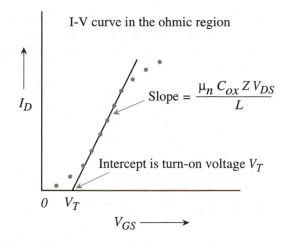

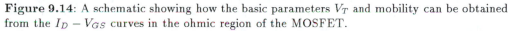

Figure 9.14: A schematic showing how the basic parameters V_T and mobility can be obtained from the $I_D - V_{GS}$ curves in the ohmic region of the MOSFET.

and this causes non-ideal behavior in the circuit. The MOSFET (and other transistors) is also used as constant current source operating in the saturation region. This also requires the transistor to have minimum drain conductance.

The transconductance of the MOSFET is closely linked to the speed of the device and is given by

$$g_m = \left.\frac{\partial I_D}{\partial V_{GS}}\right|_{V_{DS}=\text{constant}} \tag{9.46}$$

At saturation we have from Eqn. 9.41

$$\boxed{g_m = \frac{Z\mu_n C_{ox}}{L}(V_{GS} - V_T)} \tag{9.47}$$

A high-transconductance device is produced if the channel length is small and channel mobility is high. The transconductance represents the control of the gate on the channel current and is usually quoted in millisiemens per millimeter (mS/mm) to remove the dependence on the gate width Z.

9.5.2 Substrate Bias Effects

In the analysis above we have assumed that the substrate bias is the same as the source bias. In MOSFET circuits, the source-to-substrate (or body) bias V_{SB} is an additional variable that can be exploited. In Fig. 9.15 we show an n-channel MOSFET showing the source-to-body bias, which is chosen to be zero or positive to reverse bias the source-to-substrate junction.

In the absence of V_{SB}, the inversion condition occurs when V_s, the surface potential, is equal to $-2\phi_F$ as shown in Fig. 9.15b. In case V_{SB} is positive, the surface voltage required for inversion is increased as shown in Fig. 9.15c by an amount V_{SB}, since the body is at a higher energy level.

When $V_{SB} > 0$, the depletion width is no longer W_{max} but is increased to absorb the added potential V_{SB}. As noted in Eqn. 9.21, the body bias alters the threshold voltage. The charge in the threshold voltage is given by

$$\Delta V_T = \frac{\sqrt{2e\epsilon_s N_a}}{C_{ox}} \left[\sqrt{|-2\phi_F + V_{SB}|} - \sqrt{|2\phi_F|} \right] \tag{9.48}$$

To ensure a positive shift in the threshold voltage, V_{SB} must be positive for the NMOS.

The threshold voltage of a MOSFET can also be modified by altering the doping density in the silicon region as well. This can be done by ion implantation so that an added dose of acceptors (or donors) is introduced. This changes the value of the depletion charge and consequently the threshold voltage is altered. This effect is examined in Example 9.10.

EXAMPLE 9.7 Consider a n-channel MOSFET at 300 K with the following parameters:

Channel length,	L	$=$	1.5 μm
Channel width,	Z	$=$	25.0 μm
Channel mobility,	μ_n	$=$	600 cm^2/ V $\cdot$ s
Channel doping,	N_a	$=$	1×10^{16} cm^{-3}
Oxide thickness,	d_{ox}	$=$	500 Å
Oxide charge,	Q_{ss}	$=$	10^{11} cm^{-2}
Metal-semiconductor work function difference,	ϕ_{ms}	$=$	-1.13 V

Calculate the saturation current of the device at a gate bias of 5 V.

The Fermi level position for the device is given by

$$\phi_F = -0.026 \, \ell n \left(\frac{1 \times 10^{16}}{1.5 \times 10^{10}} \right) = -0.348 \text{ V}$$

The flat band voltage is

$$V_{fb} = \phi_{ms} - \frac{Q_{ss}}{C_{ox}} = -1.13 - 0.23 = -1.35 \text{ V}$$

The threshold voltage is given by Eqn. 9.13 as

$$
\begin{aligned}
V_T &= -1.35 + 0.696 \\
&+ \frac{\left[4(1.6 \times 10^{-19})(11.9)(8.84 \times 10^{-14})(10^{16})(0.348) \right]^{1/2} (500 \times 10^{-8})}{3.9(8.85 \times 10^{-14})} \\
&= 0.04 \text{ V}
\end{aligned}
$$

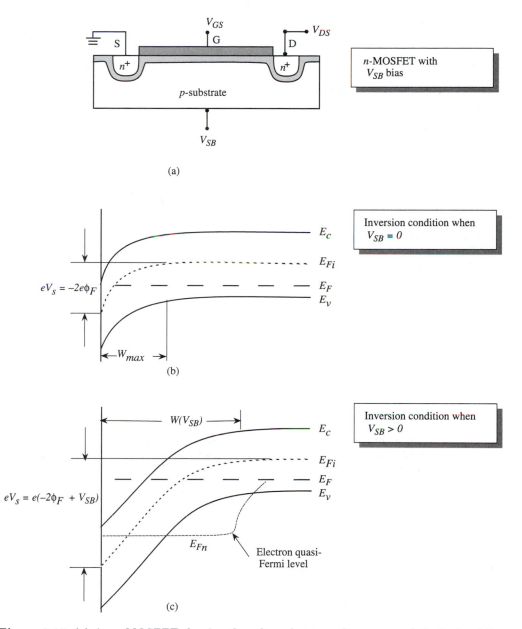

Figure 9.15: (a) An n-MOSFET showing the voltage between the source and the body of the transistor; (b) band profiles of the MOSFET with $V_{SB} = 0$ at the inversion condition; (c) band profile of the MOSFET when $V_{SB} > 0$. The depletion width increases when $V_{SB} > 0$.

The saturation current is now, from Eqn. 9.41,

$$
\begin{aligned}
I_D(sat) &= \frac{(25 \times 10^{-4})(600)}{2(1.5 \times 10^{-4})} \frac{(3.9)(8.85 \times 10^{-14})}{500 \times 10^{-8}} [5.0 - 0.04]^2 \\
&= 8.5 \text{ mA}
\end{aligned}
$$

EXAMPLE 9.8 Consider a silicon NMOS device at 300 K characterized by $\phi_{ms} = 0, N_a = 4 \times 10^{14}$ cm^{-3}, $d_{ox} = 200$ Å, $L = 1.0$ μm, $Z = 10$ μm. Calculate the drain current for a gate voltage of $V_{GS} = 5$ V and drain voltage of 4 V. The electron mobility in the channel is 700 cm^2/V·s.

We start by calculating the threshold voltage. The potential ϕ_F is given by

$$
\phi_F = -(0.026) \, \ell n \left(\frac{4 \times 10^{14}}{1.5 \times 10^{10}} \right) = -0.264 \text{ V}
$$

The threshold voltage is, from Eqn. 9.13,

$$
\begin{aligned}
V_T &= 0.528 + \frac{\left[4(1.6 \times 10^{-19})(11.9 \times 8.85 \times 10^{-14})(4 \times 10^{14})(0.264) \right]^{1/2}}{3.9 \times 8.85 \times 10^{-14}} \cdot \left(2 \times 10^{-6} \right) \\
&= 0.58 \text{ V}
\end{aligned}
$$

The saturation voltage for a gate bias of 5 V is, from Eqn. 9.38,

$$
V_{DS}(sat) = 4.42 \text{ V}
$$

The saturation current is now, from Eqn. 9.41,

$$
I_D(sat) = 11.8 \text{ mA}
$$

EXAMPLE 9.9 Consider an n-channel MOSFET with gate width $Z = 10$ μm, gate length $L = 2$ μm and oxide capacitance $C_{ox} = 10^{-7}$ F/cm^2. In the linear region, the drain current is found to have the following values at $V_{DS} = 0.1$ V:

$$
\begin{aligned}
V_{GS} &= 1.5V, I_D = 50 \text{ } \mu\text{A} \\
&= 2.5V, I_D = 80 \text{ } \mu\text{A}
\end{aligned}
$$

Calculate the mobility of electrons in silicon. How does the result compare to the electron mobility in pure silicon at 300 K? Also calculate the threshold voltage of the device.

In the linear region we have, using Eqn. 9.43 for a fixed drain bias,

$$
I_{D2} - I_{D1} = \frac{Z\mu_n C_{ox}}{L} (V_{GS2} - V_{GS1}) V_{DS}
$$

or

$$
\begin{aligned}
\mu_n &= \frac{(I_{D2} - I_{D1})L}{V_{DS}(V_{GS2} - V_{GS1})ZC_{ox}} = \frac{(30 \times 10^{-6})(2 \times 10^{-4})}{(0.1)(1.0)(10 \times 10^{-4})(10^{-7})} \\
&= 600 \text{ cm}^2/\text{V} \cdot \text{s}
\end{aligned}
$$

This mobility is almost half that of pure silicon. The decrease is primarily due to interface scattering effects since the Si-SiO$_2$ interface is not perfect.

The intercept of the $I_D - V_{GS}$ curve is at -0.16 V, which is the threshold voltage.

EXAMPLE 9.10 Consider an n-channel MOSFET with a substrate doping of $N_a = 2 \times 10^{16}$ cm^{-3} at 300 K. The SiO$_2$ thickness is 500 Å and a source-body bias of 1.0 V is applied. Calculate the shift in the threshold voltage arising from the body bias.

The potential ϕ_F is given by

$$\phi_F = -0.026 \; \ell n \left(\frac{N_a}{n_i} \right) = -0.026 \; \ell n \left(\frac{2 \times 10^{16}}{1.5 \times 10^{10}} \right) = -0.367 \text{ V}$$

The oxide areal capacitance is

$$C_{ox} = \frac{\epsilon_{ox}}{d_{ox}} = \frac{3.9(8.84 \times 10^{-14})}{500 \times 10^{-8}} = 6.9 \times 10^{-8} \text{ F/cm}^2$$

The change in the threshold voltage is

$$\Delta V_T = \frac{\left[2(1.6 \times 10^{-19})(11.9)(8.84 \times 10^{-14})(2 \times 10^{16}) \right]^{1/2}}{6.9 \times 10^{-8}}$$
$$\cdot \left\{ [2(0.367) + 1.0]^{1/2} - [2(0.367)]^{1/2} \right\}$$
$$= 0.54 \text{ V}$$

9.5.3 Depletion and Enhancement MOSFETs

In the discussions of the MOSFET so far, we saw that as the gate voltage is increased, at some positive value V_T, inversion occurs and the device starts conducting or turns ON. This, of course, is not the only configuration in which the device can operate. It is possible to design devices that are ON when no gate bias is applied or are ON when negative bias is applied. This versatility is quite important since it gives a greater flexibility to the logic designer.

A device in which the current does not flow when the gate bias is zero, and flows only when a positive or negative gate bias is applied, is called an enhancement-mode device. Conversely, if the current flow occurs when the gate bias is zero and the device turns off when the gate bias is positive or negative, the device is said to operate in the depletion mode. The device we have discussed so far is an enhancement-mode device since, in our discussions, a positive gate bias was needed to cause inversion.

To produce a depletion-mode device that is ON without any gate bias, the MOSFET fabrication is altered. As shown in Fig. 9.16, one starts with a p-type substrate and two n^+ contacts are placed. Additionally, in the depletion-mode device, one diffuses a thin layer of donors to produce a thin n-type channel between the n^+ contacts. The rest of the MOSFET is produced in the normal way by placing an oxide layer and a gate. The I-V characteristics of such a device are also shown in Fig. 9.16b.

The device discussed above can be fabricated in p-type or n-type substrates. In this device one has free carriers due to the doping and therefore the device is ON even if the gate bias is zero. The gate bias can now be used to turn the device OFF as shown.

The MOSFET can be used as a switching element in the same way as the bipolar devices or other FETs. Regardless of whether the FET is an enhancement or a depletion device, the FET carries current in one of the states of the switch. This causes power dissipation in the circuits. This is of great concern when the circuits are dense and power dissipation can cause serious heating problems. This can be avoided by using the NMOS and PMOS devices together, as will be discussed next.

9.5.4 Complementary MOSFETs

It is possible to greatly reduce the power dissipation problem if an enhancement-mode n-channel device is connected to an enhancement-mode p-channel device in series. This is the complementary MOSFET or CMOS and is fabricated on the same chip, as shown in Fig. 9.17. In the CMOS inverter shown, the drains of the n- and p-MOSFET are connected and form the output. The input is presented to the gates of the device as shown. The p-channel device has a negative threshold voltage while the n-channel device has a positive threshold voltage.

When a zero input voltage V_{in} is applied, the voltage between the source and gate of the n-channel device is zero, turning it OFF. However, the voltage between the gate and source of the p-channel device is $-V$ since the source of the p-channel device is at $+V$. This turns the p-channel device ON. Thus the p-channel is ON and the n-channel is OFF so that the output voltage is $V_{out} = V$. No current flows in the devices since they are connected in series.

When a positive gate bias is applied, the n-channel device is ON, while the p-channel device is OFF. The output voltage $V_{out} = 0$. Once again no current flow occurs since the devices are in series and one of them is OFF. As can be seen, for input of 1 (High) or 0 (Low), one of the devices remains OFF. Thus the CMOS does not consume power when it is holding the information state. Only during switching is there a current flow. This low power consumption property of the CMOS makes it very attractive for high density applications, such as for semiconductor memories. However,

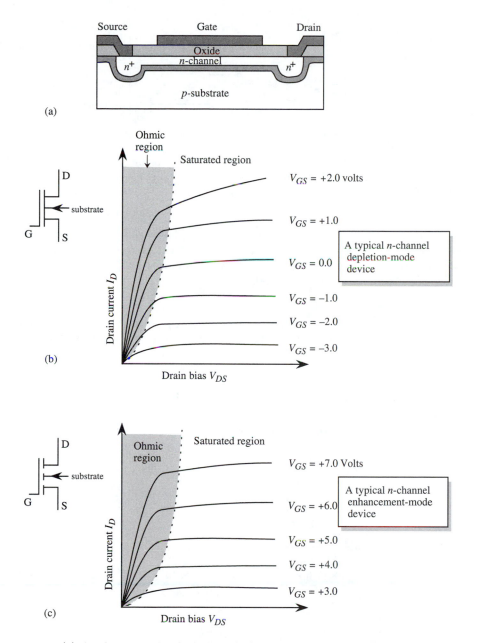

Figure 9.16: (a) A schematic of a depletion MOSFET fabricated in a *p*-type substrate, with an *n*-channel. (b) In the depletion mode, the device is ON at zero gate bias. To turn the device OFF, a negative gate bias is required as shown. The device symbol is also shown. (c) The I-V characteristics showing the device behavior in the enhancement mode. The device symbol is also shown.

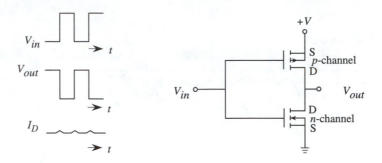

Figure 9.17: (a) A complementary MOS structure shown to function as an inverter. The circuit draws current only during the input voltage switching. (b) A schematic of the CMOS structure.

it must be noted that the device is much more complex to fabricate. Also, since the *p*-type transport is much poorer than the *n*-type transport, one has to take special care to design the two devices to have similar performances. In Chapter 10 we discuss applications of CMOS in digital and analog circuits.

EXAMPLE 9.11 An *n*-channel MOSFET is formed in a *p*-type substrate with a substrate doping of $N_a = 10^{14}$ cm^{-3}. The oxide thickness is 500 Å and $\phi_{ms} = -0.83$ V. Calculate the threshold voltage and check whether the device is an enhancement- or depletion-mode device. If the device threshold voltage is to be changed by 0.5 V by ion implanting the channel by dopants, calculate the density of dopants needed. Assume that the dopant charge is all placed near the Si-SiO$_2$ interface within a thickness of 0.1 μm. Temperature is 300 K.

The position of the Fermi level is given by

$$\phi_F = -0.026 \, \ell n \left(\frac{10^{14}}{1.5 \times 10^{10}} \right) = -0.228 \text{ V}$$

The threshold voltage is

$$
\begin{aligned}
V_T &= \phi_{ms} - 2\phi_F + \frac{\left[4e\epsilon_s N_a \, |\phi_F| \right]^{1/2}}{C_{ox}} \\
&= -0.83 + 0.456 + \frac{\left[4 \times (1.6 \times 10^{-19})(11.9 \times 8.85 \times 10^{-14})(10^{14})(0.228) \right]^{1/2} (5 \times 10^{-6})}{(3.9 \times 8.85 \times 10^{-14})} \\
&= -0.318 \text{ V}
\end{aligned}
$$

In this device there is an inversion layer formed even at zero gate bias and the device is in the depletion mode. To increase the threshold voltage by + 0.5 V, i.e., to convert the device into an enhancement-mode device, we need to place more negative charge in the channel. If we assume that the excess acceptors are placed close to the semiconductor-oxide region (i.e., within the distance W_{max}), the shift in threshold voltage is simply (N_a^{2D} is the areal density of the acceptors implanted)

$$\Delta V_T = \frac{e N_a^{2D}}{C_{ox}}$$

or

$$
\begin{aligned}
N_a^{2D} &= \frac{\Delta V_T C_{ox}}{e} = \frac{(0.5)(3.9 \times 8.85 \times 10^{-14})}{(1.6 \times 10^{-19})(5 \times 10^{-6})} \\
&= 2.16 \times 10^{11} \text{ cm}^{-2}
\end{aligned}
$$

The dopants are distributed over a thickness of 0.1 μm. The dopant density is then

$$N_a = \frac{2.16 \times 10^{11}}{10^{-5}} = 2.16 \times 10^{16} \text{ cm}^{-3}$$

The use of controlled implantation can be very effective in shifting the threshold voltage.

9.6 IMPORTANT ISSUES IN REAL MOSFETS

In the discussions above, we have made a number of simplifying assumptions. These assumptions allowed us to obtain simple analytical expressions for the I-V relationships for the device. However, in real devices a number of important effects cause the device behavior to differ from our simple results. In this section we will briefly examine the important issues that control the performance of real MOSFETs.

9.6.1 Important Effects in Long-Channel MOSFETs

We will now briefly discuss the important issues that control the performance of long-channel MOSFETs and are not covered under the simple model for I-V characteristics that we have discussed so far.

Subthreshold Conduction
In our simple analysis we assumed that the process of inversion is sudden; i.e., when the gate voltage exceeds the threshold voltage the channel is inverted and, if not, the channel is devoid of electrons. This, of course, is not really an accurate picture since the carriers in the channel change smoothly with gate bias according to the Fermi distribution function. As a result, at low applied biases we do not have the current-voltage relation given by Eqn. 9.40 (with V_{DS}^2 term ignored)

$$I_D = \frac{Z \mu_n C_{ox}}{L}(V_{GS} - V_T)V_{DS}$$

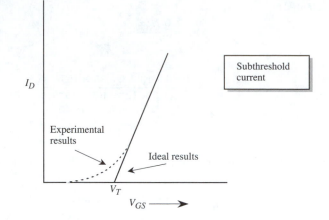

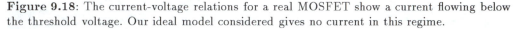

Figure 9.18: The current-voltage relations for a real MOSFET show a current flowing below the threshold voltage. Our ideal model considered gives no current in this regime.

The real current increases gradually as shown in Fig. 9.18. This phenomenon results in the undesirable situation in which the device is not turned off completely and consumes power even in the OFF state.

Mobility Variation with Gate Bias

In our simple model for carrier transport, we regarded the carrier mobility as having no dependence upon the gate bias. As the gate bias is changed the electron density in the channel changes. The electron density in turn is related to the surface field F_s by Eqn. 9.6. Thus, if the sheet charge n_s increases the surface field also increases. The increased electric field forces the electrons closer to the Si-SiO$_2$ interface. As a result, the electrons suffer a greater degree of scattering from the interface roughness and oxide impurities, and the mobility degrades.

Mobility Variation with Channel Field

The mobility of electrons (holes) in silicon is not independent of field but is high at low field and becomes smaller at high fields where the velocity saturates. As a result, the current calculated by our simple model is much larger than the current observed in real devices. More realistic device modeling approaches use a more accurate description of the velocity-field relationship. A common expression used for the velocity-field relation is (see Fig. 9.19a)

$$v(F) = \frac{\mu F}{1 + \dfrac{\mu F}{v_s}} \tag{9.49}$$

where v_s is the saturation velocity ($\sim 10^7$ cm/s). Use of this expression in calculating drain current causes a reduction in current by a factor of $\sim (1 + \mu V_{DS}/v_s L)$. In Fig. 9.19 we show a comparison of the current-voltage relations calculated using the constant-mobility model and the more accurate saturation velocity model.

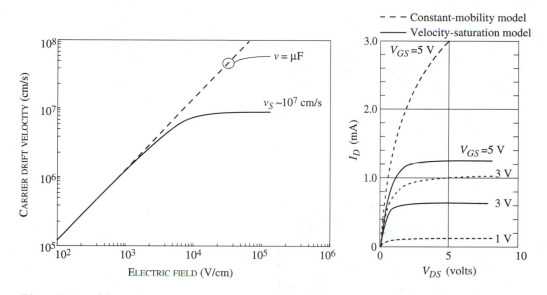

Figure 9.19: (a) Velocity-field relation for the constant-mobility model and saturation-velocity model. (b) $I_D - V_{DS}$ relations for a MOSFET using the constant-mobility model and the more accurate saturation-velocity model.

Channel Length Modulation in Saturation Region

In our simple model, once V_{DS} exceeds $V_D(sat)$ and the channel pinches off at the drain end, the current is assumed to remain independent of V_{DS}. The current in the channel is inversely proportional to the channel length. We have so far assumed that the channel length is the metallurgical channel length. However, the L that appears in the current-voltage relation represents the distance under the gate from the source side to the pinch-off point, as shown in Fig. 9.20a. As V_{DS} increases beyond $V_D(sat)$, the pinch-off point comes closer to the source side, thus effectively decreasing the channel length. This produces a change in the channel length ΔL (see Fig. 9.20b) and the current increases as

$$I_D = \frac{L}{L - \Delta L(V_{DS})} I_D(sat) \tag{9.50}$$

where $I_D(sat)$ is the current calculated assuming a fixed channel length. The effect results in an increase in the output conductance of the device. A similar effect occurs in MESFETs and JFETs, as was discussed in Section 8.4.2. It is common to represent the increase in drain current arising from channel-length modulation by an expression

$$I_D = I_D(L = \text{fixed})(1 + \lambda V_{DS}) \tag{9.51}$$

where $I_D(L = \text{fixed})$ is the current calculated assuming the channel length is fixed.

To a first approximation we can evaluate the change in effective channel length

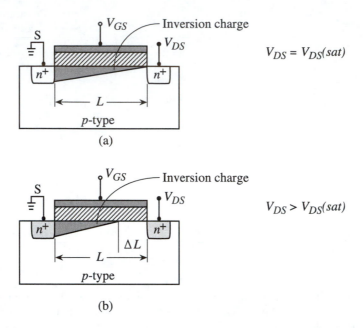

$V_{DS} = V_{DS}(sat)$

$V_{DS} > V_{DS}(sat)$

Figure 9.20: (a) A schematic of the MOSFET channel when $V_{DS} = V_{DS}(sat)$. (b) A schematic showing the decrease in the effective channel length for $V_{DS} > V_{DS}(sat)$.

by assuming that the excess potential ΔV_{DS} falls across the region L. This gives

$$\Delta L = \sqrt{\frac{2\epsilon}{eN_a}} \left[\sqrt{\phi_{fb} + V_{DS}(sat) + \Delta V_{DS}} - \sqrt{\phi_{fb} + V_{DS}(sat)} \right] \qquad (9.52)$$

where

$$\Delta V_{DS} = V_{DS} - V_{DS}(sat)$$

Radiation Effects

An important concern in real devices, once all the fabrication and packaging is finished, is the effect of radiation on device performance. MOS devices are subjected to ionizing radiation, especially when they are used in space applications. The radiation can lead to additional fixed charges as well as interface states in the MOS devices. The added charge may be due to holes generated by the radiation in the SiO_2 region and then getting trapped at impurity levels. The net result is that the threshold voltage can shift. A particularly serious problem arises if enhancement-mode devices are converted to depletion-mode devices. This could disrupt a particular circuit function and also cause unusually high power dissipation.

Breakdown Effects at High Bias

At high bias values of the gate or the drain, the channel current shows a runaway behavior and the device ceases to function as a transistor. The breakdown effect could be due to a number of causes. One cause is oxide breakdown, which can be particularly important if the oxide is very thin. The breakdown field of SiO_2 is $\sim 6 \times 10^6$ V/cm, which is a large field, but such high fields are produced since the oxide can be as narrow as 100 Å and the gate biases can approach 5-10 V.

Near the drain of the MOSFET the electric fields are quite high and an important source of breakdown is the impact ionization process where the hole carriers cause multiplication. In an n-channel MOSFET, the secondary electrons produced by impact ionization enter the drain while the holes move in the substrate and cause a substrate current. This substrate current can lead to a forward biasing of the source-substrate junction, causing electron injection into the substrate. This leads to a parasitic npn bipolar transistor action in the MOSFET. The main effect of the process is thus to limit the maximum drain voltage the MOSFET can tolerate.

A final cause for breakdown is the punchthrough effect. This occurs when the sums of the source and drain depletion regions exceed the channel length. The electrons are injected into the channel from the source and are swept by the electric field, to be collected at the drain. The important point here is that the current does not saturate and the transistor action thus depends critically on the drain bias.

9.6.2 Important Effects in Short-Channel MOSFETs

Advances in lithographic techniques are allowing MOSFET channel lengths to shrink to sizes below 1.0 μm. Experimental devices with channel lengths as small as 0.1 μm have been fabricated. The force for miniaturization is coming from the need for dense circuits for high-density memory and logic applications as well as from the need for high-frequency microwave devices. In short-channel devices the simple models we developed for the current-voltage characteristics become quite invalid for quantitative description. In addition to the effects discussed in the previous subsection for long-channel devices, specific issues relating to short-channel devices also play an important role.

Three-Dimensional Transport
In our simple model for the MOSFET current, we assumed that the current flow was one-dimensional and we could use the gradual channel approximation. For a very short-channel device, the current flow is not just parallel to the gate, but one has to consider the current flow from the source and drain side, which is highly two-dimensional. Also, if the gate width Z is small, the transport becomes truly three-dimensional, requiring enormous computations to do a proper device simulation.

Short Channel Effects
In our model for the MOSFET, we assumed that the gate area essentially covers the area of the active channel and ignored the effects of the source and drain depletion

regions. This is true for the long-channel device where the source and drain depletion lengths are small compared to the channel length. As the channel length is decreased, the source and drain depletion lengths can become a significant part of the channel length and the gate does not control the charge in these regions. The net effect is that the threshold voltage needed to cause inversion is reduced since the charge under the gate controls the threshold voltage. This causes an *n*-channel MOSFET to shift towards the depletion mode.

Hot Electron Effects

As the channel lengths shrink, the electric fields in the channel increase if the supply voltages are kept fixed. The charge becomes very "hot," i.e., acquires very high kinetic energies in such devices. These hot electrons can tunnel through the oxide barrier causing gate current. They can also cause deterioration of the device by breaking bonds in the semiconductor-oxide interface region. This damage is especially dangerous since over a period of time the device degrades and eventually the circuit based on the device loses its functionality. To avoid hot electron devices, MOSFETs are being designed so that the electric field does not become very large in any region of the device.

Radiation Effects

Radiation effects discussed in the context of long-channel MOSFETs become more critical when device dimensions decrease, and even small changes in oxide charge or interface states can cause serious effects. Radiation effects can also cause "soft errors" in which the state of memory devices can be altered by a γ-ray that generates a burst of charge.

9.6.3   Parasitic Bipolar Transistors and Latch-up in CMOS

CMOS circuits, while having the important benefit of low power consumption, have an important undesired property. This effect, known as latch-up, results from the presence of parasitic bipolar transistors present in integrated circuits. In Fig. 9.21 we show the origins of the parasitic bipolar transistors in a CMOS. We can see that in the CMOS there is an *npn* bipolar transistor and a *pnp* transistor. The resistances R_1, R_2, R_3, and R_4 are parasitic resistances associated with the *n*-substrate and *p*-well regions, as shown.

If we examine the two-terminal current between A and B as a function of bias, we find that up to a certain bias, V_L, the current is very low ($\sim \mu A$ range). However, above this critical voltage V_L (related to the punchthrough of the transistor, typically ~ 10 V), the two transistors start to conduct and the current rises abruptly to the level of milliamperes. The current is now controlled by the resistors R_3 and R_4. This phenomenon is called *latch-up*. Latch-up can occur whenever the voltages applied to input or output cause forward biasing of *pn* junctions in the devices and can cause permanent damage to the chips. To avoid latch-up it is important that device design be such that the bipolar transistor gain is low.

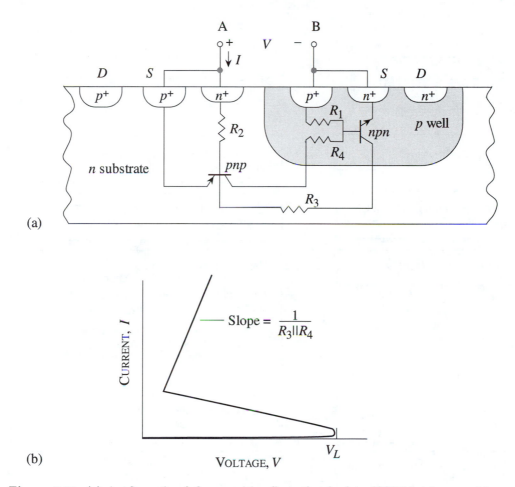

Figure 9.21: (a) A schematic of the parasitic effects that lead to CMOS latch-up problems. (b) Current versus voltage effect. The onset of latch-up is represented by a sharp rise in the parasitic current.

9.7 SUMMARY

In this chapter we have discussed the basic operating principles of one of the most important devices in solid state electronics. The MOS capacitor and the MOSFET are key devices in almost all electronic components. The summary table (Table 9.3) gives an overview of the important concepts covered in this chapter.

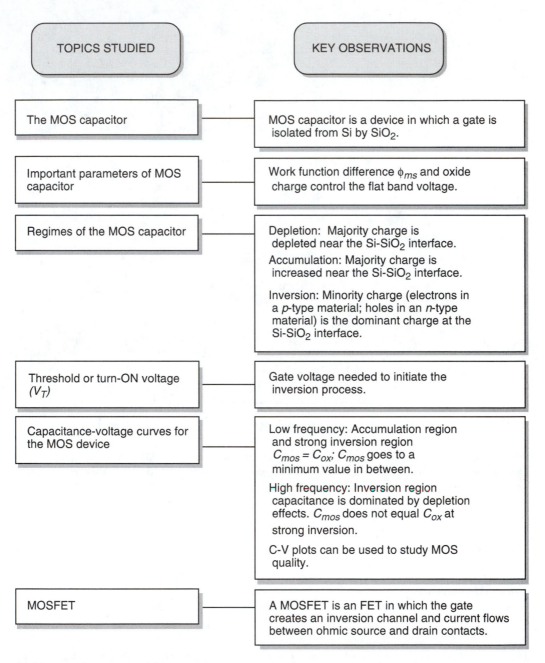

TOPICS STUDIED	KEY OBSERVATIONS
The MOS capacitor	MOS capacitor is a device in which a gate is isolated from Si by SiO_2.
Important parameters of MOS capacitor	Work function difference ϕ_{ms} and oxide charge control the flat band voltage.
Regimes of the MOS capacitor	Depletion: Majority charge is depleted near the Si-SiO_2 interface. Accumulation: Majority charge is increased near the Si-SiO_2 interface. Inversion: Minority charge (electrons in a p-type material; holes in an n-type material) is the dominant charge at the Si-SiO_2 interface.
Threshold or turn-ON voltage (V_T)	Gate voltage needed to initiate the inversion process.
Capacitance-voltage curves for the MOS device	Low frequency: Accumulation region and strong inversion region $C_{mos} = C_{ox}$; C_{mos} goes to a minimum value in between. High frequency: Inversion region capacitance is dominated by depletion effects. C_{mos} does not equal C_{ox} at strong inversion. C-V plots can be used to study MOS quality.
MOSFET	A MOSFET is an FET in which the gate creates an inversion channel and current flows between ohmic source and drain contacts.

Table 9.2: Summary table.

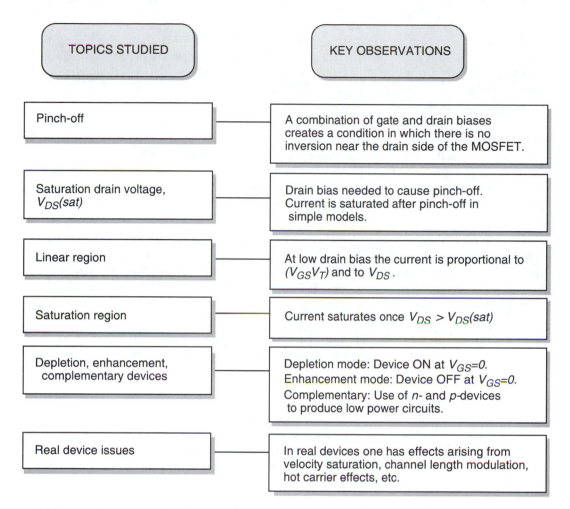

Table 9.2: Summary table (continued).

9.8 A BIT OF HISTORY

This section is common to Chapters 8 and 9.

The field-effect transistor is based on concepts much simpler than the bipolar transistor. The idea of controlling a current flowing in a device by modulating the charge density has been around for a long time. In 1926 Julius Lilienfeld patented this idea, although the technology was not available at the time to fabricate such a device. When serious work began at Bell Labs after the Second World War to develop an amplifying device based on semiconductors, the aim of the research group was to develop a field-effect device. However, the group ended up inventing the bipolar device. Research at

Bell Labs showed that it was not possible to build a field-effect transistor simply by putting a metal directly on germanium or silicon. The surface state density was simply too high. The need to place an insulator between the metal and the semiconductor was also identified.

Before the field-effect transistor could become a reality at Bell Labs, the spirit of entrepreneurship started to overtake the Bell Lab scientists. In the 1950s it was possible to start a transistor fabrication company without much capital, and many researchers were doing just that. In 1954 Shockley left Bell Labs to start his own company. He chose to be near his home town of Palo Alto and was among the first to establish a company in what is now known as Silicon Valley.

The material of choice for the bipolar transistor of the times ('50s) gradually shifted from germanium to silicon. Silicon was harder to grow as a single crystal and was more difficult to process, but the devices were more robust. Texas Instruments scientists W. A. Adcock, M. E. Jones, J. W. Thornhill, and E. D. Jackson were the first to successfully develop *npn* silicon transistors.

When Shockley left Bell Labs, he attracted a number of outstanding scientists and engineers to his new firm, among them Gordon Moore and Robert Noyce. However, Shockley's partners in this enterprise soon got tired of his research-focused approach and abandoned him. Noyce approached Fairchild Camera Company, which was a prominent company in the area, and Fairchild decided to fund the new venture into semiconductor devices. Fairchild developed the concept of mesa transistors, and to protect the transistors they used a discovery of scientists at Siemens and Western Electric (the development arm of Bell Labs). Fairchild started using the semiconductor oxide as a device-protection layer. This was a major advantage for silicon since germanium could not be oxidized. Fairchild's contribution eventually led to the planar technology.

With the development of the silicon dioxide as an insulator, the ingredients of a field-effect transistor were finally available. Steven Hofstein and Frederic Heimann, who worked at the RCA Labs, produced the MOS transistor in 1962. The concept of integrated circuits was already established by that time. The simplicity of the MOSFET was perfect for integrated circuits and RCA could, in 1962, place 16 MOSFETs on a 2.5 mm square chip. Also, self-aligned gate technology was developed at Hughes Research Labs, solving another key problem. The race for high-density chips was on!

The MOSFET, while simple, depended critically on the silicon-silicon oxide interface and it took many years of intense research to understand how to grow this interface reliably with very low trap states. With improvements in this interface, the MOSFET found a special place in the electronics industry.

Silicon has had a virtual monopoly on solid state devices, but there continues to be growing interest in other semiconductors. In 1951 Heinrich Welker showed that compound semiconductors such as GaAs and InP could do what Si and Ge were doing. He also realized that these new materials had mobilities superior to those of Si or Ge. These new compound semiconductors remained a curiosity for several decades, but as the limitations of silicon devices started to become evident, interest in them grew. The FETs made from GaAs started to find use in the microwave domain and gradually in logic applications for high speed-applications. Of course, these new materials could emit light, something that silicon could not do.

9.9 PROBLEMS

Note: Please examine problems in the Conceptual/Design Problems section as well.
(Assume a temperature of 300 K unless explicitly stated otherwise.)

Section 9.3
Problem 9.1 Calculate the maximum space charge width W_{max} in p-type silicon doped at $N_a = 10^{16}$ cm^{-3} at temperatures of 300 K and 77 K.
Problem 9.2 A p-type silicon has a uniform doping of $N_a = 10^{17}$ cm^{-3}. Calculate the surface potential needed to cause strong inversion.
Problem 9.3 Using the results given in Fig. 9.5, calculate the metal semiconductor work function for a p-type silicon with a gate of (a) n^+ polysilicon; (b) aluminum. Assume that $N_a = 5 \times 10^{15}$ cm^{-3}.
Problem 9.4 A 500 Å oxide is grown on p-type silicon with $N_a = 5 \times 10^{15}$ cm^{-3}. Assume that the oxide charge is negligible and calculate the surface potential and gate voltage to create inversion at the surface. Calculate the value of W_{max} for the device. The flat band voltage is -0.9 V.
Problem 9.5 An Al-gate MOS capacitor has an oxide thickness of 500 Å and an oxide charge density of 3×10^{11} cm^{-2}. The charge is positive. Calculate (a) the flat band voltage, (b) the turn-on voltage. Also, draw the energy band diagram and electric field profile of the structure at the onset of inversion. $N_a = 5 \times 10^{14}$ cm^{-3}.
Problem 9.6 An Al-gate n-channel MOS capacitor has a doping of $N_a = 10^{16}$ cm^{-3}. The oxide thickness is 500 Å and the flat band voltage is found to be $V_{fb} = -1.0$ V. Calculate the fixed oxide charge.
Problem 9.7. An Al-gate transistor is fabricated on a p-type substrate with an oxide thickness of 600 Å. The measured threshold voltage is $V_T = 1.0$ V, and the p-type doping is 5×10^{16} cm^{-3}. Calculate the fixed charge density in the oxide.

Section 9.4
Problem 9.8 An n-channel MOS capacitor has a doping of $N_a = 10^{15}$ cm^{-3}. The gate oxide thickness is 500 Å. Calculate the capacitances C_{ox}, C_{fb}, and C_{min} for the capacitor.
Problem 9.9 Show that if $\rho(x)$ is the distribution of charge density in the SiO$_2$ region

of thickness d_{ox}, the shift in the flat band voltage is given by

$$\Delta V_{fb} = -\frac{1}{C_{ox}} \int_0^{d_{ox}} \frac{x \rho(x) dx}{d_{ox}}$$

(Use Gauss' law for electric field due to a thin sheet of charge density. Then use the superposition principle.)

Problem 9.10 Calculate the shift in the flat band voltage using the result of Problem 9.9 for the following oxide charge distributions: (a) $Q'_{ss} = 10^{11}$ cm^{-2} is at the Si-SiO$_2$ interface; (b) the same charge is uniformly distributed in the oxide; (c) the charge is at the gate-SiO$_2$ interface. The oxide thickness is 500 Å. Assume that the charge is positive.

Section 9.5

Problem 9.11 Consider an n-channel MOSFET with a Z/L ratio of 15, a threshold voltage of 0.5 volt, mobility, $\mu_n = 500$ cm^2/V·s, and $d_{ox} = 700$ Å. Calculate the drain current and transconductance of the device (a) at $V_{DS} = 0.2$ V; (b) in the saturation region. The gate voltage is 1.5 V for both cases. Assume that the p-type doping is small.

Problem 9.12 Consider an ideal n-channel MOSFET with the following parameters:

$$
\begin{aligned}
\text{Flat band voltage,} \quad & V_{fb} & = & \quad -0.9 \text{ V} \\
\text{Channel width,} \quad & Z & = & \quad 25 \ \mu\text{m} \\
\text{Channel mobility,} \quad & \mu_n & = & \quad 450 \text{ cm}^2/ \text{ V} \cdot \text{s} \\
\text{Channel length,} \quad & L & = & \quad 1.0 \ \mu\text{m} \\
\text{Oxide thickness,} \quad & d_{ox} & = & \quad 500 \text{ Å} \\
\text{Channel doping,} \quad & N_a & = & \quad 5 \times 10^{14} \text{ cm}^{-3}
\end{aligned}
$$

Calculate and plot I_D versus V_{DS} for $0 \le V_{DS} \le 5$ V and for V_{GS} values of 0, 1, 2, 3 volts. Also, draw the locus of the $V_D(sat)$ points for each curve.

Problem 9.13 Consider an ideal p-channel MOSFET with the following parameters:

$$
\begin{aligned}
\text{Channel width,} \quad & Z & = & \quad 25 \ \mu\text{m} \\
\text{Channel mobility,} \quad & \mu_p & = & \quad 250 \text{ cm}^2/ \text{ V} \cdot \text{s} \\
\text{Channel length,} \quad & L & = & \quad 1.0 \ \mu\text{m} \\
\text{Oxide thickness,} \quad & d_{ox} & = & \quad 500 \text{ Å} \\
\text{Threshold voltage,} \quad & V_T & = & \quad -0.8 \text{ V}
\end{aligned}
$$

Calculate and plot I_D vs. V_{DS} for $-0.5 \le V_{DS} \le 0$ V for a gate bias of $V_{GS} = 0, -1, -2, -3$ V. Assume that the background doping is very small.

Problem 9.14 In the text we used the criterion that inversion occurs when $V_s = 2\phi_F$. Calculate the channel conductivity near the Si-Si0$_2$ interface for two MOS devices with the following parameters:

$$
\begin{aligned}
N_a & = & 5 \times 10^{13} \text{ cm}^{-3} & \quad\quad & N_a = 5 \times 10^{15} \text{ cm}^{-3} \\
\mu_n & = & 600 \text{ cm}^2/ \text{ V} \cdot \text{s} & \quad\quad & \mu_n = 600 \text{ cm}^2/ \text{ V} \cdot \text{s}
\end{aligned}
$$

The problem shows the rather arbitrary way of defining the inversion condition.

Problem 9.15 An n-channel and a p-channel MOSFET have to be designed so that they both have a saturated current of 5 mA when the gate-to-source voltage is 5 V for the n-MOS and -5 V for the p-MOS. The other parameters of the devices are:

$$
\begin{array}{llll}
\text{Oxide thickness,} & d_{ox} & = & 500 \text{ Å} \\
\text{Electron mobility,} & \mu_n & = & 500 \text{ cm}^2/\text{ V} \cdot \text{s} \\
\text{Hole mobility,} & \mu_p & = & 300 \text{ cm}^2/\text{ V} \cdot \text{s} \\
V_T \text{ for the } n\text{-MOS,} & & = & +0.7 \text{ V} \\
V_T \text{ for the } p\text{-MOS,} & & = & -0.7 \text{ V}
\end{array}
$$

What is the Z/L ratio for the n-MOSFET and the p-MOSFET?

Problem 9.16 An n-channel MOSFET has the following parameters:

$$
\begin{array}{llll}
\text{Oxide thickness,} & d_{ox} & = & 500 \text{ Å} \\
p\text{-type doping,} & N_a & = & 10^{16} \text{ cm}^{-3} \\
\text{Flat band voltage,} & V_{fb} & = & -0.5 \text{ V} \\
\text{Channel length,} & L & = & 1.0 \text{ } \mu\text{m} \\
\text{Channel width,} & Z & = & 15 \text{ } \mu\text{m} \\
\text{Channel mobility,} & \mu_n & = & 500 \text{ cm}^2/\text{ V} \cdot \text{s}
\end{array}
$$

Plot $\sqrt{I_D(sat)}$ versus V_{GS} over the range $0 \le I_D(sat) \le 1 \; mA$ for the source-to-body voltage of $V_{SB} = 0, 1, 2$ V.

Problem 9.17 Consider a p-channel MOSFET with oxide thickness of 500 Å, and $N_d = 10^{16}$ cm^{-3}. Calculate the body-to-source voltage needed to shift the threshold voltage from the $V_{BS} = 0$ results by -1.0 V.

Problem 9.18 A NMOS with V_T of 1.5 V is operated at $V_{GS} = 5$ V and $I_D S = 100 \; \mu$A. Determine if the device is in linear or saturation regime.

$$
k = \frac{\mu Z C_{ox}}{L} = 20 \; \mu A/V^2
$$

Problem 9.19 In the text we considered a criterion for inversion $V_s = 2\phi_F$. Consider another criterion that asserts that inversion occurs when the channel conductivity near the interface is $0.1(\Omega\text{cm})^{-1}$. Calculate the surface potential bending needed to satisfy this criterion when the channel has a p-type doping of: (a) 10^{14} cm^{-3}; (b) 10^{15} cm^{-3}; (c) 10^{16} cm^{-3}. Compare the surface bandbending arising from this new criterion to be a value of V_s given by the criterion used in the text ($\mu_n = 600$ cm^2/V·s).

Problem 9.20 Threshold bias for an n-channel MOSFET: In the text we used a criterion that the inversion of the MOSFET channel occurs when $V_s = \phi_s = 2\phi_F$ where $e\phi_F = (E_{Fi} - E_F)$. Consider another criterion in which we say that inversion occurs when the electron density at the Si/SiO$_2$ interface becomes 10^{16} cm^{-3}. Calculate the gate threshold voltage needed for an MOS device with the following parameters for the two different criteria:

$$
\begin{array}{lll}
d_{ox} & = & 500 \text{ Å} \\
\phi_{ms} & = & 1.0 \text{ V} \\
N_a & = & 10^{13} \text{ cm}^{-3}
\end{array}
$$

9.10 CONCEPTUAL/DESIGN PROBLEMS

Problem 1   Consider an n-MOSFET made from Si-doped p-type at $N_a = 10^{16}$ cm^{-3} at 300 K. The source and drain contacts are ohmic (negligible resistance) and are made from n^+-doped regions. The other parameters for the device are the following:

$$
\begin{aligned}
V_{fb} &= -1.0 \text{ V} \\
\mu_n &= 500 \text{ cm}^2 \text{ V}^{-1} \text{ s}^{-1} \\
\mu_p &= 100 \text{ cm}^2 \text{ V}^{-1} \text{ s}^{-1} \\
\text{Gate length} &= 2.0 \ \mu\text{m} \\
\text{Gate width} &= 20.0 \ \mu\text{m} \\
d_{ox} &= 500 \text{ Å}
\end{aligned}
$$

(a) Calculate the channel conductivity near the Si-SiO$_2$ interface under flat band condition and at inversion. Use the condition $V_s = 2\phi_F$ for inversion.

(b) Calculate the electron and hole densities at the Si-SiO$_2$ interface on the source and drain side of the gate when the gate bias is $V_T + 0.5$ V and $V_{DS} = 1.0$ V.

(c) Calculate the saturation current in the channel for the gate bias specified above.

(d) If the gate voltage is such that the Si bands are flat, *estimate* the current density in in the channel for a drain bias of 1.0 V.

Problem 2   Consider an n-MOSFET made from Si-doped p-type at $N_a = 5 \times 10^{16}$ cm^{-3} at 300 K. The other parameters for the device are the following:

$$
\begin{aligned}
V_{fb} &= -0.5 \text{ V} \\
\mu_n &= 600 \text{ cm}^2 \text{ V}^{-1} \text{ s}^{-1} \\
\mu_p &= 100 \text{ cm}^2 \text{ V}^{-1} \text{ s}^{-1} \\
\text{Gate length} &= 1.5 \ \mu\text{m} \\
\text{Gate width} &= 50.0 \ \mu\text{m} \\
d_{ox} &= 500 \text{ Å}
\end{aligned}
$$

The inversion condition is $V_s = 2\phi_F$.

(a) Calculate the threshold voltage V_T.

(b) Calculate the channel current when the gate bias is $V_T + 1.5$ V and the drain bias is 1.0 V.

(c) Estimate the *ratio of the electron velocities* in the channel on the source side and the drain side of the gate for the biasing in part (ii).

Problem 3   Consider an n-MOSFET made from Si doped p-type at $N_a = 5 \times 10^{16}$ cm^{-3} at 300 K. The other parameters for the device are the following:

$$
\begin{aligned}
V_{fb} &= -0.5 \text{ V} \\
\mu_n &= 600 \text{ cm}^2 \text{ V}^{-1} \text{ s}^{-1}
\end{aligned}
$$

$$\mu_p = 200 \text{ cm}^2 \text{ V}^{-1} \text{ s}^{-1}$$
$$\text{Gate length} = 1.5 \ \mu\text{m}$$
$$\text{Gate width} = 50.0 \ \mu\text{m}$$
$$d_{ox} = 500 \text{ Å}$$

The inversion condition is $V_s = 2\phi_F$. Assume that the electrons induced under inversion are in a region 200 Å wide near the Si/SiO$_2$ interface.

(a) Calculate the channel conductivity near the Si-SiO$_2$ interface under flat band condition and at inversion. Use the condition $V_s = 2\phi_F$ for inversion.

(b) Calculate the threshold voltage.

Problem 4 Consider an n-MOSFET at room temperature made from Si-doped p-type. To characterize the device C-V measurements are done for the MOS capacitor. It is found from the low-frequency measurements that the maximum and minimum capacitances per unit area are 1.72×10^{-7} F/cm^2 and 2.9×10^{-8} F/cm^2. The other parameters for the device are the following:

$$\mu_n = 600 \text{ cm}^2 \text{ V}^{-1} \text{ s}^{-1}$$
$$\text{Gate length} = 1.5 \ \mu\text{m}$$
$$\text{Gate width} = 50.0 \ \mu\text{m}$$

(a) Calculate the oxide thickness.

(b) *Estimate* the p-doping level in the channel.

(c) Calculate the channel current at saturation when the gate bias is $V_T + 1.5 \ V$.

9.11   READING LIST

- **General**

    - E. H. Nicollian and J. R. Brews, *MOS Physics and Technology* (Wiley, New York, 1982).

    - D. A. Neamen, *Semiconductor Physics and Devices: Basic Principles* (Irwin, Boston, MA, 1997).

    - R. F. Pierret, *Field Effect Devices* (Vol. 4 of the Modular Series on Solid State Devices, Addison-Wesley, Reading, MA, 1990).

    - S. M. Sze, *Physics of Semiconductor Devices* (Wiley, New York, 1981).

    - W. M. Werner, "The Work Function Difference of the MOS System with Aluminum Field Plates and Polycrystalline Silicon Field Plates" *Solid State Electronics*, **17**, *769-75* (1974).

    - M. Zambuto, *Semiconductor Devices* (McGraw-Hill, New York, 1989).

Chapter

10

MOSFET: TECHNOLOGY DRIVER

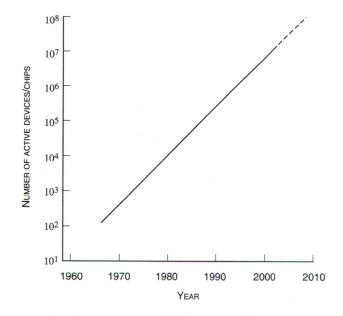

Figure 10.1: Complexity of MOSFET-based chips versus time.

10.1 INTRODUCTION

Since the mid-1960s MOS technology has been improving at an incredible rate. It has become the most important technology for information-processing systems, making up over ninety percent of the semiconductor market share. The remarkable saga of the MOSFET is captured very descriptively by two laws identified by Gordon Moore, the co-founder of Intel. According to Moore's first law, the complexity of (number of active elements on) a chip doubles every eighteen months. The second law states that the cost of a fabrication facility grows on a semi-log scale. The experimental realization of this law is shown in Fig. 10.1. Electrical engineers around the world have succeeded in keeping the first law valid—a remarkable fifty-nine percent per year increase in complexity has been achieved over three decades! Engineers are also trying their hardest to avoid the implications of the second law, according to which semiconductor factories would cost $250 billion by the year 2010. The ability to scale the MOSFET affordably and to decrease the power consumption in CMOS is responsible for the dominance MOSFETs enjoy in integrated circuits. Throughout its history, the MOS industry has achieved a cost reduction of 25-30 percent per year per function—a remarkable achievement. Is MOS technology versatile enough to continue at the pace required for Moore's first law for the next twenty years? Unfortunately, the answer seems to be "no." It is quite clear from simple physics that drastic changes in material and technology are needed by the year 2002 if MOSFET is to avoid running out of steam.

In this chapter we will examine the use of the MOSFET in digital and analog

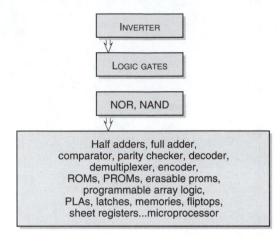

Figure 10.2: A flow chart showing how inverters are used to build higher-level circuits for digital applications.

applications. We will also discuss MOSFET scaling laws and the technology roadmap outlined by Semiconductor Industry Associates (SIA).

10.2   MOSFET IN THE DIGITAL WORLD

In this section we will make a brief survey of important uses of the MOSFET in the digital world. The MOSFET is by far the leader as far as device technologies for digital applications are concerned. Other than very high-speed applications, where GaAs technology is dominant or BJTs and HBTs are used, most digital devices are based on the MOSFET. Digital technology relies on the ON and OFF state of the transistor. These two states can be exploited using binary mathematics to do complex arithmetic operations. The basic building block of the inverter technology is the inverter in which a high input (1) to the gate results in a low output voltage (0) and vice versa. As shown in Fig. 10.2, inverters can be combined to produce logic gates, which can be used to produce various operations like NOR, NAND, etc., upon which Boolean algebra is based. Entire arrays of circuits used in digital technology can be designed using proper combinations of logic gates.

In Fig. 10.3 we show the input-output relations for an inverter. The transfer characteristics are in general represented by five critical points as shown in the figure. These are:

- V_{OH}: The output high value
- V_{OL}: The output low value
- V_{IH}: The high input value at which the slope of the $V_{out} - V_{in}$ curve becomes -1

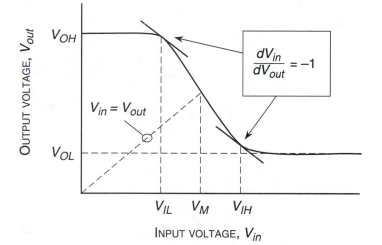

Figure 10.3: Input-output relation for an inverter. The various terms are explained in the text.

- V_{IL}: The low input value at which the slope of the $V_{out} - V_{in}$ curve becomes -1
- V_M: The point on the transfer curve where $V_{in} = V_{out}$

In Example 10.1 we discuss how one calculates these critical points for a device.

The voltage difference $V_{IH} - V_{IL}$ is an important performance parameter for the inverter. This value should be as small as possible so that fluctuations in the maximum and minimum input values would not result in possible errors in the output state. This means the output value should drop sharply from high to low over a small range of input signal. Similarly, $V_{OH} - V_{OL}$ should be as large as possible so that the output voltage, which is often used as an input to other devices, has a distinct state.

10.2.1   MOSFET as a Load

One great advantage of the MOSFET in IC technology is that it can be used not only as a transistor, but also as a resistor and a capacitor. Before discussing the use of a MOSFET as an inverter, let us consider how a MOSFET can be used as a load for the inverter circuit. Use of a MOSFET as a load has advantages in IC technology. It is difficult to make resistors with values of more than ~ 10 KΩ without taking up too much chip area. In Fig. 10.4 we show an approach in which the gate is tied to the drain to produce a MOSFET that can then be used as a resistive load. The resistance is given by the locus of points on the device I-V characteristics for which $V_{GS} = V_{DS}$. As can be seen from Fig. 10.4, the resistance is nonlinear—a fact that is quite useful in circuit

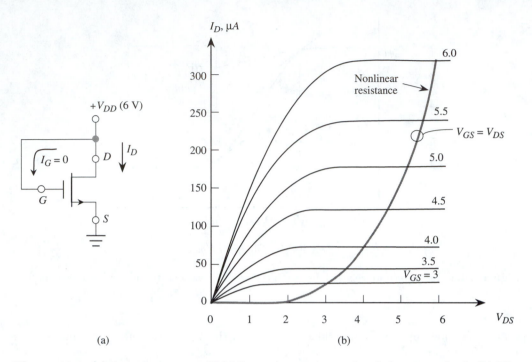

Figure 10.4: (a) An enhancement NMOS transistor connected to behave as a load. (b) The nonlinear resistance characteristic of the circuit in part (a).

applications. Note that the device always operates in the saturation region, since

$$V_{GS} - V_T < V_{DS} \tag{10.1}$$

In Example 10.2 we discuss how such a load is useful in an inverter circuit. A depletion MOSFET can also be used as a load resistance in the configuration shown in Fig. 10.5. Here the source and gate are tied so that $V_{GS} = 0$ defines the resistance of the device. Once again, we can see that the resistance of the MOSFET is quite nonlinear and, most importantly, it can be made to be quite large.

10.2.2   MOSFET as an Inverter

The most important application of the MOSFET is as a switch in digital technology. In Fig. 10.6a we show a circuit in which a resistive load is used to design a switch. As noted above, in IC technology the load is usually another MOSFET. In an ideal case we can consider an input signal to the gate of the MOSFET switching as shown in Fig. 10.6a. To understand the signal at the output we draw the load line on the I-V characteristics of the transistor. The load line in a graphical representation of the equation

$$I_D R_L + V_{DS} = V_{DD} \tag{10.2}$$

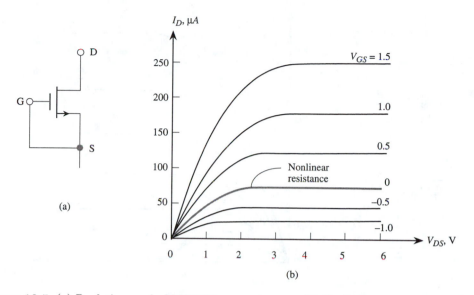

Figure 10.5: (a) Depletion-mode MOSFET as a resistance. (b) Nonlinear characteristics (V_{GS} = 0).

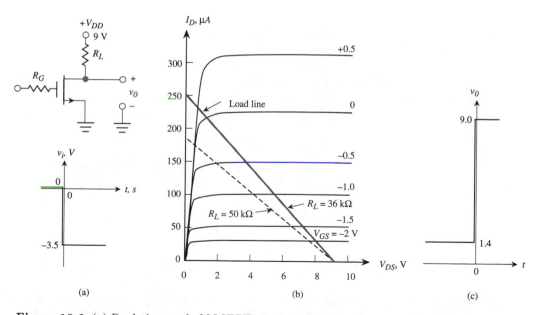

Figure 10.6: (a) Depletion-mode MOSFET circuit and input voltage waveform. (b) MOSFET output characteristics (the load lines correspond to V_{DD} = 9 V and R_L = 36 KΩ (solid) and R_L = 50 KΩ (dashed)). (c) The output voltage waveform.

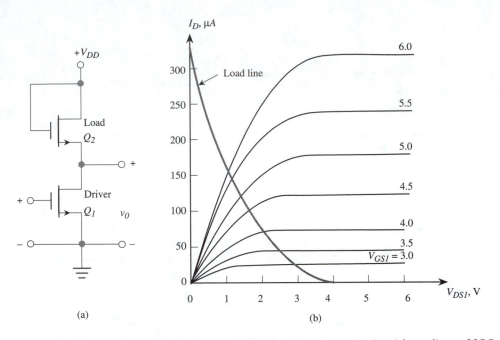

Figure 10.7: (a) Metal-oxide semiconductor field-effect transistor circuit with nonlinear MOS-FET load resistance. (b) MOSFET output characteristics showing nonlinear load line.

The load line is just linear, intersecting the V_{DS} axis at $(I_D = 0)$ V_{DD} and the I_D axis $(V_{DS} = 0)$ at $I_D = V_{DD}/R_L$.

We note that the MOSFET of Fig. 10.6 is a depletion MOSFET. When the input voltage is 0 the device is turned ON and the output voltage from the load line is 1.4 volt. When the input voltage is -3.5 volt, the device is turned OFF and the output voltage is V_{DD} or 9.0 volt.

In Figs. 10.7 and 10.8 we show the MOSFET switch with a MOSFET load. In these cases the load line must be computed from the nonlinear resistance. In Examples 10.2 and 10.3 we consider how these circuits function.

10.2.3   NMOS Logic Gates

The NMOS inverters discussed here can be exploited to form logic gates by using multiple devices that feed a single load. In Fig. 10.9a we show a two-input NOR gate. The circuit consists of two identical NMOS enhancement drivers and a depletion NMOS acting as load. If either or both of the inputs is at $V(1)$, the output is $v_0 = 0$. If both inputs are at $V(0)$, the output is V_{DD}. The truth table of the NOR gate is shown in

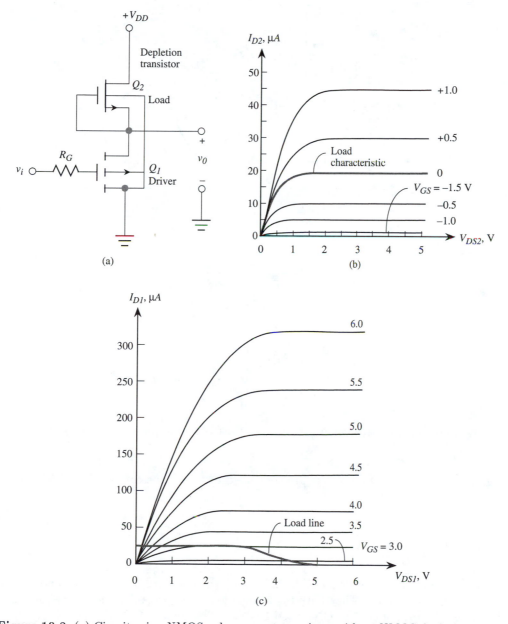

Figure 10.8: (a) Circuit using NMOS enhancement transistor with an NMOS depletion transistor connected as the load. (b) The depletion load resistance characteristic. (c) The load line corresponding to part (b) plotted on the output curves of Q_1.

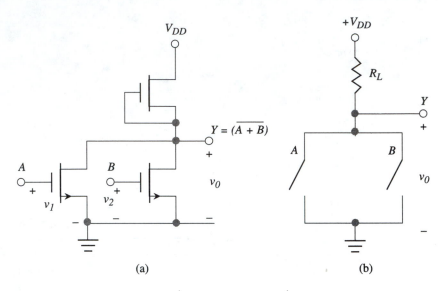

(a)   (b)

A		B		Y	
v_1	State	v_2	State	v_0	State
$\leq V_{IL}$	0	$\leq V_{IL}$	0	$\geq V_{OH}$	1
$\leq V_{IL}$	0	$\geq V_{IH}$	1	$\leq V_{OL}$	0
$\geq V_{IH}$	1	$\leq V_{IL}$	0	$\leq V_{OL}$	0
$\geq V_{IH}$	1	$\geq V_{OL}$	1	$\leq V_{OL}$	0

(c)

Figure 10.9: (a) An NMOS NOR gate. (b) Idealized representation of (a). (c) Truth table.

Fig. 10.9c. If the NMOS inverters are connected in series, one can obtain NAND logic.

10.2.4   CMOS Inverter

Complementary metal-oxide FETs have become the standard for digital technology in a wide range of systems. They have the tremendous advantage that they have virtually no power dissipation in the ON or OFF state. This makes CMOS technology especially useful for applications in mobile computing and communications, where power consumption is a critical issue.

The basic CMOS inverter is shown in Fig. 10.10. The circuit consists of an NMOS driver connected in series to a PMOS load. The drains of the two devices are

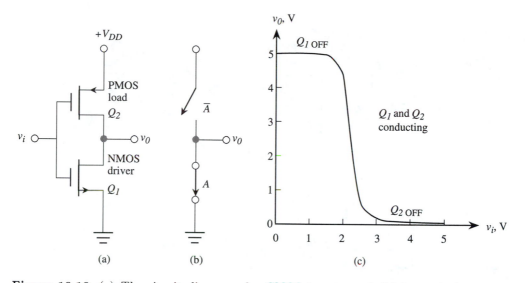

Figure 10.10: (a) The circuit diagram of a CMOS inverter and (b) its equivalent switch representation. (c) The voltage transfer characteristic for (a) with $V_{DD} = 5$ V and threshold voltages of 2 V (Q1) and -2 V (Q2).

connected. The input signal v_i is applied to both transistors and varies from $V(0) = 0$ V to $V(1) = V_{DD}$. When $v_i = 0$, the transistor Q_1 is OFF, while $V_{GS2} = -V_{DD}$ and the PMOS device is ON. Since two devices are in series, the current $I_{D1} = I_{D2} = 0$. Thus the PMOS device operates at the origin of the I-V characteristics. The input-output relation is

$$v_i = 0 \ : \ v_0 = V_{DD} \tag{10.3}$$

When $v_i = V_{DD} = V_{GS1}$, the transistor Q_1 is ON, but Q_2 is OFF, since $V_{GS2} = 0$. Thus the current is again zero and Q_1 operates near the origin of its I-V characteristics. We now have

$$v_i = V_{DD} \ ; v_0 = 0 \tag{10.4}$$

We see that there is essentially no current in the CMOS inverter in the ON or OFF state (in reality, there is a small leakage current). As a result, the power dissipation in the circuit is extremely small.

10.2.5 CMOS Logic Gates

One can use the CMOS inverter to make logic gates such as NAND and NOR. The advantage of the CMOS technology is the low power consumption, although since both

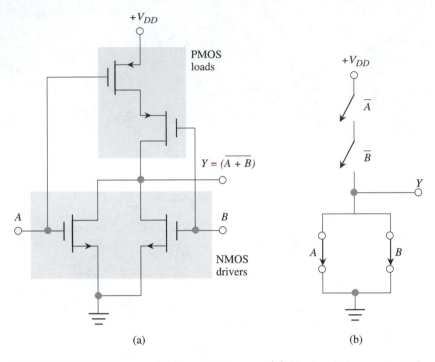

Figure 10.11: CMOS NOR gate. (a) Circuit diagram. (b) Ideal switch representation.

NMOS and PMOS devices are needed, the component density and speed are not as high as one would get in the NMOS technology.

In Fig. 10.11a we show a CMOS NOR gate. The drivers are connected in parallel, while the loads are connected in series, as shown. The circuit representation is shown in Fig. 10.11b.

If A or B are at logic 1, one or the other of the PMOS leads (which are in series) represents an open switch. The output is at ground. If both inputs are at $V(0)$, the PMOS switches are closed and $v_0 = V_{DD}$.

EXAMPLE 10.1 Consider an inverter in the configuration shown in Fig. 10.6. Find the critical voltages for the voltage transfer curve. Also find the noise margin for the inverter. Assume the following values for the device:

$$
\begin{aligned}
k' &= 25 \ \mu\text{A/V}^2 \\
V_T &= 1.0 \ \text{V} \\
Z/L &= 2.0 \\
V_{DD} &= 5.0 \ \text{V} \\
R_L &= 100 \ k\Omega
\end{aligned}
$$

To address this problem we first establish some simple relations for the critical voltages.

(1) V_{OH}: The high output is just $V_{DD} = 5.0$ V and arises when the device is OFF and no current flows in the inverter.

(2) V_{OL}: In this case the device is ON and in the ohmic region. The currents through the load and the transistor are equal. We assume that the device is receiving an input from another inverter with $V_{GS} = V_{OH} = V_{DD}$. This gives ($k = Zk'/L$)

$$k(V_{OH} - V_T)V_{OL} = \frac{V_{DD} - V_{OL}}{R_L}$$

or

$$
\begin{aligned}
V_{OL} &= \frac{V_{DD}}{1 + kR_L(V_{DD} - V_T)} \\
&= \frac{(5.0 \text{ V})}{1 + (2 \times 25 \times 10^{-6} \times 10^5)(4.0)} = 0.24 \text{ V}
\end{aligned}
$$

(3) V_{IL}: The low input value where the transistor is in the saturation region and the output resistance r_{DS} is essentially infinite. The volume of V_{IL} is defined as the point where

$$
\begin{aligned}
\frac{dV_{out}}{dV_{in}} &= -1 \\
\frac{dV_{out}}{dV_{in}} &= \frac{-dI_D}{dV_{in}} \times (R_L \| r_{DS})
\end{aligned}
$$

Using $(R_L \| r_{DS}) = R_L$ we get, using $V_{in} = V_{GS} = V_{IL}$,

$$
\begin{aligned}
k(V_{IL} - V_T)R_L &= 1 \\
V_{IL} = \frac{1}{kR_L} + V_T &= 1.2 \text{ V}
\end{aligned}
$$

(4) V_{IH}: At this point, once again the slope of the voltage transfer function is -1. The device is in the ohmic region and r_{DS} is small. Thus $(R_L \| r_{DS}) \sim r_{DS}$.

$$\frac{dV_{out}}{dV_{in}} = \frac{-dI_D}{dV_{in}} \cdot r_{DS} = -1$$

Using

$$r_{DS} = \frac{dV_{out}}{dI_D}$$

and the current-voltage relations in the ohmic region, we get

$$\frac{kV_{DS}}{k(V_{GS} - V_T - V_{DS})} = 1$$

Also, since $V_{in} = V_{GS}$ and $V_{out} = V_{DS}$, from the equation above we get the relation between V_{out} and V_{in} at this critical point:

$$V_{out} = \frac{V_{in} - V_T}{2}$$

Equating the current flowing in the MOSFET and the load, we get

$$\frac{k}{2}\left[2(V_{IH} - V_T)V_{out} - V_{out}^2\right] = \frac{V_{DD} - V_{out}}{R_L}$$

Using the relation between V_{out} and $V_{in}(= V_{IH})$, we have

$$\frac{k}{2}\left[\frac{3}{4}\left(V_{IH} - V_T\right)^2\right] + \frac{V_{IH} - V_T}{2R_L} - \frac{V_{DD}}{R_L} = 0$$

Solving for V_{IH} for the device in this example, we get

$$V_{IH} - V_T = \frac{-0.2 \pm 2.45}{1.5}\,\text{V}$$

The positive (physically correct) solution is

$$V_{IH} = 1.0 + 1.5 = 2.5 \text{ V}$$

(5) V_M: At this point $V_{in} = V_{out}$. Equating the current in the load and the transistor yields

$$\frac{k}{2}\left(V_M - V_T\right)^2 = \frac{V_{DD} - V_M}{R_L}$$

or

$$V_M^2 - 2V_M\left(V_T - \frac{1}{kR_L}\right) + \left(V_T^2 - \frac{2V_{DD}}{kR_L}\right) = 0$$

For our device we get (keeping the positive solution)

$$V_M = 2.08 \text{ V}$$

We finally have for the noise margins of the device

$$NM_L = V_{IL} - V_{OL} = 1.2 - 0.24 \quad = \quad 0.96 \text{ V}$$
$$NM_H = V_{OH} - V_{IH} = 5.0 - 2.5 \quad = \quad 2.5 \text{ V}$$

An important point to note in this example is that the large 100 kΩ resistance (needed for low power consumption) would require a very large chip area. A resistor of this value could easily take up an area 100 times that of the transistor. This is the reason one avoids using a resistor as a load in an inverter.

EXAMPLE 10.2 Consider the inverter transistor of Example 10.1. Assume that an enhancement MOSFET is used in load as shown in Fig. 10.7. Discuss the device parameters of the load needed to provide the same current for $V_{DD} = 5$ V and V_{OL} value as in the inverter of Example 10.1.

In the inverter of Example 10.1 we considered a load of 100 kΩ and discussed the area requirement for such a resistor. If we use a MOSFET with $V_{GS} = V_{DS}$, we have (the device is in saturation or cutoff), at $V_{out} = V_{OL}$ and an equivalent $R_L = 100$ kΩ,

$$\frac{k_L}{2}\left(V_{GS} - V_T\right)^2 = \frac{V_{DD} - V_{OL}}{R_L} = \frac{5 - 0.24}{10^5} = 47.6 \text{ }\mu\text{A}$$

Here k_L is the parameter for the load transistor. Using the value of V_{GS} for the local device of $V_{DD} - V_{OL} \sim 4.7$ V, we get

$$k_L = \frac{2(47.6 \times 10^{-6} \text{ A})}{(4.7 - 1.0 \text{ V})^2} = 6.95 \text{ }\mu\text{A/V}^2$$

In Example 10.1, the k value for the active transistor is 50 $\mu A/V^2$. Since the value of k' is the same (25 $\mu A/V^2$) in the load transistor, we have for the load

$$\frac{Z}{L} = \frac{6.95}{25} = 0.28$$

For the active device discussed in Example 10.1, Z/L is 2. Consider a case where the smallest dimension possible is 3 μm. Then we could have, for the inverter MOSFET,

$$L = 3 \ \mu\text{m}; \ Z = 6 \ \mu\text{m}$$

and for the load (approximately)

$$L = 10 \ \mu\text{m}; \ Z = 3 \ \mu\text{m}$$

The load area is about twice that of the inverter. On the other hand, if a resistor were used, the resistive load can have an area 100 times that of the transistor.

EXAMPLE 10.3 Consider an NMOS inverter with a saturated enhancement load. Assume that $V_{TO} = 1.5$ V, $2\phi_F = 0.7$ V, and $\gamma = 0.4 \ V^{1/2}$. Calculate the output high of the inverter if $V_{DD} = 5$ V. The body (of both devices) is grounded.

   The output high of the inverter V_{OH} is limited by the condition that the load is conducting. We have

$$
\begin{aligned}
V_{OH} &= V_{DD} - V_T(V_{OH}) \\
&= V_{DD} - \left[V_{TO} + \gamma \left(\sqrt{V_{OH} + 2\phi_F} - \sqrt{2\phi_F} \right) \right]
\end{aligned}
$$

Using the values of the parameters given, we get

$$V_{OH} = 3.16 + 0.4\sqrt{V_{OH} + 1.4}$$

Solving for V_{OH}, we get

$$V_{OH} = 4.1 \text{ V}$$

We see that the high value of the output in this device is not V_{DD}, but 4.1 V. This is a disadvantage of the enhancement load.

   If a depletion device is used as the load along with an enhancement transistor (as shown in Fig. 10.8), the value of V_{OH} is V_{DD}. Thus by combining a depletion and enhancement device, one is able to improve the inverter performance.

10.3   MOSFET AS AN AMPLIFIER

One of the important applications of the MOSFET (and other transistors, such as BJTs, MESFETs, etc.) is the amplification of weak signals. In the MOSFET the input signal is applied to the gate and controls the current output of the device. In order for the

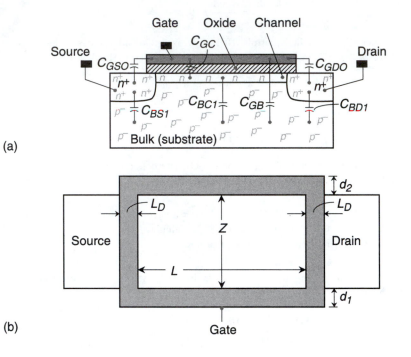

Figure 10.12: (a) Parasitic capacitances in MOS transistors. (b) A top view showing the overlap of the gate, source, and drain regions. (c) Values for parasitic capacitances under various operating conditions.

transistor to operate as an amplifier, it has to be biased in the saturation region so that the current output has minimal dependence on the drain-source potential.

In the high-frequency regime parasitic capacitances present in the MOSFET structure play a dominant role. In Fig. 10.12a we show a schematic of a MOSFET along with the parasitic capacitances that dominate device performance at high frequencies. The capacitances C_{BD1}, C_{BC1}, and C_{BS1} represent the capacitances of the junctions between the body and the drain, channel, and source, respectively, as shown. Important capacitances are C_{GSO} and C_{GDO}, which are the gate-to-source and gate-to-drain overlap capacitances. They arise because, as shown in Fig. 10.12b, there is a diffusion of the dopants from the source and drain region under the gate. If L_D represents the diffusion under the gate, then, as shown in Fig. 10.12b, the capacitance due to this overlay is

$$C_{GDO} = C_{GSO} = C_{ox} Z L_D \qquad (10.5)$$

In the cutoff region gate-to-channel capacitance C_{GC} is zero. In the ohmic region the channel is quite uniform and the gate-to-channel capacitance is

$$C_{GC} = C_{ox} Z L \qquad (10.6)$$

Half of this capacitance is distributed between the gate and source and the other half is included in the gate-to-drain capacitance. Once the device goes to saturation, the gate-to-channel capacitance on the drain side is neglected, while the gate on the source side, typically $2/3\ C_{GC}$, is added to the gate capacitance (see Fig. 10.12c).

When the device is in the cutoff region there is no inversion layer so that the bulk region extends to the Si/SiO$_2$ interface and the gate-to-body capacitance is

$$C_{GB} = C_{ox} Z L \qquad (10.7)$$

In the saturation and ohmic regions the inversion layer is formed and C_{GB} becomes negligible.

The capacitance C_{BC1} is the bulk-to-channel junction capacitance. In the cutoff region there is no inversion layer so that C_{BC1} is zero. In the ohmic and saturation regime, the bulk-to-channel capacitance exists and is given by the junction capacitances of the p-n region. The bulk-to-channel capacitance is distributed across the bulk-to-source and bulk-to-drain capacitance, as shown in Fig. 10.12c.

Finally, the capacitances C_{BS1} and C_{BD1} are capacitances associated with the bulk-source and bulk-drain junctions. The total bulk-source and bulk-drain capacitances include contributions from the bulk-channel junction, as shown in Fig. 10.12c. In Section 10.4 we present some typical values of these capacitances. In Appendix D the capacitance values are calculated for a typical device process.

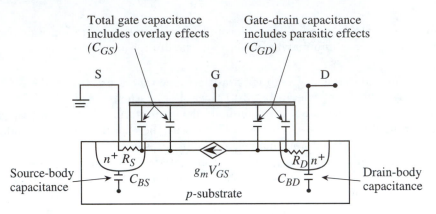

Figure 10.13: The source of various resistances and capacitances in a MOSFET. The gate-source capacitance C_{GS} and the gate-drain capacitance G_{GD} include the capacitances due to the oxide overlay as shown. The overlay capacitance is a serious source of degradation of high-frequency performance and must be minimized.

The high-frequency, high-speed issues of the MOSFET are controlled by the capacitance charging times and the transit time of carriers through the channel. To describe the small-signal response of a MOSFET we need to develop an equivalent circuit for the device. The various components of such a circuit are shown in Fig. 10.13. The various capacitances have been discussed above. R_S and R_D are the series resistances associated with the source and the drain. The voltage V'_{GS} is the internal voltage appearing between the source and the gate and is related to the external gate voltage V_{GS} by the IR drop across the source resistance.

The equivalent circuit for a common source MOSFET is shown in Fig. 10.14a. The parameter g_D^{-1} represents the slope of the $I_D - V_{DS}$ curves of the MOSFET. For an ideal MOSFET, in saturation this slope is zero and g_D^{-1} is infinite. However, as discussed in the previous chapter, gate length modulation effects cause g_D^{-1} to be finite.

Let us consider a simple model of the MOSFET to examine the high-frequency cutoff frequency. We ignore R_S, R_D, and C_{DS} to develop a simpler equivalent circuit shown in Fig. 10.14b. If we examine the currents at the input gate node, we get (for a signal to the gate of frequency ω)

$$I_i = j\omega C_{GS} V_{GS} + j\omega C_{GD} \left(V_{GS} - V_{DS}\right) \tag{10.8}$$

where I_i is the small-signal input current. Summing the currents at the output drain node, we get

$$\frac{V_{DS}}{R_L} + g_m V_{GS} + j\omega C_{GD} \left(V_{DS} - V_{GS}\right) = 0 \tag{10.9}$$

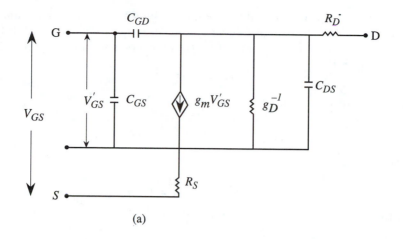

(a)

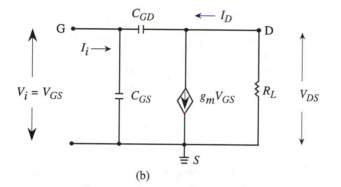

(b)

Figure 10.14: (a) An equivalent circuit for the MOSFET incorporating the elements shown in Fig. 10.15. The quantity g_D^{-1} is the resistance associated with the slope of the $I_D - V_{DS}$ curves in saturation. (b) A simple model for the high-frequency equivalent circuit of a common source n-MOSFET. A load resistance R_L is connected to this simpler equivalent circuit.

Eliminating V_{DS} from Eqns. 10.8 and 10.9, we get

$$I_i = j\omega \left\{ C_{GS} + C_{GD} \left[\frac{1 + g_m R_L}{1 + j\omega R_L C_{GD}} \right] \right\} V_{GS} \qquad (10.10)$$

For realistic devices $j\omega R_L C_{GD}$ is small and we can neglect it in the denominator of Eqn. 10.10. This gives

$$I_i = j\omega \left\{ C_{GS} + C_{GD}(1 + g_m R_L) \right\} V_{GS} \qquad (10.11)$$

The second term in the parentheses is the total gate-to-drain capacitance multiplied by $(1 + g_m R_L)$ and is called the Miller capacitance:

$$C_M = C_{GD}(1 + g_m R_L) \qquad (10.12)$$

The capacitance C_{GD} is dominated by the gate-drain overlay capacitance and thus the parasitic effects cause serious high-frequency-response limitations in the MOSFET. The cutoff frequency is the frequency where $I_i = I_D$, i.e. there is no current gain. The output current I_D is given by

$$I_D = g_m V_{GS} \tag{10.13}$$

Also from Eqns. 10.11 and 10.12 we can write

$$I_i = j\omega(C_{GS} + C_M)V_{GS} \tag{10.14}$$

The current gain is

$$\left| \frac{I_D}{I_i} \right| = \frac{g_m}{2\pi f(C_{GS} + C_M)} \tag{10.15}$$

Equating the gain to unity, we get for the cutoff frequency

$$f_T = \frac{g_m}{2\pi(C_{GS} + C_M)} \tag{10.16}$$

This expression shows how important it is to reduce C_M. This can be done by reducing the overlay capacitance by using "self-aligned gate technology" in which the drain contact is placed by using the gate as a mask. This greatly reduces the overlap between the gate and the drain. In an ideal MOSFET, the parasitic capacitance is zero, as is the intrinsic gate-to-drain capacitance in the saturation region, so that $C_M = 0$. Also the gate-to-source capacitance is (note that C_{ox} is capacitance per unit area)

$$C_{GS} \cong C_{ox}ZL = \frac{\epsilon_{ox}ZL}{d_{ox}} \tag{10.17}$$

In the model developed in Section 9.5 based on constant mobility, the transconductance in the saturation region is given by Eqn. 9.47 and we get (assuming C_M is negligible)

$$f_T = \frac{g_m}{2\pi C_{GS}} = \frac{\mu_n(V_{GS} - V_T)}{2\pi L^2} \tag{10.18}$$

The cutoff frequency can be improved by reducing the gate length or improving the carrier mobility. Since it is difficult to improve μ_n for Si, the greatest improvements come from the reduction of the gate length. The expression for the cutoff frequency given above is only approximate since it depends upon the constant-mobility model. For short channel devices the velocity of the device saturates to v_s so that f_T is

$$f_T = \frac{1}{2\pi t_{tr}} \tag{10.19}$$

where t_{tr} is the transit time through the channel $(= L/v_s)$.

EXAMPLE 10.4 Consider an n-channel MOSFET with mobility $\mu_n = 700$ cm^2/V·s and a channel length of 1.0 μm. Assume that the threshold voltage V_T is 1.5 V and the gate bias is 3.0 V for a small-signal application. Calculate the cutoff frequency using a constant-mobility model. Compare the results with a model based on the transit time for electrons moving with a saturation velocity of 10^7 cm/s. The cutoff frequency for the constant-mobility model is

$$f_T = \frac{\mu_n(V_G - V_T)}{2\pi L^2} = \frac{700(3.0 - 1.5)}{2\pi(10^{-4})^2} = 16.7 \text{ GHz}$$

If we obtain the cutoff frequency from the transit time under the saturation velocity model, we get

$$f_T = \frac{1}{2\pi t_{tr}} = \frac{v_s}{2\pi L} = \frac{10^7}{2\pi \times 10^{-4}} = 15.9 \text{ GHz}$$

The two results are quite similar in this example. Remember, however, that we have ignored the parasitic effects in these simple calculations.

10.4 FROM DEVICE PHYSICS TO CIRCUIT SIMULATORS

The MOSFET is the most important building block in modern microelectronics. Its low cost, high yield, and reasonable performance have made it the device of choice for memories, logic gates, microprocessors, digital signal processors, etc. In order to design and understand the performance of very complex circuits using MOSFETs, it is essential to have modeling tools that can represent the behavior of active devices as well as passive elements such as resistors, capacitors, etc. This requires that the device be represented by input-output relations which typically give the drain current as a function of drain bias. In this chapter and the previous one we have developed a simple physical model for the MOSFET I-V characteristics.

An important development in the microelectronic industry is the role being played by *fabless design companies*. These companies specialize in design only and contract out to other manufacturing houses for the final product. To succeed at what they do, it is critical to have very accurate device models. This is becoming increasingly difficult as the MOSFET shrinks to gate lengths below 0.1 μm. As a result circuit simulators increasingly are based not on physical models (i.e., something that can be derived from device dimensions and the transport properties of the material) but on equations with parameters that are *fitted to simulate actual experiments*.

Within the integrated circuit industry there are a number of simulator packages. Some of these are proprietary while others are publicly available. Most of these packages are based on SPICE (Simulation Program with Integrated Circuit Emphasis) originally developed at the University of California at Berkeley in the late 1960s. Modeling of FETs (particularly MOSFETs) within SPICE has become an important research and development challenge for industrial and academic scientists.

As shown in Fig. 10.15, during the early phase of development of circuit simulators (when devices were long) the basic device equations were based on physical models.

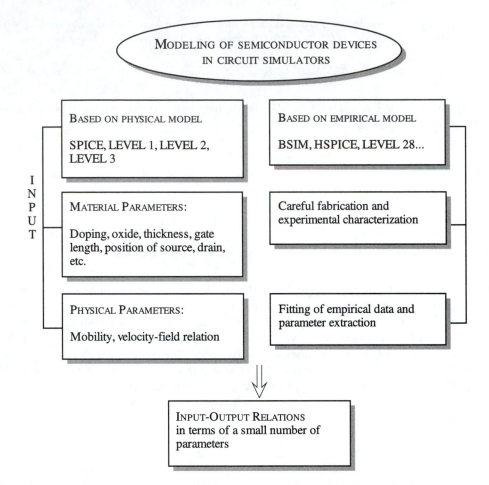

Figure 10.15: Approaches used to develop device models used in circuit simulators.

As devices became smaller, the underlying physics became more complex and physical models were unable to give results that agreed with experiments. An approach was then taken that is more empirical in nature. As shown in Fig. 10.15, the new approach depends on using experimental data to fit equations with a number of adjustable parameters. The parameters have some link with the physical parameters but this link is not entirely clear. For example the current-voltage relation for the MOSFET is written as

Linear regime (see Eqns. 9.36 and 9.51):

$$I_D = k \left[(V_{GS} - V_T)V_{DS} - \frac{V_{DS}^2}{2} \right] (1 + \lambda V_{DS}) \qquad (10.20)$$

Saturation regime (see Eqns. 9.41 and 9.51):

$$I_D = \frac{k}{2}(V_{GS} - V_T)^2(1 + \lambda V_{DS}) \tag{10.21}$$

where λ is the parameter to account for the change in the effective gate length, as discussed in Section 9.6. In our simple model the value of the parameter k is

$$k = \frac{\mu Z C_{ox}}{L} = \frac{k'Z}{L} \tag{10.22}$$

The parameter k is called the *device transconductance parameter*, while k' is called the *process transconductance parameter*. Such a simple (or even more involved physics-based) description for k and λ does not give good agreement for modern devices. As a result, in advanced simulators k is expressed in terms of a mobility that has adjustable parameters obtained from a fit to experimental results.

To understand the equivalent circuit used in SPICE to represent a MOSFET, we show in Fig. 10.16a a cross-sectional view of a typical MOSFET. The various capacitances and resistances associated with the device are shown on the structure and in Fig. 10.12. In Section 10.3 we discussed the origin of the parasitic capacitances of the device. In Fig. 10.16b we show a simple equivalent circuit used to develop a MOS model for SPICE LEVEL 1.

It is very instructive to go through an exercise in which the device process parameters are related to the parameters that form a SPICE input file. With this in mind we show a typical CMOS structure in Fig. 10.17a. Typical process parameters are shown in Fig. 10.17b. It is reasonably straightforward to obtain the various SPICE parameters from the process parameters. In Appendix D we go through this exercise. The resulting SPICE parameters are shown in Table 10.1. The resistance values (RD, RS, RSH) are simple estimates, as are the saturation current values for the junctions.

10.5 MOSFET TECHNOLOGY ROADMAP

Is MOS technology in its present form versatile enough to allow us to satisfy Moore's law over the next ten to twenty years? Unfortunately, the answer is no. It is quite clear from simple physics that drastic changes in materials and technology have to be made, starting around the year 2002. In order to understand some of the important challenges faced by the MOSFET technology, let us take a brief overview of how devices are scaled.

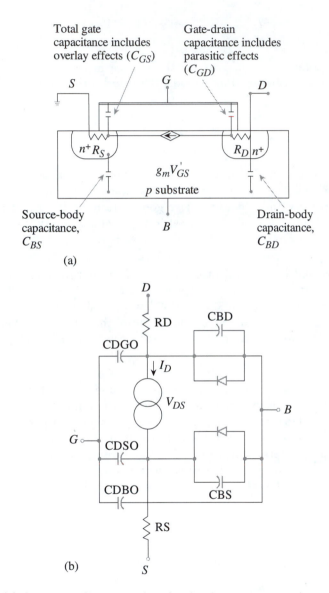

Figure 10.16: (a) A structural cross-section showing important capacitances and resistances asociated with a MOSFET. (b) LEVEL 1 SPICE model based on the structural model shown.

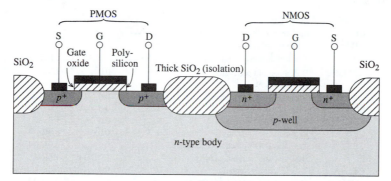

Parameter	Value	Units
n^+ active doping	5×10^{18}	cm^{-3}
p^+ active doping	5×10^{17}	cm^{-3}
p-well doping	5×10^{16}	cm^{-3}
n-well doping	1×10^{16}	cm^{-3}
Lateral diffusion		
n-channel	0.5	μm
p-channel	0.6	μm
oxide thickness	500	Å
Channel mobility		
n-channel	700	$cm^2/V{\cdot}s$
p-channel	230	$cm^2/V{\cdot}s$
Source area		
NMOS	100	$μm^2$
PMOS	100	$μm^2$
Drain area		
NMOS	100	$μm^2$
PMOS	100	$μm^2$
V_T	±1.0	V

(a)

(b)

Figure 10.17: (a) A cross-section of a CMOS circuit. (b) Process parameters typical of a CMOS circuit.

SPICE MOS TRANSISTOR MODEL PARAMETERS

Symbol	Name	Parameter	Units	Default	Example	
Electrical Parameters					NMOS	PMOS
	LEVEL	Model index		1		
μ	UO	Surface mobility	cm^2/V•s	600	700	230
V_{T0}	VTO	Zero-bias threshold voltage	V	0.0	1.0	−1.0
k'	KP	Transconductance parameter	A/V^2	2.0E–5	4.83E–5	1.59E–5
γ	GAMMA	Bulk threshold parameter	V$^{1/2}$	0.0	1.88	0.84
$2|\phi_F|$	PHI	Surface potential	V	0.6	0.78	0.7
λ	LAMBDA	Channel length modulation	V^{-1}	0.0	0.027	0.02
r_D	RD	Drain ohmic resistance	Ω	0.0	1.0	1.0
r_S	RS	Source ohmic resistance	Ω	0.0	1.0	1.0
C_{BD}	CBD	Zero-bias B-D junction capacitance	F	0.0	6.8E–14	3.2E–14
C_{BS}	CBS	Zero-bias B-DS junction capacitance	F	0.0	6.8E–14	3.2E–14
	IS	Bulk junction saturation current	A	1.0E–14	1.0E–15	1.0E–15
ϕ_0	PB	Bulk junction potential	V	0.8	0.8	0.8
C_{GSO}	CGSO	Gate-source overlap capacitance	F	0.0	6.9E–15	8.3E–15
C_{GDO}	CGDO	Gate-drain overlap capacitance	F	0.0	6.9E–15	8.3E–15
	RSH	Drain and source diffusion sheet resistance	Ω/square	0.0	10.0	10.0

Table 10.1: SPICE model parameters for a typical NMOS and CMOS device. The results given in the EXAMPLE columns are for the device of Fig. 10.17 and are calculated in Appendix D.

10.5.1   Scaling of MOS Circuits

In the introduction to this chapter we discussed how the continual shrinking of the transistor size has led to higher-density chips with greater complexity. This has been the key to the success of IC industry and the technologies that depend upon it. In this section we will briefly discuss considerations that determine how various transistor parameters must change as the device gate length shrinks.

One can consider two kinds of scaling possibilities. In one, as the device size shrinks by S, the voltages V_T, V_D also shrink by the same factor. In the second case, the voltages are kept constant. Usually the second case is of greater practical importance, since it is common to retain compatibility in supply voltages and logic values with previous-generation chips.

Full Scaling

In full scaling, each vertical dimension, surface dimension, and voltage level is scaled by the same factor S. As shown in Table 10.2, the oxide capacitance for the device decreases by $1/S$, although C_{ox} (capacitance per unit area) increases by S. The device current decreases by $1/S$, since

$$I_D \propto C_{ox} \frac{Z}{L} V^2 \propto \frac{1}{S} \tag{10.23}$$

The power consumption thus falls by $1/S^2$. The device switching time scales as $1/S$ and

PARAMETER	FULL SCALING	CONSTANT-VOLTAGE SCALING
Gate width, Z Gate length, L Oxide thickness, d_{ox}	1/S	1/S
Drain bias, V_{DS} Threshold bias, V_T	1/S	1
Oxide capacitance for the device	1/S	1/S
Drain current, I_D	~1/S	~S
DC power consumption	~1/S²	~S
Device switching time	~1/S	~1/S²
Power-delay product	~1/S³	~1/S

Table 10.2: Scaling of MOSFET structural and performance parameters for full scaling and constant-voltage scaling.

the power-delay product goes as $1/S^3$.

Full scaling rules are usually not followed, since that would make the new product incompatible with the older bias voltages.

Constant-Voltage Scaling

In this approach the supply voltage values are kept constant, while the device dimensions are reduced by S. In Table 10.2 we show the scaling factors for various device parameters.

In order to understand the difficulty that is anticipated in MOS scaling at gate lengths below 0.1 μm, we notice that as the gate length decreases, one has to decrease the gate oxide thickness in order to maintain the same control of the gate on the channel. For example, in Intel's microprocessor NMOS devices a relation between gate length L_G and oxide thickness is

$$L_G \sim 45 \, d_{ox} \tag{10.24}$$

Advances in oxide growth and control have allowed engineers to deposit 15 Å of oxide over 200 mm wafers. It is expected that even thinner oxides can be grown. This would suggest that one should be able to go down to gate lengths of ~ 500 Å without any problem. Unfortunately, the limit in this case is set by quantum-mechanical tunneling of electrons. The strong electric field between the channel and the gate causes electrons to tunnel through the oxide barrier, especially under inversion. Once the tunneling current approaches $\sim 10^2$ A/cm², the gate leakage current becomes unacceptably high. This occurs for SiO₂ thickness of ~ 16 Å.

10.5.2   Challenges Ahead

To appreciate the challenges faced by the semiconductor industry (the MOSFET industry), it is important to consider the following issues. The semiconductor industry has consistently been able to achieve an annual cost reduction of 25-30% per year per function during its history. This is a result of design innovation, shrinking devices, wafer size increase, yield improvement, and equipment improvement. To continue such a cost reduction is becoming increasingly difficult, however, as problems inherent in lithography and device physics emerge for the standard MOSFET technology. In Table 10.3 we show a projected roadmap complied by Semiconductor Industry Associates (SIA; http://www.sia.com) on how the MOSFET-based industry is expected to progress through the year 2012. One can see that a very aggressive struggle has to be waged to meet the projection shown. The challenges to be overcome are derived from the following:

- **Lithography**: Current devices are based on optical lithography, which limits the device dimensions to 0.19 μm (190 nm). As can be seen from Table 10.3, this causes a serious problem in meeting the challenges of 1999 devices. The limitations of 0.19 μm arises because present materials used for masks are not transparent below this optical wavelength. Thus, new materials are needed for masks. It is possible to use 0.19 μm light and draw somewhat smaller features, but this clearly is not a feasible approach for the future. Alternate choices for small feature sizes are *e*-beam lithography or X-ray lithography. These techniques are more expensive.

- **Gate Dielectric**: As the gate lengths for MOSFETs shrink, the gate oxide must shrink as well, in order for the gate to have an effective control of the channel. Intel, the world's biggest microprocessor manufacturer, reports that for its devices, gate length is approximately 45 times the gate oxide thickness. According to this relation, once the gate length shrinks below 100 nm, the gate oxide will have to be ~2 nm. Generating uniform oxide with this thickness is a formidable challenge, but it has already been met! The more serious and devastating challenge comes from basic physics. When d_{ox} is 2 nm, the tunneling current between the gate and channel becomes too large. This produces a very "leaky" FET (like a leaky faucet where water dribbles out of the faucet head). We can see from Table 10.3 that the gate current problem will become serious around the year 2005. Clearly, one needs either new insulator materials or new device designs.

- **Interconnect Delays**: It is normally expected that as gate lengths shrink, the overall speed of the circuit (for a given function) will improve. This is certainly true for gate delay, i.e., the time taken for the FET to switch. However, in a chip, the signal travels from one point to another, creating an "interconnect delay." For gate lengths of about 0.1 μm, the dominant delay is the gate delay. However, as devices shrink and packing

Year	1997	1999	2001	2003	2006	2009	2012
DRAM cell half-pitch (nm)	250	180	150	130	100	70	50
Gate length for MPU (nm)	200	140	120	100	70	50	35
Maximum substrate diameter (mm)	200	300	300	300	300	450	450
Acceptable defect density at 60% yield for DRAM	2080	1455	1310	1040	735	520	370
Defect density for MPU	1940	1710	1510	1355	1120	940	775
Power supply (V)	1.8-2.5	1.5-1.8	1.2-1.5	1.2-1.5	0.9-1.2	0.6-0.9	0.5-0.6
Power dissipation with heat sink (W)	70	90	110	130	160	170	175
Power dissipation without heat sink (for portable electronics) (W)	1.2	1.4	1.7	2.0	2.4	2.8	3.2
Cost per function DRAM (μcents/function)	120	60	30	15	5.3	1.9	0.66
Cost per function MPU (μcents/function)	3000	1735	1000	580	255	110	49

Table 10.3: Silicon technology roadmap as presented by Semiconductor Industry Associates (website: http://www.sia.com).

densities increase, the interconnect lines have to shrink as well. This increases their resistance and increases the interconnect delay. To appreciate this issue, consider this. For a gate length of 0.1 μm the gate delay is 25 ps and interconnect delay is ~1.5 ps. For a gate length of 100 nm, the gate delay is 3 ps, while the interconnect delay is ~38 ps. If the interconnects are made from copper instead of aluminum, the gate delay for the 100 nm device falls to 10 ps.

In 1998-1999 copper interconnect technology was perfected by IBM and the first products based on this technology were shipped. This is one of the important steps in

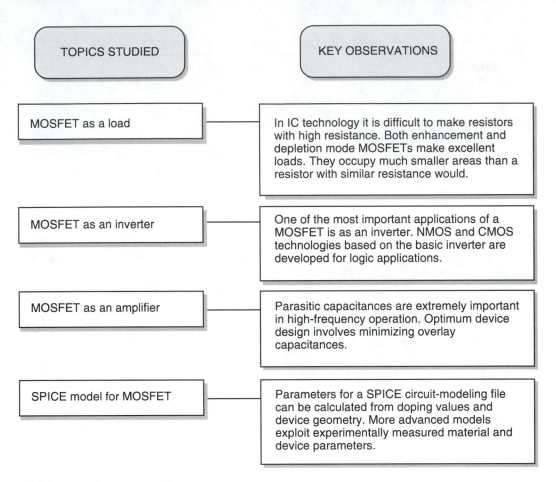

Table 10.4: Summary table.

meeting the challenges of the semiconductor roadmap. In Chapter 6 we have discussed issues related to copper-based interconnects.

• **Metrology and Testing**: As device dimensions shrink and packing densities increase, it is critical that defects on the chip be minimized. As can be seen from Table 10.3, the defects have to be dramatically reduced to ensure a constant yield as packing density increases. This requires that purity of crystal growth and wafer monitoring equipment also must become more sophisticated.

10.6 SUMMARY

In Table 10.4 we present an overview of the topics discussed in this chapter.

10.7 PROBLEMS

Section 10.2

Problem 10.1 Assume that in an IC technology one can make resistors with a width of 5.0 μm from a material with sheet resistance of 100 Ω per square. Calculate the area needed to make a resistor of value 100 $k\Omega$. Compare this area with the area of a MOSFET of gate length 3.0 μm and a width of 10.0 μm.

Problem 10.2 An NMOS device is needed for an inverter with load resistance of 10 $k\Omega$ and $V_{DD} = 5$ V. Design a device (i.e., find the dimensions, doping, and oxide thickness) that has noise margins of at least

$$NM_L = V_{IL} - V_{OL} = 1.0 \text{ V}$$
$$NM_H = V_{OH} - V_{IL} = 2.5 \text{ V}$$

Problem 10.3 Design an NMOS for a resistive load inverter with load resistance of 10 $k\Omega$ and $V_{DD} = 5$ V that would have the following noise margins:

$$NM_L \geq 0.7 \text{ V}$$
$$NM_H = \geq 1.2 \text{ V}$$

Problem 10.4 Consider a resistive load inverter with the following device parameters:

$$
\begin{aligned}
k' &= 20 \ \mu A/V^2 \\
\frac{Z}{L} &= 20 \\
V_{TB} &= 1 \text{ V} \\
\gamma &= 0.4 \text{ V}^{1/2} \\
\lambda &= 0
\end{aligned}
$$

For $V_{DD} = 5$ V, calculate V_{OL}, V_{OH}, NM_L and NM_H.

Problem 10.5 Consider an NMOS inverter with a saturated enhancement load as shown in Fig. 10.7. The devices have the following parameters:

$$
\begin{aligned}
\left(\frac{Z}{L}\right)_{inverter} &= 20 \\
\left(\frac{Z}{L}\right)_{load} &= 2 \\
V_{TO} &= 1.5 \text{ V} \\
2\phi_F &= 0.6 \text{ V} \\
\gamma &= 0.4 \text{ V}^{1/2}
\end{aligned}
$$

Find the critical points $V_{OH}, V_{DL}, V_{IL}, V_{IH}$, and V_M for the inverter.

Section 10.3

Problem 10.6 An n-channel MOSFET has the following parameters:

$$
\begin{aligned}
\text{Channel mobility,} \quad \mu_n &= 600 \ \text{cm}^2/ \text{V} \cdot \text{s} \\
\text{Oxide thickness,} \quad d_{ox} &= 500 \ \text{Å} \\
\text{Channel length,} \quad L &= 1.0 \ \mu\text{m} \\
\text{Channel width,} \quad Z &= 20.0 \ \mu\text{m} \\
\text{Turn-on voltage,} \quad V_T &= 0.8 \ \text{V}
\end{aligned}
$$

Assume an ideal model for the transistor if the device is biased at $V_G = 2.0$ V. Calculate the cutoff frequency. In the above problem the gate oxide overlaps the source and the drain by 0.2 μ m and the external load has a resistance of $R_L = 10$ kΩ. Calculate the Miller capacitance and the cutoff frequency at $V_G = 2.0$ V.

Problem 10.7 Consider a 1.0 μm channel length n-channel MOSFET in which the drain voltage is 5.0 V. Assume that the electrons see an average electric field in the channel given by V_D/L. Calculate the cutoff frequency of the device if the transit time limits the performance. Use the velocity-field relations for Si to calculate the answer.

Problem 10.8 Consider an n-type 1.0 μm FET made from GaAs and Si. The device cutoff frequency is limited by transit time effects. Using the velocity-field relations for electrons in GaAs and Si, estimate the cutoff frequency for the devices when (a) $V_D = 0.5V$, i.e., for low-power applications and (b) $V_D = 10$ V, i.e., for high-power applications. Assume that the average field in the channel is V_D/L.

Problem 10.9 Consider an NMOS with $d_{ox} = 200$ Å, $L = 2.0$ μm, $L_D = 0.15$ μm, $Z = 10$ μm, $C_{SBO} = C_{DBO} = 40$ $fF, V_0 = 0.8$ V, and $|V_{SB}| = |V_{DB}| = 2$ V. Calculate the following capacitances when the transistor is operating in saturation: $C_{ox}, C_{GS}, C_{GD}, C_{SB}, C_{DB}$. In the list of parameters given above, V_0 is the built-in voltage for the p-n junction between the source (or drain) and the body.

Problem 10.10 Consider a MOSFET used as an amplifier. The device has the following parameters:

$$
\begin{aligned}
C_{GS} = C_{GD} &= 1 \ \text{pF} \\
I_D(\text{sat}) &= 2 \ \text{mA} \\
V_{GS} &= 2 \ \text{V} \\
V_T &= 1 \ \text{V}
\end{aligned}
$$

The effective load resistance of the amplifier is 5 kΩ. Calculate the cutoff frequency f_T of the amplifier.

10.8 READING LIST

- **General**

    - A. S. Sedra and K. C. Smith, *Microelectronic Circuits* (Oxford University Press, New York, 1998).

    - D. A. Hodges and H. G. Jackson, *Analysis and Design of Digital Integrated Circuits* (McGraw-Hill, New York, 1988).

    - R. L. Geiger, P. E. Allen, and N. R. Strader, *VLSI Design Techniques for Analog and Digital Circuits* (McGraw-Hill, New York, 1990).

Chapter

11

SEMICONDUCTOR OPTOELECTRONICS

Chapter at a Glance

11.1 INTRODUCTION

So far in this text we have discussed electronic devices that are made from semiconductors. An important and growing application of semiconductors is in the area of optoelectronics. Semiconductors are able to "marry" useful properties of optics and electronics, spawning new technologies. Light has many properties that make it very attractive for information processing. Some of these properties, as shown in Fig. 11.1, are:

(i) *Immunity to electromagnetic interference*: Since light particles carry no charge, electromagnetic activities such as lightning and other potential discharges that can play havoc with electrical signals have essentially no effect on optical signals;

(ii) *Non-interference of crossing light signals*: Two unrelated light beams can cross one another and emerge with little effect on each other, a property that can be exploited in very high-density information processing. In electronic signals, two crossing signals will have serious effects and cause loss of information;

(iii) *Promise of high parallelism*: The benefits of optics as far as parallelism is concerned are obvious to us when we see an image and are able to process it in parallel to make real-time decisions like crossing a busy road;

(iv) *High speed and high bandwidth*: Optical pulses have been produced with widths of only a few femtoseconds! In principle, such short pulses could be exploited for a variety of high-speed applications;

(v) *Signal (beam) steering*: Optical beams can be steered quite easily by use of lens or holograms. This is difficult or impossible to do for electron beams in a reasonable manner. The beam-steering phenomenon can, in principle, allow one to reconfigure interconnections in very short times and thus generate circuits that can be flexible (or functional) in real time;

(vi) *Special-function devices*: This is a most exciting property of optical devices and has great potential in high-speed information processing. An important example in this case is the lens, which, when used with a proper object-to-image relationship, can produce a Fourier transform of the object image. This property is exploited in numerous recognition-based systems. Another example is the use of optics in spectrum analyzers, which exploits the special diffraction properties of light;

(vii) *Wave nature of light*: Since light suffers little scattering over long distances (compared to electrons), its wave nature can be readily exploited for special-purpose devices. In electronics, the wave nature of electrons comes into the picture only when the device dimensions are below $\sim$500 Å, since scattering effects smear out the electron's phase over longer distances;

(viii) *Nonlinear interactions*: A number of materials have a strongly nonlinear response to optical intensity and can be exploited for devices;

(ix) *Ease of coupling with electronics*: This is one of the most important features of optics and the one that has the paid most dividends so far. Optical and electronic interactions can easily be merged in semiconductor devices. This has led to the most important optoelectronic devices, viz. the laser, the detector, and the modulator.

It is obvious from the above brief discussion that optics has the potential of playing a major role in the information-processing age. Optics has been very important in a number of areas. These areas include:

(i) *Memory*: Information is stored digitally on optical discs (compact discs or CDs) as tiny "bumps" that can be read by a solid state laser. This has greatly revolutionized the music industry as well as the general information-storage industry. However, the laser in the CD player still has to be backed up by an electronic chip that does all the signal processing and controls the audio output;

(ii) *Optical communication*: This has been the most important area where optoelectronic devices have made inroads into modern technology. This area has also given impetus to compound semiconductor research and development. Optical fibers are rapidly replacing the traditional copper cables for carrying long-distance telephone conversations;

(iii) *Local area networks (LANs)*: This is another area where optical interconnects between local computers, telephones, etc., are making office buildings and factories more efficient and capable of handling high-volume information;

(iv) *Printing and desktop publishing*: This area has received a great boost with the availability of laser printers;

(v) *Guidance and control*: Laser-guided weapons and unmanned flying craft have become important components of modern armies;

(vi) *Photonic switching and interconnects*: The use of optical devices in chip-to-chip interconnects is becoming increasingly feasible.

In this chapter we will discuss the basic principles of several important optoelectronic devices. In semiconductors, the interaction between light or photons and the semiconductor is mediated by electrons. When light with an appropriate energy impinges upon a semiconductor, it can generate electrons and hole pairs. These electrons and holes can be used to convert the photon energy into electrical energy. Solar cells and optical detectors are based upon this interaction. Light can also be generated in semiconductors by the recombination of an electron and hole. Excess electrons and holes

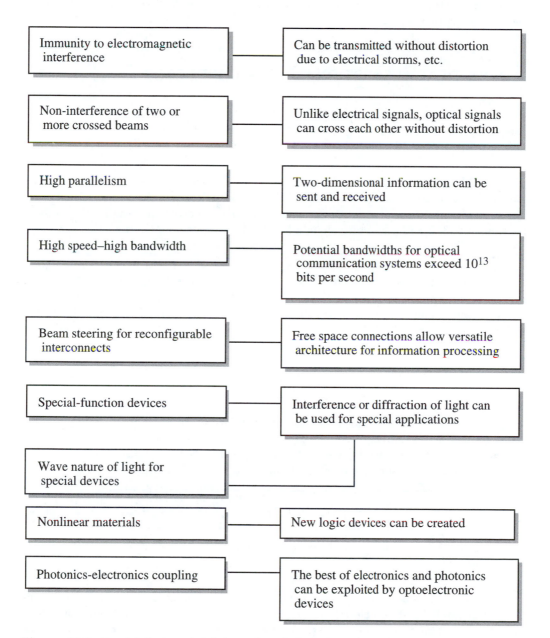

Figure 11.1: Special features of light and optical devices that make optics an attractive medium for information processing. Currently, semiconductor optoelectronic devices are playing a key role in communication, memory storage, guidance systems, etc.

can be injected by a power supply into a semiconductor and one can convert electronic signals into optical signals.

11.2   OPTICAL ABSORPTION IN A SEMICONDUCTOR

In order for a semiconductor device to be useful as a detector, some property of the device should be affected by radiation. The most commonly used property is the conversion of light into electron-hole pairs, which can be detected in a properly chosen electric circuit.

When light impinges on a semiconductor, it can scatter an electron in the valence band into the conduction band. This process, called the absorption of a photon, has been discussed in Chapter 3. We will summarize some of the results discussed in Sections 3.6, 3.7, and 3.8, which the reader should review to understand this chapter.

The photon-absorption process is strongest when the photon can directly cause an electron in the valence band to go into the conduction band. Since the photon momentum is extremely small on the scale of the electron momentum, the conservation of momentum requires that the electron-hole transitions are *vertical in k-space*, as shown in Fig. 11.2a (see Section 3.8.1). Such transitions are possible only near the bandedge for direct bandgap semiconductors. For such semiconductors one can write the absorption coefficient as

$$\alpha(\hbar\omega) = \frac{\pi e^2 \hbar}{2n_r c m_o^2 \epsilon_o} \frac{|p_{cv}|^2}{\hbar\omega} \frac{\sqrt{2}(m_r^*)^{3/2}(\hbar\omega - E_g)^{1/2}}{\pi^2 \hbar^3} \tag{11.1}$$

where m_r^* is the reduced *e-h* mass, n_r the refractive index, $\hbar\omega$ the photon energy, E_g the band gap and p_{cv} a momentum matrix element that allows the transition to take place. For directgap semiconductors, when the various values for the constants are plugged into Eqn. 11.1, the absorption coefficient turns out to be ($\hbar\omega$, E_g in eV, $\hbar\omega \geq E_g$)

$$\boxed{\alpha(\hbar\omega) \cong (3.5 \pm 0.5) \times 10^6 \left(\frac{m_r^*}{m_o}\right)^{3/2} \frac{(\hbar\omega - E_g)^{1/2}}{\hbar\omega} \ \text{cm}^{-1}} \tag{11.2}$$

The variation in α is due to small changes in the constant p_{cv} of Eqn. 11.1 for different semiconductors.

When a semiconductor does not have a direct bandgap, vertical *k* transitions are not possible, and the electrons can absorb a photon only if a phonon (or lattice vibration) participates in the process, as shown in Fig. 11.2b. Such processes are not as strong as the ones that do not involve a phonon. The absorption coefficient for indirect gap materials is typically a factor of 100 smaller than in the direct gap case for the same value of photon energy above bandgap ($\hbar\omega - E_g$).

As can be seen from Eqn. 11.1, the absorption coefficient is zero above a cutoff

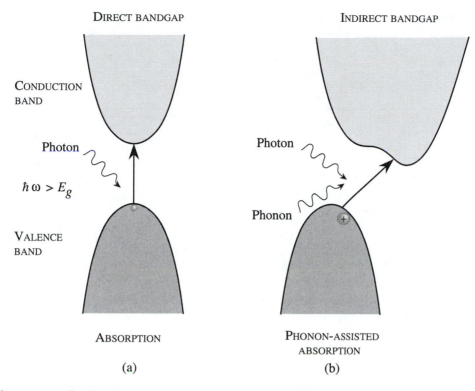

Figure 11.2: Band-to-band absorption in direct and indirect semiconductors. (a) An electron in the valence band "absorbs" a photon and moves into the conduction band. Momentum conservation ensures that only vertical transitions are allowed. (b) In indirect semiconductors a phonon or lattice vibration must participate to take an electron from the top of the valence band to the bottom of the conduction band. This is a comparatively weak process.

wavelength given by λ_c, where

$$\boxed{\lambda_c = \frac{hc}{E_g} = \frac{1.24}{E_g(\text{eV})} \ (\mu\text{m})}$$

(11.3)

where E_g is the semiconductor bandgap. In Fig. 11.3 we show the bandgap and cutoff wavelengths for several semiconductors along with the relative response of the human eye. In Fig. 11.4 we show the absorption coefficients for several different semiconductors. Materials like GaAs, InP, InGaAs, etc., have strong optical absorption at the bandedges because the optical absorption can occur without phonon participation. On the other hand, Si and Ge have an indirect bandgap, and the absorption strength is weak near the bandedge. However, this does not mean that these materials cannot be used as detectors (unfortunately they cannot be used as lasers, as we will see in Section 11.5). For detection of an optical signal, the light should be absorbed. If L is the length of the sample, the fraction of incident light absorbed in the sample is

$$1 - \exp\left(-\alpha L\right)$$

(11.4)

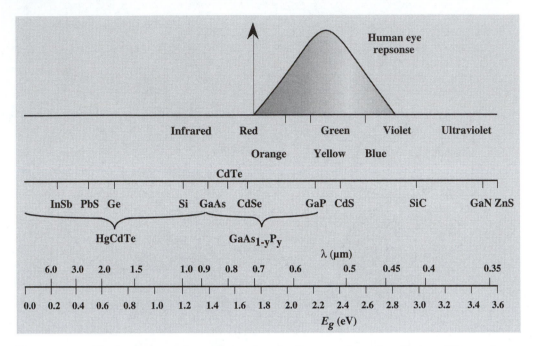

Figure 11.3: The bandgap and cutoff wavelengths for several semiconductors. The semiconductor bandgaps range from 0 (for $Hg_{0.84}Cd_{0.15}Te$) to well above 3 eV, providing versatile detection systems.

Thus, for strong absorption, we must have

$$L > \frac{1}{\alpha(\hbar\omega)} \tag{11.5}$$

Thus, if Si is to absorb at GaAs laser emission ($\hbar\omega \sim 1.45$ eV), we can see from Fig. 11.4 that one needs a material thickness of 10-20 μm. On the other hand, a Ge detector would require an interaction length of only ~ 1 μm, even though Ge is an indirect gap material.

Once the absorption coefficient for a semiconductor is known, one needs to know the rate at which electron-hole pairs will be generated. To calculate the rate of e-h pair generation, consider an optical beam with intensity $P_{op}(0)$ impinging upon a semiconductor per unit area. The intensity at a point x is given by (intensity has units of W/cm^2)

$$P_{op}(x) = P_{op}(0) \exp(-\alpha x) \tag{11.6}$$

The energy absorbed per second per unit area in a region of thickness dx between points x and $x + dx$ is (dx is very small)

$$P_{op}(x) - P_{op}(x + dx) = P_{op}(0)\left[\exp(-\alpha x) - \exp(-\alpha(x + dx))\right]$$

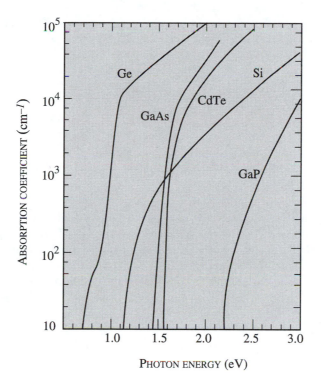

Figure 11.4: Absorption coefficients in different semiconductors. Notice the slow increase in absorption coefficients in indirect gap semiconductors (Si, Ge).

$$= P_{op}(0) \left[\exp\left(-\alpha x \right) \right] \alpha dx = P_{op}(x) \alpha dx \qquad (11.7)$$

If this absorbed energy produces *e-h* pairs of energy $\hbar\omega$, the rate of the carrier generation is G_L (rate per unit volume):

$$\boxed{G_L = \frac{\alpha P_{op}(x)}{\hbar\omega} = \alpha J_{ph}(x)} \qquad (11.8)$$

where J_{ph} is the photon flux density impinging at point x (flux has units of $cm^{-2}\ s^{-1}$).

When light impinges upon a semiconductor and generates *e-h* pairs, the detector performance depends upon collecting these carriers and thus changing the conductivity of the material or generating a voltage signal. In the absence of an electric field or a concentration gradient, the *e-h* pairs will recombine with each other and not generate a detectable signal. An important property of the detector is described by its responsivity, which gives the current, I_L, produced by a certain optical power. The responsivity R_{ph} is defined by

$$\boxed{R_{ph} = \frac{I_L/A}{P_{op}} = \frac{J_L}{P_{op}}} \qquad (11.9)$$

where I_L is the photocurrent produced in a device of area A and J_L is the photocurrent density. The quantum efficiency of the detector is defined by

$$\eta_Q \;=\; \frac{J_L/e}{P_{op}/\hbar\omega}$$

$$=\; R_{ph}\,\frac{\hbar\omega}{e} \qquad\qquad (11.10)$$

The quantum efficiency essentially tells us how many carriers are collected for each photon impinging on the detector.

The responsivity of a detector has a strong dependence upon the wavelength of the impinging photons. If the wavelength is above the cutoff wavelength, the photons will not be absorbed and no photocurrent will be generated. When the wavelength is smaller than λ_c, the photon energy will be larger than the bandgap energy and the difference will be released as heat. Thus, even though the photon energy increases above the bandgap, it still produces the same number of *e-h* pairs. Thus the responsivity starts to decrease.

To collect the electron-hole pairs generated by light one needs an electric field. This can be generated either by simply applying a bias across an undoped semiconductor or by using a *p-n* diode. The former choice leads to the photoconductive detector, in which the *e-h* pairs change the conductivity of the semiconductor. The *p-n* (or *p-i-n*) diode is widely used as a detector and exploits the built-in electric fields present at the junction together with an applied reverse bias to collect electrons and holes.

In addition to the diodes, transistors can also be used to detect optical signals. Phototransistors are widely used in optoelectronic technology. Phototransistors offer high gain due to the transistor gain.

EXAMPLE 11.1 Calculate the absorption coefficient for GaAs for an incident optical beam with energy of 1.7 eV. Assume that $E_g = 1.43$ eV. The absorption coefficient for GaAs is

$$\alpha(\hbar\omega) = 5.6 \times 10^4 \frac{(\hbar\omega - E_g)^{1/2}}{\hbar\omega} \;\; \text{cm}^{-1}$$

where $\hbar\omega$ and E_g are expressed in eV. For our case we get

$$\alpha(\hbar\omega = 1.7 \text{ eV}) = 4.21 \times 10^4 \frac{(0.27)}{1.7} = 6.7 \times 10^3 \;\; \text{cm}^{-1}$$

EXAMPLE 11.2 A Ge detector is to be used for an optical communication system using a GaAs laser with emission energy of 1.43 eV. Calculate the depth of the detector needed to be able to absorb 90% of the optical signal entering the detector.

From Fig. 11.4 we see that at $\hbar\omega = 1.43$ eV, $\alpha \cong 2.5 \times 10^4$ cm^{-1}. The length of material required to absorb 90% of the light is given by

$$L = -\frac{1}{\alpha}\ell n\ (1 - 0.9) = \frac{2.3}{(2.5 \times 10^4\ \mathrm{cm}^{-1})} = 0.92\ \mu\mathrm{m}$$

Thus a rather thin region of Ge can absorb a large fraction of light emitted from a GaAs laser. Of course, if the light was emitted by an $\mathrm{In}_{0.53}\mathrm{Ga}_{0.47}\mathrm{As}$ laser ($\hbar\omega \sim 0.8$ eV), Ge would not be as suitable a material.

EXAMPLE 11.3 An optical intensity of 10 W/cm^2 at a wavelength of 0.75 μm is incident on a GaAs detector. Calculate the rate at which electron-hole pairs will be produced at this intensity at 300 K. If the *e-h* recombination time is 10^{-9} s, calculate the excess carrier density.

From Fig. 11.4, the absorption coefficient of GaAs at this wavelength is $\sim 7 \times 10^3$ cm^{-1}. The *e-h* generation rate is (0.75 μm wavelength is equivalent to a photon of 1.65 eV)

$$G_L = \frac{\alpha P_{op}}{\hbar\omega} = \frac{(7 \times 10^3\ \mathrm{cm}^{-1})(10\ \mathrm{Wcm}^{-2})}{(1.65 \times 1.6 \times 10^{-19}\ \mathrm{J})} = 2.65 \times 10^{23}\ \mathrm{cm}^{-3}\ \mathrm{s}^{-1}$$

In steady state the excess carrier density is

$$\begin{aligned}
\delta n = \delta p = G_L \tau &= (2.65 \times 10^{23}\ \mathrm{cm}^{-3}\ \mathrm{s}^{-1})(10^{-9}\ \mathrm{s}) \\
&= 2.65 \times 10^{14}\ \mathrm{cm}^{-3}
\end{aligned}$$

11.3 PHOTOCURRENT IN A *P-N* DIODE

When light impinges upon a semiconductor to create electron-hole pairs, some of the carriers are collected at the contact and lead to the photocurrent. Let us consider a long *p-n* diode in which excess carriers are generated uniformly at a rate G_L. Fig. 11.5 shows a *p-n* diode with a depletion region of width W. The electron-hole pairs generated in the depletion region are swept out rapidly by the electric field existing in the region. Thus the electrons are swept into the *n*-region while the holes are swept into the *p*-region. The photocurrent arising from the photons absorbed in the depletion region is thus

$$I_{L1} = A \cdot e \int_0^{x'} G_L \cdot dx = A \cdot e G_L W \tag{11.11}$$

where A is the diode area and we have assumed a uniform generation rate in the diode. *Since the electrons and holes contributing to I_{L1} move under high electric fields, the response is very fast, and this component of the current is called the prompt photocurrent.*

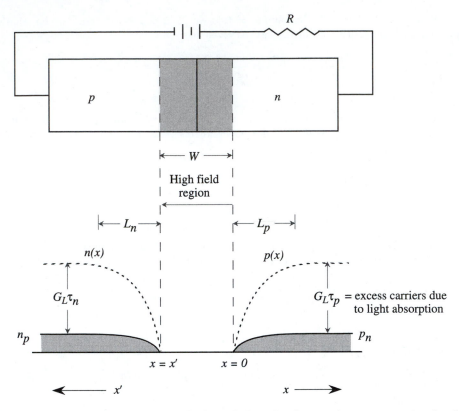

Figure 11.5: A schematic of a p-n diode and the minority carrier concentration in absence and presence of light. The minority charge goes to zero at the depletion region edge due to the high field, which sweeps the charge away. The equilibrium minority charge is p_n and n_p in the n- and p-sides, respectively.

In addition to the carriers generated in the depletion region, e-h pairs are generated in the neutral n- and p-regions of the diode. On physical grounds, we may expect that holes generated within a distance L_p (the diffusion length) of the depletion region edge ($x = 0$ of Fig. 11.5) will be able to enter the depletion region, from whence the electric field will sweep them into the p-side. Similarly, electrons generated within a distance L_n of the $x' = 0$ side of the depletion region will also be collected and contribute to the current. Thus the photocurrent should come from all carriers generated in a region $(W + L_n + L_p)$. A quantitative analysis reaches the same conclusion.

The total current due to carriers in the neutral region and the depletion region is

$$I_L = I_{nL} + I_{pL} + I_{L1} = eG_L(L_p + L_n + W)A \qquad (11.12)$$

It must be noted that the e-h pair generation is not uniform with penetration

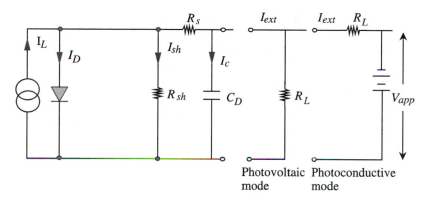

Figure 11.6: The equivalent circuit of a photodiode. The device can be represented by a photocurrent source I_L feeding into a diode. The device's internal characteristics are represented by a shunt resistor R_{sh} and a capacitor C_D. R_s is the series resistance of the diode. In the photovoltaic mode (used for solar cells and other devices) the diode is connected to a high-resistance R_L, while in the photoconductive mode (used for detectors) the diode is connected to a load R_L and a power supply.

depth but decreases with it. Thus G_L has to be replaced by an average generation rate for an accurate description. *It is also important to note that the photocurrent flows in the direction of the reverse current of the diode.*

The total current in the diode connected to the external load, as shown in Fig. 11.6, is given by the light-generated current and the diode current in the absence of light. In general, if the voltage across the diode is V, the total current is (note that the photocurrent flows in the opposite direction to the forward-bias diode current)

$$I = I_L + I_0 \left[1 - \exp\left\{\frac{e(V + R_s I)}{n k_B T}\right\}\right] \tag{11.13}$$

where R_s is the diode series resistance, n the ideality factor, and V the voltage across the diode. As shown in Fig. 11.6, the photodiode can be used in one of two configurations. In the photovoltaic mode, used for solar cells, there is no external bias applied. The photocurrent passes through an external load to generate power. In the photoconductive mode, used for detectors, the diode is reverse biased and the photocurrent is collected.

11.3.1   Application to a Solar Cell

An important use of the *p-n* diode is to convert optical energy to electrical energy as in a solar cell. The solar cell operates without an external power supply and relies on the optical power to generate current and voltage. To calculate the important parameters of a solar cell, consider the case where the diode is used in the open circuit mode so that the current I is zero. This gives, for Eqn. 11.13,

$$I = 0 = I_L - I_0 \left[\exp \left(\frac{eV_{oc}}{nk_BT} \right) - 1 \right] \tag{11.14}$$

where V_{oc} is the voltage across the diode, known as the open circuit voltage. We get for this voltage

$$V_{oc} = \frac{nk_BT}{e} \ln \left(1 + \frac{I_L}{I_0} \right) \tag{11.15}$$

At high optical intensities the open-circuit voltage can approach the semiconductor bandgap. In the case of Si solar cells for solar illumination (without atmospheric absorption), the value of V_{oc} is roughly 0.7 eV.

A second limiting case in the solar cell is the one where the output is short-circuited, i.e., $R = 0$ and $V = 0$. The short-circuit current is then

$$I = I_{sc} = I_L \tag{11.16}$$

A plot of the diode current in the solar cell as a function of the diode voltage then provides the curve shown in Fig. 11.7. In general, the electrical power delivered to the load is given by

$$P = I \times V = I_L V - I_0 \left[\exp \left(\frac{eV}{k_BT} \right) - 1 \right] V \tag{11.17}$$

The maximum power is delivered at voltage and current values of V_m and I_m, as shown in Fig. 11.7.

The conversion efficiency of a solar cell is defined as the rate of the output electrical power to the input optical power. When the solar cell is operating under maximum power conditions, the conversion efficiency is

$$\eta_{conv} = \frac{P_m}{P_{in}} \times 100 \text{ (percent)} = \frac{I_m V_m}{P_{in}} \times 100 \text{ (percent)} \tag{11.18}$$

Another useful parameter in defining solar cell parameters is the fill factor F_f, defined as

$$F_f = \frac{I_m V_m}{I_{sc} V_{oc}} \tag{11.19}$$

In most solar cells the fill factor is ~ 0.7.

In the solar-cell conversion efficiency, it is important to note that photons that have an energy $\hbar\omega$ smaller than the semiconductor bandgap will not produce any electron-hole pairs. *Also, photons with energy greater than the bandgap will produce electrons and holes with the same energy (E_g), regardless of how large ($\hbar\omega - E_g$) is. The*

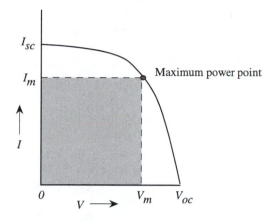

Figure 11.7: The relationship between the current and voltage delivered by a solar cell. The open-circuit voltage is V_{oc} and the short-circuit current is I_{sc}. The maximum power is delivered at the point shown.

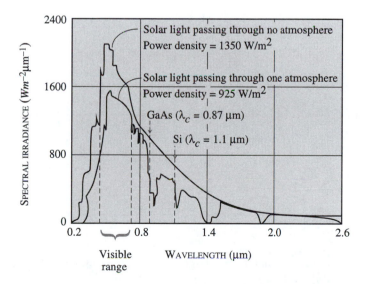

Figure 11.8: The spectral irradiance of the solar energy. The spectra are shown for no absorption in the atmosphere and for the sea-level spectra. Also shown are the cutoff wavelengths for GaAs and Si.

excess energy $\hbar\omega - E_g$ is simply dissipated as heat. Thus the solar cell efficiency depends quite critically on how the semiconductor bandgap matches the solar energy spectra. In Fig. 11.8 we show the solar energy spectra. Also shown are the cutoff wavelengths for silicon and GaAs. GaAs solar cells are better matched to the solar spectra and provide greater efficiencies. However, the technology is more expensive than Si technology. Thus GaAs solar cells are used for space applications while silicon (or amorphous silicon) solar cells are used for applications where cost is a key factor.

EXAMPLE 11.4 Consider a long Si p-n junction that is reverse biased with a reverse bias voltage of 2 V. The diode has the following parameters (all at 300 K):

Diode area,	A =	$10^4 \, \mu m^2$
p-side doping,	N_a =	$2 \times 10^{16} \, cm^{-3}$
n-side doping,	N_d =	$10^{16} \, cm^{-3}$
Electron diffusion coefficient	D_n =	$20 \, cm^2/s$
Hole diffusion coefficient,	D_p =	$12 \, cm^2/s$
Electron minority carrier lifetime,	τ_n =	$10^{-8} \, s$
Hole minority carrier lifetime,	τ_p =	$10^{-8} \, s$
Electron-hole pair generation rate by light,	G_L =	$10^{22} \, cm^{-3} \, s^{-1}$

Calculate the photocurrent.

The electron length diffusion length is

$$L_n = \sqrt{D_n \tau_n} = \left[(20)(10^{-8}) \right]^{1/2} = 4.5 \, \mu m$$

The hole diffusion length is

$$L_p = \sqrt{D_p \tau_p} = \left[(12)(10^{-8}) \right]^{1/2} = 3.46 \, \mu m$$

To calculate the depletion width, we need to find the built-in voltage,

$$V_{bi} = \frac{k_B T}{e} \, \ell n \, \left(\frac{N_a N_d}{n_i^2} \right) = 0.026 \, \ell n \, \left(\frac{(2 \times 10^{16})(10^{16})}{(1.5 \times 10^{10})^2} \right) = 0.715 \, V$$

The depletion width is now

$$
\begin{aligned}
W &= \left\{ \frac{2\epsilon_s}{e} \left(\frac{N_a + N_d}{N_a N_d} \right) (V_{bi} + V_R) \right\}^{1/2} \\
&= \left\{ \frac{2(11.9)(8.85 \times 10^{-14})}{(1.6 \times 10^{-19})} \left(\frac{(2 \times 10^{16} + 10^{16})}{(2 \times 10^{16})(10^{16})} \right) (2.715) \right\}^{1/2} \\
&= 0.73 \, \mu m
\end{aligned}
$$

We see in this case that L_n and L_p are larger than W. The prompt photocurrent is thus a small part of the total photocurrent. The photocurrent is now

$$
\begin{aligned}
I_L &= eAG_L(W + L_n + L_p) \\
&= (1.6 \times 10^{-19} \, C)(10^4 \times 10^{-8} \, cm^2)(10^{22} \, cm^{-3} \, s^{-1}) \\
&\quad (0.73 \times 10^{-4} \, cm + 4.5 \times 10^{-4} \, cm + 3.46 \times 10^{-4} \, cm) \\
&= 0.137 \, mA
\end{aligned}
$$

The photocurrent is much larger than the reverse saturation current I_0 and its direction is the same as the reverse current.

EXAMPLE 11.5 Consider an Si solar cell at 300 K with the following parameters:

Area,	A	$=$	$1.0\ \text{cm}^2$
Acceptor doping,	N_a	$=$	$5 \times 10^{17}\ \text{cm}^{-3}$
Donor doping,	N_d	$=$	$10^{16}\ \text{cm}^{-3}$
Electron diffusion coefficient,	D_n	$=$	$20\ \text{cm}^2/\text{s}$
Hole diffusion coefficient,	D_p	$=$	$10\ \text{cm}^2/\text{s}$
Electron recombination time,	τ_n	$=$	$3 \times 10^{-7}\ \text{s}$
Hole recombination time,	τ_p	$=$	$10^{-7}\ \text{s}$
Photocurrent,	I_L	$=$	$25\ \text{mA}$

Calculate the open-circuit voltage of the solar cell.

To find the open-circuit voltage, we need to calculate the saturation current I_0, which is given by

$$I_0 = A\left[\frac{eD_n n_p}{L_n} + \frac{eD_p p_n}{L_p}\right] = Aen_i^2\left[\frac{D_n}{L_n N_a} + \frac{D_p}{L_p N_d}\right]$$

Also,

$$L_n = \sqrt{D_n\tau_n} = \left[(20)(3 \times 10^{-7})\right]^{1/2} = 24.5\ \mu\text{m}$$

$$L_p = \sqrt{D_p\tau_p} = \left[(10)(10^{-7})\right]^{1/2} = 10.0\ \mu\text{m}$$

Thus,

$$I_0 = (1)(1.6 \times 10^{-19})(1.5 \times 10^{10})^2\left[\frac{20}{(24.5 \times 10^{-4})(5 \times 10^{17})} + \frac{10}{(10 \times 10^{-4})(10^{16})}\right]$$

$$= 3.66 \times 10^{-11}\ \text{A}$$

The open-circuit voltage is now

$$V_{oc} = \frac{k_B T}{e}\ell n\left(1 + \frac{I_L}{I_0}\right) = (0.026)\,\ell n\left(1 + \frac{25 \times 10^{-3}}{3.66 \times 10^{-11}}\right) = 0.53\ \text{V}$$

EXAMPLE 11.6 A single solar cell of area 1 cm^2 has a photocurrent of $I_L = 25$ mA and a diode saturation current of 3.66×10^{-11} A at 300 K. (a) Calculate the open-circuit voltage and short-circuit current of the solar cell; (b) calculate the power extracted from each cell if the fill factor is 0.8; (c) if a solar power system requires a power of 10 W at a voltage level of 10 V, calculate the number of solar cells needed in series and the number of rows in parallel for such a solar cell array. (This diode has the same features as the diode considered in Example 11.5.)

The open-circuit voltage was calculated in Example 11.5 and is 0.53 V. The short-circuit current is simply $I_L = 25$ mA. The power per solar cell is

$$P = 0.8I_{sc}V_{oc} = 0.8(25 \times 10^{-3})(0.53) = 1.06\ \text{mW}$$

The number of solar cells needed in series to produce an output voltage of 10 V is (each cell produces approximately $V_M \sim (F_f)^{1/2} \sim 0.9 V_{oc}$)

$$N(\text{series}) = \frac{10}{0.9 \times 0.53} \sim 24 \text{ cells}$$

The number of rows needed to produce a power of 10 W is now ($I_m \sim 0.9 I_{sc}$)

$$N(\text{parallel}) = \frac{10W}{10V(25 \times 10^{-3} \times 0.9A)} = 45 \text{ rows}$$

Thus the system needs a total of 1080 solar cells to meet the specifications.

11.4 *P-I-N* PHOTODETECTOR

An important mode of operation of the *p-n* (or *p-i-n*) diode under illumination is when the diode is under reverse-bias conditions. The reverse bias is, however, not so strong that there are breakdown effects, as in the avalanche photodiode to be discussed in the next section.

A schematic of the band profile of a *p-i-n* detector is shown in Fig. 11.9. Since the device is in reverse bias, the diode current in dark is I_0 and is independent of the applied bias. The photocurrent I_L is essentially due to the carriers generated in the depletion region (*i*-region) that are collected. The diode is reverse biased so that the entire *i*-region is depleted and has a strong electric field. The device response is fast since the photocurrent is primarily due to the prompt photocurrent discussed in Section 11.3.1. The maximum current that can be collected is (we assume that the intrinsic region is larger than the diffusion lengths of the electrons or holes)

$$I_L = eA \int_0^W G_L(x) dx \tag{11.20}$$

where W is the depletion width. In this expression we will take into account the fact that as photons penetrate a material, their intensity decreases through absorption. The generation rate at a point x is given from Eqn. 11.8 as

$$G_L(x) = \alpha J_{ph}(0) \ \exp(-\alpha x) \tag{11.21}$$

where $J_{ph}(0)$ is the photon flux (number per cm^2 per second) at $x = 0$. The photocurrent is then from Eqns. 11.20 and 11.21

$$I_L = eA J_{ph}(0) \left[1 - \exp\left(-\alpha W\right)\right] \tag{11.22}$$

If R is the reflectivity of the surface (i.e., the fraction of photons that actually go into the device is $1 - R$), the photocurrent is

$$I_L = eA J_{ph}(0)(I - R) \left[1 - \exp(-\alpha W)\right]$$

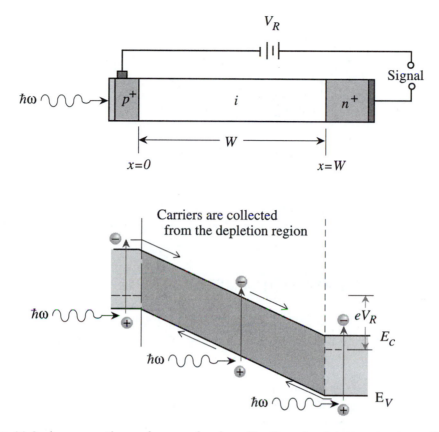

Figure 11.9: A cross-section and energy band profile of a *p-i-n* detector structure. Carriers generated in the depletion region are collected and contribute to the current. If the intrinsic region is thick, the photocurrent is dominated by carriers collected from the depletion region, since the carriers generated in the neutral regions contribute a smaller fraction of photocurrent. Since the photocurrent is dominated by the prompt photocurrent, the device response is fast.

One measure of the detector efficiency is the ratio of the photocurrent density to the incident flux,

$$\eta_{det} = \frac{I_L}{A J_{ph}(0)} = (I - R)\left[1 - \exp(-\alpha W)\right] \qquad (11.23)$$

For high efficiency one must have a small R (by placing anti-reflective coatings) and a long W. However, if W is too long the electron transit time, which controls the device speed, becomes too large, reducing the device speed. High-speed devices have W of about a micron or less and can operate at speeds in excess of 10 GHz.

EXAMPLE 11.7 Consider a silicon *p-i-n* photodiode with an intrinsic region of width 10 μm. Light from a GaAs laser at energy $\hbar\omega = 1.43$ eV impinges upon the diode. The optical power is 1 W/cm^2. Calculate the photocurrent density in the detector.

The photon flux incident on the detector is

$$\Phi_0 = \frac{P_{op}}{\hbar\omega} = \frac{1 \text{ W/cm}^2}{1.43(1.6 \times 10^{-19} \text{ J})} = 4.37 \times 10^{18} \text{ cm}^{-2} \text{ s}^{-1}$$

The absorption coefficient for Si at GaAs wavelength (i.e., photons with energy 1.43 eV) is $\sim$700 cm^{-1}. The photocurrent density is (assuming no reflection losses)

$$\begin{aligned}
J_L &= e\Phi_o \left\{ 1 - \exp -(\alpha W) \right\} \\
&= (1.6 \times 10^{-19})(4.37 \times 10^{18}) \left\{ 1 - \exp\left(-700 \times 10^{-3}\right) \right\} \\
&= 0.352 \text{ A/cm}^2
\end{aligned}$$

One can see from this example that Si detectors are capable of producing acceptable response to GaAs photons. Since Si technology is so advanced, one uses Si detectors for GaAs lasers.

11.5   LIGHT EMISSION: LIGHT-EMITTING DIODE

The simplicity of the light-emitting diode (LED) makes it a very attractive device for display and communication applications. The basic LED is a *p-n* junction that is forward biased to inject electrons and holes into the *p-* and *n*-sides respectively. The injected minority charge recombines with the majority charge in the depletion region or the neutral region. *In direct band semiconductors, this recombination leads to light emission since radiative recombination dominates in high-quality materials. In indirect gap materials, the light emission efficiency is quite poor and most of the recombination paths are nonradiative, which generates heat rather than light.* In the following subsections we will examine the important issues that govern the LED operation.

We will briefly outline some of the important considerations in choosing a semiconductor for LEDs or laser diodes.

Emission Energy: The light emitted from the device is very close to the semiconductor bandgap, since the injected electrons and holes are described by quasi-Fermi distribution functions. The desire for a particular emission energy may arise from a number of motivations. In Fig. 11.3 we show the response of the human eye to radiation of different wavelengths. Also shown are the bandgaps of some semiconductors. If a color display is to be produced that is to be seen by people, one has to choose an appropriate semiconductor. Very often one has to choose an alloy, since there is a greater flexibility in the bandgap range available. In Fig. 11.10 we show the loss characteristics of an optical fiber. As can be seen, the loss is least at 1.55 μm and 1.3 μm. If optical communication sources are desired, one must choose materials that can emit at these wavelengths. This is especially true if the communication is long haul, i.e., over hundreds or even thousands of kilometers. Materials like GaAs that emit at 0.8 μm can still be used for local area networks (LANs), which involve communicating within a building or local areas.

Substrate Availability: Almost all optoelectronic light sources depend upon epitaxial crystal growth techniques where a thin active layer (a few microns) is grown on a substrate (which is ∼ 200 μm). The availability of a high-quality substrate is extremely important in epitaxial technology. If a substrate that lattice-matches to the active device layer is not available, the device layer may have dislocations and other defects in it. These can seriously hurt device performance. The important substrates that are available for light-emitting technology are GaAs and InP. A few semiconductors and their alloys can match these substrates. The lattice constant of an alloy is the weighted mean of the lattice constants of the individual components, i.e., the lattice constant of the alloy $A_x B_{1-x}$ is

$$a_{all} = x a_A + (1 - x) a_B \qquad (11.24)$$

where a_A and a_B are the lattice constants of A and B. Semiconductors that cannot lattice-match with GaAs or InP have an uphill battle for technological success. The crystal grower must learn the difficult task of growing the semiconductor on a mismatched substrate without allowing dislocations to propagate into the active region.

Important semiconductor materials exploited in optoelectronics are the alloy $Ga_x Al_{1-x} As$, which is lattice-matched very well to GaAs substrates; $In_{0.53}Ga_{0.47}As$ and $In_{0.52}Al_{0.48}As$, which are lattice-matched to InP; InGaAsP, which is a quaternary material whose composition can be tailored to match with InP and can emit at 1.55 μm; and GaAsP, which has a wide range of bandgaps available. Recently there has been considerable interest in large bandgap materials such as ZnSe, ZnS, InN, GaN, and AlN to produce devices that emit green and blue light. The motivation is for superior display technology and high-density optical memory applications (a shorter wavelength allows reading of smaller features).

In general, the electron-hole recombination process can occur by radiative and nonradiative channels. Under the condition of minority carrier recombination or high injection recombination, as shown in Sections 3.8.1 and 3.8.2, one can define a lifetime for carrier recombination. If τ_r and τ_{nr} are the radiative and nonradiative lifetimes, the total recombination time is (for, say, an electron)

$$\frac{1}{\tau_n} = \frac{1}{\tau_r} + \frac{1}{\tau_{nr}} \qquad (11.25)$$

The internal quantum efficiency for the radiative processes is then defined as

$$\eta_{Qr} = \frac{\frac{1}{\tau_r}}{\frac{1}{\tau_r} + \frac{1}{\tau_{nr}}} = \frac{1}{1 + \frac{\tau_r}{\tau_{nr}}} \qquad (11.26)$$

In high-quality direct gap semiconductors, the internal efficiency is usually close to unity. In indirect materials the efficiency is of the order of 10^{-2} to 10^{-3}.

Before starting the discussion of light emission, let us remind ourselves of some

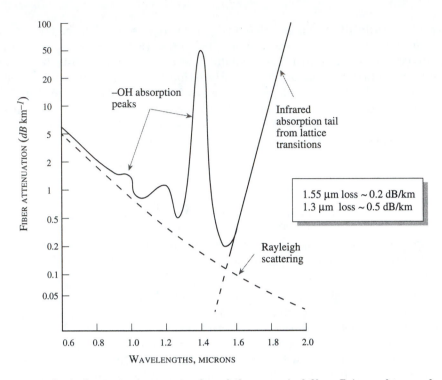

Figure 11.10: Optical attenuation vs. wavelength for an optical fiber. Primary loss mechanisms are identified as absorption and scattering.

important definitions and symbols used in this chapter:

I_{ph} : photon current = number of photons passing a cross-section/second.

J_{ph} : photon current density = number of photons passing a unit area/second.

P_{op} : optical power intensity = energy carried by photons per second per area.

11.5.1 Carrier Injection and Spontaneous Emission

The LED is essentially a forward-biased *p-n* diode, as shown in Fig. 11.11. Electrons and holes are injected as minority carriers across the diode junction and they recombine by either radiative recombination or nonradiative recombination. The diode must be designed so that the radiative recombination can be made as strong as possible.

The theory of the *p-n* diode was discussed in detail in Chapter 5. In the forward-bias conditions the electrons are injected from the *n*-side to the *p*-side while holes are injected from the *p*-side to the *n*-side. As noted in Chapter 5, the forward-bias current is dominated by the minority charge diffusion current. The diffusion current, in general, has three components: (i) minority carrier electron diffusion current; (ii) minority carrier hole diffusion current; and (iii) trap-assisted recombination current in the depletion region of width W. These current densities have the following forms, respectively (see Eqns. 5.42, 5.43, and 5.55):

$$J_n = \frac{eD_n n_p}{L_n}\left[\exp\left(\frac{eV}{k_B T}\right) - 1\right] \tag{11.27}$$

$$J_p = \frac{eD_p p_n}{L_p}\left[\exp\left(\frac{eV}{k_B T}\right) - 1\right] \tag{11.28}$$

$$J_{GR} = \frac{en_i W}{2\tau}\left[\exp\left(\frac{eV}{2k_B T}\right) - 1\right] \tag{11.29}$$

where τ is the recombination time in the depletion region and depends upon the trap density. *The LED is designed so that the photons are emitted close to the top layer and not in the buried layer, as shown in Fig. 11.11. The reason for this choice is that photons emitted deep in the device have a high probability of being reabsorbed. Thus one prefers to have only one kind of carrier injection for the diode current.* Usually the top layer of the LED is p-type, and for photons to be emitted in this layer one must require the diode current to be dominated by the electron current (i.e., $J_n \gg J_p$). The ratio of the electron current density to the total diode current density is called the injection efficiency γ_{inj}. Thus we have

$$\boxed{\gamma_{inj} = \frac{J_n}{J_n + J_p + J_{GR}}} \tag{11.30}$$

If the diode is pn^+, $n_p \gg p_n$ and, as can be seen from Eqns. 11.27 and 11.28, J_n becomes much larger than J_p. If, in addition, the material is high quality so that the recombination current is small, the injection efficiency approaches unity.

Once the minority charge (electrons) is injected into the doped neutral region (p-type), the electrons and holes will recombine to produce photons. They may also recombine nonradiatively via defects or via phonons. The radiative recombination process was discussed in Chapter 3, and we will briefly review it for the direct bandgap semiconductors.

As discussed in Section 3.8, the radiative process is "vertical," i.e., the k-value of the electron and that of the hole are the same in the conduction and valence bands, respectively. From Fig. 11.12 we see that the photon energy and the electron and hole energies are related by

$$\hbar\omega - E_g = \frac{\hbar^2 k^2}{2}\left[\frac{1}{m_e^*} + \frac{1}{m_h^*}\right] = \frac{\hbar^2 k^2}{2m_r^*} \tag{11.31}$$

where m_r^* is the reduced mass for the e-h system.

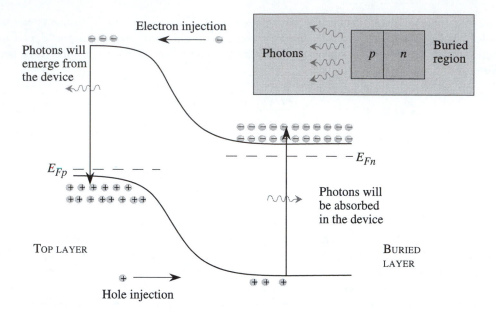

Figure 11.11: In a forward-biased p-n junction, electrons and holes are injected as shown. In the figure, the holes injected into the buried n region will generate photons that will not emerge from the surface of the LED. The electrons injected will generate photons that are near the surface and have a high probability to emerge.

If an electron is available in a state k and a hole is also available in the state k (i.e., if the Fermi functions for the electrons and holes satisfy $f^e(k) = f^h(k) = 1$), the radiative recombination rate is found to be

$$W_{em} \sim 1.5 \times 10^9 \hbar\omega(eV) \ s^{-1} \tag{11.32}$$

and the recombination time becomes ($\hbar\omega$ is expressed in electron volts)

$$\tau_o = \frac{0.67}{\hbar\omega(eV)} \ ns \tag{11.33}$$

The recombination time discussed above is the shortest possible spontaneous emission time since we have assumed that the electron has a unit probability of finding a hole with the same k-value.

When carriers are injected into the semiconductors, the occupation probabilities for the electron and hole states are given by the appropriate quasi-Fermi levels. The emitted photons leave the device volume so that the photon density never becomes high in the e-h recombination region. In a laser diode the situation is different, as we shall see later. The photon emission rate is given by integrating the emission rate W_{em} over all the electron-hole pairs after introducing the appropriate Fermi functions.

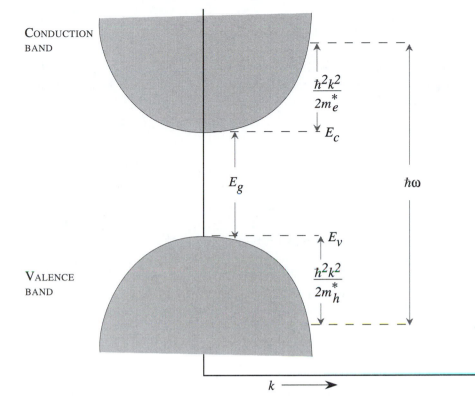

CONDUCTION
BAND

$$\frac{\hbar^2 k^2}{2m_e^*}$$

E_c

E_g

$\hbar\omega$

E_v

$$\frac{\hbar^2 k^2}{2m_h^*}$$

VALENCE
BAND

$k \longrightarrow$

Figure 11.12: A schematic of the *E-k* diagram for the conduction and valence bands. Optical transitions are vertical; i.e., the *k*-vector of the electron in the valence band and in the conduction band is the same.

There are several important limits of the spontaneous rate:

(i) *In the case where the electron and hole densities n and p are small* (nondegenerate case), the Fermi functions have a Boltzmann form ($\exp(-E/k_B T)$). The recombination rate is found to be

$$R_{spon} = \frac{1}{2\tau_o} \left(\frac{2\pi\hbar^2 m_r^*}{k_B T m_e^* m_h^*} \right)^{3/2} np \qquad (11.34)$$

The rate of photon emission depends upon the product of the electron and hole densities. If we define the lifetime of a single electron injected into a lightly doped ($p = N_a \leq 10^{17} \text{cm}^{-3}$) *p*-type region with hole density p, it would be given from Eqn. 11.34 by

$$\boxed{\frac{R_{spon}}{n} = \frac{1}{\tau_r} = \frac{1}{2\tau_o} \left(\frac{2\pi\hbar^2 m_r^*}{k_B T m_e^* m_h^*} \right)^{3/2} p} \qquad (11.35)$$

The time τ_r in this regime is very long (hundreds of nanoseconds), as shown in Fig. 11.13, and becomes smaller as p increases.

(ii) *In the case where electrons are injected into a heavily doped p-region (or holes are injected into a heavily doped n-region), the function $f^h(f^e)$ can be assumed to be unity.* The spontaneous emission rate is

$$R_{spon} \sim \frac{1}{\tau_o} \left(\frac{m_r^*}{m_h^*} \right)^{3/2} n \qquad (11.36)$$

for electron concentration n injected into a heavily doped p-type region and

$$R_{spon} \sim \frac{1}{\tau_o} \left(\frac{m_r^*}{m_e^*} \right)^{3/2} p \qquad (11.37)$$

for hole injection into a heavily doped n-type region.

The minority carrier lifetimes (i.e., n/R_{spon}) play a very important role not only in LEDs but also in diodes and bipolar devices. In this regime the lifetime of a single electron (hole) is independent of the holes (electrons) present since there is always a unity probability that the electron (hole) will find a hole (electron). The lifetime is now essentially τ_o, as shown in Fig. 11.13.

(iii) *Another important regime is that of high injection, where $n = p$ is so high that one can assume $f^e = f^h = 1$ in the integral for the spontaneous emission rate.* The spontaneous emission rate is

$$\boxed{R_{spon} \sim \frac{n}{\tau_o} \sim \frac{p}{\tau_o}} \qquad (11.38)$$

and the radiative lifetime ($n/R_{spon} = p/R_{spon}$) is τ_o.

(iv) *A regime that is quite important for laser operation is one where sufficient electrons and holes are injected into the semiconductor to cause "inversion."* As will be discussed later, this occurs if $f^e + f^h \geq 1$. If we make the approximation $f^e \sim f^h = 1/2$ for all the electrons and holes at inversion, we get the relation

$$R_{spon} \sim \frac{n}{4\tau_o} \qquad (11.39)$$

or the radiative lifetime at inversion is

$$\boxed{\tau \sim \frac{\tau_o}{4}} \qquad (11.40)$$

This value is a reasonable estimate for the spontaneous emission rate in lasers near threshold.

The radiative recombination depends upon the radiative lifetime τ_r and the non-radiative lifetime τ_{nr}. To improve the efficiency of photon emission one needs a value

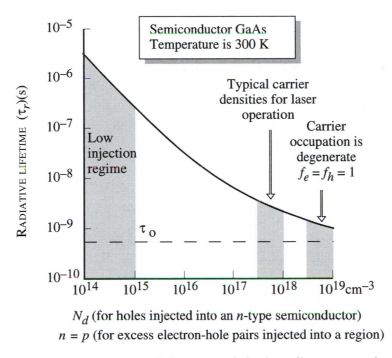

N_d (for holes injected into an n-type semiconductor)

$n = p$ (for excess electron-hole pairs injected into a region)

Figure 11.13: Radiative lifetimes of electrons or holes in a direct gap semiconductor as a function of doping or excess charge. The figure gives the lifetimes of a minority charge (a hole) injected into an n-type material. The figure also gives the lifetime behavior of electron-hole recombination when excess electrons and holes are injected into a material as a function of excess carrier concentration.

of τ_r as small as possible and τ_{nr} as large as possible. To increase τ_{nr} one must reduce the material defect density. This includes improving surface and interface qualities.

EXAMPLE 11.8 Calculate the e-h recombination time when an excess electron and hole density of $10^{15}\,\mathrm{cm}^{-3}$ is injected into a GaAs sample at room temperature.

Since 10^{15} cm^{-3} or $10^{21}\ m^{-3}$ is a very low level of injection, the recombination time is given by Eqn. 11.35 as

$$\frac{1}{\tau_r} = \frac{1}{2\tau_o}\left(\frac{2\pi\hbar^2 m_r^*}{k_B T m_e^* m_h^*}\right)^{3/2} p$$

$$= \frac{1}{2\tau_o}\left(\frac{2\pi\hbar^2}{k_B T m_e^* + m_h^*}\right)^{3/2} p$$

Using $\tau_o = 0.6$ ns and $k_B T = 0.026$ eV, we get for $m_e^* = 0.067\,m_o, m_h^* = 0.45\,m_o,$

$$\frac{1}{\tau_r} = \frac{10^{21}\ \mathrm{m}^{-3}}{2\times(0.6\times10^{-9}\ \mathrm{s})}\left[\frac{2\times3.1416\times(1.05\times10^{-34}\ \mathrm{Js})^2}{(0.026\times1.6\times10^{-19}\ \mathrm{J})\times(0.517\times9.1\times10^{-31}\ \mathrm{kg})}\right]^{3/2}$$

$$\tau_r = 5.7 \times 10^{-6} s \cong 9.5 \times 10^3 \tau_o$$

We see from this example that at low injection levels, the carrier lifetime can be very long. Physically, this occurs because at such a low injection level, the electron has a very small probability of finding a hole to recombine with.

EXAMPLE 11.9 In two n^+p GaAs LEDs, $n^+ \gg p$ so that the electron injection efficiency is 100% for both diodes. If the nonradiative recombination time is 10^{-7}s, calculate the 300 K internal radiative efficiency for the diodes when the doping in the p-region for the two diodes is 10^{16} cm^{-3} and 5×10^{17} cm^{-3}.

When the p-type doping is 10^{16} cm^{-3}, the hole density is low and the e-h recombination time for the injected electrons is given by Eqn. 11.35 as

$$\frac{1}{\tau_r} = \frac{1}{2\tau_o} \left(\frac{2\pi\hbar^2 m_r^*}{k_B T m_e^* m_h^*} \right)^{3/2} p$$

From the previous example, we can see that for p equal to 10^{16} cm^{-3}, we have (in the previous example the value of p was ten times smaller)

$$\tau_r = 5.7 \times 10^{-7} \text{ s}$$

In the case where the p-doping is high, the recombination time is given by the high-density limit (see Eqn. 11.36) as

$$\frac{1}{\tau_r} = \frac{R_{spon}}{n} = \frac{1}{\tau_o} \left(\frac{m_r^*}{m_h^*} \right)^{3/2}$$

$$\tau_r = \frac{\tau_o}{0.05} \sim 20\tau_o \sim 12 \text{ ns}$$

For the low-doping case, the internal quantum efficiency for the diode is

$$\eta_{Qr} = \frac{1}{1 + \frac{\tau_r}{t_{nr}}} = \frac{1}{1 + (5.7)} = 0.15$$

For the more heavily doped p-region diode, we have

$$\eta_{Qr} = \frac{1}{1 + \frac{10^{-7}}{20 \times 10^{-9}}} = 0.83$$

Thus there is an increase in the internal efficiency as the p doping is increased.

EXAMPLE 11.10 Consider a GaAs p-n diode with the following parameters at 300 K:

Electron diffusion coefficient,	D_n	=	30 cm^2/V $\cdot$ s
Hole diffusion coefficient,	D_p	=	15 cm^2/V $\cdot$ s
p-side doping,	N_a	=	5×10^{16} cm^{-3}
n-side doping,	N_d	=	5×10^{17} cm^{-3}
Electron minority carrier lifetime,	τ_n	=	10^{-8} s
Hole minority carrier lifetime,	τ_p	=	10^{-7} s

Calculate the injection efficiency of the LED assuming no recombination due to traps.

The intrinsic carrier concentration in GaAs at 300 K is 1.84×10^6 cm^{-3}. This gives

$$n_p = \frac{n_i^2}{N_a} = \frac{(1.84 \times 10^6)^2}{5 \times 10^{16}} = 6.8 \times 10^{-5} \text{ cm}^{-3}$$

$$p_n = \frac{n_i^2}{N_d} = \frac{(1.84 \times 10^6)^2}{5 \times 10^{17}} = 6.8 \times 10^{-6} \text{ cm}^{-3}$$

The diffusion lengths are

$$L_n = \sqrt{D_n \tau_n} = \left[(30)(10^{-8}) \right]^{1/2} = 5.47 \ \mu m$$

$$L_p = \sqrt{D_p \tau_p} = \left[(15)(10^{-7}) \right]^{1/2} = 12.25 \ \mu m$$

The injection efficiency is now (assuming no recombination via traps)

$$\gamma_{inj} = \frac{\frac{e D_n n_{po}}{L_n}}{\frac{e D_n n_{po}}{L_n} + \frac{e D_p p_{no}}{L_p}} = 0.98$$

EXAMPLE 11.11 Consider the p-n^+ diode of the previous example. The diode is forward biased with a forward-bias potential of 1 V. If the radiative recombination efficiency $\eta_{Qr} = 0.5$, calculate the photon flux and optical power generated by the LED. The diode area is 1 mm^2.

The electron current injected into the p-region will be responsible for the photon generation. This current is

$$\begin{aligned}
I_n &= \frac{A e D_n n_{po}}{L_n} \left[\exp \left(\frac{eV}{k_B T} \right) - 1 \right] \\
&= \frac{(10^{-2} \text{ cm}^2)(1.6 \times 10^{-19} \text{ C})(30 \text{ cm}^2/s)(6.8 \times 10^{-5} \text{ cm}^{-3})}{5.47 \times 10^{-4} \text{ cm}} \left[\exp \left(\frac{1}{0.026} \right) - 1 \right] \\
&= 0.30 \text{ mA}
\end{aligned}$$

The photons generated per second are

$$\begin{aligned}
I_{ph} = \frac{I_n}{e} \cdot \eta_{Qr} &= \frac{(0.30 \times 10^{-3} \text{ A})(0.5)}{1.6 \times 10^{-19} \text{ C}} \\
&= 9.38 \times 10^{14} \text{ s}^{-1}
\end{aligned}$$

Each photon has an energy of 1.41 eV (= bandgap of GaAs). The optical power is thus

$$\begin{aligned}
\text{Power} &= (9.38 \times 10^{14} \text{ s}^{-1})(1.41)(1.6 \times 10^{-19} \text{ J}) \\
&= 0.21 \text{ mW}
\end{aligned}$$

11.6   SEMICONDUCTOR LASER: BASIC PRINCIPLES

The semiconductor laser diode is a forward-bias p-n junction, just like the LED studied in the previous sections. *However, while the structure appears similar to the LED as far as the electron and holes are concerned, it is quite different from the point of view of the photons.*

As in the case of the LED, electrons and holes are injected into an active region by forward biasing the laser diode. At low injection, these electrons and holes recombine radiatively via the spontaneous emission process to emit photons. However, the laser structure is so designed that at higher injections emission occurs by a process known as stimulated emission. As we will discuss below, the stimulated emission process provides spectral purity to the photon output, provides coherent photons, and offers high-speed performance. *Thus the key difference between the LED and the laser diode arises from the difference between spontaneous and stimulated emission.*

11.6.1   Spontaneous and Stimulated Emission:
The Need for an Optical Cavity

The key to understanding the semiconductor laser diode is the physics behind spontaneous and stimulated emission. Let us develop this understanding using Fig. 11.14. Consider an electron with wave vector k and a hole with a wave vector k in the conduction and valence bands, respectively, of a semiconductor. In the case shown in Fig. 11.14a, initially there are no photons in the semiconductor. The electron and hole recombine to emit a photon as shown, and this process is the spontaneous emission. The spontaneous emission rate was discussed in the context of the LED.

In Fig. 11.14b, we show the electron-hole pair along with photons of energy $\hbar\omega$ equal to the electron-hole energy difference. In this case, in addition to the spontaneous emission rate, one has an additional emission rate called the stimulated emission process. The stimulated emission process is proportional to the photon density (of photons with the correct photon energy to cause the *e-h* transition). *The photons that are emitted are in phase (i.e., same energy and wave vector) with the incident photons.* Quantum-mechanical calculations show that the rate for stimulated emission is

$$\boxed{W_{em}^{st}(\hbar\omega) = W_{em}(\hbar\omega) \cdot n_{ph}(\hbar\omega)} \qquad (11.41)$$

where $n_{ph}(\hbar\omega)$ is the photon density and W_{em} is the spontaneous emission rate we have already discussed. Thus if $n_{ph}(\hbar\omega) \sim 0$, there is no stimulated emission process. In the LED, when photons are emitted by spontaneous emission, they are either lost by reabsorption or simply leave the structure. Thus $n_{ph}(\hbar\omega)$ remains extremely small and stimulated emission cannot get started.

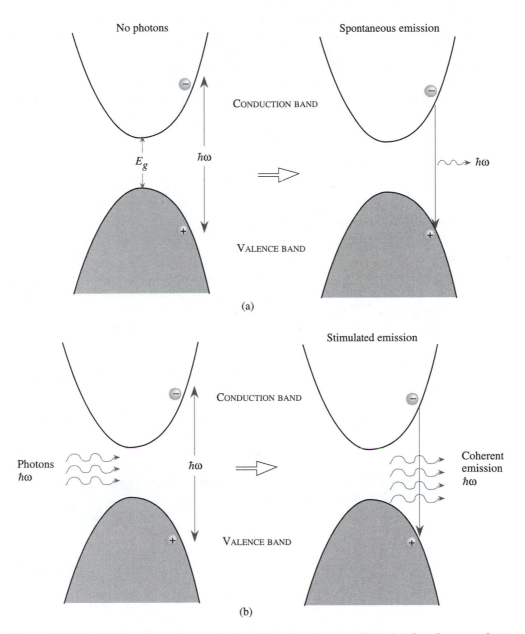

Figure 11.14: (a) In spontaneous emission, the *e-h* pair recombines in the absence of any photons present to emit a photon. (b) In simulated emission, an *e-h* pair recombines in the presence of photons of the correct energy $\hbar\omega$ to emit coherent photons. In coherent emission the phase of the photons emitted is the same as the phase of the photons causing the emission.

Consider now the possibility that, when photons are emitted via spontaneous emission, an optical cavity is designed so that photons with a well defined energy are selectively confined in the semiconductor structure. If this is possible, two important effects occur: (i) the photon emission for photons with the chosen energy becomes stronger due to stimulated emission; (ii) the e-h recombination rate increases, as can be seen from Eqn. 11.41. These two effects are highly desirable since they produce an optical spectrum with very narrow emission lines and the light output can be modulated at high speeds.

The challenge for the design of the laser is, therefore, to incorporate an optical cavity that ensures that the photons emitted are allowed to build up in the semiconductor device so that stimulated emission can occur.

11.6.2   The Laser Structure: Optical Cavity

While both the LED and the laser diode use a forward-biased *p-n* junction to inject electrons and holes to generate light, the laser structure is designed to create an "optical cavity" that can "guide" the photons generated. The optical cavity is essentially a resonant cavity in which the photons have multiple reflections. Thus, when photons are emitted, only a small fraction is allowed to leave the cavity. As a result, the photon density in the cavity starts to build up. A number of important cavities are used for solid state lasers. These are the Fabry-Perot cavity, cylindrical cavity, rectangular cavity, etc. For semiconductor lasers, the most widely used cavity is the Fabry-Perot cavity shown in Fig. 11.15a. The important ingredient of the cavity is a polished mirror surface that assures that resonant modes are produced in the cavity as shown in Fig. 11.15b. These resonant modes are those for which the wave vectors of the photon satisfy the relation

$$L = q\lambda/2 \qquad (11.42)$$

where q is an integer, L is the cavity length, and λ is the light wavelength in the material and is related to the free-space wavelength by

$$\lambda = \frac{\lambda_o}{n_r} \qquad (11.43)$$

where n_r is the refractive index of the cavity. The spacing between the stationary modes is given by

$$\Delta k = \frac{2\pi}{L} \qquad (11.44)$$

As can be seen from Fig. 11.15a, the Fabry-Perot cavity has mirrored surfaces on two sides. The other sides are roughened so that photons emitted through these sides are not reflected back and are not allowed to build up. Thus only the resonant modes are allowed to build up and participate in the stimulated emission process.

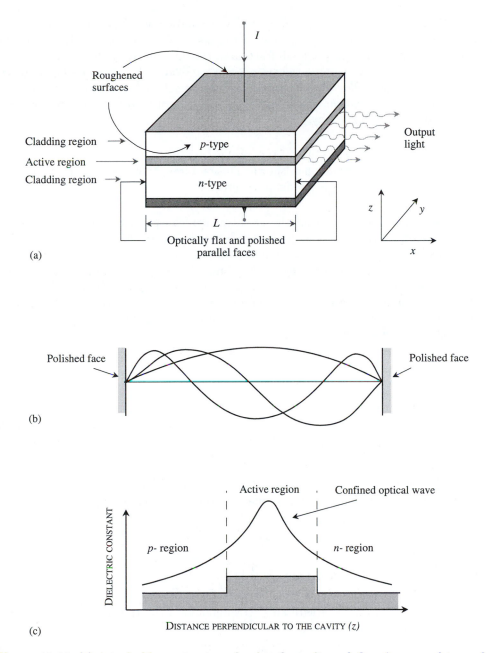

Figure 11.15: (a) A typical laser structure showing the cavity and the mirrors used to confine photons. The active region can be quite simple, as in the case of double heterostructure lasers, or quite complicated, as in the case of quantum-well lasers. (b) The stationary states of the cavity. The mirrors are responsible for these resonant states. (c) The variation in dielectric constant is responsible for the optical confinement. The structure for the optical cavity shown in this figure is called the Fabry-Perot cavity.

While the optical cavity can confine the photons with certain characteristics, it must be noted that the active region of the laser in which electron-hole pairs are recombining may occupy only a small fraction of the optical cavity. *It is important that a large fraction of the optical waveform overlap with the active region since only this fraction will be responsible for stimulated emission. As a result, it is important to design the laser structure so that the optical wave has a high probability of being in the region where e-h pairs are recombining.*

If a planar heterostructure of the form shown in Fig. 11.15c is used, the optical wave is confined in the z-direction as shown. This requires the light-confining or cladding layers to be made from a large bandgap material.

Many advances in laser physics are being driven by superior optical cavities. In the above discussion the optical confinement is improved by the heterostructure cladding layers. This is straightforward to do in an epitaxial process. It is somewhat difficult to produce dielectric constant variation in the plane of the laser, i.e., in the y-direction. Thus, the laser is usually fabricated as a strip of width $\sim 10~\mu m - 50~\mu m$. The strip is produced by etching. It is also possible to produce "buried" lasers where the y-direction optical confinement is produced by doping or defect introduction since these processes can also change the dielectric constant.

An important parameter of the laser cavity is the optical confinement factor Γ, which gives the fraction of the optical wave in the active region,

$$\Gamma = \frac{\int_{\text{active region}} |F(z)|^2 dz}{\int |F(z)|^2 dz} \tag{11.45}$$

This confinement factor is almost unity for "bulk" double heterostructure lasers where the active region is $\gtrsim 1.0~\mu m$, while it is as small as 1% for advanced quantum-well lasers.

11.6.3   Optical Absorption, Loss, and Gain

The photon current associated with an electromagnetic wave traveling through a semi-conductor is described by

$$I_{ph} = I_{ph}^0 \exp(-\alpha x) \tag{11.46}$$

where α (the absorption coefficient) is usually positive and I_{ph}^0 is the incident light intensity at $x = 0$. The optical intensity, which is the photon current multiplied by the photon energy $\hbar\omega$, falls as the wave travels if α is positive. However, if electrons are pumped in the conduction band and holes in the valence band, the electron-hole recombination process (photon emission) can be stronger than the reverse process of electron-hole generation (photon absorption). In general, the gain coefficient is defined by gain = emission coefficient−absorption coefficient. If $f^e(E^e)$ and $f^h(E^h)$ denote the electron and hole occupation, the emission coefficient depends upon the product of $f^e(E^e)$ and $f^h(E^h)$ while the absorption coefficient depends upon the product of $(1-f^e(E^e))$ and $(1-f^h(E^h))$. Here the energies E^e and E^h are related to the photon energy by the condition of vertical k-transitions. For these transitions we have

$$
\begin{aligned}
E^e &= E_c + \frac{m_r^*}{m_e^*}(\hbar\omega - E_g) \\
E^h &= E_v - \frac{m_r^*}{m_h^*}(\hbar\omega - E_g)
\end{aligned}
\tag{11.47}
$$

The occupation probabilities f^e and f^h are determined by the quasi-Fermi levels for electrons and holes, as discussed in Section 3.7.1.

The gain, which is the difference of the emission and absorption coefficient, is now proportional to

$$
g(\hbar\omega) \sim f^e(E^e)\cdot f^h(E^h) - \{1 - f^e(E^e)\}\{1 - f^h(E^h)\} = \{f^e(E^e) + f^h(E^h)\} - 1 \tag{11.48}
$$

The optical wave has a general spatial intensity dependence:

$$
I_{ph} = I_{ph}^0 \ \exp\ (g(\hbar\omega)x) \tag{11.49}
$$

and *if g is positive, the intensity grows because additional photons are added by emission to the intensity.* The condition for positive gain requires "inversion" of the semiconductor system, i.e., from Eqn. 11.48,

$$
\boxed{f^e(E^e) + f^h(E^h) > 1} \tag{11.50}
$$

The quasi-Fermi levels must penetrate their respective bands for this condition to be satisfied. It is found that gain is approximately given by

$$
\boxed{g(\hbar\omega) \cong 5.7 \times 10^4 \frac{(\hbar\omega - E_g)^{1/2}}{\hbar\omega}\left[f^e(E^e) + f^h(E^h) - 1\right]\ \mathrm{cm}^{-1}} \tag{11.51}
$$

To evaluate the actual gain in a material as a function of carrier injection $n\ (= p)$, one has to find the electron and hole quasi-Fermi levels and the occupation probabilities $f^e(E^e)$ and $f^h(E^h)$, where E^e and E^h are related to $\hbar\omega$ by Eqn. 11.47. The procedure is described in more detail through Examples 11.13 and 11.14.

It must be noted that the laser operates under conditions where f^e and f^h are larger than 0.5. In this high-injection limit, the occupation probabilities are not given accurately by the Boltzmann statistics. A useful approach is to use the Joyce-Dixon

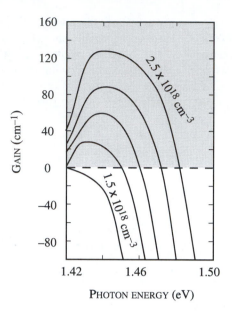

Figure 11.16: Gain vs. photon energy curves for a variety of carrier injections for GaAs at 300 K. The electron and hole injections are the same.

approximation for the position of the Fermi levels. For a given injection density $n \, (= p)$, the position of the quasi-Fermi levels is given by

$$E_{Fn} = E_c + k_B T \left[\ell n \frac{n}{N_c} + \frac{1}{\sqrt{8}} \frac{n}{N_c} \right] \tag{11.52}$$

$$E_{Fp} = E_v - k_B T \left[\ell n \frac{p}{N_v} + \frac{1}{\sqrt{8}} \frac{p}{N_v} \right] \tag{11.53}$$

where N_c and N_v are the effective density of states at the conduction and valence bands.

With these expressions the gain can be calculated as a function of photon energy for various levels of injection densities $n \, (= p)$. At low injections, f^e and f^h are quite small and the gain is negative. However, as injection is increased, for electrons and holes near the bandedges, f^e and f^h increase and gain can be positive. However, even at high injections, for $\hbar\omega \gg E_g$, the gain is negative. The general form of the gain-energy curves for different injection levels is shown in Fig. 11.16.

The gain discussed above is called the material gain and comes only from the active region where the recombination is occurring. Often this active region is of very small dimensions. In this case, one needs to define the cavity gain, which is given by

$$\boxed{\text{Cavity gain} = g(\hbar\omega)\Gamma} \tag{11.54}$$

where Γ is the fraction of the optical intensity overlapping with the gain medium, as discussed in Section 11.6.2. The value of Γ is almost unity for double heterostructure lasers and ~ 0.01 for quantum-well lasers. In quantum-well lasers, the overall cavity gain can still be very high since the gain in the quantum well is very large for a fixed injection density when compared to bulk semiconductors.

The optical wave suffers two important sources of photon loss in the cavity. Free carrier loss where light is absorbed by free carriers in the conduction or the valence band is one important source. We denote by α_{loss} the absorption coefficient due to this and other defect-related absorption in the cavity. The other important loss is due to the loss of photons from the cavity by escape to the outside. We can represent this photon loss also as a loss coefficient α_R. If R is the reflection coefficient of the mirror in the cavity, the fraction of photons lost after traveling a distance length L is $(1 - R)$. The loss coefficient α_R is thus defined as

$$1 - \exp\left(-\alpha_R L\right) = 1 - R$$

or

$$\boxed{\alpha_R = -\frac{1}{L}\ (\ell n R)} \tag{11.55}$$

For the GaAs-air interface, the value of the reflection coefficient is

$$R = \frac{\left(n_r(\text{GaAs}) - 1\right)^2}{\left(n_r(\text{GaAs}) + 1\right)^2} \sim 0.33 \tag{11.56}$$

since the refractive index n_r of GaAs is 3.66.

EXAMPLE 11.12 Consider a GaAs Fabry-Perot laser cavity. The absorption loss in the cavity is given by an absorption coefficient of 20 cm^{-1}. Calculate the cavity length at which the absorption loss and the mirror loss become equal.

The length of the cavity is given by

$$\alpha_R = \alpha_{loss} = -\frac{1}{L}\ \ell n R$$

or

$$L = \frac{-1}{20}\ \ell n(0.33) = 554\ \mu m$$

11.6.4   The Laser Below and Above Threshold

In Fig. 11.17 we show the light output as a function of injected current density in a laser diode (LD). If we compare this with the output from an LED shown in Fig. 11.8 we notice an important difference. The light output from a laser diode displays a rather abrupt change in behavior below the "threshold" condition and above this condition. The threshold condition is usually defined as the condition where the cavity gain overcomes the cavity loss for any photon energy, i.e., when

$$\boxed{\Gamma g(\hbar\omega) = \alpha_{loss} - \frac{\ell n R}{L}} \qquad (11.57)$$

In high-quality lasers $\alpha_{loss} \sim 10 \text{ cm}^{-1}$ and the reflection loss may contribute a similar amount. Another useful definition in the laser is the condition of transparency when the light suffers no absorption or gain, i.e.,

$$\boxed{\Gamma g(\hbar\omega) = 0} \qquad (11.58)$$

When the *p-n* diode making up the semiconductor laser is forward biased, electrons and holes are injected into the active region of the laser. These electrons and holes recombine to emit photons. It is important to identify two distinct regions of operation of the laser. Referring to Fig. 11.18, when the forward-bias current is small, the number of electrons and holes injected is small. As a result, the gain in the device is too small to overcome the cavity loss. The photons that are emitted are either absorbed in the cavity or lost to the outside. Thus, in this regime there is no buildup of photons in the cavity. However, as the forward bias increases, more carriers are injected into the device until eventually the threshold condition is satisfied for some photon energy. As a result, the photon number starts to build up in the cavity. As the device is further biased beyond threshold, stimulated emission starts to occur and dominates the spontaneous emission. The light output in the photon mode for which the threshold condition is satisfied becomes very strong.

Below the threshold, the device essentially operates as an LED except that there is a higher cavity loss in the laser diode since photons cannot escape from the device due to the mirrors. Let β_{loss} be the fraction of photons that cannot escape from the device. The photon current output is given by

$$
\begin{aligned}
I_{ph} &= (1 - \text{loss}) \text{ (total } e\text{-}h \text{ recombination per second)} \\
&= (1 - \text{loss}) \text{ (electron particle current)} \qquad (11.59)
\end{aligned}
$$

i.e.,

$$I_{ph} = (1 - \beta_{loss}) (R_{spon} A d_{las}) = (1 - \beta_{loss})\frac{I}{e} \qquad (11.60)$$

where A is the laser cavity area, d_{las} is the thickness of the active layer where the recombination is occurring, and I is the injected current. The light output I_{ph} is lower than a corresponding output in an LED due to the high value of the photon loss term β_{loss}. This situation is shown schematically in Fig. 11.17. Figure 11.18a shows this regime of operation.

Once the carrier density of the electrons and holes is high enough that the threshold condition given by Eqn. 11.57 is met, the photons generated in the laser

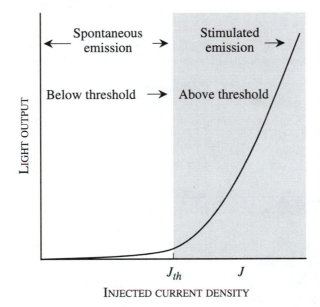

Figure 11.17: The light output as a function of current injection in a semiconductor laser. Above threshold, the presence of a high photon density causes stimulated emission to dominate.

cavity grow in intensity after emission. Of course, out of all the optical modes that are allowed in the cavity, one or two will have the highest gain since the gain curves have a peak at some energy, as seen from Fig. 11.18b. Since the gain is positive, the photon density in the laser cavity starts to increase rapidly. As a result, the stimulated emission process starts to grow. As noted in Section 11.6.1, the stimulated emission rate is related to the spontaneous emission rate by

$$W_{em}^{st}(\hbar\omega) = W_{em}(\hbar\omega) \cdot n_{ph}(\hbar\omega) \tag{11.61}$$

where $n_{ph}(\hbar\omega)$ is the photon density in the mode.

In order to study the laser characteristics around and above threshold, let us establish a simple relation between the injected current density, radiative lifetime, dimensions of the active region where recombination occurs, and the carrier density ($n = p$) in the active region. The rate of arrival of electrons (holes) into the active region is

$$\frac{JA}{e}$$

The rate at which the injected e-h pairs recombine is

$$\frac{nAd_{las}}{\tau_r(J)}$$

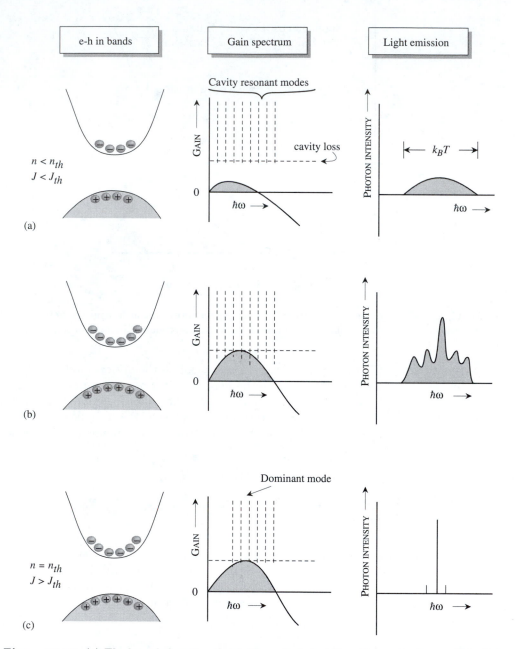

Figure 11.18: (a) The laser below threshold. The gain is less than the cavity loss and the light emission is broad as in an LED. (b) The laser at threshold. A few modes start to dominate the emission spectrum. (c) The laser above threshold. The gain spectrum does not change but, due to the stimulated emission, a dominant mode takes over the light emission.

where $\tau_r(J)$ is the current-density-dependent radiative lifetime. Assuming a radiative efficiency of unity, we equate the two results given above to get

$$n = \frac{J\tau_r(J)}{ed_{las}} \qquad (11.62)$$

At threshold we have

$$n_{th} = \frac{J_{th}\tau_r(J_{th})}{ed_{las}} \qquad (11.63)$$

As discussed in Section 11.5.1, at threshold $\tau_r(J_{th}) \sim 4\tau_o$ ($\sim$2 ns for a GaAs laser).

As the current density exceeds J_{th}, the photon density in the dominant mode builds up as discussed above, and the value of τ_r starts to become smaller. *As a result, even though the injected charge density increases, the carrier density in the active region saturates close to the threshold density n_{th}.*

The light output is given by ($n = n_{th}$; use Eqn. 11.62 for the current density)

$$I_{ph} = \frac{I}{e} = \frac{n_{th}Ad_{las}}{\tau_r} \qquad (11.64)$$

Upon comparing this equation with the results for light output from an LED, it may be seen that the photon current is similar for the same injected current for an LED and LD (biased above threshold). However, in the case of the laser diode, the entire photon output emerges only in one or two photon modes rather than in a broad spectrum of width k_BT. This spectral purity that arises because of the importance of stimulated emission distinguishes the LD from an LED. Also, the light output is highly collimated and coherent for similar reasons.

An important relationship is given by Eqn. 11.63 among n_{th}, J_{th}, and the active region thickness d_{las}. From Eqn. 11.63 it is clear that for low threshold current density, the active layer thickness should be decreased. However, from Eqn. 11.64 it is seen that for higher optical output the active layer thickness should be large. Depending upon the application, one designs a semiconductor laser according to these results. It has been found that lasers with active regions made from very thin layers called quantum wells ($\sim$100 Å) produce the lowest threshold current devices.

EXAMPLE 11.13 According to the Joyce-Dixon approximation, the relation between the Fermi level and carrier concentration is given by

$$E_F - E_c = k_BT\left[\ell n\frac{n}{N_c} + \frac{1}{\sqrt{8}}\,\frac{n}{N_c}\right]$$

where N_c is the effective density of states for the band. Calculate the carrier density needed for the transparency condition in GaAs at 300 K and 77 K. The transparency condition is defined at the situation where the maximum gain is zero (i.e., the optical beam propagates without loss or gain).

At room temperature the valence and conduction band effective density of states are

$$N_v = 7 \times 10^{18} \text{ cm}^{-3}$$
$$N_c = 4.7 \times 10^{17} \text{ cm}^{-3}$$

The values at 77 K are

$$N_v = 0.91 \times 10^{18} \text{ cm}^{-3}$$
$$N_c = 0.61 \times 10^{17} \text{ cm}^{-3}$$

In the semiconductor laser, an equal number of electrons and holes are injected into the active region. We will look for the transparency conditions for photons with energy equal to the bandgap. The approach is very simple: (i) choose a value of n or p; (ii) calculate μ from the Joyce-Dixon approximation; (iii) calculate $f^e + f^h - 1$ and check if it is positive at the bandedge. The same approach can be used to find the gain as a function of $\hbar\omega$.

For 300 K we find that the material is transparent when $n \sim 1.1 \times 10^{18}$ cm^{-3} at 300 K and $n \sim 2.5 \times 10^{17}$ cm^{-3} at 77 K. Thus a significant decrease in the injected charge occurs as temperature is decreased.

EXAMPLE 11.14 Consider a GaAs laser at 300 K. The optical confinement factor is unity. Calculate the threshold carrier density assuming that it is 20% larger than the density for transparency. If the active layer thickness is 2 μm, calculate the threshold current density.

From the previous example we see that at transparency

$$n = 1.1 \times 10^{18} \text{ cm}^{-3}$$

The threshold density is then

$$n_{th} = 1.32 \times 10^{18} \text{ cm}^{-3}$$

The radiative recombination time is approximately four times τ_o, i.e., $\sim$2.4 ns. The current density then becomes

$$J_{th} = \frac{e \cdot n_{th} \cdot d}{\tau_r} = \frac{(1.6 \times 10^{-19} \text{ C})(1.32 \times 10^{18} \text{ cm}^{-3})(2 \times 10^{-4} \text{ cm})}{2.4 \times 10^{-9} \text{ s}}$$
$$= 1.76 \times 10^4 \text{ A/cm}^2$$

11.7   SUMMARY

In this chapter we have examined optoelectronic devices that can convert an optical signal to an electrical signal. Most of these devices depend upon band-to-band transitions in which electrons are transferred from the valence band to the conduction band by the photons. The key findings of this chapter are summarized in the chapter summary table (Table 11.1).

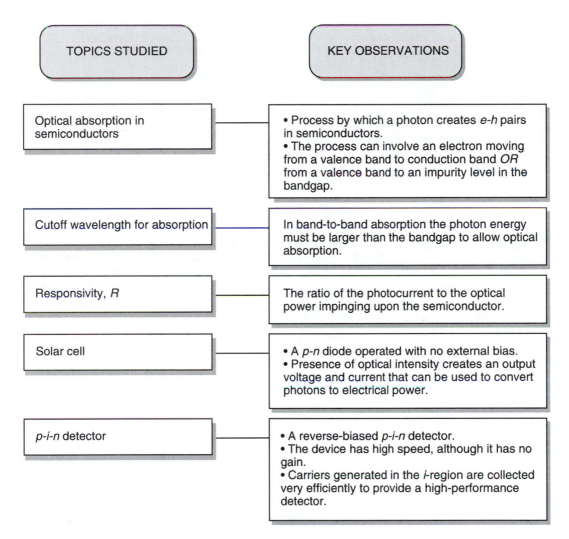

Table 11.1: Summary table.

11.8 A BIT OF HISTORY

The interaction of light with crystals has been a field of study for a long time. Some of the earliest uses of semiconductors exploited optical effects. Optical effects in materials were of key importance in ushering in the new age of quantum mechanics. This occurred when Einstein studied the photoelectric effect, which proved that light acted as a particle of energy $\hbar\omega$. This allowed one to experimentally determine Planck's constant.

Optical interactions with semiconductors were an important area of study in Prof. Pohl's laboratory in Göttingen during the early part of this century. He and his

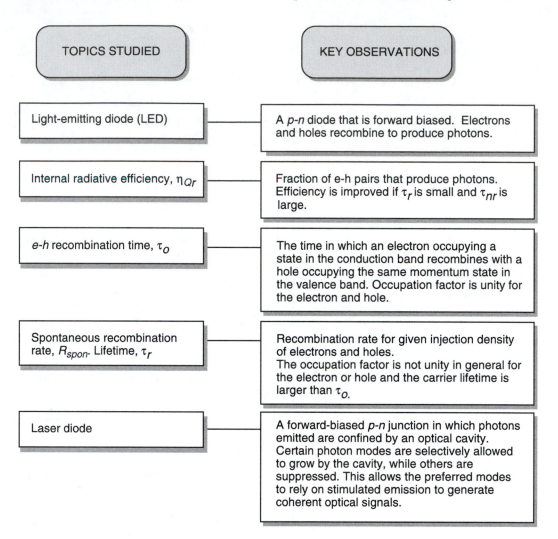

Table 11.1: Summary table (continued).

co-workers found that the conductivity of certain crystals was altered when they were irradiated with light. Industries were using semiconductors for scintillation counters as early as 1915.

When semiconductor technology blossomed in the '50s, the main applications were in the area of electronic devices. All of the intelligent decision-making devices were electronic in nature. Optoelectronic devices based on semiconductors were developed either as display devices or as devices used to convert electromagnetic signals into electrical signals. An important application was in the area of solar energy conversion in the space program. However, the solar energy conversion has not proceeded as its proponents expected, partly because of the politics involved and partly because of the difficulty in competing with fossil fuels. However, a renewed sense of environmental concern may push solar cells back into the forefront of energy alternatives.

The detector has certainly been acquiring an important role due to the increasing importance of optical communication. The optical detector is still not at a level where it can compete with microwave detectors. Coherent detection of optical signals is still not a perfected technology, so that the full benefits of optical communications are far from being realized. In fact, we may argue that the best of optoelectronics is still to come, since the device needs are quite apparent, but device technology is unable to meet these needs at the moment.

Stimulated and spontaneous emission processes, the two underlying processes behind optical sources, caused a great deal of difficulty for scientists in the early part of this century. As we have discussed, the spontaneous emission of a photon is supposed to occur *without* the presence of photons, a concept that is not consistent with classical physics or even semi-classical quantum physics. How can you disturb a system without a perturbation? And if there are no photons, there is no perturbation. The dilemma was resolved by Dirac, and now we know that even in the absence of photons there is a "zero point" photon energy that is responsible for spontaneous emission.

The concept that spontaneous and stimulated emission could be used to generate very narrow bandwidth power was first demonstrated, not in the optical regime, but in the microwave regime. For this work, which started in the 1940s, the 1964 Nobel prize was awarded to the American scientist Charles H. Townes and to the Russian scientists A. M. Prokorov and N. Basov. Following the success of the maser (microwave amplification by stimulated emission of radiation), many scientists attempted to generalize the results to the optical domain. In 1960, T. H. Maiman was able to build the first ruby laser at the Hughes Research Laboratories.

As the importance of semiconductors increased, it was realized that lasing could be achieved in these materials as well. In 1962, W. P. Dumke showed that laser action should be possible in direct bandgap materials such as GaAs, and scientists at IBM, GE, and Bell Labs were able to show lasing in semiconductors. The initial homojunction lasers had very poor efficiencies, and only after H. Kroemer suggested the use of heterostructures did the efficiency improve to levels where room-temperature laser operation was routinely possible.

Optoelectronic devices such as LEDs and lasers have become increasingly important, thanks to developments in optical fiber technology. With the availability of optical fibers with losses approaching 0.2 dB/km, optical communication has become the obvious choice for voice and data transmission. This technology is one of the fastest growing components in our "high-tech" society.

Optoelectronics continues to have a lot of success stories, but it has also suffered from overoptimistic scenarios. The attraction of the term "computation at the speed of light" has caused many people to expect that "optical computers" will replace conventional electronic computers. This has not happened, and closer examination suggests that it is not likely to happen for several decades. However, as intelligent optoelectronic devices become available, they will be part of the growing optoelectronic family, which will supplement the electronic family in information processing. Most of these intelligent devices are yet to be invented, making this area an extremely exciting one.

11.9 PROBLEMS

Section 11.2
Problem 11.1 The bandgap of the $Hg_{1-x}Cd_xTe$ alloy is given by the expression

$$E_g(x) = -0.3 + 1.9x \ (\text{eV})$$

Calculate the composition of an alloy that gives a cutoff wavelength of (a) 10 μm; (b) 5.0 μm.

Problem 11.2 Calculate the cutoff wavelength for a GaAs detector. If the cutoff wavelength is to be decreased to 0.7 μm, how much AlAs must be added to a GaAs? Assume that the bandgap of $Ga_{1-x}Al_xAs$ is given by

$$E_g(x) = 1.43 + 1.25x \ (\text{eV})$$

as long as $x \leq 0.4$.

Problem 11.3 Calculate the absorption coefficient for GaAs for photons with energy 1.8 eV. Calculate the fraction of this light absorbed in a GaAs sample of thickness of 0.5 μm.

Problem 11.4 An optical power density of 1 W/cm^2 is incident on a GaAs sample. The photon energy is 2.0 eV and there is no reflection from the surface. Calculate the excess electron-hole carrier densities at the surface and 0.5 μm from the surface. The e-h recombination time is 10^{-8} s.

Problem 11.5 Assume that all the photons in an optical beam produce an electron-hole pair in a Ge detector. If all the carriers are collected, calculate the responsivity for photon energies of (a) 0.7 eV; (b) 1.0 eV; (c) 2.0 eV.

Section 11.3
Problem 11.6 Consider a long Si *p-n* junction with a reverse bias of 1 V at 300 K.

The diode has the following parameters:

Diode area,	A	$= 1 \text{ cm}^2$
p-side doping,	N_a	$= 3 \times 10^{17} \text{ cm}^{-3}$
n-side doping,	N_d	$= 10^{17} \text{ cm}^{-3}$
Electron diffusion coefficient,	D_n	$= 12 \text{ cm}^2/\text{s}$
Hole diffusion coefficient,	D_p	$= 8 \text{ cm}^2/\text{s}$
Electron minority carrier lifetime,	τ_n	$= 10^{-7} \text{ s}$
Hole minority carrier lifetime,	τ_p	$= 10^{-7} \text{ s}$
Optical absorption coefficient,	α	$= 10^3 \text{ cm}^{-1}$
Optical power density,	P_{op}	$= 10 \text{ W/cm}^2$
Photon energy,	$\hbar\omega$	$= 1.7 \text{ eV}$

Calculate the photocurrent in the diode.

Problem 11.7 Consider a long Si p-n junction solar cell with an area of 4 cm^2 at 300 K. The solar cell has the following parameters:

n-type doping,	N_d	$= 10^{18} \text{ cm}^{-3}$
p-type doping,	N_a	$= 3 \times 10^{17} \text{ cm}^{-3}$
Electron diffusion coefficient,	D_n	$= 15 \text{ cm}^2/\text{s}$
Hole diffusion coefficient,	D_p	$= 7.5 \text{ cm}^2/\text{s}$
Electron minority carrier lifetime,	τ_n	$= 10^{-7} \text{ s}$
Hole minority carrier lifetime,	τ_p	$= 10^{-7} \text{ s}$
Photocurrent,	I_L	$= 1.0 \text{ A}$
Diode ideality factor,	m	$= 1.25$

Calculate the open circuit voltage of the diode. If the fill factor is 0.75, calculate the maximum power output.

Problem 11.8 Consider the solar cell of problem 11.7. A solar system is to be developed from such cells to deliver a power of 15 W at a voltage level of 5 V. Calculate the total number of solar cells needed.

Section 11.4

Problem 11.9 Consider a long silicon p-n photodiode at 300 K on which light from a GaAs laser ($\hbar\omega = 1.43$ eV) is impinging. The optical power density is 10^{-2} W/cm^2. The diode has the following parameters:

Device area,	A	$= 10^{-6} \text{ cm}^2$
n-type doping,	N_d	$= 5 \times 10^{16} \text{ cm}^{-3}$
p-type doping,	N_a	$= 10^{17} \text{ cm}^{-3}$
Electron diffusion coefficient,	D_n	$= 20 \text{ cm}^2/\text{s}$
Hole diffusion coefficient,	D_p	$= 12 \text{ cm}^2/\text{s}$
Electron minority carrier lifetime,	τ_n	$= 1.5 \times 10^{-7} \text{ s}$
Hole minority carrier lifetime,	τ_p	$= 10^{-7} \text{ s}$
Absorption coefficient,	α	$= 700 \text{ cm}^{-1}$

A reverse bias voltage of 5 V is applied to the diode. Assume that the carrier generation rate is uniform. Calculate the prompt photocurrent and the total photocurrent in the detector.

Problem 11.10 Consider a GaAs *p-i-n* detector with an intrinsic layer width of $1.0\mu m$. Optical power density (photon energy 1.6 eV) of 0.1 watt/cm^2 impinges upon the detector. The absorption coefficient for the active region is 10^4cm^{-1}. Calculate the prompt photocurrent of the device. The device area is 10^{-4} cm^2.

Problem 11.11 Consider a silicon *p-i-n* photodetector in which the *i* layer is 10.0 μm thick. Calculate the maximum quantum efficiency of this detector if only light absorbed in the undoped region contributes to the photocurrent. The absorption coefficient is 10^3 cm^{-1}. Also calculate the minimum thickness of the *i*-region needed to ensure a quantum efficiency of 0.8. There are no reflection losses.

Problem 11.12 Consider an In$_{0.53}$Ga$_{0.47}$As sample at 300 K in which excess electrons and hole are injected. The excess density is 10^{15} cm^{-3}. Calculate the rate at which photons are generated in the system. The bandgap is 0.8 eV and carrier masses are m_e^* = $0.042m_o$; m_h^* = $0.4m_o$. Also calculate the photon emission rate if the same density is injected at 77 K. (Use the low injection approximation).

Section 11.5

Problem 11.13 Determine the error in the position of the Fermi levels E^{Fn} and E^{Fp} calculated from the Boltzmann and Joyce-Dixon approximations for GaAs when the free carrier density $n = p$ ranges from 10^{15} cm^{-3} to 10^{18} cm^{-3}. Calculate the results at 77 K and 300 K.

Problem 11.14 Consider a GaAs *p-n*$^+$ junction LED with the following parameters at 300 K:

Electron diffusion coefficient,	D_n =	25 cm^2/s
Hole diffusion coefficient,	D_p =	12 cm^2/s
n-side doping,	N_d =	5×10^{17} cm^{-3}
p-side doping,	N_a =	10^{16} cm^{-3}
Electron minority carrier lifetime,	τ_n =	10 ns
Hole minority carrier lifetime,	τ_p =	10 ns

Calculate the injection efficiency of the LED assuming no trap-related recombination.

Problem 11.15 The diode in Problem 11.14 is to be used to generate an optical power of 1 mW. The diode area is 1 mm^2 and the external radiative efficiency is 20%. Calculate the forward bias voltage required.

Problem 11.16 Consider the GaAs LED of Problem 11.14. The LED is to be used in a communication system. The binary data bits 0 and 1 are to be coded so that the optical pulse output is 1 nW and 50 μW. If the external efficiency factor is 10%, calculate the forward bias voltages required to send the 0s and 1s. The LED area is 1 mm^2.

Section 11.6

Problem 11.17 Consider the semiconductor alloy InGaAsP with a bandgap of 0.8 eV.

The electron and hole masses are 0.04 m and 0.35 m, respectively. Calculate the injected electron and hole densities needed at 300 K to cause inversion for the electrons and holes at the bandedge energies. How does the injected density change if the temperature is 77 K? Use the Joyce-Dixon approximation.

Problem 11.18 Consider a GaAs-based laser at 300 K. Calculate the injection density required at which the inversion condition is satisfied at (a) the bandedges; (b) an energy of $\hbar\omega = E_g + k_B T$. Use the Joyce-Dixon approximation.

Problem 11.19 Consider a GaAs-based laser at 300 K. A gain of 30 cm^{-1} is needed to overcome cavity losses at an energy of $\hbar\omega = E_g + 0.026$ eV. Calculate the injection density required. Also, calculate the injection density if the laser is to operate at 400 K.

Problem 11.20 Consider the laser of the previous problem. If the time for *e-h* recombination is 2.0 ns at threshold, calculate the threshold current density at 300 K and 400 K. The active layer thickness is 20 μm and the optical confinement is unity.

Problem 11.21 Two GaAs/AlGaAs double heterostructure lasers are fabricated with active region thicknesses of 2.0 μm and 0.5 μm. The optical confinement factors are 1.0 and 0.8, respectively. The carrier injection density needed to cause lasing is 1.0×10^{18} cm^{-3} in the first laser and 1.1×10^{18} cm^{-3} in the second one. The radiative recombination times are 1.5 ns. Calculate the threshold current densities for the two lasers.

Problem 11.22 A GaAs laser has a threshold current density of 500 A/cm^2. The laser has dimensions of 10 μm $\times$ 200 μm. The active region is $d_{las} = 100$ Å. The electron-hole recombination time at threshold is 1.5 ns. Calculate the optical power emitted when a current density of $5J_{th}$ is injected into the laser. The emitted photons have an energy of 1.43 eV.

11.10 READING LIST

- **General**

  - J. Gowar, *Optical Communication Systems* (Prentice-Hall, Englewood Cliffs, NJ, 1989).

  - J. I. Pankove, *Optical Processes in Semiconductors* (Dover Publications, New York, 1977).

  - J. Singh, *Physics of Semiconductors and Their Heterostructures* (McGraw-Hill, New York, 1993).

  - J. Wilson and J. F. B. Hawkes, *Optoelectronics: An Introduction* (Prentice-Hall, Englewood Cliffs, NJ, 1983).

  - H. Kressel and J. K. Butler, *Semiconductor Lasers and Heterojunction LEDs* (Academic Press, New York, 1977).

  - R. Baets, "Heterostructures in III-V Optoelectronic Devices," *Solid State Electronics*, **30**, *1175* (1987).

– W. T. Tsang, "High Speed Photonic Devices" in *High Speed Semiconductor Devices*, edited by S. M. Sze (Wiley-Interscience, New York, 1990).

– J. Singh, *Optoelectronics—An Introduction to Materials and Devices* (McGraw-Hill, New York, 1996).

– P. K. Bhattacharya, *Semiconductor Optoelectronic Devices* (Prentice-Hall, Englewood Cliffs, NJ, 1994).

Appendix

A

LIST OF SYMBOLS

a lattice constant (edge of the cube for the semiconductor fcc lattice)

B base transport factor in a bipolar transistor

c velocity of light

C_{BD}, C_{BS} body-to-drain and body-to-source capacitance in a MOSFET
C_{ox} oxide capacitance per unit area
C_{mos} capacitance (per area) of an MOS capacitor
$C_{mos(min)}$ minimum capacitance (per area) of an MOS capacitor
$C_{mos(fb)}$ capacitance (per area) of an MOS capacitor under flatband conditions
C_{GS}, C_{GD} gate-to-source and gate-to-drain capacitance in a FET
C_{DS} drain-to-substrate capacitance in an FET
C_j, C_d junction, diffusion capacitance in a p-n diode

D_n electron diffusion coefficient
D_p hole diffusion coefficient
D_b diffusion coefficient in the base of a bipolar transistor

D_e	diffusion coefficient in the emitter of a bipolar transistor
D_c	diffusion coefficient in the collector of a bipolar transistor

e	magnitude of the electron charge

E	energy of a particle
E_F	Fermi level
E_{Fi}	intrinsic Fermi level
E_{Fn}	electron quasi-Fermi level
E_{Fp}	hole quasi-Fermi level
$E^e(E^h)$	energy of an electron (hole) in an optical absorption or emission measured from the bandedges
$E_c(E_v)$	conduction (valence) bandedge

$f(E)$	occupation probability of an electron state with energy E at equilibrium. This is the Fermi-Dirac function
$f^e(E)$	occupation function for an electron in non-equilibrium state. This is the quasi-Fermi function
$f^h(E)$	occupation function for a hole $= 1 - f^e(E)$
f_T	cutoff frequency for unit current gain
f_{max}	frequency at which available power gain is unity

F	electric field
F_{ext}	external force such as an electric or magnetic force

g_m	transconductance of a transistor
g_D	output conductance of a transistor

G_L	electron-hole generation rate due to a light beam

$\hbar$	Planck's constant divided by 2π
h	channel thickness of a JFET or an MESFET
$h(x)$	depletion region thickness in an FET at position x along the source-to-drain channel

H	magnetic field

I_{ph}	photon particle current
I_E, I_B, I_C	emitter, base, and collector current in a BJT
I_{En}, I_{Ep}	electron, hole part of the emitter current in an *npn* BJT

I_D drain current in an FET

I_o reverse-bias saturation current in a p-n diode

I_s reverse-bias saturation current in a Schottky diode

I_{GR} generation recombination current in a diode

I_{GR}^o prefactor for the generation recombination current

J current density

J_L photocurrent density

J_{ph} photon particle current density

k device transconductance parameter for a MOSFET

k' process transconductance parameter for a MOSFET

ℓ mean free path between successive collisions

L_n diffusion length for electron

L_p diffusion length for holes

m grading index for doping in a p-n structure

m_o free electron mass

m_e^* electron mass

m_h^* hole mass

m_{dos}^* density of states mass

m_σ^* conductivity mass

m_{hh}^* mass of the heavy hole

$m_{\ell h}^*$ mass of the light hole

m_r^* reduced mass of the electron-hole system

M, M_e, M_h multiplication factor, multiplication factor for electrons, multiplication factor for holes

n diode ideality factor

n electron concentration in the conduction band

n_i intrinsic electron concentration in the conduction band

n_d electrons bound to the donors

$n_p(n_p)$ equilibrium electron density in the p-side (n-side) of a p-n junction

N_{cv} joint density of states for electrons and holes

$N_e(E)$ density of states of electrons in the conduction band

$N_h(E)$ density of states of holes in the valence band

$N_c(E)$ effective density of states in the conduction band

$N_v(E)$ effective density of states in the valence band

N_d donor density

N_a acceptor density

N_t density of impurity states (trap states)

N_{ab} acceptor concentration in the base of an *npn* BJT

N_{de} donor concentration in the emitter of an *npn* BJT

N_{dc} donor concentration in the collector of *npn* BJT

p momentum of a particle

p hole concentration in the valence band

p_i intrinsic hole concentration in the valence band

p_a holes bound to acceptors

p_{cv} momentum matrix element for an optical transition between the valence and conduction band

$p_n(p_p)$ equilibrium hole density in the *n*-side (*p*-side) of a *p-n* junction

P_{op} optical power density (energy flow/sec/area)

Q_s total charge (per area) in an MOS channel

Q_n total mobile charge (per area) in an MOS channel

Q_{ss} surface charge density in an MOS capacitor

Q_{dep} depletion charge (per area) in an MOS channel

R_{spon} total rate at which an electron-hole system recombines to emit photons by spontaneous recombination

R_s, R_G, R_D parasitic resistances associated with the source, gate, and drain of a transistor respectively

R_L load resistance

R^* Richardson constant in a Schottky barrier

t_{tr} transit time of a carrier through a channel

T tunneling probability

$U(r)$ position-dependent potential energy

v velocity of the electron

v_s saturation velocity of the carrier (electron, hole)

V_{bi}	built-in voltage
V_{GS}	gate bias (referred to the source)
V_{DS}	drain bias
$V_{DS}(sat)$	drain bias at which pinch-off occurs at the drain
V_p	pinch-off voltage to deplete the channel of an FET
V_T	threshold gate bias for pinch-off
V_{fb}	flat band voltage. Voltage needed to make the semiconductor bands flat in an MOS capacitor
V_{SB}	source-to-body (substrate) potential
$V_r(V_f)$	reverse (forward) bias voltage in a diode
V_{BE}, V_{BC}	base-to-emitter, base-to-collector bias in a bipolar transistor
V_{pt}	punchthrough voltage
$W_n(W_p)$	depletion region edge on the n-side (p-side) of a p-n junction
W	depletion region width
W_b, W_{bn}	base width, neutral base width of a bipolar transistor
α	optical absorption coefficient
α	current transfer ratio in a bipolar transistor
α_R	reflection loss coefficient in an optical cavity
α_{imp}	impact ionization coefficient for electrons
β	base-to-collector current amplification factor in a BJT
β_{imp}	impact ionization coefficient for holes
δn	excess electron density in a region. This is the density above the equilibrium density
δp	excess hole density in a region
ΔE_g	bandgap difference between two materials
$\Delta E_c, \Delta E_v$	band discontinuity in the conduction, valence band in a heterostructure
ϵ_o	free space permittivity
ϵ	product of the relative dielectric constant and ϵ_o
γ_e	emitter efficiency of a bipolar transistor
γ_{inj}	injection efficiency of a p-n diode for electron (hole) current
λ	channel length modulation parameter
μ	mobility of a material
$\mu_n(\mu_p)$	electron (hole) mobility
$\sigma_n(\sigma_p)$	electron (hole) capture cross-section for an impurity
σ	conductivity of a material
τ_{sc}	scattering time between successive collisions. Also called relaxation time
τ_o	rate at which an electron recombines radiatively with a hole at the same momentum value
τ_r	radiative recombination time for e-h pair

τ_{nr}	nonradiative recombination time for a *e-h* pair
τ_n	lifetime of an electron to recombine with a hole
τ_p	lifetime of a hole to recombine with an electron
τ_{sd}	storage delay time in a diode
ϕ_m	metal work function
ϕ_s	work function of a semiconductor
ϕ_{ms}	difference between a metal and semiconductor work function
ϕ_b	barrier height seen by electrons coming from a metal towards a semiconductor
χ_s	electron affinity of a semiconductor
ψ	electron wavefunction
ω	frequency

Appendix

B

SOME QUANTUM-MECHANICS PROBLEMS

In this text we have used several important concepts from quantum mechanics. These include the concepts of valence band, conduction band, and bandgap, as well as effective mass, density of states, and mobility. Without these concepts we would not be able to understand and exploit phenomena on which modern microelectronics is based. In this appendix we will present solutions to a few important quantum-mechanics problems.

B.1 DENSITY OF STATES

We will start with the problem of density of states of electrons in free space or in perfectly crystalline materials. We have noted that in perfect semiconductors, electrons can be regarded as "free" electrons.

Let us consider the Schrödinger equation for free electrons. The time-independent

509

equation is

$$\frac{-\hbar^2}{2m_o} \left(\frac{\partial^2}{\partial x^2} + \frac{\partial^2}{\partial y^2} + \frac{\partial^2}{\partial z^2} \right) \psi(r) = E\psi(r) \tag{B.1}$$

A general solution of this equation is

$$\psi(r) = \frac{1}{\sqrt{V}} e^{\pm ik \cdot r} \tag{B.2}$$

where the factor $\frac{1}{\sqrt{V}}$ comes because we wish to have one electron per volume V, or

$$\int_V d^3r \mid \psi(r) \mid^2 = 1 \tag{B.3}$$

We assume that the volume V is a cube of side L.

The corresponding energy of the electron is obtained from Eqn. B.1 and is

$$E = \frac{\hbar^2 k^2}{2m_o} \tag{B.4}$$

The momentum of the electron is (replacing $\boldsymbol{p}$ by the differential operator)

$$\boldsymbol{p}\psi = -i\hbar \frac{\partial}{\partial \mathbf{r}} \psi \rightarrow \hbar \mathbf{k} \text{ or } \boldsymbol{p} = \hbar \boldsymbol{k} \tag{B.5}$$

while the velocity is

$$\mathbf{v} = \frac{\hbar \mathbf{k}}{m_o} \tag{B.6}$$

In classical mechanics the energy-momentum relation for the free electron is $E = p^2/2m_o$, and p can be a *continuous variable*. The quantity $\hbar k$ appearing above seems to be replacing p, in quantum mechanics. Due to the wave nature of the electron, the quantity k is not continuous but discrete. For describing moving electrons, the boundary condition used is known as a periodic boundary condition. Even though we focus our attention on a finite volume V, the wave can be considered to spread in all space as we conceive the entire space is made up of identical cubes of sides L, as shown in Fig. B.1. Then,

$$\begin{aligned}
\psi(x, y, z + L) &= \psi(x, y, z) \\
\psi(x, y + L, z) &= \psi(x, y, z) \\
\psi(x + L, y, z) &= \psi(x, y, z)
\end{aligned} \tag{B.7}$$

The boundary conditions impose certain restrictions in the k of the wavefunction.

Because of the boundary conditions the allowed values of k are (n are integers— positive and negative)

$$k_x = \frac{2\pi n_x}{L}; \ k_y = \frac{2\pi n_y}{L}; \ k_z = \frac{2\pi n_z}{L} \tag{B.8}$$

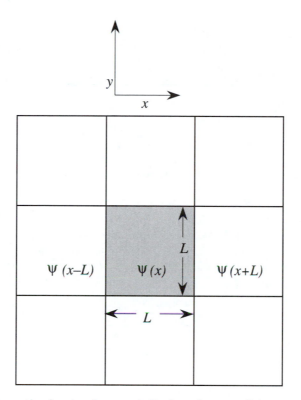

Figure B.1: A schematic showing how periodic boundary conditions are applied. A large volume is considered to be made up of identical cubic volumes.

If L is large, the spacing between the allowed k values is very small. It is useful to discuss the *volume in k-space that each electronic state occupies*. As can be seen from Fig. B.2, this volume is

$$\left(\frac{2\pi}{L}\right)^3 = \frac{8\pi^3}{V} \tag{B.9}$$

If Ω is a volume of k-space, the number of electronic states in this volume is

$$\boxed{\frac{\Omega V}{8\pi^3}} \tag{B.10}$$

We will now use the discussion provided above to derive the concept of density of states. The concept of density of states is extremely powerful, and important physical properties such as optical absorption, transport, etc., are intimately dependent upon it. Density of states is the number of available electronic states *per unit volume per unit energy* around an energy E. If we denote the density of states by $N(E)$, the number of states in a unit volume in an energy interval dE around an energy E is $N(E)dE$.

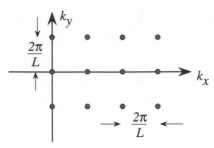

Figure B.2: k-space volume of each electronic state. The separation between the various allowed components of the k-vector is $\frac{2\pi}{L}$.

The energies E and $E + dE$ are represented by surfaces of spheres with radii k and $k + dk$ as shown in Fig. B.3. In a three-dimensional system, the k-space volume between vector k and $k + dk$ is $4\pi k^2 dk$. We have shown in Eqn. B.9 that the k-space volume per electron state is $(\frac{2\pi}{L})^3$. Therefore, the number of electron states in the region between k and $k + dk$ is

$$\frac{4\pi k^2 dk}{8\pi^3} V = \frac{k^2 dk}{2\pi^2} V \tag{B.11}$$

Denoting the energy and energy interval corresponding to k and dk as E and dE, we see that the number of electron states between E and $E + dE$ per unit volume is

$$N(E)dE = \frac{k^2 dk}{2\pi^2} \tag{B.12}$$

and since

$$E = \frac{\hbar^2 k^2}{2m_o} \tag{B.13}$$

$$k^2 dk = \frac{\sqrt{2}m_o^{3/2} E^{1/2} dE}{\hbar^3} \tag{B.14}$$

and

$$N(E)dE = \frac{m_o^{3/2} E^{1/2} dE}{\sqrt{2}\pi^2 \hbar^3} \tag{B.15}$$

An electron can have two possible states with a given energy. These states are called the spin states. The electron can have a spin state $\hbar/2$ or $-\hbar/2$. To take spin into account, the density of states obtained above is simply multiplied by 2,

$$\boxed{N(E) = \frac{\sqrt{2}m_o^{3/2} E^{1/2}}{\pi^2 \hbar^3}} \tag{B.16}$$

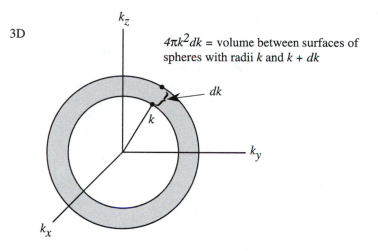

Figure B.3: Geometry used to calculate density of states. By finding the k-space volume in an energy interval between E and $E+dE$, one can find out how many allowed states there are.

In Fig. B.4, we show the density of states. Note that in the derivation given here, we have assumed that the electron energy starts at $E = 0$. If the electron energy is given by

$$E = \frac{\hbar^2 k^2}{2m_0} + V_0 \tag{B.17}$$

i.e., if there is a background potential V_0, the density of states becomes

$$N(E) = \frac{\sqrt{2}m_0^{3/2}(E - V_0)^{3/2}}{\pi^2 \hbar^3} \tag{B.18}$$

In this case, the density of states is zero for $E < V_0$.

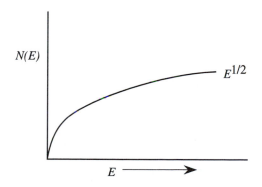

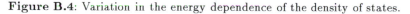

Figure B.4: Variation in the energy dependence of the density of states.

B.2   TUNNELING THROUGH BARRIERS

Quantum mechanics often produces situations in nature that are intuitively hard to grasp. An important phenomenon that occurs in nature is tunneling of a "particle" with energy E through a potential barrier with barrier height U_0 *greater than* E. Classically, this is forbidden, but again one has such an experience with, say, light waves. Light waves are known to penetrate "forbidden" regions by what are known as evanescent waves. This phenomenon is well known in optics and is based on the classical wave description of light.

   Tunneling is a very important phenomenon in semiconductor devices. Apart from special devices based upon this phenomenon, the functioning of the ohmic contacts discussed in Chapter 6 depends upon this effect. Ohmic contacts provide a connection between the outside world and a semiconductor and exploit the tunneling behavior. Let us briefly review the tunneling phenomenon.

   Consider a potential barrier as shown in Fig. B.5a. We have divided the space in three regions as shown. Let's say an electron has an energy E and impinges from region I onto the barrier. We are interested in the fraction of the electron that tunnels through into region III.

   The general solutions of the Schrödinger equation in the three regions considered independently are:

$$\text{Region I}: \quad \psi_1(x) \;=\; Ae^{ikx} + Be^{-ikx}; \quad k^2 = \frac{2m_o E}{\hbar^2} \tag{B.19}$$

$$\text{Region II}: \quad \psi_2(x) \;=\; Ce^{-\alpha x} + De^{\alpha x}; \quad \alpha^2 = \frac{2m_o}{\hbar^2}(U_o - E) \tag{B.20}$$

$$\text{Region III}: \quad \psi_3(x) \;=\; Fe^{ikx}; \quad k^2 = \frac{2m_o E}{\hbar^2} \tag{B.21}$$

In region I, the wavefunction represents a wave proceeding towards the barrier (Ae^{ikx}) and away from the barrier (Be^{-ikx}), representing the incident and reflected waves. In region III we consider only the wave going away from the barrier, representing the wave that is tunneling out.

   We now use the fact that the wavefunction and its derivative are continuous at the boundaries. This gives us the following conditions:

$$\text{at } x = 0 \begin{cases} A + B &= C + D \\ ikA - ikB &= -\alpha C + \alpha D \end{cases}$$

$$\text{at } x = d \begin{cases} Ce^{-\alpha d} + De^{+\alpha d} &= Fe^{ikd} \\ -\alpha Ce^{-\alpha d} + \alpha De^{\alpha d} &= ikFe^{ikd} \end{cases}$$

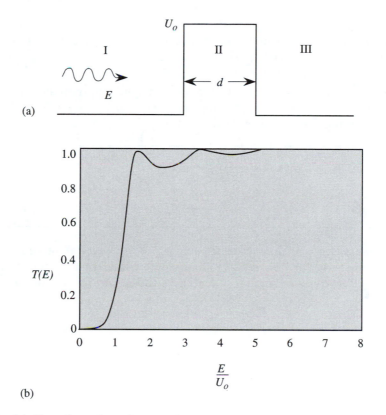

(a)

(b)

Figure B.5: (a) Tunneling of an electron through a potential barrier. (b) The transmission coefficients of a square barrier as a function of particle energy for $U_o = 1.0$ $eV, d = 7.77$ Å. The results are typical of transmittance through a barrier. Notice that even for $E > U_o$, the probability goes through values less than unity, i.e., some reflection occurs.

Eliminating B, C, D from these four equations, we can obtain the tunneling probability T:

$$T = \left|\frac{F}{A}\right|^2 = \frac{4}{4\cosh^2(\alpha d) + [(\alpha/k) - (k/\alpha)]^2 \sinh^2(\alpha d)} \qquad (B.22)$$

A typical tunneling probability is plotted in Fig. B.5b. Note that as the barrier height or width increases, the tunneling probability decreases.

Often the barrier that an electron has to tunnel through does not have the nice rectangular form assumed here. In such cases it is useful to use the Wentzel-Kramers-Brillouin expression for the tunneling rate. The general form of this expression is

$$T \cong \exp\left[-2\int_{x_1}^{x_2} |\alpha(x)| dx\right] \qquad (B.23)$$

where $\alpha(x)$ is given by Eqn. B.20 with U_o replaced by a position-dependent potential energy, and x_1 and x_2 represent the limits of the classically forbidden region where the potential energy is larger than the electron energy.

Two special cases of barriers are the triangular barrier and the trapezoidal barriers shown in Fig. B.6. The tunneling rates for such barriers are (F is the applied electric field causing the formation of the barriers)

$$
\text{Triangular} \quad : \quad T = \exp\left\{ \frac{-4(2m_o)^{1/2}}{3eF\hbar}(U_o - E)^{3/2} \right\}
$$

$$
\text{Trapezoidal} \quad : \quad T = \exp\left\{ \frac{-4(2m_o)^{1/2}}{3eF\hbar}\left[(U_o - E)^{3/2} - (U_o - E - eFd)^{3/2}\right] \right\}
$$

$$\tag{B.24}$$

These expressions for tunneling are extremely useful in understanding many important aspects of semiconductor devices, including ohmic contacts and Zener tunneling. An ohmic contact involves a triangular potential as shown in Fig. B.6 and is discussed in Chapter 6.

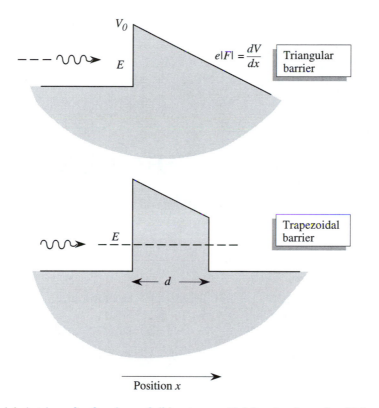

Figure B.6: (a) A triangular barrier and (b) a trapezoidal barrier through which an electron with energy E can tunnel. Such barriers are encountered in many electronic device structures. The potential shape is described by an electric field F.

Appendix

C

PROPERTIES OF SEMICONDUCTORS

In this appendix we include some important properties of semiconductors. As we have seen in this text, material properties like bandgap, electron and hole masses, mobilities, etc. determine how devices perform. We have also included breakdown field values for some semiconductors.

EXPERIMENTAL ENERGY GAP E_g (eV)				
Semi-conductor	Type of energy gap	0 K	300 K	Temperature dependence of energy gap (eV)
AlAs	Indirect	2.239	2.163	$2.239 - 6.0 \times 10^{-4}T^2/(T + 408)$
GaP	Indirect	2.338	2.261	$2.338 - 5.771 \times 10^{-4}T^2/(T + 372)$
GaAs	Direct	1.519	1.424	$1.519 - 5.405 \times 10^{-4}T^2/(T + 204)$
GaSb	Direct	0.810	0.726	$0.810 - 3.78 \times 10^{-4}T^2/(T + 94)$
InP	Direct	1.421	1.351	$1.421 - 3.63 \times 10^{-4}T^2/(T + 162)$
InAs	Direct	0.420	0.360	$0.420 - 2.50 \times 10^{-4}T^2/(T + 75)$
InSb	Direct	0.236	0.172	$0.236 - 2.99 \times 10^{-4}T^2/(T + 140)$
Si	Indirect	1.17	1.11	$1.17 - 4.37 \times 10^{-4}T^2/(T + 636)$
Ge	Indirect	0.66	0.74	$0.74 - 4.77 \times 10^{-4}T^2/(T + 235)$

Table C.1: Energy gaps of some semiconductors along with their temperature dependence.

Material	Electron mass (m_o)	Hole mass (m_o)
AlAs	0.1	
AlSb	0.12	$m_{dos} = 0.98$
GaN	0.19	$m_{dos} = 0.60$
GaP	0.82	$m_{dos} = 0.60$
GaAs	0.067	$m_{lh} = 0.082$ $m_{hh} = 0.45$
GaSb	0.042	$m_{dos} = 0.40$
Ge	$m_l = 1.64$ $m_t = 0.082$	$m_{lh} = 0.044$ $m_{hh} = 0.28$
InP	0.073	$m_{dos} = 0.64$
InAs	0.027	$m_{dos} = 0.4$
InSb	0.13	$m_{dos} = 0.4$
Si	$m_l = 0.98$ $m_t = 0.19$	$m_{lh} = 0.16$ $m_{hh} = 0.49$

Table C.2: Electron and hole masses for several semiconductors. Some uncertainty remains in the value of hole masses for many semiconductors.

Semiconductor	Bandgap (eV) 300 K	Mobility at 300 K (cm^2/V·s)	
		Electrons	Holes
C	5.47	800	1200
Ge	0.66	3900	1900
Si	1.12	1500	450
α-SiC	2.996	400	50
GaSb	0.72	5000	850
GaAs	1.42	8500	400
GaP	2.26	110	75
InSb	0.17	8000	1250
InAs	0.36	33000	460
InP	1.35	4600	150
CdTe	1.56	1050	100
PbTe	0.31	6000	4000

Table C.3: Bandgaps, electron and hole mobilities of some semiconductors.

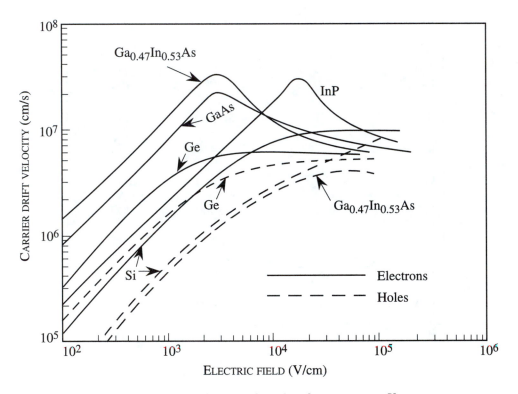

Figure C.1: Velocity-field relations for several semiconductors at 300 K.

Material	Bandgap (eV)	Breakdown electric field(V/cm)	
GaAs	1.43	4×10^5	
Ge	0.664	10^5	
InP	1.34		
Si	1.1	3×10^5	
$In_{0.53}Ga_{0.47}As$	0.8	2×10^5	
C	5.5	10^7	
SiC	2.9	$2\text{-}3 \times 10^6$	
SiO_2	9	10^7	
Si_3N_4	5	10^7	

Table C.4: Breakdown electric fields in some materials.

Appendix

D

FROM DEVICE PHYSICS TO CIRCUIT MODELING

In this text we have focused our attention on fundamental physics- and technology-related issues that control device performance. It must be realized, however, that most electrical engineering students eventually use devices to design and optimize circuits. In circuit design one represents devices by equivalent circuit models—models that capture, to varying degrees of accuracy, how the device behaves. Thus there are circuit models for large-signal applications, for small-signal applications, and for switching applications.

Computer simulations of circuits have become an integral part of undergraduate (and graduate) electrical engineering programs. The circuit simulation program SPICE has emerged as a standard in academia and in industry (although many industries have their own simulation packages). SPICE, developed by researchers at the University of California at Berkeley, has evolved over the last 25 years and can perform circuit simulations for a variety of devices and applications. It can be used for nonlinear dc, nonlinear transient, and linear-ac analysis. Additionally, it can study noise, temperature dependence, etc.

There are varying levels of device models in SPICE. The number of parameters describing a particular device usually increases as the level of the model increases. In this text we have discussed SPICE models used for a *p-n* diode (Section 5.7), bipolar junction transistor (Section 7.9), and a MOSFET (Section 10.4). It is very useful for students to see how basic parameters such as doping, mobility, diffusion coefficients, etc., relate to SPICE model parameters. It is also important to see how technology parameters such as length scales, areas, overlap between contacts, etc., influence SPICE model parameters.

Understanding this connection—even on a semi-quantitative level—helps develop a better ability as a circuit designer. It also helps one understand what parameters are important in different circuits and to what extent they can be optimized.

In this appendix we present the details of the calculations for SPICE parameters—one for a *p-n* diode and one for a CMOS structure. The calculation for a BJT are similar in many respects to the *p-n* diode case discussed here.

D.1   SPICE PARAMETERS FOR A *P-N* DIODE

In Fig. 5.18 we presented a set of process parameters for a *p-n* diode. We will now calculate the input file (the results are presented in Fig. 5.19). For coherence of this section we reproduce Fig. 5.18 as Fig. D.1.

(1) V_{bi}: The built-in voltage.

$$V_{bi} = \frac{k_B T}{e} \ln \left(\frac{5 \times 10^{31} \text{ cm}^{-6}}{2.25 \times 10^{20} \text{ cm}^{-6}} \right) = 0.68 \text{ V}$$

(2) I_o: The saturation current for the ideal diode.

$$
\begin{aligned}
n_p &= 2.25 \times 10^4 \text{ cm}^{-3} \\
p_n &= 4.5 \times 10^4 \text{ cm}^{-3} \\
L_n &= 17.32 \ \mu m \gg W_{\ell p} \rightarrow \text{narrow diode} \\
L_p &= 10.0 \ \mu m \gg W_{\ell m} \rightarrow \text{narrow diode} \\
I_o &= 1.6 \times 10^{-19} \left[\frac{10 \times 4.5 \times 10^4}{5 \times 10^{-4} \text{ cm}} + \frac{30 \times 2.25 \times 10^4}{5 \times 10^{-4} \text{ cm}} \right] \times (10^{-4} \text{ cm}^2) \\
&= 3.6 \times 10^{-14} \text{ A}
\end{aligned}
$$

(3) C_{jo}: Zero-bias junction capacitance.

$$
\begin{aligned}
W(V = 0) &= \left[\frac{2 \times 11.9 \times 8.84 \times 10^{-14} \times 0.68}{1.6 \times 10^{-19}} \left(\frac{1.5 \times 10^{16}}{5 \times 10^{31}} \right) \right]^{1/2} \\
&= 5.18 \times 10^{-5} \text{ cm}
\end{aligned}
$$

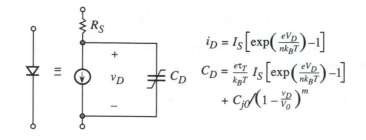

$$i_D = I_S\left[\exp\left(\frac{eV_D}{nk_BT}\right)-1\right]$$

$$C_D = \frac{e\tau_T}{k_BT}I_S\left[\exp\left(\frac{eV_D}{nk_BT}\right)-1\right]$$

$$+ C_{jo}\Big/\left(1-\frac{v_D}{V_0}\right)^m$$

(a)

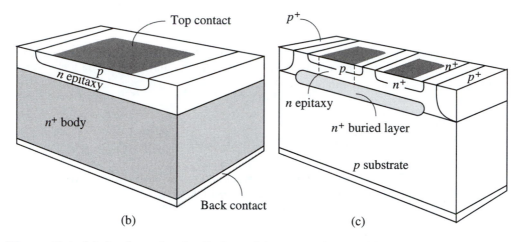

(b)   (c)

Figure D.1: (a) A schematic of a diode model used in SPICE. (b) A typical input file for SPICE simulation of a *p-n* diode.

$$C_{jo} = \frac{11.9 \times 8.84 \times 10^{-14} \times 10^{-4}}{5.18 \times 10^{-5}} = 2 \text{ pF}$$

(4) T_T: Transit times of holes across the *n*-region.

$$T_T = \frac{\left(5 \times 10^{-4} \text{ cm}\right)^2}{2 \times \left(10 \text{ cm}^2/\text{s}\right)} = 12.5 \text{ ns}$$

(5) I_{GR}^o: Generation-recombination prefactor. We calculate this at several bias points.

$$I_{GR}^o(V=0) = \frac{\left(1.6 \times 10^{-19} \text{ C}\right)\left(10^{-4} \text{ cm}\right)\left(5.18 \times 10^{-5} \text{ cm}\right)\left(1.5 \times 10^{10} \text{ cm}^{-3}\right)}{2 \times 10^{-7} \text{ s}}$$

$$= 6.2 \times 10^{-11} \text{ A}$$

$$I_{GR}^o(V=0.5) = 3.16 \times 10^{-11} \text{ A}$$

$$I_{GR}^o(V=0.6 \text{ V}) = 2.1 \times 10^{-11} \text{ A}$$

(6) n: Ideality factor. We calculate the current at several voltages and estimate n.

$$I(0.5 \text{ V}) = 8.6 \times 10^{-6} \text{ A}$$

$$I(0.6 \text{ V}) = 3.8 \times 10^{-4} \text{ A}$$

Using

$$ln\left(\frac{I(0.6 \text{ V})}{I(0.5 \text{ V})}\right) = \frac{e\Delta V}{nk_B T}$$

we get

$$n = 1.015$$

(7) I_S: Saturation current. We get this value by fitting the current (say at $V = 0.6$ V) to

$$I(0.6 \text{ V}) = I_S \exp\left(\frac{0.6 \text{ V}}{nk_B T}\right)$$

This gives

$$I_S = 5 \times 10^{-14} \text{ A}$$

(8) V_{BD}: Breakdown voltage.

$$V_{BD} \cong \frac{\epsilon F_{crit}^2}{2eN_a}$$

$$= \frac{\epsilon \left(4 \times 10^5 \text{ V/cm}\right)^2}{2e\left(10^{16} \text{ cm}^{-3}\right)} \cong 50 \text{ V}$$

(9) I_{BV}: Current at breakdown voltage (dominated by recombination-generation current).

$$I_{BV} = 5.35 \times 10^{-10} \text{ A}$$

(10) R_S: Series resistance from the neutral n and p regions of the diode. The carrier mobilities are, from the Einstein relation,

$$\mu_p = 384 \text{ cm}^2/\text{V} \cdot \text{s}; \ \mu_n = 1154 \text{ cm}^2/\text{V} \cdot \text{s}$$

$$R_S(p\text{-side}) = \frac{\left(5 \times 10^{-4} \text{ cm}\right)}{\left(10^{16} \text{ cm}^{-3}\right)\left(1.6 \times 10^{-19} \text{ C}\right)\left(384 \text{ cm}^2/\text{V} \cdot \text{s}\right)\left(10^{-4} \text{ cm}^2\right)}$$

$$= 8.1 \ \Omega$$

$$R_S(n\text{-side}) = \frac{\left(5 \times 10^{-4} \text{ cm}\right)}{\left(5 \times 10^{15} \text{ cm}^{-3}\right)\left(1.6 \times 10^{-19} \text{ C}\right)\left(1154 \text{ cm}^2/\text{V} \cdot \text{s}\right)\left(10^{-4} \text{ cm}^2\right)}$$

$$= 5.4 \ \Omega$$

The total resistance is $R_S \sim 13.5 \ \Omega$.

D.2   SPICE PARAMETERS FOR A CMOS

We will now present results for a typical CMOS SPICE parameter. Once again, in Fig. D.2 we reproduce Fig. 10.17. The calculation of the SPICE parameters is given below.

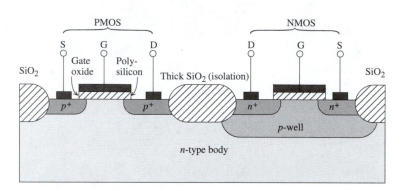

(a)

Parameter	Value	Units
n^+ active doping	5×10^{18}	cm^{-3}
p^+ active doping	5×10^{17}	cm^{-3}
p-well doping	5×10^{16}	cm^{-3}
n-well doping	1×10^{16}	cm^{-3}
Lateral diffusion		
n-channel	0.5	µm
p-channel	0.6	µm
oxide thickness	500	Å
Channel mobility		
n-channel	700	cm^2/V·s
p-channel	230	cm^2/V·s
Source area		
NMOS	100	µm^2
PMOS	100	µm^2
Drain area		
NMOS	100	µm^2
PMOS	100	µm^2
V_T	±1.0	V

(b)

Figure D.2: (a) A cross-section of a CMOS circuit. (b) Process parameters typical of a CMOS circuit.

(1) Device dimensions assumed:

$$L(\text{NMOS}) = 3.0 \ \mu\text{m} \quad Z(\text{NMOS}) \quad = \quad 20 \ \mu\text{m}$$
$$L(\text{PMOS}) = 9.0 \ \mu\text{m} \quad Z(\text{PMOS}) \quad = \quad 20 \ \mu\text{m}$$

(2) k': Process transconductance parameter.

$$
\begin{aligned}
k'(\text{NMOS}) \quad &= \quad \mu_n C_{ox} = \frac{\mu_n \epsilon_{ox}}{d_{ox}} \\
&= \quad \frac{\left(700 \ \text{cm}^2/\text{V} \cdot \text{s}\right)\left(3.9 \times 8.85 \times 10^{-14} \ \text{F/cm}\right)}{\left(500 \times 10^{-8} \ \text{cm}\right)} \\
&= \quad 4.83 \times 10^{-5} \ \text{A/V}^2 \\
k'(\text{PMOS}) \quad &= \quad 1.59 \times 10^{-5} \ \text{A/V}^2
\end{aligned}
$$

(3) γ: Body factor.

$$
\begin{aligned}
\gamma(\text{NMOS}) \quad &= \quad \frac{1}{C_{ox}}\sqrt{2e\epsilon_{si}N_A} \\
&= \quad \frac{1}{6.9 \times 10^{-8} \ \text{F/cm}^2} \\
&\qquad \sqrt{2\left(1.6 \times 10^{-19} \ \text{C}\right)\left(11.9 \times 8.85 \times 10^{-14} \ \text{F/cm}\right)\left(5 \times 10^{16} \ \text{cm}^{-3}\right)} \\
&= \quad 1.88 \ \text{V}^{1/2} \\
\gamma(\text{PMOS}) \quad &= \quad 0.84 \ \text{V}^{1/2}
\end{aligned}
$$

(4) $2\phi_F$: Surface potential.

$$
\begin{aligned}
\phi_F(\text{NMOS}) \quad &= \quad \frac{k_B T}{e} \ ln \frac{ni}{N_a} \\
&= \quad (0.026 \ \text{V}) ln\left(\frac{1.5 \times 10^{10} \ \text{cm}^{-3}}{5 \times 10^{16} \ \text{cm}^{-3}}\right) = -0.39 \ \text{V} \\
2\phi_F(\text{NMOS}) \quad &= \quad -0.78 \ \text{V} \\
2\phi_F(\text{PMOS}) \quad &= \quad 0.7 \ \text{V}
\end{aligned}
$$

(5) λ: Channel length modulation parameter.

$$
\begin{aligned}
\Delta L(\text{NMOS}) \quad &= \quad \sqrt{\frac{2\epsilon_{si}}{eN_a}}\left[\sqrt{\phi_{fb} + V_{DS}(\text{sat}) + \Delta V_{DS}} - \sqrt{\phi_{fb} + V_{DS}(\text{sat})}\right] \\
&\simeq \quad \sqrt{\frac{2\epsilon_{si}}{eN_a}}\sqrt{\phi_{fb} + V_{DS}(\text{sat})}\frac{\Delta V_{DS}}{2}
\end{aligned}
$$

Using

$$I'_D \quad = \quad I_D\left(1 + \frac{\Delta L}{L}\right) \equiv I_D\left(1 + \lambda V_{DS}\right)$$

$$\lambda \equiv \frac{\Delta L}{L \Delta V_{DS}} = \frac{1}{2L}\sqrt{\frac{2\epsilon_{si}}{eN_a}}\sqrt{\phi_{fb} + V_{DS}(\text{sat})}$$

$$\lambda(\text{NMOS}) \cong \frac{1}{2 \times 3 \times 10^{-4}\ \text{cm}}\sqrt{\frac{2 \times 11.9 \times 8.85 \times 10^{-14}\ \text{F/cm}}{(1.6 \times 10^{-19}\ \text{C})(5 \times 10^{16}\ \text{cm}^{-3})}}$$

$$= \ 2.7 \times 10^{-2}\ \text{V}^{-1}$$

$$\lambda(\text{PMOS}) \ = \ 2 \times 10^{-2}\ \text{V}^{-1}$$

(6) C_{BD}: Zero-bias body-to-drain capacitance.

$$V_{bi}(\text{NMOS}) \ = \ (0.026\ \text{V})\ell n\left(\frac{(5 \times 10^{18}\ \text{cm}^{-3})(5 \times 10^{16}\ \text{cm}^{-3})}{(2.25 \times 10^{20}\ \text{cm}^{-6})}\right)$$

$$= \ 0.9\ \text{V}$$

$$V_{bi}(\text{PMOS}) \ = \ 0.8\ \text{V}$$

$$C_{BD}(\text{NMOS}) \ = \ \frac{A}{2}\left[\frac{2e\epsilon_{si}}{V_{bi}}\frac{N_A N_D}{N_A + N_D}\right]^{1/2}$$

$$= \ (100 \times 10^{-8}\ \text{cm}^2)(6.8 \times 10^{-8}\ \text{F/cm}^2)$$

$$= \ 6.8 \times 10^{-14}\ \text{F}$$

$$C_{BD}(\text{PMOS}) \ = \ 3.2 \times 10^{-14}\ \text{F}$$

(7) C_{BS}: Zero bias body-source junction capacitance. Same as above.

(8) C_{GSO}: Gate-source overlap capacitance.

$$C_{GSO}(\text{NMOS}) \ = \ C_{ox}L_d Z = (6.9 \times 10^{-8}\ \text{Fcm}^{-2})(0.5 \times 10^{-4}\ \text{cm})(20 \times 10^{-4}\ \text{cm})$$

$$= \ 6.9 \times 10^{-15}\ \text{F}$$

$$C_{GSO}(\text{PMOS}) \ = \ 6.9 \times 10^{-8} \times 0.6 \times 10^{-4} \times 20 \times 10^{-4}$$

$$= \ 8.3 \times 10^{-15}\ \text{F}$$

(9) C_{GBO}: Gate-to-body capacitance.

$$C_{GBO}(\text{NMOS}) \ = \ (6.9 \times 10^{-8}\ \text{Fcm}^{-2})(3.0 \times 10^{-4}\ \text{cm})(20 \times 10^{-4}\ \text{cm})$$

$$= \ 4.14 \times 10^{-14}\ \text{F}$$

$$C_{GBO}(\text{PMOS}) \ = \ 6.9 \times 10^{-8} \times 9.0 \times 10^{-4} \times 20 \times 10^{-4}$$

$$= \ 1.24 \times 10^{-13}\ \text{F}$$

As we noted in Section 10.4, in advanced SPICE models, the parameters for the SPICE file are usually obtained from a more empirical approach. However, the exercise given in this appendix is very useful in illustrating how various process parameters control the SPICE file parameter.

Index